Evolutionary Dynamics

Evolutionary Dynamics: Exploring the Interplay of Selection, Accident, Neutrality, and Function

Editors

James P. Crutchfield
Santa Fe Institute

Peter Schuster
University of Vienna

Santa Fe Institute
Studies in the Sciences of Complexity

2003

OXFORD
UNIVERSITY PRESS

Oxford New York
Auckland Bangkok Buenos Aires Cape Town Chennai
Dar es Salaam Delhi Hong Kong Istanbul Karachi Kolkata
Kuala Lumpur Madrid Melbourne Mexico City Mumbai Nairobi
São Paulo Shanghai Taipei Tokyo Toronto

Published by Oxford University Press, Inc.
198 Madison Avenue, New York, New York 10016

www.oup.com

Library of Congress Cataloging-in-Publication Data
Evolutionary dynamics : exploring the interplay of selection, accident, neutrality,
and function / editors, James P. Crutchfield, Peter Schuster.
p. cm.—(Santa Fe Institute studies in the sciences of complexity)
Includes bibliographical references and index.
ISBN 0-19-514264-0; ISBN 0-19-514265-9 (pbk.)
1. Evolution (Biology) I. Crutchfield, James P. (James Patrick)
II. Schuster, Peter, 1939– III. Proceedings volume in the Santa Fe Institute
studies in the sciences of complexity.
QH366.2 .E867 2002
576.8—dc21 2002025267

9 8 7 6 5 4 3 2 1

Printed in the United States of America
on acid-free paper

About the Santa Fe Institute

The *Santa Fe Institute* (SFI) is a private, independent, multidisciplinary research and education center, founded in 1984. Since its founding, SFI has devoted itself to creating a new kind of scientific research community, pursuing emerging science. Operating as a small, visiting institution, SFI seeks to catalyze new collaborative, multidisciplinary projects that break down the barriers between the traditional disciplines, to spread its ideas and methodologies to other individuals, and to encourage the practical applications of its results.

All titles from the *Santa Fe Institute Studies in the Sciences of Complexity* series carry this imprint which is based on a Mimbres pottery design (circa A.D. 950–1150), drawn by Betsy Jones. The design was selected because the radiating feathers are evocative of the outreach of the Santa Fe Institute Program to many disciplines and institutions.

Santa Fe Institute
Studies in the Sciences of Complexity

Lecture Notes Volume

Author	*Title*
Eric Bonabeau, Marco Dorigo, and Guy Theraulaz	Swarm Intelligence: From Natural to Artificial Systems

Proceedings Volumes

Editor	*Title*
James H. Brown and Geoffrey B. West	Scaling in Biology
Timothy A. Kohler and George J. Gumerman	Dynamics in Human and Primate Societies
Lee A. Segel and Irun Cohen	Design Principles for the Immune System and Other Distributed Autonomous Systems
H. Randy Gimblett	Integrating Geographic Information Systems and Agent-Based Modeling Techniques
James P. Crutchfield and Peter Schuster	Evolutionary Dynamics: Exploring the Interplay of Selection, Accident, Neutrality, and Function

Contributors List

Lionel Barnett, *University of Sussex, Centre for Computational Neuroscience and Robotics, School of Cognitive and Computing Sciences, Brighton BN1 9QG, United Kingdon; e-mail: lionelb@cogs.susx.ac.uk*

Aviv Bergman, *Stanford University, Center for Computational Genetics and Biological Modelling, 385 Serra Herrin 428, Stanford, CA 94305-5020; e-mail: aviv@stanford.edu*

Stefan Bornholdt, *Universität Kiel, Inst. für Theoretische Physik, Leibnizstr. 15, D-24098 Kiel, Germany; e-mail: bornholdt@theo-physik.uni-kiel.de*

James P. Crutchfield, *Santa Fe Institute, 1399 Hyde Park Road, Santa Fe, NM 87501; e-mail: chaos@santafe.edu*

Rajarshi Das, *IBM T.J.Watson Research Center, P.O. Box 704, Yorktown Heights, NY 10598; e-mail: rajarshi@watson.ibm.com*

Gunther Eble, *National Museum of Natural History, Smithsonian Institution, MRC-122, Washington, DC 20560 and Santa Fe Institute, 1399 Hyde Park Road, Santa Fe, NM 87501. Present Address: Centre National de la Recherche Scientific, UMR 5561, 6 Boulevard Gabriel, 21000 Dijon, France; e-mail: eble@santafe.edu*

Niles Eldredge, *American Museum of Natural History, Central Park West at 79th Street, New York, NY 10024-5192; e-mail: epunkeek@amnh.org*

Marcus W. Feldman, *Stanford University, Department of Biological Sciences, Stanford, CA 94305-5020; e-mail: marc@charles.stanford.edu*

Sergey Gavrilets, *University of Tennessee, Ecology and Evolutionary Biology, Knoxville, TN 37996; e-mail: gavrila@tiem.utk.edu*

Johan Johansson, *Chalmers University, Department of Microelectronics and Nanoscience, SE-41296, Göteborg, Sweden; e-mail: tfkjj@fy.chalmers.se*

Michael Lachmann, *Santa Fe Institute, 1399 Hyde Park Road, Santa Fe, NM 87501; e-mail: dirk@santafe.edu*

Kristian Lindgren, *Chalmers University, Department of Physical Resource Theory, SE-41296, Goteborg, Sweden; e-mail: frtkl@fy.chalmers.se*

Melanie Mitchell, *Santa Fe Institute, 1399 Hyde Park Road, Santa Fe, NM 87501; e-mail: mm@santafe.edu*

Martin Nilsson, *Los Alamos National Laboratory, Geology and Subsurface, Mail Stop D450, Los Alamos, NM 87545; e-mail: martin@santafe.edu*

Tomoko Ohta, *National Institute of Genetics, Division of Population Genetics, Yata 1111 Mishima, Shizuoka 411-8540, Japan; e-mail: tohta@lab.nig.ac.jp*

Peter Schuster, *University of Vienna, Waehringerstrasse 17, A-1090 Wien, Austria; e-mail: pks@tbi.univie.ac.at*

Guy Sella, *Stanford University/Tel Aviv University, Department of Biological Sciences, Stanford, CA 94305; e-mail: guy@charles.stanford.edu*

Nigel Snoad, *Nuix, Level 8, 433 York Street, Sydney 2000, Australia; e-mail: nigel@santafe.edu*

Peter Stadler, *University of Vienna, Inst. für Theoretische Chemie und Molekulare Strukfurbiologie, Währingerstraßbe 17, A-1090 Wien, Austria; e-mail: stadler@santafe.edu*

Contents

POPULATION GENETICS, DYNAMICS, AND OPTIMIZATION

EVOLUTION OF COOPERATION

Preface: Dynamics of Evolutionary Processes

James P. Crutchfield
Peter Schuster

Starting with a brief historical introduction, we review recent results in the theory of evolutionary dynamics, emphasizing new mathematical and simulation methods that promise to provide experimental access to evolutionary phenomena. One impetus for these developments comes from a fresh look at evolutionary population dynamics from the computer science perspectives of stochastic search and adaptive computation. These applications of evolutionary processes are complemented by new theoretical approaches to Wright's concept of *adaptive landscapes*—analyses that borrow from statistical physics and dynamical systems theory. We also discuss extensions of basic neo-Darwinian dynamics that include, for example, morphological and functional aspects of phenotypes, as well as cooperative interactions between individuals and between species. We argue that understanding the behavioral and structural richness these extensions engender requires new levels of mathematical and theoretical inventiveness that are appropriate for the massive amounts of data soon to be produced by automated experimental evolutionary systems.

1 INTRODUCTION

Evolution unfolds over a stunningly wide range of temporal and spatial scales—from seconds and nanometers at the molecular level to geologic epochs and continents at the macroevolutionary level. The mathematical theory of *evolutionary dynamics* attempts to articulate a consistent conceptual description of these processes—what is similar and what is different across these scales. It even asks if such consistency is at all possible. It focuses especially on how the component mechanisms of selection, genetic variation, population dynamics, the spontaneous emergence of structure, morphological constraints, environmental variation, and so on interact to produce the huge diversity of biological structure, function, and behavior that we observe. Since analysis always has limitations, in its attempt to understand the interplay of these components evolutionary dynamics integrates modern analytical methods from statistics and nonlinear physics and mathematics with current software engineering techniques and simulation methods.

This book presents an up-to-date, but selected, overview of results in the theory and practice of evolutionary dynamics. Starting with a brief historical introduction, here we set the context for the collection by reviewing new developments that promise to provide experimental access to evolutionary phenomena. One impetus for these developments comes from a fresh look at evolutionary population dynamics from the perspective of computer science: stochastic search and adaptive computation. These applications of evolutionary processes are complemented by new theoretical approaches to Wright's concept of *adaptive landscapes*—analyses that borrow heavily from the theories of phase transitions, critical phenomena, and self-organization in statistical physics and of bifurcations, pattern formation, and chaos in dynamical systems theory. These new investigations adopt a rather different stance toward evolutionary dynamics, as compared to (say) mathematical population genetics. To give a sense of this, we review one recent research thread that attempts to address the evolutionary emergence of structural complexity. The new approaches also depend heavily on simulation—partly, to test analytical results but also to push into regimes where analysis fails or cannot yet be employed. When viewed as a whole, the new developments promise to provide deeper insights into the mechanisms of Darwinian optimization through variation and selection and into the roles of stochasticity and nonadaptive evolution—processes beyond the optimization metaphor. The new approaches are particularly important in teasing out how evolutionary mechanisms interact nonlinearly to produce such a wide range of biological organization and dynamical behavior.

There are important extensions of basic neo-Darwinian evolution that include morphological and functional aspects of phenotypes, as well as cooperative interactions between individuals and between species. Cooperative interactions, for example, are now believed to play a predominant role in the major evolutionary transitions. Understanding the behavioral and structural richness these extensions engender will require, we believe, new levels of mathematical and theoretical inventiveness that address the coming deluge of data produced by automated experimental evolutionary systems. Cognizant of these trends, the contributions in this volume on evolutionary dynamics are dedicated to stimulating a balanced integration of creativity, theoretical analysis, and experimental exploration.

2 A BRIEF HISTORY OF EVOLUTIONARY DYNAMICS

For decades after Darwin laid down its basic principles, evolution was the domain of biologists and paleontologists. When the synthetic theory brought the successful union of Darwinian principles with Mendelian genetics [64, 81] at the turn of the nineteenth century, most biologists were confident that they had a solid conceptual basis for biology. The mathematical theory of evolution came to be dominated by population genetics, which was commonly thought to provide a sufficiently deep theoretical framework for analyzing the constituent mechanisms driving evolutionary processes. The synthetic theory—*neo-Darwinian evolution*—came to be viewed as a universal paradigm for biology.

By the mid-twentieth century, however, the successes of the molecular life sciences [47] introduced new perspectives into evolutionary biology. First, a great number of empirical facts and time-worn rules—such as, the Mendelian laws of inheritance, the structure of cellular metabolism, and the mechanisms underlying mutation—found straightforward and satisfactory explanations at the molecular scale. Second, the concepts and methods of chemistry and physics slowly began to make their way into biological thought. The result today is that molecular reasoning has become an indispensable part of biology. Chemical kinetics, as one example, was successfully integrated with population genetics [23, 24], resulting in a single framework for analyzing prokaryotic and, in particular, viral evolution [22]. Finally, and perhaps predictably, the successes of molecular biology have revealed that the fundamental concept of the *gene*—in many ways responsible for this progress—is now in need of an overhaul [52].

As a result, evolution has come to be modeled as an intrinsically stochastic and (nonlinear) dynamical system in which a population of structured individuals, monitored as a set of genotypes, diffuses through the space of all possible genotypes. The diffusion is far from random, but instead is driven by genetic variation and environmental fluctuations and guided by constraints imposed by developmental processes and selection according to phenotypic fitness, for example. Genotype space is now formulated as a sequence space of genes [97], proteins [62], or, most appropriately, polynucleotides [23].

This blueprint is incomplete, however. For example, the discovery of gene sequence variation having no apparent effect on fitness led to the idea of non-adaptive, or *neutral*, evolution [53]. The result of the many-to-one nature of the mappings from genotype to phenotype and phenotype to fitness, neutrality radically modifies the effective architecture of genotype space and so, too, the resulting evolutionary dynamics. At the macroscopic scale, phenotypes appear unchanged for long periods of time, while at the microscopic scale genotypes constantly vary as they diffuse across large, selectively neutral networks in genotype space. In this view, selection acts mainly to eliminate deleterious mutations. The result is that the majority of mutants reaching fixation in a population consists of individuals with selectively neutral genotypes. The flip side is that advantageous mutants are rare and can have little influence on the genetic sequences recorded in phylogenetic trees [53].

Certainly of equal significance with neutrality, the theory of evolutionary dynamics must also confront the difficult questions of how developmental processes interact with, constrain, and drive the evolution of biological complexity [39]. Considering even the simplest class of model, one runs into daunting technical problems. Populating an evolutionary system with pattern-forming individuals leads to a process that operates on two basic time scales. The first is evolutionary and relatively long, the second developmental and markedly faster. As a prerequisite to analyzing this type of two-scale stochastic dynamical system, one must first analyze the nonlinear pattern formation that occurs during development and then identify which of the emergent features become evolutionarily relevant so that selection can act on them. Topping off these complications, a developmental evolutionary theory needs a measure of the emergent features' structure and function, if the theory is to make quantitative predictions.

These observations only serve to emphasize the magnitude of the task of building a theory of evolutionary dynamics that naturally integrates selection, accident, neutrality, structure, and function. Fortunately, biology is not alone in facing many of the attendant theoretical problems. Over the same period that witnessed the flourishing of evolutionary science, starting in the mid- to late-nineteenth century, new concepts and methods were developed in mathematics and the physical sciences that now promise to remove several of the roadblocks to an integrative theory of evolutionary dynamics.

3 ORIGINS OF NOVELTY AND STRUCTURE

In parallel with the rise of evolutionary science, phenomena were investigated in physics, chemistry, and other areas of science outside biology that turned out to be intimately related to biological evolution. From the mid-nineteenth to the mid-twentieth centuries, a number of mechanisms underlying the emergence of randomness and structure in nature were discovered and mathematically analyzed. Today, we now appreciate that phase transitions and critical phenomena, pattern formation, bifurcations, and deterministic chaos occur in both inanimate and animate nature and are implicated in fundamental ways with evolutionary population dynamics and self-organization in biological development. Perhaps somewhat surprisingly, even optimization through variation and selection turned out not to be restricted to biological systems. By way of elucidating several parallels and possible future tools for evolutionary dynamics theory, let us consider these phenomena in turn.

One of the most widely applicable lessons from other disciplines is that when systems consist of competing elementary forces the tensions that arise create structural complexity. Some of the earliest examples of the spontaneous formation of order are found in the equilibrium statistical mechanics of interacting spins on a lattice. At low temperatures, the energy of spin interactions dominates and, in a ferromagnet for example, the neighboring spins align, creating a global order of all "up" spins. At high temperatures, though, thermal fluctuations overcome the local ordering force of spin alignment and configurations consist of an array of randomly oriented spins. There must be an intermediate temperature, it was argued [77], at which the tendencies to order and to disorder balance, producing a new kind of structured *critical state* with long-range correlations and aligned-spin clusters of all sizes. In other words, at these *phase transitions* the forces leading to order and disorder compete, resulting in states more complex than those away from the transition. Analogous tensions are well known in evolution: the selection-mutation balance, the balance between replication fidelity and mutation (the *error threshold*), and the interaction between gene stability, which is required for heredity, and genetic diversity, which is necessary for species adaptability, are only three examples. The emergence of complexity, an apparently common phenomenon, and the parallels between the architecture of physical and biological phenomena only hint at the beginnings of a synthetic theory that describes how evolution generates structure. Fortunately, however, there is a much broader constellation of ideas that can be brought to bear on the difficult conceptual problems of evolutionary dynamics.

Novel, and rather straightforward, interpretations of common phenomena initiated the development of more general theories of self-organization as it occurs in *nonequilibrium* systems. The latter describe *open systems* and so are often more appropriate models for biological processes, which are sustained by fluxes of energy and resources, than equilibrium systems. One example is the spontaneous formation of spatial patterns in systems *far from equilibrium*. Interestingly, this phenomenon had been predicted and developed as a model of embryological morphogenesis in the 1950s by Alan Turing, one of the founders of theoretical computer science [87]. Twenty years later, these ideas were turned into a comprehensive model of pattern formation in biological development [65]. It was only about ten years ago, though, that Turing's predictions were verified in an experimental chemical reaction-diffusion system [15]. There have been even more direct arguments—for example, [41, 69]—that biological evolution itself is an example of far-from-equilibrium self-organization.

Another, complementary approach to the evolution of biological complexity originates from the observation that rich dynamical behavior and intricate structures emerge when a few simple rules are applied over and over again. This is the domain of *dynamical systems theory* [48, 74, 86, 94], which classifies temporal behavior into four categories: (i) equilibrium or *fixed-point* behavior, (ii) oscillations or *limit-cycle* behavior, (iii) *deterministic chaos*, and (iv) *transients* (that relax onto stable behaviors (i), (ii), or (iii)). *Bifurcation theory*, a branch of dynamical systems, analyzes and classifies the structural changes that can occur when one kind of behavior makes a transition to another, as a system control parameter is varied. A result that typifies the kind of general principle available from dynamical systems theory is that the dominant signature of a system undergoing a bifurcation is the enhancement of its transient behavior. That is, the system takes longer and longer to settle down as a transition nears.

A hybrid approach to the emergence of complex behaviors and structures, combining ideas from equilibrium and nonequilibrium statistical mechanics and dynamical systems, is found in the study of *cellular automata*, a class of spatial system consisting of a lattice of locally coupled finite-state machines. Although invented in the 1940s by John von Neumann in part to formally investigate the minimal requirements of self-reproduction [92], the study of their spatiotemporal behavior was rekindled in the early 1980s with the introduction of a classification scheme that mimics the four dynamical categories above [96]. For example, in addition to periodic and "chaotic" behavior, those cellular automata expressing the richest kinds of self-organization, those in "class IV," are associated with transient behavior. It was suggested that their behavior is a product of a dynamical interplay between regularity and disorder [57, 75].

One of the most popular examples of self-organization in cellular automata is John Conway's two-dimensional cellular automaton, the *Game of Life*. The Game of Life produces a wide diversity of intricate static and propagating structures, despite the fact that its behavior is entirely specified by a simple rule that operates on local neighborhoods of "live" or "dead" cells. One notable indication of its behavioral richness is that a universal Turing machine, the most powerful kind of discrete computational device, can be embedded in the Game of Life by carefully programming the initial configuration of live and dead cells [8, 78].

Conway's Game of Life was introduced and popularized in the early 1970s, but it was not until the mid-1980s that a more systematic investigation of the dynamics and structures generated by cellular automata was begun using the methods of dynamical systems, information and computation theories, and statistical mechanics. The general goal was to develop a thorough appreciation of the possible behaviors and structures that systems with demonstrably simple architectures could generate. Implicit in this agenda was the belief that, if one could not develop a consistent vocabulary and set of analytical tools for cellular automata, then systems with more complicated architectures, such as those found in biological processes, would remain forever inaccessible. Exhaustive surveys of cellular automata in one and two spatial dimensions were carried out [96]. The surveys suggested, for example, that universal computation could also be performed by even one-dimensional cellular automata, and constructive proofs of particularly simple examples were then produced [60].

At the time, it was believed that these results suggested an alternative and novel view of the evolution of biological complexity [28, 56]. By varying dynamical properties, it appeared possible that qualitatively different levels of computational structure could emerge in pattern-forming systems and that these levels could become a substrate for novel forms of biological structure and information processing [17, 20, 49, 57]. The emergence of this kind of structural complexity was investigated in an evolutionary setting to test if evolution could find cellular automata with increased computational power [18, 68, 75] or find formal logic systems with increased organization [30, 50, 80].

At about the same time, a new mechanism for self-organization was discovered through simulation studies of critical phenomena in nonequilibrium systems, specifically, simple models of sand-pile avalanching. It turned out that the size distribution of avalanches exhibits a *power-law* scaling indicating that there were no characteristic temporal or spatial scales; there was structure at all scales. More notably, for systems expressing this *self-organized criticality*, it was proposed that systems naturally tend to these complex states, and that, in turn, these states are stable [6]. This kind of self-regulating, stable complexity could play a role in the evolutionary maintenance of biological structure and function.

Note that all of these investigations—some theoretical, many using simulations—turn on the idea that between the extremes of pure order and utter disorder lie behavioral regimes that produce structural complexity. Moreover, it appears that processes, evolutionary or not, can naturally move to structured states—an adaptive behavior that sustains complexity. Very recently, similar ideas have reinvigorated the analysis of the structure and dynamics of networks. It has been suggested that highly structured networks, such as those found in a wide range of natural and artificial systems, lie between regular and purely random topologies [93]. Despite a sometimes checkered history, the notion that complexity arises at the order-disorder border has highlighted an important interplay between dynamics, structure, information processing, and computation in pattern formation and in evolutionary processes.

This brief history of the origins of novelty and structure emphasizes that one of the overriding problems in all fields concerned with self-organization—whether with its phenomenology, emergence analysis, application, biological function, and so on—is the issue of complexity. How does one detect that a system has become organized? For that matter, what does one mean by "organization" in the first place? Where is the "self" in self-organization? These questions have stimulated a substantial, though disparate, body of research that addresses how to define and quantify structural complexity [16]. It seems fair to say, though, that as things stand today the implications for evolutionary dynamics have yet to be fully exploited. This observation leads us to think more broadly about the future.

4 EVOLUTION OF STRUCTURE AND FUNCTION?

Self-organization of the general kinds mentioned here sets the stage for evolution at two levels. First, self-organization guides the processes that produce the structured entities on which variation and selection operate [39, 50]. Second, self-organization can emerge spontaneously in evolutionary dynamics itself, either as complex temporal population dynamics or as spatially structured populations. Of course, these two levels interact, and this interaction greatly complicates mathematical modeling and analysis and controlled experimentation.

What lessons for evolutionary dynamics should one take away from the parallel developments outside biology? First and foremost, they force one to recognize the essential tension between selection, accident, organization, and neutrality in evolutionary processes and that these tensions can be an intrinsic driving force in the generation of structure. One also comes to appreciate the sheer complication that can result from this interplay, which will certainly outstrip that found in simple chaotic dynamical systems and cellular automata. Second and more concretely, they provide a new set of conceptual and analytical tools with which to begin modeling complex evolutionary and developmental processes. An abiding question, however, presents itself, How do we integrate these results and tools into a comprehensive whole? We believe many key pieces (such as the few mentioned above) are now in place. We also believe that their integration has only just begun.

Looking even further ahead, what would be the goal, beyond success in this integration? Perhaps the most important would be that the resulting synthesis leads to *predictive* theories that bear directly on experimental observations. It would appear that there needs to be a new balance between evolutionary theory and experiment that bridges the gulf that now exists. We believe recent developments indicate that such a balance—one more familiar in the physical sciences—is now possible.

In addition to the constant need to revisit and reformulate the mathematical foundations of evolutionary dynamics, as we just argued, we believe that a new conceptual framework for the evolution of structure and function is a pragmatic priority. It is clear, even passé today, that the exploration of molecular genetics proceeds at a breathtaking pace and has led to a rapidly growing number of fully sequenced genomes. Together with an impressive array of other molecular data and new laboratory-scale biological evolutionary systems, the available information represents a vast and untapped wealth that waits to be exploited by theorists with new concepts and analytical methods especially as it highlights our lack of firm principles, definition of structure, and function.

To put some flesh on these bones, in the following two sections we discuss recent experimental and theoretical approaches to evolutionary dynamics. The choice of subjects was guided by a desire to highlight a few stepping stones that may play a role in articulating an integrative and experimentally relevant theory of evolutionary dynamics. The final section briefly introduces the contributions to this volume within this setting. The chapters are intended as a collection of recent ideas in evolutionary dynamics. We will consider the collection a success if it brings some of the current conceptual challenges to the attention of a wide range of theorists and, in this way, is sufficiently provocative to stimulate novel synthetic approaches. We also hope that experimentalists and engineers will find the reviews and the diversity of topics and results a stimulus to new experimental directions.

5 EXPERIMENT AND DESIGN

5.1 BIOLOGICAL AND MOLECULAR MODEL SYSTEMS

Research on biological evolution suffers from the fact that until recently no direct experimental studies have been possible on the dynamics of evolutionary adaptation. The time scales of many evolutionary processes are simply not compatible with experimentalists' lifetimes. Hence, one has to extract information from, for example, the fossil record or from comparisons of genetic sequence data of (almost exclusively) contemporary organisms [14, 59]. Moreover, running control experiments in these cases is out of the question, a situation reminiscent of that found in astrophysics and cosmology.

Fortunately, new biological and molecular model systems have begun to break down the barriers of time and control. Experiments with rapidly multiplying bacteria and molecules replicating in vitro have recently led to a marked reduction in generation times to less than one hour. At this time scale, evolutionary phenomena become observable over days, weeks, and years. To date, populations of the eubacterium *Escherichia coli* have been studied for thousands of generations under precisely controlled conditions [25, 58, 76]. Two findings from these experiments are of particular interest for evolutionary dynamics. First, the evolution of bacterial phenotypes, monitored through recording cell size (a more or less direct correlate of fitness), does not show a gradual adaptation toward an optimum. Rather, innovations in fitness occur in jumps interrupted by rather long metastable epochs [25]. Second, genetic evolution recorded in terms of DNA sequences does not stop during the epochs of phenotypic stasis but proceeds at least at the same pace, if not faster than, during the adaptive innovations [76]. Punctuation in bacterial evolution thus occurs without external triggers, and there is clear evidence for neutral evolution, which manifests itself in genetic changes despite observationally constant phenotypes. Although occurring at the level of single-cell organisms, this intermittent population dynamics reminds one, of course, of the punctuated equilibria proposed to explain the long periods of morphological constancy found in the fossil record [40].

The first attempts to study the evolution of molecules in the test tube date back to the 1960s [84]. In RNA evolution, to take one example, the rate of RNA synthesis is a proxy for the mean fitness of a population of RNA molecules. In RNA evolution experiments, this fitness was optimized in serial transfer experiments. Later on in vitro evolution of RNA was investigated and analyzed in great detail [10], and the mechanism of optimization through mutation and selection is now fully understood at the level of chemical reaction kinetics [9]. One take-home lesson of in vitro molecular evolution concerns the entities that can be subjected to evolutionary optimization: Evolution is not restricted to cells or higher-level organisms. The operation of a Darwinian mechanism can be observed with free molecules in solution, provided that they are capable of replication and that the reaction medium sustains replication.

5.2 EVOLUTIONARY ENGINEERING

Recent engineering applications of in vitro molecular evolution to the production of molecules with predefined properties—called *applied molecular evolution* or *evolutionary biotechnology*—has generated additional key insights on the mechanisms underlying optimization through variation and selection [95]. In these applications, desired molecular properties often do not use high replication efficiency as a proxy for fitness. One illustration of this is found in *molecular breeding*, which plays off the analogy between an evolutionary "producer" of molecules and an animal breeder or gardener in a plant nursery. In contrast to natural selection, in *artificial selection* the experimentalist interrupts the process of optimizing fitness by picking out suitable candidates from the molecular progeny, discarding the remaining variants irrespective of their potential reproductive success. In this way, a molecular breeder implements a modified fitness through its interventions.

Most molecular breeding experiments to date have been performed with RNA or DNA molecules, since they are readily amplified through replication without requiring other molecules as intermediates. (Protein evolution, in contrast, depends on DNA or RNA genes and so requires translation and its attendant complex molecular machinery.) Genetic diversity of molecular populations can be controlled in a more or less straightforward way either by replication with properly adjusted mutation rates or by random chemical synthesis of oligonucleotides. Selection, though, requires ingenious chemical or physical devices. Unfortunately, it would take us too far afield to discuss these here. Instead, we mention two successes in molecular breeding. In the first, RNA or DNA molecules—*aptamers*, which bind specifically to predefined targets—were produced for almost all classes of known (bio)molecules. Binding constants were then optimized through mutation and selection [12, 61]. In the second example, catalysts based on RNA or DNA, known as *(deoxy)ribozymes*, were evolved for a wide variety of natural reactions, as well as for some chemical processes. Interestingly, some of these processes have no counterpart in biochemistry or molecular biology [13, 45].

Molecular breeding illustrates an important feature of evolution in general: The evolutionary process creates or produces "solutions" to "problems" *without* (intentionally) designing them, in contrast to what an engineer would attempt to do. In order to produce molecules with desired properties or functions by means of variation and selection, one need not know the exact molecular structure that solves a task or that expresses a given property or function. Biochemical engineers performing rational design, in contrast, start by constructing a structure

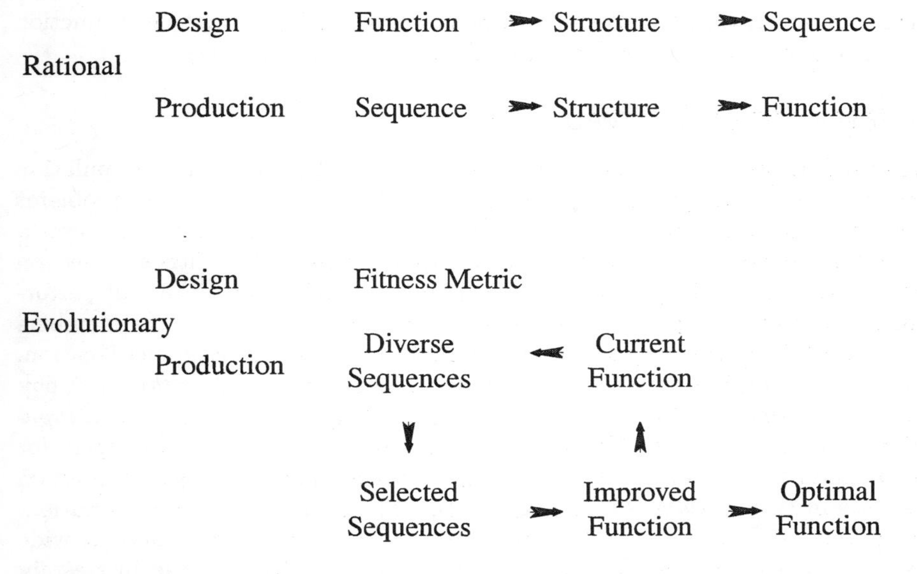

FIGURE 1 Engineering paradigms: Comparing rational and evolutionary design and production methodologies.

that they expect will serve a predefined purpose. Then, they try to find biopolymer sequences that form the structure in question. In the production process, the sequence is synthesized and transformed into the structure with the desired property. Clearly, rational engineering is a different kind of process from evolution.

To draw out this difference and so emphasize the role of dynamics in evolutionary processes, figure 1 compares these two strategies for engineering biomolecules with predefined functions. What the figure illustrates is the dynamical nature of the evolutionary approach: The inclusion of the feedback loop in an evolutionary process is crucial. It introduces an implicit temporal component (dynamics). Most importantly, feedback allows for rapid convergence—even exponentially fast near an optimum.

With the current state of the art, rational design often encounters great difficulties because the contemporary knowledge of the relationships between the sequence, structure, and function of biomolecules is simply not adequate for the requirements of top-down engineering. Given this, it is an advantage in evolutionary production that prior knowledge of structure is not necessary. In fact, one can even forego a focus on reaching a unique optimum. Often two or more structures serving a purpose equally well have been found by evolutionary dynamics. In some ways, this multiplicity is not so unexpected. Different solutions for the same task are frequently observed in macroscopic biology, since only function and efficiency count in the selection-mediated evaluation of fitness.

5.3 EVOLUTIONARY COMPUTATION

Over the last two decades the theory of evolution and, in particular, population genetics have found application to general combinatorial optimization problems as various kinds of population-based stochastic search algorithm. In these, a population consists of candidate solutions that compete via a fitness based on how well they solve the problem. Those (partial) solutions with the best performances are selected as parents of the next generation's population—the members of which are formed from the parents by replication with genetic modification. Variations on this basic theme go by different names: *genetic algorithms* [38, 66], *genetic programming* [54], *evolutionary strategies* [4], and *evolutionary computation* [46], among others. These search methods differ in their preference for one or another type of problem encoding—such as binary strings or function trees—or in their emphasis on one or another mechanism for genetic variation, recombination, and selection. Evolutionary search methods have found a wide range of application; see, for example, [7, 21, 27, 34]. In fact, the interest in adapting evolutionary ideas to problems in computer science—such as artificial intelligence—has a long history [29, 43]. One can view the design of adaptive and evolutionary algorithms as examples of engineering's attempt to take advantage of nature's methods for problem solving [5, 44].

6 CONCEPTS, MODELS, AND METHODS

6.1 "LANDSCAPES"

The concept of *adaptive landscapes* was introduced by Sewall Wright in the 1930s, initially as a metaphor to visualize evolution as a hill-climbing (local optimization) process [79, 97]. Much later, his geographic analogy was revived with the development of formal methods to handle optimization problems in complex physical systems [35]—such as finding configurations with minimum free energy in spin glasses. In simple evolutionary dynamics models, one makes an analogy between a physical system minimizing energy as its state moves down an *energy surface* and an evolving population maximizing fitness as it climbs a "landscape" to an adaptive peak. The *fitness landscape* (or, more properly, *fitness function*) is defined by assigning a reproduction rate to every point in genotype space.

Though specified by simple interaction rules, it turns out that the spin-configuration energy functions of spin glasses are interesting models for a class of evolutionary dynamics in which there are many adaptive peaks. Their main feature is that, as fitness functions, they have a large number of local energy minima [11]. In slightly modified form, spin-glass-like fitness functions are used in the popular NK model of "rugged" fitness landscapes [51], which have been used as simple models of gene-to-gene coupling in genetic networks.

The central question to be answered before one uses the geographic metaphor of landscapes as a starting point for modeling evolutionary dynamics is whether or not a direct analog of an energy function exists. As has been appreciated for some time in mathematical population genetics, this is typically not the case. Except in fairly restricted settings, there is no energy function whose gradient determines the dynamics of evolutionary processes. One generally must employ the theory of dynamical systems [86] or, when the fluctuations due to finite-population sampling are dominant, stochastic extensions of this theory when constructing and analyzing evolutionary models.

Concern over how to best model evolutionary behavior and how to analyze the mechanisms that drive adaptive dynamics has stimulated recent work on a number of different kinds of fitness function that are not "landscapes"—i.e., that do not specify gradient dynamical systems. For example, in the theory of evolutionary computation, the *Royal Road* fitness functions, which require specific blocks of "genes" to be correctly set before a unit of fitness is given to a genotype, were invented to test the "building block" hypothesis [43] that genetic algorithms using crossover during replication preferentially assemble functional subsets of genes. The building-block hypothesis turned out not to hold in general [67], but the study led to a detailed mathematical analysis of the finite-population dynamics of the Royal Road genetic algorithm [91] and the discovery of how neutrality leads to epochal evolution [19].

Another interesting example of investigating the behavioral consequences of structured fitness functions, the *holey adaptive landscape*, was developed in population biology [36, 37] to probe the effects of neutrality on evolutionary dynamics. There, a network of genotypes of (almost equally) high fitness percolates sequence space, leaving holes of low fitness. A similar model, with a neutral genotypic plateau above genotypes of low fitness, was shown to lead spontaneously to the emergence of phenotypes with increased resilience to mutational variation—phenotypes that have *mutational robustness* or, using an older terminology, phenotypes that can sustain a higher *genetic load* [89]. Finally, there have been attempts to investigate evolutionary dynamics produced by molecularly realistic fitness functions based on folding RNA sequences into RNA secondary structures [2, 33]. These studies predicted a high degree of neutrality for properties of RNA molecules [83], which was confirmed recently by an elegant experimental study [82].

A wide class of fitness functions can be compared and studied analytically with respect to the nature of their "ruggedness" using methods based on the algebra of linear operators [85]. They can be compared also in terms of the statistics of adaptive walks or of optimization processes taking place on them [51].

In many of these studies, simulations play an important role, initially giving access to a basic appreciation of the diverse behaviors and properties that can emerge and finally providing confirmation of theoretical predictions. For example, simulation studies of evolutionary dynamics have given direct insights into the mechanisms that promote and inhibit optimization on a spin glass landscape [1], on the Royal Road genetic algorithm [90], and in RNA evolution [31, 32].

Time and again in these mathematical and simulation studies, one is confronted by the sheer complication and richness of evolutionary dynamics. The result is that it is difficult to make blanket statements about how evolution "works." For example, population dynamics depends, critically sometimes, on parameters—such as population size and mutation rate. Varying a parameter, even a small amount, can lead to very different population dynamics. In addition, subtle variations in fitness are simply not seen by selection and so do not control, even local, optimization [73, 88]. One of the overriding lessons is that specifying the fitness function is only one, and sometimes not the dominant, contributor to evolutionary behavior. These observations simply serve to emphasize that the interplay of selection, accident, neutrality, and function is multifaceted. This, of course, points to the challenges and also the opportunities in evolutionary dynamics research.

6.2 COEVOLUTIONARY DYNAMICS AND COOPERATION

Stepping back a bit, note that we have been talking here largely about modeling and analysis *within* the neo-Darwinian framework. One important area of research on evolutionary dynamics that, strictly speaking, lies outside the conventional Darwinian paradigm of variation and selection concerns coevolution and cooperation within and between species. In symbiosis, for example, competition is suppressed because the long-term benefits of cooperation outweigh short-term competitive advantages. The evolution of cooperation that combines competitors into a new functional unit has been invoked as an explanation of the occurrence of major evolutionary transitions [24, 63]. Periods of cooperation, in which Darwinian survival is suppressed, are thought to be implicated in the transition to more complex and hierarchically organized entities. Examples of these include transitions from unicellular to multicellular organisms and from solitary individuals to societies. Notably, cooperation is possible even between molecular species in cell-free assays. Experiments studying the emergence of cooperative molecular assemblies are under way [26]. A mathematical framework to model adaptive dynamics in such non-Darwinian systems has already been developed [42]. Particular attention has also been directed to the evolution of cooperation in animal and human societies [3, 70, 71, 72].

7 A SNAPSHOT OF EVOLUTIONARY DYNAMICS

The contributions to this volume were initiated by a conference, **Towards a Comprehensive Dynamics of Evolution—Exploring the Interplay of**

Selection, Accident, Neutrality, and Function, held October 5–9, 1998 at the Santa Fe Institute. We have loosely grouped them into a few major headings: macroevolution; epochal evolution; population genetics, dynamics, and optimization; and evolution of cooperation. The chapters should be seen as pointing to an enriched perspective on evolutionary dynamics, one that appreciates the sheer diversity of behavior and that acknowledges that this diversity emerges from, and is often not directly determined by, evolution's elementary operations.

Macroevolution: Niles Eldredge starts with an up-to-date presentation on external triggers in biological evolution. Next, Gunther Eble describes the interaction between developmental morphology and evolution using a space of parametrized organism shapes. His analysis of the fossil record reveals different evolutionary stages of morphological divergence and convergence. Then, Stefan Bornholdt reviews the present state of the art in modeling the dynamics of macroscopic biological systems. Despite an overwhelming amount of new data, a comprehensive model of macroevolution is still out of reach, but individual questions—such as the origin of punctuated equilibria—can already be addressed successfully, as several of the other contributors relate.

Epochal Evolution: Aviv Bergman and Marcus Feldman review the classical analysis of punctuation and stasis in evolution. They show that the occurrence of epochal or stepwise evolution need not depend on external triggering. It can be intrinsic to the stochastic sampling dynamics of intermittent fixation that is induced by finite populations. Moreover, punctuation happens independently of many details in the governing evolution equations; in particular, it occurs in asexual as well as in Mendelian populations. James Crutchfield reviews Erik van Nimwegen's and his analysis of the mechanisms leading to epochal evolution via sudden adaptive innovations. They use the methods of maximum entropy and self-averaging from statistical physics to show how evolutionary innovations arise via a series of phase transitions: a population dynamical system can discover (via genetic variation) and then stabilize (through selection) new levels of structural complexity. Sergey Gavrilets introduces his concept of holey landscapes and analyzes the evolutionary dynamics on them. An impressive number of applications of the concept ranging from molecular evolution to organismic evolution and speciation are discussed. Peter Schuster addresses the evolutionary dynamics of asexual reproduction. He reviews the theory of molecular quasispecies and presents a comprehensive theory of evolution for molecular phenotypes, which leads to new definitions of continuity and discontinuities in evolution.

Population Genetics, Dynamics, and Optimization: Tomoko Ohta presents the most recent developments in the nearly neutral theory of evolution, one of the most important extensions of Motoo Kimura's neutral theory. It explicitly considers weakly selected and slightly deleterious mutations. Peter Stadler then presents a formal and mathematically quite demanding theory of adaptive landscapes. Basic to his approach is an analysis of the spectra of linear operators on "rugged landscapes." The Fourier transform provides a useful tool for a classification of linearly decomposable landscapes according to the hardness of optimization dynamics. Next, Nigel Snoad and Martin Nilsson extend the concept of quasispecies to dynamic fitness functions. They find two thresholds for the copying fidelity: The lower bound is given by the well-known error threshold, which itself sets an upper limit on mutation rate. The upper fidelity limit expresses the fact that a population with exact replication is doomed to die out on a dynamic landscape since it cannot adapt to a changing environment. Lionell Barnett studies evolutionary dynamics in finite populations with recombination. He models the system as a birth-and-death process under the assumptions of the Moran model. Interestingly, he observes bistability: different initial conditions give rise to different stationary populations.

Evolution of Cooperation: Guy Sella and Michael Lachmann study the evolutionary dynamics of a population of agents interacting via the Prisoners' Dilemma. They investigate spatial cooperation with a population of agents on a lattice. A common scenario they find consists in life cycles of populations that are established by spreading from single cooperators that then die after invasion by agents with parasitic "defect" strategies. A dynamical steady state with persistent cooperation is encountered when the global birth rate of populations founded by dispersed cooperators is balanced by the death rate of populations caused by invading defectors. Kristian Lindgren and Johan Johansson present results on the evolutionary dynamics of a population of finite-state agents playing the N-person Prisoners' Dilemma—a well-known model for competitive game-theoretic interactions. They analyze a difference equation that models the resulting population dynamics in the case of nonoverlapping generations and asexual reproduction. One class of their models exhibits a predominance of cooperation through a dynamics that avoids less cooperative stable fixed points.

In the final chapter, James Crutchfield, Raja Das, and Melanie Mitchell analyze the evolutionary emergence of global computation in spatial lattices of finite-state machines (cellular automata). They focus particularly on the interaction between a series of evolutionary innovations (ultimately producing high computational performance) and the structural aspects of spatial cooperation that convey high fitness to the best cellular automata.

ACKNOWLEDGMENTS

The workshop **Towards a Comprehensive Dynamics of Evolution—Exploring the Interplay of Selection, Accident, Neutrality, and Function** in October 1998 was hosted and sponsored by the Santa Fe Institute, including core funding from the the National Science Foundation (PHY 9600400). The Keck Foundation's Evolutionary Dynamics Program at SFI provided financial support. JPC's support comes from DARPA contract F30602-00-2-0583, NSF grant IRI-9705830, and AFOSR via NSF grant PHY-9970158. The organizers wish to thank the workshop participants for their generous contributions and stimulating discussions. The authors thank Jennifer Dunne for helpful comments on this chapter and Victoria Alexander for comments and editorial assistance on the book.

RESOURCES

A number of resources for evolutionary dynamics are available at the workshop's website: http://www.santafe.edu/~jpc/evdyn.html. We also recommend the book *Evolution as Computation* [55] as a companion to this one.

REFERENCES

[1] Amitrano, C., L. Peliti, and M. Saber. "A Spin-Glass Model of Evolution." In *Molecular Evolution on Rugged Landscapes*, edited by A. S. Perelson and S. A. Kauffman, 27–38. Santa Fe Institute Studies in the Sciences of Complexity, Proc. Vol. IX. Redwood City, CA: Addison-Wesley, 1991.

[2] Ancel, L. W., and W. Fontana. "Plasticity, Evolvability and Modularity in RNA." *J. Exp. Zool. (Mol. & Dev. Evol.)* **288** (2000): 242–283.

[3] Axelrod, R. *The Evolution of Cooperation.* New York, NY: Basic Books, 1984.

[4] Bäck, T. *Evolutionary Algorithms in Theory and Practice: Evolution Strategies, Evolutionary Programming, Genetic Algorithms.* New York: Oxford University Press, 1996.

[5] Bäck, T., D. Fogel, and Z. Michalewicz. *Handbook of Evolutionary Computation.* Oxford: Oxford University Press, 1996.

[6] Bak, P., C. Tang, and K. Wiesenfeld. "Self-Organized Criticality." *Phys. Rev. A* **38** (1988): 364–374.

[7] Belew, R. K., and L. B. Booker, ed. *Proceedings of the Fourth International Conference on Genetic Algorithms.* San Mateo, CA: Morgan Kaufmann, 1991.

[8] Berlekamp, E. R., J. H. Conway, and R. K. Guy. *Winning Ways for Your Mathematical Plays*, vol. 2. New York: Academic Press, 1984.

[9] Biebricher, C. K., and M. Eigen. "Kinetics of RNA Replication by Qβ Replicase." In *RNA Genetics I: RNA-Directed Virus Replication*, edited by E. Domingo, J. J. Holland, and P. Ahlquist, 1–21. Boca Raton, FL: Plenum Publishing, 1987.

[10] Biebricher, C. K., and W. C. Gardiner. "Molecular Evolution of RNA *in vitro*." *Biophys. Chem.* **66** (1997): 179–192.

[11] Binder, K. "Spin Glasses: Experimental Facts, Theoretical Concepts, and Open Questions." *Rev. Mod. Phys.* **58** (1986): 801–976.
[12] Breaker, R. R. "DNA Aptamers and DNA Enzymes." *Curr. Opin. Chem. Biol.* **1** (1997): 26–31.
[13] Breaker, R. R. "Making Catalytic DNAs." *Science* **290** (2000): 2095–2096.
[14] Carporale, L. H., ed. *Molecular Strategies in Biological Evolution.* Annals of the New York Academy of Sciences, vol. 80. New York: The New York Academy of Sciences, 1999.
[15] Castets, V., E. Dulos, J. Boissonade, and P. De Kepper. "Experimental Evidence of a Sustained Standing Turing-Type Nonequilibrium Chemical Pattern." *Phys. Rev. Lett.* **64** (1990): 2953–2956.
[16] Cowan, G., D. Pines, and D. Melzner, eds. *Complexity: Metaphors, Models, and Reality.* Santa Fe Institute Studies in the Sciences of Complexity, Proc. Vol. XIX. Reading, MA: Addison-Wesley, 1994.
[17] Crutchfield, J. P. "The Calculi of Emergence: Computation, Dynamics, and Induction." *Physica D* **75** (1994): 11–54.
[18] Crutchfield, J. P., and M. Mitchell. "The Evolution of Emergent Computation." *Proc. Natl. Acad. Sci.* **92** (1995): 10742–10746.
[19] Crutchfield, J. P., and E. van Nimwegen. "The Evolutionary Unfolding of Complexity." In *Evolution as Computation*, edited by L. F. Landweber, E. Winfree, R. Lipton, and S. Freeland. Lecture Notes in Computer Science. New York: Springer-Verlag, 2000.
[20] Crutchfield, J. P., and K. Young. "Inferring Statistical Complexity." *Phys. Rev. Lett.* **63** (1989): 105–108.
[21] Davis, L. D., ed. *The Handbook of Genetic Algorithms.* New York: Van Nostrand Reinhold, 1991.
[22] Domingo, E., and J. J. Holland. "RNA Virus Mutations and Fitness for Survival." *Ann. Rev. Microbiol.* **51** (1997): 151–178.
[23] Eigen, M. "Self-Organization of Matter and the Evolution of Biological Macromolecules." *Naturwissenschaften* **58** (1971): 465–523.
[24] Eigen, M., and P. Schuster. *The Hypercycle. A Principle of Natural Self-Organization.* Berlin: Springer-Verlag, 1978.
[25] Elena, S. F., V. S. Cooper, and R. E. Lenski. "Punctuated Evolution Caused by Selection of Rare Beneficial Mutations." *Science* **272** (1996): 1802–1804.
[26] Ellinger, T., R. Ehricht, and J. S. McCaskill. "*In vitro* Evolution of Molecular Cooperation in CATCH, a Cooperatively Coupled Amplification System." *Chem. & Biol.* **5** (1998): 729–741.
[27] Eshelman, L., ed. *Proceedings of the Sixth International Conference on Genetic Algorithms.* San Mateo, CA: Morgan Kaufmann, 1995.
[28] Farmer, J. D., and S. Rasmussen, eds. *Evolution, Games, and Learning: Models for Adaptation in Machines and Nature.* Proceedings of the Fifth Annual International Conference of the Center for Nonlinear Studies, Los Alamos, NM 87545, USA, May 20–24, 1985. Amsterdam: North Holland, 1986.

[29] Fogel, L. J., A. J. Owens, and M. J. Walsh, eds. *Artificial Intelligence through Simulated Evolution.* New York: Wiley, 1966.

[30] Fontana, W. "Algorithmic Chemistry." In *Artificial Life II*, edited by C. Langton, C. Taylor, J. D. Farmer, and S. Rasmussen, 159–209. Santa Fe Institute Stuides in the Sciences of Complexity, Proc. Vol. XI. Redwood City, CA: Addison-Wesley, 1991.

[31] Fontana, W., and P. Schuster. "A Computer Model of Evolutionary Optimization." *Biophys. Chem.* **26** (1987): 123–147.

[32] Fontana, W., and P. Schuster. "Continuity in Evolution. On the Nature of Transitions." *Science* **280** (1998): 1451–1455.

[33] Fontana, W., P. F. Stadler, E. G. Bornberg-Bauer, T. Griesmacher, I. L. Hofacker, M. Tacker, P. Tarazona, E. D. Weinberger, and P. Schuster. "RNA Folding and Combinatory Landscapes." *Phys. Rev. E* **47** (1993): 2083–2099.

[34] Forrest, S., ed. *Proceedings of the Fifth International Conference on Genetic Algorithms.* San Mateo, CA: Morgan Kaufmann, 1993.

[35] Frauenfelder, H., A. R. Bishop, A. Garcia, and A. Perelson, eds. *Landscape Concepts in Physics and Biology.* Amsterdam: Elsevier Science, 1997. Special Issue of Physica D, vol. 107: 2–4.

[36] Gavrilets, S. "Evolution and Speciation on Holey Adaptive Landscapes." *TREE* **12** (1997): 307–312.

[37] Gavrilets, S., and J. Gravner. "Percolation on the Fitness Hypercube and the Evolution of Reproductive Isolation." *J. Theor. Biol.* **184** (1997): 51–64.

[38] Goldberg, D. E. *Genetic Algorithms in Search, Optimization, and Machine Learning.* Reading, MA: Addison-Wesley, 1989.

[39] Goodwin, B. C. *How the Leopard Changed Its Spots: The Evolution of Complexity.* New York: C. Scribner's Sons, 1994.

[40] Gould, S. J., and N. Eldredge. "Punctuated Equilibria: The Tempo and Mode of Evolution Reconsidered." *Paleobiology* **3** (1977): 115–251.

[41] Haken, H. *Synergetics.* Berlin: Springer-Verlag, 1977.

[42] Hofbauer, J., and K. Sigmund. "Adaptive Dynamics and Evolutionary Stability." *Appl. Math. Lett.* **3** (1990): 75–79.

[43] Holland, J. H. *Adaptation in Natural and Artificial Systems.* Ann Arbor: University of Michigan Press, 1975.

[44] Holland, J. H. "Escaping Brittleness: The Possibilities of General Purpose Learning Algorithms Applied to Parallel Rule-Based Systems." In *Machine Lerning*, edited by R. S. Michalski, J. G. Carbonell, and T. M. Mitchell, vol. 2, 593–623. Los Altos, CA: Morgan Kaufmann, 1986.

[45] Jäschke, A., and B. Seelig. "Evolution of DNA and RNA as Catalysts for Chemical Reactions." *Curr. Opin. Chem. Biol.* **4** (2000): 257–262.

[46] Jong, K. D. "Editorial Introduction." *Evol. Comp.* **1** (1993): 1–3.

[47] Judson, H. F. *The Eighth Day of Creation. The Makers of the Revolution in Biology.* London: Jonathan Cape, 1979.

[48] Kaplan, D., and L. Glass. *Understanding Nonlinear Dynamics.* New York, NY: Springer-Verlag, 1995.

[49] Kauffman, S. A. "Metabolic Stability and Epigenesis in Randomly Connected Nets." *J. Theor. Biol.* **22** (1969): 437–467.
[50] Kauffman, S. A. *The Origins of Order. Self-Organization and Selection in Evolution.* New York: Oxford University Press, 1993.
[51] Kauffman, S. A., and S. Levin. "Towards a General Theory of Adaptive Walks on Rugged Landscapes." *J. Theor. Biol.* **128** (1987): 11–45.
[52] Keller, E. F. *The Century of the Gene.* Cambridge, MA: Harvard University Press, 2000.
[53] Kimura, M. *The Neutral Theory of Evolution.* Cambridge, UK: Cambridge University Press, 1983.
[54] Koza, J. R. *Genetic Programming: On the Programming of Computers by Means of Natural Selection.* Cambridge, MA: MIT Press, 1992.
[55] Landweber, L. F., and E. Winfree, eds. *Evolution as Computation.* Natural Computing Series. New York: Springer-Verlag, 2000.
[56] Langton, C. G., ed. *Artificial Life.* Redwood City, CA: Addison-Wesley, 1989.
[57] Langton, C. G. "Computation at the Edge of Chaos: Phase Transitions and Emergent Computation." In *Emergent Computation*, edited by S. Forrest, 12–37. Amsterdam: North-Holland, 1990.
[58] Lenski, R. E. "Evolution in Experimental Populations of Bacteria." In *Population Genetics of Bacteria*, edited by S. Baumberg, J. P. W. Young, E. M. H. Wellington, and J. R. Saunders, 193–215. Cambridge, UK: Cambridge University Press, 1974.
[59] Li, W. H., and D. Graur. *Fundamentals of Molecular Evolution.* Sunderland, MA: Sinauer Associates, 1991.
[60] Lindgren, K., and M. G. Nordahl. "Universal Computation in a Simple One-Dimensional Cellular Automaton." *Complex Systems* **4** (1990): 299–318.
[61] Marshall, K. A., and A. D. Ellington. "*In vitro* Selection of RNA Aptamers." *Methods Enzymol.* **318** (2000): 193–214.
[62] Maynard Smith, J. "Natural Selection and the Concept of a Protein Space." *Nature* **225** (1970): 563–564.
[63] Maynard Smith, J., and E. Szathmàry. *The Major Transitions in Evolution.* Oxford, UK: W. H. Freeman, 1995.
[64] Mayr, E., and W. B. Provine, eds. *The Evolutionary Synthesis. Perspectives of the Unification of Biology.* Cambridge, MA: Harvard University Press, 1980.
[65] Meinhardt, H. *Models of Biological Pattern Formation.* London: Academic Press, 1982.
[66] Mitchell, M. *An Introduction to Genetic Algorithms.* Cambridge, MA: MIT Press, 1996.
[67] Mitchell, M., J. H. Holland, and S. Forrest. "When Will a Genetic Algorithm Outperform Hillclimbing?" In *Advances in Neural Information Processing Systems*, edited by J. D. Cowan, G. Tesauro, and J. Alspector, vol. 6, 51–56, San Mateo, CA: Morgan Kauffman, 1993.

[68] Mitchell, M., P. Hraber, and J. P. Crutchfield. "Revisiting the Edge of Chaos: Evolving Cellular Automata to Perform Computations." *Complex Systems* **7** (1993): 89–130.

[69] Nicolis, G., and I. Prigogine. *Self-Organization in Non-equilibrium Systems.* New York: Wiley-Interscience, 1977.

[70] Nowak, M. A., R. M. May, and K. Sigmund. "The Arithmetics of Mutual Help." *Sci. Am.* **272** (1995): 50–55.

[71] Nowak, M. A., and K. Sigmund. "The Dynamics of Indirect Reciprocity." *J. Theor. Biol.* **194** (1998): 561–574.

[72] Nowak, M. A., and K. Sigmund. "Evolution of Indirect Reciprocity by Image Scoring." *Nature* **393** (1998): 573–577.

[73] Ohta, T., and J. H. Gillespie. "Development of Neutral and Nearly Neutral Theories." *Theor. Pop. Biol.* **49** (1996): 128–142.

[74] Ott, E. *Chaos in Dynamical Systems.* New York: Cambridge University Press, 1993.

[75] Packard, N. H. "Adaptation Toward the Edge of Chaos." In *Dynamic Patterns in Complex Systems*, edited by A. J. M. J. A. S. Kelso and M. F. Shlesinger, 293–301. Singapore: World Scientific, 1988.

[76] Papadopoulos, D., D. Schneider, J. Meier-Eiss, W. Arber, R. E. Lenski, and M. Blot. "Genomic Evolution during a 10,000-Generation Experiment with Bacteria." *Proc. Natl. Acad. Sci.* **96** (1999): 3807–3812.

[77] Peierls, R. "On Ising's Model of Ferromagnetism." *Proc. Cambridge Phil. Soc.* **32** (1936): 477–481.

[78] Poundstone, W. *The Recursive Universe: Cosmic Complexity and the Limits of Scientific Knowledge.* New York, NY: Morrow, 1985.

[79] Provine, W. B. *Sewall Wright and Evolutionary Biology*, 304–317. Chicago, IL: University of Chicago Press, 1986.

[80] Ray, T. S. "An Approach to the Synthesis of Life." In *Artificial Life II*, edited by C. Langton, C. Taylor, J. D. Farmer, and S. Rasmussen, 371–408. Santa Fe Institute Stuides in the Sciences of Complexity, Proc. Vol. XI. Redwood City, CA: Addison-Wesley, 1991.

[81] Reif, W. E., T. Junker, and U. Hoßfeld. "The Synthetic Theory of Evolution: General Problems and the German Contribution to the Synthesis." *Theor. Biosci.* **119** (2000): 41–91.

[82] Schultes, E. A., and D. P. Bartel. "One Sequence, Two Ribozymes: Implications for the Emergence of New Ribozyme Folds." *Science* **289** (2000): 448–452.

[83] Schuster, P., W. Fontana, P. F. Stadler, and I. L. Hofacker. "From Sequences to Shapes and Back: A Case Study in RNA Secondary Structures." *Proc. Roy. Soc. Lond. B* **255** (1994): 279–284.

[84] Spiegelman, S. "An Approach to the Experimental Analysis of Precellular Evolution." *Quart. Rev. Biophys.* **4** (1971): 213–253.

[85] Stadler, P. F. "Landscapes and Their Correlation Functions." *J. Math. Chem.* **20** (1996): 1–45.

[86] Strogatz, S. H. *Nonlinear Dynamics and Chaos: With Applications to Physics, Biology, Chemistry, and Engineering.* Reading, MA: Addison-Wesley, 1994.

[87] Turing, A. M. "The Chemical Basis of Morphogenesis." *Philos. Trans. Roy. Soc. Lond. B* **237** (1952): 37–72.

[88] van Nimwegen, E., and J. P. Crutchfield. "Optimizing Epochal Evolutionary Search: Population-Size Dependent Theory." *Machine Learning* (2001): to appear.

[89] van Nimwegen, E., J. P. Crutchfield, and M. A. Huynen. "Neutral Evolution of Mutational Robustness." *Proc. Natl. Acad. Sci. USA* **96** (1999): 9716–9720.

[90] van Nimwegen, E., J. P. Crutchfield, and M. Mitchell. "Finite Populations Induce Metastability in Evolutionary Search." *Phys. Lett. A* **229** (1997): 144–150.

[91] van Nimwegen, E., J. P. Crutchfield, and M. Mitchell. "Statistical Dynamics of the Royal Road Genetic Algorithm." *Theor. Comp. Sci.* **229** (1999): 41–102.

[92] von Neumann, J. *Theory of Self-Reproducing Automata.* Urbana: University of Illinois Press, 1966.

[93] Watts, D. J., and S. H. Strogatz. "Collective Dynamics of 'Small-World' Networks." *Nature* **393** (1998): 440–442.

[94] Wiggins, S. *Introduction to Applied Nonlinear Dynamical Systems and Chaos.* New York: Springer-Verlag, 1990.

[95] Wilson, D. S., and J. W. Szostak. "*In vitro* Selection of Functional Nucleic Acid." *Ann. Rev. Biochem.* **68** (1999): 611–647.

[96] Wolfram, S. *Theory and Applications of Cellular Automata.* Singapore: World Scientific Publishers, 1986.

[97] Wright, S. "The Roles of Mutation, Inbreeding, Crossbreeding and Selection in Evolution." In *Intl. Proceedings of the Sixth International Congress on Genetics*, edited by D. F. Jones, vol. 1, 356–366. Brooklyn, NY: Brooklyn Botanic Garden, 1932.

Macroevolution

The Sloshing Bucket: How The Physical Realm Controls Evolution

Niles Eldredge

What drives evolution? Is the history of life deeply contingent as some (most notably, perhaps, S. J. Gould—see e.g., Gould [33]) would have it? Or is the dominant signal in evolution a sort of stochastic determinism, with natural selection constantly coming up with similar adaptive "solutions" to the two classes of "problems" all organisms face: (1) obtaining nutrients and energy to differentiate, grow, and simply stay alive, and (2) reproducing? This latter alternative—one that sees natural selection as not only necessary but entirely in itself *sufficient* to explain the history of life—and thus the process of evolution—has long been the dominant view. Its origin, of course, lies in Darwin's[1] [8] original exposition, but it clearly lives on, albeit in starkly extreme form, in such formulations as Dawkins's "selfish gene."

[1] I find the teleology of this take on the evolutionary process downright baffling—as other theorists closely tied with ultradarwinism (e.g., especially George Williams [62, 63]) have been particularly articulate and persuasive on the inability of selection to "see" into the future. How could a gene, or an organism, care about seeing versions of its genetic information in place in future generations?

As I have pointed out elsewhere, "selfish gene" models of natural selection in effect invert Darwin's original concept: Darwin [8] (see p. 5 for a very clear statement) saw differential success (vis-à-vis conspecifics in a local population) in the economic side of an organism's life as having a differential effect on what we now call "reproductive success." He contrasted such "natural" selection with "sexual" selection (especially clearly articulated in Darwin [9, p. 256])—which he saw as differential reproductive success arising purely from reproductive adaptations themselves. Dawkins and other modern "ultradarwinians," in contrast see selection in all forms as arising initially from competition for reproductive success—be it among organisms themselves, or in an even more starkly reductive mode, among organisms' genes themselves. Though the biotic and abiotic environment may determine the statistical "winners" of such a competitive race, causality in the evolutionary process stems, at base, from the competition for reproductive success itself.

Such gene-centered views of the very heart and soul of the evolutionary process transform natural selection from its traditional conception as, in effect, an information filter (where what gets passed along to succeeding generations is the genetic information underlying differential economic success) to something resembling an "active force" shaping biological evolution. The older version of natural selection, in contrast, lets us understand evolution as simply the historical fallout of what I take to be the *real* dynamics of life: relative success in the economic sphere (within populations, or *avatars* [7] in the context of local ecosystems, as explained in section 2) impinging on reproductive success within *demes.*

In this chapter, I develop a model of the evolutionary process that starts from the other end-member extreme—as far away from "competition among genes for reproductive success" as can possibly be articulated. My aim is to specify the causal relations between physical environmental events and biotic response, taking the overarching view that environmental factors (including biotic factors) generally impinge on cross-genealogical biotic systems—"ecosystems"—and ultimately on over to the genealogical systems (genomes, demes, species, higher taxa) that are the normal and traditional purview of evolutionary biology.

The great virtue of such an approach is that it yields a vision of the evolutionary process much more readily matched with events in the history of life—especially, though not exclusively, as seen in the fossil record. The downside, of course, is that the approach—consciously and deliberately—is indeed an extreme end member and does not explicitly fold in the myriad microevolutionary dynamics that have been so well established in population genetics and ecology. Fortunately, growing numbers of paleontologists and systematists, ecologists, and geneticists (both population and molecular) realize that biological systems, hence processes, are hierarchically arrayed—and the results of one discipline are by no means automatically to be held as either antithetical—or strictly reducible—to results obtained in other disciplines. For example, stasis within species (see below for discussion) by no means negates the very commonly encountered result that portions, at least, of a given species' collective genome are in constant flux—if not total chaos.

Thus ultimate understanding of the evolutionary process, at least in my opinion, comes from a more precise understanding of the ontological nature of biological systems and their relations to one another. The scheme presented below is but one of several that have been mooted—and is not presented as the ultimate word. But I do believe the hierarchical approach is bound to be more successful than George Gaylord Simpson's [52] abortive attempt to synthesize the data of paleontology with the then-understanding of what he called the "determinants" of the evolutionary process in the genetics of his day. In the preface of his *Tempo and Mode in Evolution* (Simpson [52, p. xv]), he writes:

> "Not long ago paleontologists felt that a geneticist was a person who shut himself in a room, pulled down the shades, watched small flies disporting themselves in bottles, and thought he was studying nature. A pursuit so removed from the realities of life, they said, had no significance for the true biologist. On the other hand, the geneticists said that paleontology had no further contributions to make to biology, that its only point had been the completed demonstration of the truth of evolution, and that it was a subject too purely descriptive to merit the name 'science.' The paleontologist, they believed, is like a man who undertakes to study the principles of the internal combustion engine by standing on a street corner and watching the motor cars whiz by."

Simpson's approach was frankly reductionist—in that he used his understanding of contemporary genetics to explain some recurring phenomena he delineated in the fossil record. He developed a model—"Quantum Evolution"—that used Sewall Wright's [64, 65] adaptive landscape imagery and relied heavily on Wright's notion of genetic drift as a means whereby small populations of organisms might rapidly "lose" their adaptations, go through a phase that can only be described as "adaptive uncertainty," and finally developing sufficient embryonic adaptations that selection could rapidly drive the population up a different adaptive peak. The model—indeed, Simpson's entire enterprise to bridge the gap between paleontology and genetics, foundered largely because Wright [66] lambasted Simpson's use of his imagery, models, and mathematics, and because Simpson himself, in his lengthy revision of *Tempo and Mode in Evolution* (renamed *The Major Features of Evolution* [53]), revised his three-step notion of quantum evolution to simply one of rapid phyletic evolution—entirely under the guidance of natural selection throughout (see Gould [32], on Simpson's modification of his earlier views as an aspect of the "hardening" of the "synthesis").

But what Simpson did leave as a vibrant intellectual legacy—one that informs the entire middle section of this chapter—is his recognition that the history of life is fraught with hauntingly similar, repeated patterns. And he was right to assume that process, consistent with (but, I would add, not necessarily reducible to) the "determinants" of evolution as seen, for example, in traditional population genetics, can be inferred from these very patterns themselves. Patterns lie at the very heart of all scientific work: they suggest the problems and the solutions that scientists investigate—be these patterns the traces of evanescent subatomic particles generated in cyclotrons, the patterns of kinship linking up all of life, or the sorts of patterns that Simpson himself singled out: the relatively abrupt appearance of higher taxa, stamped with highly divergent novel adaptations implying rapid bursts of evolution as typical of the origin of most higher taxa.

Thus, in the Simpsonian spirit of taking patterns in the history of life seriously—in the utter conviction that such patterns must at the very least be explained by a complete evolutionary theory, and in the further belief that such patterns actually have much to tell us about how the evolutionary process actually operates—I devote the next section to an examination of three utterly typical patterns in the history of life. These patterns are additive, in the sense that Pattern 1 is a part of Pattern 2, which is itself a part of Pattern 3. In considering them in their proper order, we obtain a first-order approximation of how genealogical systems—the traditional purview of evolutionary theory, and rightly so—connect with cross-genealogical, ecological systems—a connection that further suggests how the physical environment is related to—and, I am convinced, to a great extent "controls"—the evolutionary process.

1 THREE PATTERNS IN THE HISTORY OF LIFE

PATTERN 1: STASIS

Darwin [8, chs. 9 and 10] attributed the lack of examples of "imperceptibly graded series" of fossils to the infancy of paleontology and the incompleteness of the fossil record itself. Darwin's vision of the workings of natural selection on heritable variation extrapolated naturally to a picture of gradual incremental change. New species arose from old, but only by the gradual wholesale transformation of an entire lineage. Were the fossil record complete, we would not be able to discern where the old species left off and the new one began. Thus it was Darwin who initiated the mantra of thanksgiving that the fossil record is *not* complete—otherwise we would not be able to draw definitive boundaries around species nor classify living systems with any ease. Darwin attributed the spottiness of the fossil record to vagaries of preservation and discovery—in so doing, performing the very useful task of founding the field of taphonomy.

Nevermind that the common paleontological experience of the day already knew there was pattern in the history of species's existence. Instead of seeing a swiss-cheese record distorting a true signal of gradual phyletic change, however, several prominent paleontologists of the day were quick to point out that the dominant signal was strong, despite the gaps inherent in the record itself. In his very useful compendium of reviews of the first edition of Darwin's *Origin*, David

Hull [35] reprints several written by experienced paleontologists. All complained that Darwin had failed to note the strongest pattern of all: the manifest stability of species once they first appear in the record. For example, F. J. Pictet writes:

> "The theory of Mr. Darwin supposes that species were modified during thousands of generations by very minimal changes in their characters. Thus, all species must be connected with their common ancestors and, in consequence, with each other in all possible degrees. Why don't we find these gradations in the fossil record, and why, instead of collecting thousands of identical individuals, do we not find more intermediary forms?" (Pictet [45], as reprinted in Hull [35, p. 149].)

Spotty as the record may be (and its spottiness varies greatly from environment, taxon, and to some extent age, of the example in question), some signal is bound to get through. These early paleontologists, without having any reliable basis for assigning actual lengths of time to events and patterns of lineage persistence in the fossil record,[2] nonetheless saw that species tend to persist recognizably the same (with little net evolutionary change) for periods far longer than the events that brought them into existence and that caused their eventual demise—if, that is, their typically abrupt appearances and disappearances in the record accurately mirrored true evolutionary and extinction episodes.

We now know that these early paleontologists were correctly perceiving a dominant signal in the history of life (fig. 1). For reasons difficult to fathom completely, paleontologists themselves fell into "the fossil record is too spotty" mantra in the latter half of the nineteenth century—though some did "look for evolution in the rocks," and proclaimed that their patterns upheld Darwin's extrapolated vision of slow steady transformation of ancestral species into descendants [2, 48]. Thus it is not, in any real sense, the fault of early evolutionary biologists unschooled in the data of paleontology that stasis was lost as a dominant pattern in the history of life: paleontologists themselves for the most part stopped discussing stasis and its implications for understanding the evolutionary process.

Since its "rediscovery" in the early 1970s (see Eldredge [16, 23], where it was given its name), stasis has come to be nearly universally recognized among both paleontologists and biologists whose focus is on the Recent. Paleontologists once intransigently opposed to recognizing stasis as the dominant signal [31] now concede its reality. As do prominent evolutionary biologists, who themselves nonetheless find much in "punctuated equilibria" with which to disagree [42, 63]. The questions become, What exactly is stasis? and What *causes* stasis?

---

[2]Ironically, it was Charles Darwin, probably more than any other mid-nineteenth-century scientist, who most contributed to the realization of the true dimensions of geological time. Darwin needed long periods of time for his extrapolated vision of evolution to produce modern diversity—and it was he who most visibly spoke in terms of hundreds of millions of years—an order of magnitude more than most of his geological colleagues were prone to use.

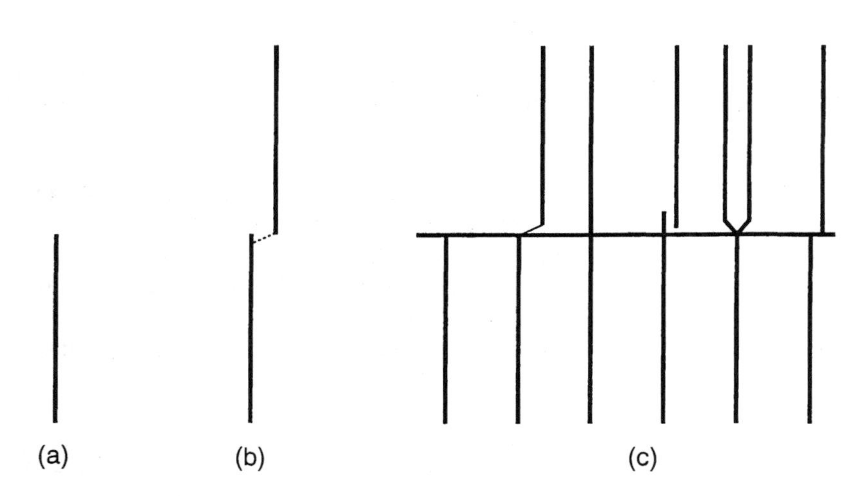

FIGURE 1 Three additive evolutionary patterns. (a) Pattern 1: Stasis. Depicted as a simple vertical line, stasis of species actually encompasses within- and among-population variation through time, with little or no cumulative evolutionary change. (b) Pattern 2: Adaptive change correlated with speciation. Little or no stratigraphic overlap (as depicted here) might be involved or there may be significant temporal overlap between ancestral and descendant species. (c) Pattern 3: "Coordinated stasis" or "turnover pulse." The dominant pattern of extinction and evolution in the last half billion years of the history of life, incorporating elements of both Patterns 1 and 2.

The answer to the first question is relatively simple: Species in the fossil record show very much the same sorts of within- and among-population phenotypic variation as is routinely documented in extant species. Some vary a lot, some relatively little. Some occur over vast expanses and inhabit a variety of environments; while others are restricted to narrow geographic ranges and habitats. Mean values of quantitative characteristics, as well as the less precise assessment of variation in qualitative characteristics, tend naturally enough to vary as one samples through time—again, both within and among populations. Some of those shifts in mean value are pure sampling error; but some of them, surely, represent true evolutionary change. But, contrary to a modern, restated version of Darwin's original extrapolated predictions, seldom if ever does the variation within an entire species become translated in any one particular direction for very long within the entire history of a species. This is as clear in the quantitative data graphed and statistically analyzed in purported cases of phyletic gradualism [30, 31, 51] (to name two of the more celebrated instances) as it is for graphs of examples whose original interpretation was stasis. The point is simply that the phenotypes of organisms within most species end up looking pretty much like they did when they started—and that is true, in the case of

Paleozoic marine invertebrates, at any rate, for species durations of 7–10 million years or even more.

Thus stasis is a dynamic, nonregular oscillation of within- and among-population variation of great empirical generality. It is truly a major pattern in the history of life that must be acknowledged before a complete understanding of the evolutionary process can be reached. The next question, then, is: what *causes* stasis?

George Williams [63, p. 127ff.] has looked at stasis and concluded that, with all the variation going on within the genomes of populations, hence of species, there must be some sort of pruning mechanism that regularly erases true evolutionary change, leaving for the fossil record just the conservative lineage that keeps spawling off new species. This theoretical gambit has some attractive qualities—not least that Williams is positing a higher-level form of selection ("species selection") to explain stasis. But Williams's suggested solution fails because it neither (1) explains the stability of the core conservative lineage nor does it acknowledge (2) the stability encountered among the short-lived, highly modified evolutionary spin-offs of the core conservative lineage. Something more is clearly needed.

Consider almost any species of the modern biota, for example, the American robin (*Turdus migratorius*) living on the North American continent. This particular species has a great geographic range: Virtually all of the North American continent—if summer and winter ranges are added to the year-round range that encompasses all but the northernmost tier of states. It inhabits an equally impressive spectrum of habitats (to name but two—from the deep moist woods of the Adirondack Mountains to the high-elevation arid scrub of Santa Fe—with a patchwork quilt of habitats, not to mention suburban lawns and gardens, in between). Think about this species and its vast range, and then think about how natural selection can possibly take this entire melange and "march" it in any one particular evolutionary direction.

Such a "gedanken" experiment easily boggles the mind, as it seems on the face of it unlikely that selection could effect such evolutionary change on such a far-flung collection of organisms. Further thought reveals why. Robins, like all other species, are broken up into local populations that are integrated into—play concerted roles within—local ecosystems. Life for a robin on the grounds of the Santa Fe Institute in August is very different than life for a robin in the Adirondack deep woods at the same time. Food resources are different, as are water availability, predators, and pathogens. Two related considerations immediately follow, both of which shed light on why stasis should be the *expected* pattern of most species's histories:

1. With the possible exception of some pandemic pathogenic infections, the only "selection force" that might reasonably be expected to exert an influence on all members of such a far-flung species would be physical climate. Even among climatic factors, only concerted (global) cooling or warming might

be imagined to affect all members of a species—though even in the case of temperature change the local manifestation of such events (just like changes in rainfall pattern) is unevenly spotty.

Now, traditional models tended to view selection as tracking physical environmental change. Provided the requisite genetic variation were present, selection, for example, might modify the physiologies and anatomies of organisms subjected to concerted bouts of global cooling. What really happens, we now know (e.g., Coope [6]; Davis [10]), is that extreme conditions of global cooling, where continental glaciers expand southward in the northern continents as they have done four times in the past 1.65 million years, causes a collapse toward the equator of tundra, evergreen (boreal) forest, mixed hardwood forests, etc. Not that ecosystems move smoothly and in lock-step fashion. But communities are reassembled further south that are composed of populations of more-or-less the same species previously occurring further north.

In other words, species track displaced habitats much more than selection tracks environmental change in any one place. Everything can move, including plants through their propagules. The literature on Pleistocene movements—and indeed on migrations northward in the last century as global warming has set in—is resoundingly clear on this point. And as long as species can find and inhabit "recognizable" habitat (i.e., recognizable in terms of adaptations already in place), they will tend to persist, and to do so more or less unchanged. There were lions in Trafalgar Square during an exceptionally warm interglacial some 120,000 years ago, as well as hippos in the Thames and a particular species of beetle living nearby. The beetle (same species, based on male genitalia) persists today in southern Italy [6], and the lions and hippos were the very same species known today only from Africa (and India in the case of the lions).

Successful habitat tracking amounts to stabilizing selection in the face of environmental change. Extinction is the second most likely outcome—as when climatic change goes too far too fast and recognizable habitats cannot be found (a topic to which we return under Pattern 3 below). Directional selection modifying adaptations, as per the original model of adaptive modification of an entire species, is the least likely outcome and only occurs in the exceptional circumstances discussed below.

2. Consider, again, the fact that all species—even those with only moderately broad distributions—are broken up into local populations and integrated into ecosystems that differ in detail among themselves: from the point of view of the niche of the focal populations of conspecifics (meaning the robins on the grounds of the Santa Fe Institute versus the robins in the Adirondack deep woods), to some degree or other different food resources, predators, climatic regimes, etc. If even a pandemic selection factor such as climate change does not take an entire species and modify it to fit changing conditions, what is the probability that the initial sampling differences of the genomes in all these

different populations, plus different mutational histories, drift and selection regimes peculiar to each ecological setting, will propel *an entire species* in any one particular evolutionary direction?

Recently, I (e.g., Eldredge [20]; Eldredge and Gould [25]) have been suggesting that Wright's very notion of the adaptive landscape (e.g., Wright [64, 65]) ought to have been sufficient to predict the ubiquity of stasis. These statements need clarification: I had in mind the version of Wright's landscape that sees local populations of a species occupying peaks of varying heights on the landscape (itself an extension of the primordial concept—which saw, instead, different "gene combinations" within a species as occupying peaks of varying heights). Thanks in no small measure to discussions following presentation of this chapter at the Santa Fe Institute (October, 1998), I am reminded that Wright's original intent was to explain how genes conveying superior fitness in one population may spread to another and, thus, provide an effective model for precisely what I am claiming is seldom encountered in the history of life: a smoothly gradational evolutionary change affecting an entire species!

Wright's shifting balance, however, can work only for situations where genes conveying higher fitness do so without respect to the local climatic and biotic environments in which all the various "demes" ("avatars" is better in an ecological context) exist. Though such "general adaptations" can readily be imagined, the original and still the core problem of evolutionary theory is to explain the phenotypic diversity of life—and that diversity in the main reflects economic adaptations to specific environmental conditions. That's what I had in mind when I originally said that the adaptive landscape itself should have led to expectations of stasis and not to transformation of an entire species. When I look at a landscape depicting different avatars/demes, I automatically think of how disparate the environmental conditions are for each deme in the real world and imagine the different quasi-autonomous evolutionary histories each deme will undergo. The net effect, when summing up the evolutionary change over an interval of time over an entire species range, would have to be a zero-sum game: just looking at the internal organization of species broken up into avatars within local ecosystems should lead to the null hypothesis of stasis for the entire species.

But there are further corollary predictions. If selection should be expected not to produce patterns of directional phyletic change of entire species, we would nonetheless predict (see Eldredge [19, p. 88]) such examples of directional phyletic change that might be documented in the fossil record to come from samples drawn either from (1) restricted regions within a species distribution or (2) samples drawn from the "same" communities that can be shown to be tracking habitats. For example, Gingerich [31] has documented instances of directional phyletic change (areal increase of the surface of first lower molars M_1) in several lineages of Eocene mammals. All of his samples (meticulously collected, documented, measured, and analyzed as they were) were drawn from the Bighorn Basin—but one of several basins of deposition preserving fossils of these various species lineages. Indeed, Schankler [50] (see Prothero [46] for review) has shown at least for some of these lineages that the histories of M_1 size were different within the same lineage even in different parts of the Bighorn Basin.

Lieberman and colleagues [40, 41] studied two different species of brachiopods from the Middle Devonian Hamilton formation, representing an interval of time of some 6 or 7 million years. Both species occurred in a variety of community settings. Though the net history of both species was stasis, samples from one of the more widespread, open-marine settings did show some gradual phyletic change. Teasing the ecological settings in which a species is found into finer categories and tracing the evolutionary histories of portions of species restricted to each of the communities types typically reveals gradual change. Though even in these two examples (as in Gingerich's mammals and Sheldon's trilobites from the Builth Inlier in Wales—likewise in all likelihood a restricted aspect of the species' geographic ranges) the phyletic changes seen in these subsets of species eventually reversed—amounting to no net change: stasis.

If stasis is the norm in evolutionary history and if stasis, upon serious reflection, is a none-too-surprising outcome of (1) habitat tracking and (2) the internal structure of species vis-à-vis the distribution of avatars in different ecosystems, the next question becomes: when does adaptive change come in evolution?

PATTERN 2: SPECIATION AND ADAPTIVE CHANGE

Perhaps the greatest surprise among the repeated empirical patterns in the history of life is not stasis (though it counters Darwinian extrapolated predictions), but the second pattern, left by two species, both in stasis and in presumed ancestral-descendant relationship (fig. 1(b)).[3] For the simple fact is that, with

[3] I say "presumed" ancestral-descendant relationship because such evolutionary relationships among species-level taxa are notoriously difficult to test—i.e., falsify. (See Schaeffer, Hecht, and Eldredge [49]; Eldredge and Tattersall [28]; and Eldredge and Crancraft [22], for early discussions of the relative testability of sister group relations in cladograms, versus ancestral-descendant relationships of evolutionary trees and even more complex statements of evolutionary history.) The problem lies in the necessity that true ancestors must by definition be plesiomorphic in all respects to descendants, a form of negative evidence: testability in cladistics relies on positive evidence in the form of (syn-)apomorphy. However, the ubiquity of multiple, cross-genealogical patterns (Pattern 3), which are multiple examples of Pattern 2 affecting independent lineages at the same time and place as discussed in the next section—where presumed ancestors are in fact plesiomorphic and stratigraphically older than their presumed descendants—make it clear that, if any one hypothesis of ancestry and descent may be in doubt, when many putative cases are considered together, we may be more confident that at least some of them are correct.

On a related matter, the patterns and theories of process discussed in this chapter are every bit as testable as are cladograms, and in the very same manner. We test, for example, stasis (Pattern 1) by continuing to collect further samples from within the time range of a species-level taxon. We test Pattern 2 by adding further data to the within-lineage histories of putative ancestor and putative descendant and by gathering stratigraphic data pertaining to their overlap (or not) in time. We test Pattern 3 by doing the same for many independent lineages. And we test the generality of these patterns by examining other biotas from different parts of the geologic column.

stasis being the norm, adaptive change is ipso facto correlated with the origin of new species.

Darwin, of course, would not have been surprised at this since he defined species as collections of morphologically distinguishable organisms—collections that evolve one from another through a cumulative process of adaptive change generated by natural selection. But with the advent of "the biological species concept" (see especially Dobzhansky [13, 14, 43]), the (still) dominant view of the nature of species in evolutionary biology became one of reproductive communities. Dobzhansky [14], for example, argued that Darwin solved the problem of generation of adaptive phenotypic diversity, but ignored completely a second major problem: the origin of discontinuities evident among species. In other words, gaps between species are as much a direct outcome of evolution as they are a reflection of either a spotty fossil record or simple extinction of intermediate forms. (Darwin preferred to think that advanced species would routinely, and for the most part, necessarily drive their less-highly adapted progenitors to extinction via competition.) Mayr, for his part, charged Darwin with not even addressing the "origin" of species in his book of that very name.

Dobzhansky and Mayr changed the concept of species from (a) a collection of similar organisms who owed their similarity to a shared gene pool to (b) a shared gene pool (i.e., reproductive community) that set the outer reproductive boundaries of a species. This insured that organisms within a species largely resemble each other more than they do members of other species.

They did this because they understood the origin of discontinuity to be triggered by enforced reproductive "isolation"—itself almost always allopatric, as that is the easiest scenario to imagine—and this arguably fits the distributional patterns of most closely related species in the modern biota.

But that leaves us with a striking paradox. Why would most economic adaptive change tend to accrue in relatively brief (a few thousands, perhaps at most tens of thousands, of years) spurts *in conjunction with the origin of new reproductive adaptations*—which is what the speciation process actually is? Especially with the work of H. E. H. Patterson [44], it is clear that speciation is not just reproductive isolation borne of an accumulation of genetic change (i.e., underlying any and all sorts of adaptive change + genetic drift), but rather most often reflects divergence in reproductive isolation pure and simple.

In other words, a 2×2 contingency table, of the sort first proposed by Vrba [58] (see fig. 2) contrasting within and among species levels of adaptive differences, would predict that all squares would be filled, as a null hypothesis, with equal frequencies. There should be as many instances of reproductively isolated, well-differentiated (i.e., phenotypically, in terms of economic adaptations) closely related species as instances of sibling species (isolated, but not economically highly differentiated) and as many instances of noneconomically differentiated ("varying") single species as examples of species with great internal variation in terms of the economic adaptations of their component organisms.

		Speciation	
		Within species	Between species
Morphology	Discernible differentiation abscent	A	B
	Discernible differentiation present	C	D

FIGURE 2 Redrawn version of Vrba's [58, p. 68, fig. 5] (see Eldredge [19, p. 118, fig. 4.2]) 2 × 2 contingency table depicting the "relationship between morphological differentiation and speciation." For fuller explanation, see text.

In short, there should be no predicted, necessary correlation between reproductive isolation and degree of economic phenotypic evolutionary diversity. Yet the fossil record—Pattern 2—proclaims loudly and clearly that there is such a relationship. It says that economic adaptive evolutionary change is virtually restricted to instances where lineages split—presumably via factors such as those analyzed initially by Dobzhansky and Mayr—and subsequently developed far further by a host of ecologists and geneticists.

Pattern 2—two species in stasis and in putative ancestral-descendant evolutionary relationship—constitutes the original repeated empirical observations underlying the notion of "punctuated equilibria" [16, 23]. There are two versions of this pattern, one showing nontrivial stratigraphic overlap between ancestor and presumed descendant—a pattern I have long insisted must be there to be more nearly certain that true speciation has taken place. Alternatively, as shown in figure 1(b), there may be a "replacement" pattern that could result either from true evolution (though it raises the specter of "saltationism"—one of the red herrings originally urged against punctuated equilibria) or from biogeographic range alterations. I have long felt this latter pattern to be too ambiguous to accept under the rubric of punctuated equilibria (i.e., stasis + speciation). Consideration, however, of Pattern 3, relaxes this particular constraint—as discussed immediately below.

If economic adaptive change is overwhelmingly correlated with the origin of new reproductive communities (new species, via reproductive adaptive change), how does this process work? Why this correlation? Though the more complete answer lies in consideration of Pattern 3, there are at least two general considerations pertaining to genealogical lineages which, at least in part, shed some light on the paradox.

First, it was Darwin himself [9] who, in a relatively rare moment of considering species as mutually isolated entities more "permanent" than varieties, had the insight that within-species variation may easily be lost, but once a species goes its own separate way, what we would now call the "genetic information" has been, so to speak, injected into the phylogenetic mainstream. Simpson, in his very definition of species [54, p. 153], had a similar vision when he spoke of a species as a lineage "with its own unitary role and tendencies." And Futuyma [29], in a brief note (all the while disavowing any credence in punctuated equilibria!), in effect remarked that, if there is anything general about what I have been calling "Pattern 2" in this chapter, it must be that speciation acts to partition genetic information: subsets of genetic information are isolated and have a better chance to survive—a modern expression of Darwin's insight.

But beyond this undoubtedly important conservation of already-generated genetically-based adaptive variation afforded as a side effect of speciation, I believe that there is at least one under-appreciated aspect of the speciation process which itself would foster the generation of adaptive change. Though developed fleetingly 25 years ago [24], the model has received little further attention. But, I believe, it may have some significance, particularly in light of interpretations of Pattern 3 below.

Briefly, the model is predicated on the assumption that the geographic limits of the distribution of species reflects the adaptations of their component organisms. In general, optimal habitats (in terms especially of physical environmental parameters and energy and nutrient sources, etc.) occur nearer the center of a species' range and its outer limits are in fact imposed by more marginally favorable habitat.[4] Caveats aside, if this model linking adaptations and habitat optimality has some useful generality for understanding the causes of distributional limits of species, then, under the rubrics of parapatric or microallopatric speciation theory, it should follow that isolation of populations at the periphery of a species range may lead to rapid diversification of economic adaptations.

[4] I realize that exceptions abound—and am especially aware that favorable habitat may well lie outside the current distributional boundaries of a given species. Habitat, in other words, that would in principle be "recognizable" to component organisms' adaptations, but habitat that, for one reason or another, has not—at least yet—been reached and successfully colonized.

Given the requisite genetic variation, natural selection would be expected to modify economic adaptations—in effect "redefining" marginal habitat as now optimal for the isolated population.

Thus, we have at least one model that plausibly links reproductive isolation with rapid economic differentiation—where reproductive isolation can actually act as a trigger for such rapid adaptive evolutionary change. There are, undoubtedly, other plausible such models. But all such models still suffer generically from what I have recently come to believe is the greatest lacuna in all of evolutionary biological thinking since the founding days of Charles Darwin. For, quite sensibly, Darwin saw the primary signal of evolution as genealogical—and the idea that evolution produces ancestral-descendant series of packages of genetic information has only been reinforced by the molecular genetics revolution.

Indeed, had I been writing this chapter ten years ago, "Pattern 3" would not have been the same "Pattern 3" addressed immediately below. In keeping with the purely genealogical perspective traditional in evolutionary biology, since the early days of punctuated equilibria discussion and debate, the prevailing tendency was to extend the analysis to larger-scale genealogical systems. Thus the next pattern after Patterns 1 and 2 above, typically concerned large-scale monophyletic taxa—especially the (likewise paradoxical) generation of trends—patterns of directional accumulation of adaptive change *among* species, where the prevailing pattern *within* species is one of relative stasis.

Viewing evolution as a purely genealogical concern, however—a concern in which the physical environment plays a cameo role as the background to which adaptations must be matched—implies a seldom explored additional null hypothesis: that speciation (and, for that matter, extinction of species) in one lineage is random with respect to speciation and extinction in other lineages. Evolution is like a ticking clock, and rates within and among lineages are known to vary—so the presumption, never fully examined, has always been that speciation itself is like a ticking clock, with events in one lineage having nothing to do with events in other lineages (whether in the same general area, or, of course, in far-flung regions). The apotheosis of this generally unstated assumption is that there is a steady "background" ticking of statistically constant amounts of speciation and extinction going on all the time throughout the history of life. Nothing could be further from the truth.

PATTERN 3: CROSS-GENEALOGICAL PATTERNS OF SYNCHRONOUS SPECIATION AND EXTINCTION

To a remarkable—if not yet fully explored—degree, the history of life looks like Pattern 3 (fig. 1(c)). It is a pattern that amalgamates Patterns 1 and 2, with the notable additional feature that such diagrams pertain typically to broad regions *and include all species-level taxa known, whether closely related or not.* There is a preponderance of the form of Pattern 2 that shows little or no documented stratigraphic overlap between putative ancestors and their descendants, and with some of the pattern reflecting biogeographic distributional change, rather than extinction and evolution. Moreover, some taxa survive unchanged across the boundaries—Pattern 1, in other words, prevailing even when other lineages are variably becoming extinct, speciating—or moving in or out of a region.

Before considering some of the details of ecological and evolutionary dynamics at play in these situations of rapid biotic turnover, a few examples to buttress the claim that these patterns are the rule, not the exception, in the history of life (at least the history of eukaryotic, multicellular life over the past half-billion years) are in order.

First, I call attention simply to the geological time scale—elements of which have been in place since pioneer efforts by the likes of Cuvier, Brongniart, Smith, Sedgwick, and Murchison from the late eighteenth into the early decades of the nineteenth century. Long before Darwin, geologists and paleontologists utilized natural "breaks" in the fossil record of life to recognize divisions of geological time. The larger the scale of these breaks—meaning the larger the scale taxonomically (and thereby, as a natural if not entirely automatic corollary, the broader the geographic scale)—the higher up a hierarchical scale of divisions they would recognize. The "Paleozoic," "Mesozoic," and "Cenozoic" (ancient, middle, and recent life) divisions were well entrenched by the mid-1840s and the divisions between them correspond to two of the five greatest global mass extinction events so far to have struck the earth's biota.

I return to mass extinctions in the final section of this chapter. For the moment, the main point is that there is indeed a nested set of such biotic breaks that corresponds exactly to the nested sets of divisions of geological time. Lesser events are more numerous, more closely spaced—and less globally encompassing—than the global mass extinction events. But they are nonetheless real—and, I think, if anything even more important for understanding dynamics of interacting ecological and evolutionary processes *at the primary level where adaptive change normally and typically occurs in evolutionary history.*

It is these regional-level phenomena on which I shall focus here. One very well worked out series of examples are documented primarily by Carlton Brett and colleagues (e.g., Brett and Baird [1]; see Ivany and Schopf [36], for an entire volume devoted to data and analysis pertaining to this pattern) from the Paleozoic of the eastern Appalachian Basin of North America and, in some instances, adjacent epeiric seas on the continental interior. Brett and Baird [1] call Pattern 3 "coordinated stasis," basing their name, understandably, on the image of stability within each of the succeeding intervals they studied. In a nutshell, Brett and Baird [1] document a succession of eight such faunas, in which the vast majority of known component species[5] remains in stasis throughout

[5]There are from a few or several dozen, to over several hundred, described species from these eight intervals. With the exception of a few vertebrate and plant species documented primarily from the younger (Devonian) biotas, all are hard-shelled marine invertebrates: brachiopods, arthropods, mollusks, bryozoans, echinoderms, etc. Though some have openly worried that the hard parts of shelled invertebrates is a poor sampling of all species present (and offer

the entire interval. Each interval of stability (and the stability refers as much to monotonously repeated ecological assemblages as it does to the overall stasis exhibited by the vast majority of component species) lasts from 5 to 7 million years—an aggregate total time of some 45–55 million years. Brett and colleagues estimate that between 70% and 85% of the taxa (species) are present throughout their respective intervals—and only some 20%, again on average, make it through to the next succeeding biotic interval. Thus, the situation Brett and Baird describe fits the pattern of figure 1(c)—Pattern 3—exactly.

Though much is known about at least some of the events that triggered the turnovers in the Paleozoic faunas (e.g., eustatic [sea level] oscillations caused by plate tectonic movements or glaciation episodes; depositional regimes changed by episodes of mountain building—themselves caused by continental collisions which, among other things, caused the introduction of new species from outside the Appalachian faunal province, etc.), I defer discussion of causation underlying Pattern 3 events for the discussion of Pliocene African events, where the data are clearer.

Though the simple existence of a detailed geological time scale suggests that this picture is general for the entirety of Phanerozoic time, I'll mention two more especially well worked out examples. Beginning with the work of Cheetham [3], and together with Jackson [4, 5, 37, 38], the richly fossiliferous sediments of the Neogene of the Dominican Republic have recently emerged as a strikingly cross-genealogical Pattern 3 example. Just as my original "Pattern 2"-punctuated equilibria—trilobite example (the *Phacops rana* species complex), drawn as a single clade among what turn out to be many cross-genealogical examples in one of Brett and Baird's Paleozoic intervals (Middle Devonian), Cheetham's initial work on Neogene Dominican bryozoans has been greatly expanded by other workers and it has long since become clear that the mollusks, echinoderms, and other marine invertebrates all, together, clearly left a lock-step Pattern 3-mode of evolutionary history.

But it is the examples documented by Elisabeth Vrba from the Pliocene of eastern (and now southern) Africa which perhaps shed the most light on Pattern 3-style phenomena. Unlike Brett and Baird, Vrba chose the biotic breaks as inspiration for her name for what I have been more blandly calling Pattern 3: her term is "turnover pulse" [60]. Though there are data pertaining to biotic turnover some 5 million years ago, it is the events between 2.8 and 2.5 million years which have attracted the most attention. Based on the analysis of oxygen isotopes from marine microfossils, it is apparent that there was a global cooling event that saw the earth's mean temperature drop between 10° and 15° C over a period of about 300,000 years. According to Vrba, the early stages of this drop—which brought cooler and notably drier conditions to the wet woodland biome that dominated much of eastern Africa 3 million years ago—had little detectable

---

only tangential insight to the soft-tissue anatomy of the shelled creatures themselves), it is well to bear in mind that hard-part morphology also evolves—and it is the pattern of evolution of hard-part morphology that is under discussion here.

effect on the biota. But then, around 2.5 million years ago, very abruptly the wetter woodlands gave way to more open savannah grasslands.

With the radical change in the plant community came, naturally enough, an equally dramatic change in the mammalian biota. The pattern is abrupt—and it is, of course, cross-genealogical—as it was the change in composition of the plant communities that, in all likelihood, underlay the changes that quickly came to the mammalian fauna. And these changes were, according to Vrba's analysis (she herself is an expert on antelope), of two basic sorts: ecological-biogeographic and evolutionary. On the ecological-biogeographic front, new species showing up for the first time in the fossil record of eastern Africa at about 2.5 million years ago included some already adapted to open savannahs: they simply came in through habitat tracking. Likewise, among those that disappeared just about 2.5 million years ago, some presumably survived elsewhere—perhaps in western Africa where conditions similar to those to which they were adapted persisted. This latter possibility is difficult to address, as the fossil record of western Africa at that time is sparse to nonexistent.

Some species—such as impalas (a notorious "living fossil," see Vrba [59])—sailed right through the turnover. Impalas are ecological generalists and can live in virtually any African ecological setting, from woodlands to near-desert scrublands, only requiring the presence of open water for drinking.

Yet other species apparently became extinct around 2.5 million years ago. And some, without doubt, evolved. Habitat disruption through climate change appears to have been sufficiently severe that a threshold was reached, and extinction apparently claimed a number of mammalian species in eastern and southern Africa. One species that disappeared at this time was none other than *Australopithecus africanus*, sole known hominid species of the interval between 3 million years and 2.5 million years ago. *Australopithecus africanus* was a slight-of-build species; males stood (literally, this was a bipedal species) barely 4 feet tall and weighed less than 100 pounds; brain size was about 400 cc.—roughly the size of a modern chimp's brain.

Australopithecus africanus disappeared about 2.5 million years ago—but right at 2.5 million years, there appeared not one but two radically distinct hominid species. The *Paranthropus robustus* lineage was one of these. The so-called "black skull," at 2.5 million years, is so far the earliest record of the superficially retro-ape-like lineage of hominids—with massive skulls (males had sagittal crests rather like gorillas) supporting thick musculature to operate the obligate-herbivore nut-and-tuber grinding jaw and dental apparatus. This lineage has five species documented so far over its one million year known stratigraphic range, exhibiting the high rates of speciation and extinction so typical of stenotopic species-lineages in general.

The other lineage is more lightly built ("gracile")—and possessed of a brain (to judge from the earliest discovered and best preserved specimen—Richard Leakey's 1470 from East Rudolf, Kenya) expanded to ca. 750 cc. This *Homo habilis* species lineage[6] was presumably the one who made the earliest stone tools—which turn up just at 2.5 million years ago at Olduvai Gorge in Tanzania. It is difficult to avoid the conclusion that whatever was driving the extinction and evolution of new species among the antelopes, suids, etc., in eastern and southern Africa seemed to be having a comparable effect among our hominid ancestors and collateral kin.

What to make of this pattern? If it is clear that habitat tracking accounts for some of the disappearances and consequent appearances of mammalian species once the plant communities have transformed, it is equally clear that ecological disruption was sufficiently great to drive some species extinct—and perhaps to "cause" others to evolve. If so, what precisely would have been going on?

Once again, following the argument of the immediately preceding section, it is difficult to imagine how extinction and the putative consequence of "freeing niche space" can actually cause speciation—or simple phyletic evolution—to occur. Vrba [60] herself seems to solve this problem by pointing out that habitat fragmentation (as when savannahs spread at the expense of woodlands, but in so doing fragment the distribution of formerly contiguous woodlands) is the *sine qua non* of speciation itself. It seems quite likely, in other words, that habitat tracking—plus extinction and speciation—are threshold responses to accumulating environmental stress that simply goes too far for the system to take. I would further suggest that extinction, rather than driving ("causing") speciation, simply increases the probability of survival of fledgling species. Speciation (formation of reproductively isolated units) may very well be happening "all the time," but it is only after ecosystemic disruption that numbers of fledgling species (which are, after all, typically not all that differentiated ecologically from their parental species) can expect to survive in any great numbers.[7]

I believe that the threshold phenomena encapsulated especially clearly in Vrba's 2.5-million-year turnover pulse pattern—Pattern 3—has much to tell us about the constraints—the hows, whys, and whens—of adaptive change in evolution. In the final section of this chapter, I'll try to integrate these phenomena at various spatiotemporal levels to produce a thumbnail sketch of the evolutionary process—one that integrates ecological systems with genealogical systems and the whole with the physical environment. But first, a penultimate section that examines more closely and more formally the very nature, as well as hierarchical structure, of both genealogical and ecological systems. This dual hierarchical scheme of biological systems will then serve as the conceptual hatrack for the multiscalar, multilevel theory sketched out in the final section.

---

[6]Tattersall [55] argues that there are more species than *habilis* among the early habilines in east (and presumably southern) Africa around this time.

[7]A number of colleagues have suggested to me, variously, that shock proteins, transposons, and homeobox genes might be able to be expressed in times following environmental disruption and extinction—a possibility that fits with the scenario of increased probability of fledgling species' survival, but one which nonetheless is too far from my paleontological purview to comment on intelligently.

2 THE GENEALOGICAL AND ECOLOGICAL HIERARCHIES

Organisms engage in two, and only two, classes of activity. They engage in matter-energy transfer processes to differentiate, grow, and maintain the soma. And, to enable them to pursue their one other class of activity, they reproduce. Physiologists have long seen this dichotomy, as they list processes (digestion, respiration, elimination, etc.), and duly note that only one—reproduction—is not essential for an organism's very existence.

The consequences of this duality of organismic activity for the construction and maintenance of large-scale biological systems (i.e., above the organism level) are straightforward and have been discussed in detail elsewhere (see especially Eldredge and Salthe [27]; Eldredge [17, 18]; Eldredge and Grene [26]). Here, I present only a simplified summary.

As matter-energy transfer "machines," organisms interact with others of their own species in the context of the local economy of nature: organisms variously compete for, cooperate, or maintain neutral relations in their search for energy sources and nutrients. This is the local "population" in an ecological context—a collection of interacting conspecifics that Damuth [7] has usefully termed an "avatar." Use of the term "avatar" allows reservation of the term "deme" specifically for local populations of interbreeding conspecifics. Demes, in this terminology, are divisions of species, while avatars are parts of local ecosystems.

While a case can be made that local avatars and demes often do not coincide in composition (e.g., demes of African elephants, which are divided for the bulk of the nonreproductive year into loner bulls, small collections of younger males, and large hierarchically arranged matriarchal herds of females and young—each of which grouping having a decidedly different impact on the local ecosystem), it is at the next higher level that differences between the ecological and genealogical hierarchies are most clearly seen. Avatars interact with other avatars as biotic components of local ecosystems. Matter-energy transfer occurs between genealogically unrelated avatars within local ecosystems. To my mind the most compelling definition of the very term "niche" is the role each such avatar plays in the context of the local system—which consists, as well, of the physical setting and the nonorganic original sources of the energy that fuels the system.

Avatars are held together by (nonreproductive) interactions among local conspecifics; similarly, the biotic components of local ecosystems are held together by matter-energy transfer processes between avatars. At the next level—and this is a sliding spatial scale up through the global biosphere—is the matter-energy flow between adjacent ecosystems. In the ecological hierarchy, local ecosystems are interlinked (through the very same sort of moment-by-moment matter/energy transfer relation) to form larger, regional systems—which are themselves linked up into even larger-scale systems.

Moment-by-moment matter-energy transfer processes among entities at the next lower level give definition and cohesion to entities at the next higher level—

and the hierarchy consists of organisms/avatars/local ecosystems/regional ecosystems.../biosphere. One could easily point to the hierarchical arrangement of the soma of Metazoa as an extension downward of this very same hierarchy, e.g., organ systems/organs/tissues/cells/organelles/proteins.

Reproduction, likewise, has immediate consequences for the formation and definition of larger-scale biological systems. Demes, as already noted, are reproductively interactive groups of local conspecifics. Demes are parts of species—and the genetic interactions among demes has traditionally played a strong role in evolutionary theory (perhaps most notably those of Wright [64, 65], but also of many others—see especially Wade and Goodnight [61]). The splitting, merging, and extinction of demes provides the internal structure of species—defined here as the largest collectivity of organisms sharing reproductive adaptations (e.g., the "specific mate recognition system" of Patterson [44]).

Thus, species are formed and held together by simple organismic sexual reproduction, via the structure afforded by the intermediate level of the deme. Organisms in a reproductive sense, demes, and species are simply nested packages of genetic information—each one formed and held together by the on-going "more-making" activities of entities at the next lower level. There is but one higher level to add to this "genealogical" (or simply "evolutionary") hierarchy: species also "make more of themselves": they "speciate" creating ancestral-descendant skeins of species that are called "monophyletic taxa." Note that, though monophyletic taxa are also nested into a hierarchical system (i.e., the Linnaean hierarchy), genera, families, and the like do not make more entities of like kind. The reproductive buck in a purely functional sense stops at the species level.

This necessarily abbreviated summary of the natures of the ecological and genealogical hierarchies nonetheless clarifies a number of ontological issues pertaining to the relation between ecological and genealogical systems. For one thing, it is clear that species do not have "ecological niches."[8] Yet the evolutionary literature is full of claims to the contrary and the occupation by species of niches is fundamental to traditional models of the evolutionary process that extend Wright's landscape imagery to entire "adaptive (mountain) ranges" said to be occupied by monophyletic taxa (see, e.g., Dobzhansky [15, pp. 9–10]; Eldredge [19]). Demes, species, and monophyletic taxa, in a nutshell, do not actually do anything in the natural world—except to act as reservoirs of genetic information. Species and monophyletic taxa are simultaneously byproducts of the reproductive process and historical entities—fruits of the evolutionary process that reflect the interaction between the ecological and genealogical spheres.

What is the nature of this interaction? At first glance at the hierarchical scheme given in figure 3, the realms of ecology and evolution seem rather divorced from each other. The one (ecological hierarchy) is strictly about matter-energy transfer: systems are held together by such interactions on a moment-by-

[8] Unless, of course, a species consists of but a single population (avatar/deme)—a condition not likely to be encountered except perhaps at the very beginning, and again at the very end, of a species' existence.

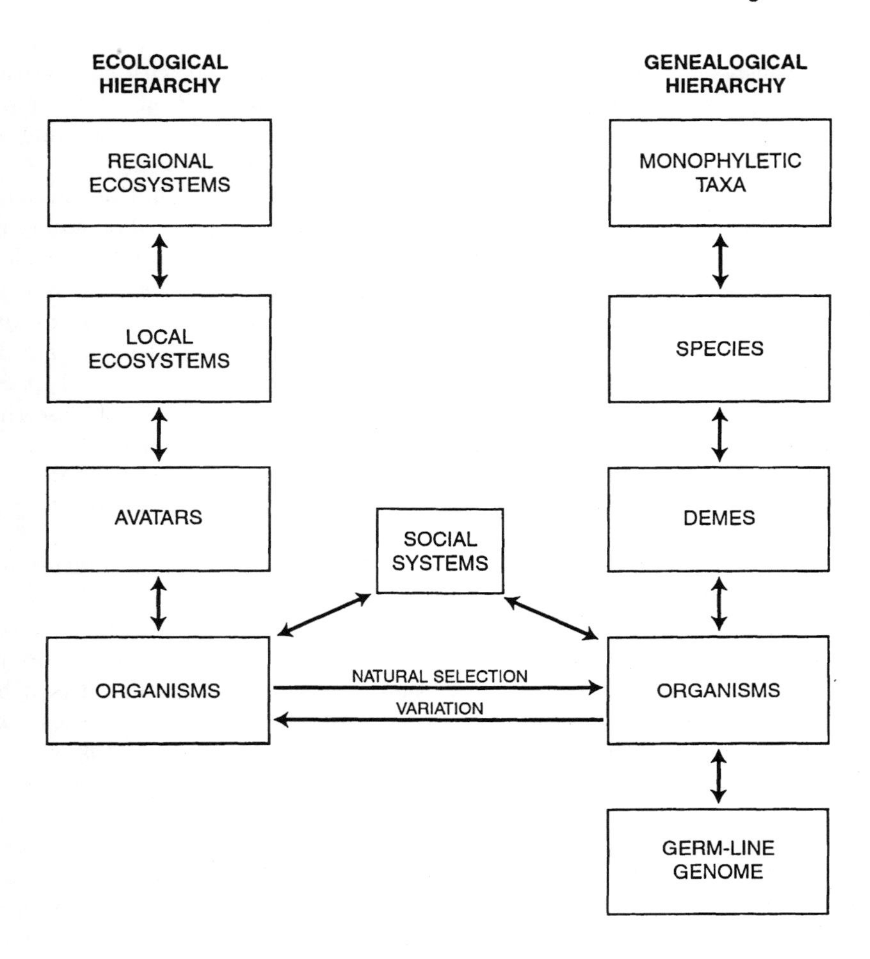

FIGURE 3 The ecological and genealogical hierarchies. For details, see text.

moment basis. The other—the genealogical hierarchy—is concerned with replication, reproduction—generic "more-making."

Yet connections between the two systems are actually rather obvious—especially since organisms are simultaneously parts of both systems. For example, the dual hierarchy scheme depicted in figure 3 is a marvelous way to read Darwin's original formulation of natural selection: natural selection (as contrasted with sexual selection) is the effect that relative economic success has on relative reproductive success, all other things being equal, among conspecifics of local avatars/demes. Under this (original Darwinian) rubric, natural selection is a passive information filter, reflecting simply what worked better among a local group of conspecifics exhibiting heritable variation and faced with competition for inevitably finite (energy) resources. (As already noted, sexual selection was later defined by Darwin as the advantage some local conspecifics of the same sex have over others in matters purely reproductive—see Darwin [9, p. 256].)

Note how starkly different this view of natural selection is from Dawkins's notion of forcing from below. That the urge to replicate (among genes) or simply reproduce (among organisms) sets in motion competition for resources: genealogical processes, in Dawkins's vision, drive ecological processes. One does not eat to live and in so doing possibly to reproduce. Rather, according to Dawkins, one lives to reproduce—which raises the necessity of procuring energy to maintain the protective soma and to enable reproduction to occur. Indeed, Dawkins [11, p. 90] has extended his reductionist hegemony of genealogy over ecology by asserting that one day ecologists will interpret the structure of ecosystems in terms of evolutionary stable strategies—rather than the interactive dynamics of matter-energy flow they persist in seeing as the very core of the dynamic structure of ecosystems.

A far more neutral view of the relation between matter-energy transfer and reproductive processes—one that agrees far better with the received wisdom of the old physiologists—sees neither as more "important" or "fundamental" than the other (unless it be the economic activities necessary for simply staying alive). Rather, there is a reciprocal relationship—where economic success biases reproductive success—but where, too, ongoing reproduction furnishes each generation's players in the ecological arena. As we shall now see, it is what is available in the genetic stocks—these demes, species, and monophyletic groups—that stocks ecosystems and determines in large measure what will happen next in evolutionary history. The dual hierarchy scheme serves as a very useful underlying conceptual framework for a multi-spatiotemporal-scalar nested hierarchy of ecological and evolutionary processes—based on the three patterns developed earlier in this chapter: the "Sloshing Bucket" model of evolution.

3 THE SLOSHING BUCKET

Integrating the three evolutionary patterns discussed earlier in this chapter with the dual genealogical and ecological hierarchy scheme discussed in the immediately preceding section yields the following conceptual model of the evolutionary process (see also Eldredge [21, Ch.6]). Each level corresponds to physical disturbance at a particular spatiotemporal scale—from essentially no disturbance (stability of the local physical environment within parameters of daily and seasonal oscillations over periods of a few years) up through massive disruptions caused by, e.g., extraterrestrial bolide impacts—events that happen rarely, take anywhere from a few days to a million years or more, and affect virtually all biomes on the earth. Also, each level forms a part of the next higher level—so that all processes and patterns of all constituent lower levels can be found within any given focal level. For this reason I start with the lowest level and work up:

LEVEL 1

Short-term, within-deme/avatar and species. As discussed above (see Pattern 1: Stasis), patterns of within- and among-deme variation are seldom if ever translated into cumulative evolutionary change. Nonetheless, as Thompson [56] has particularly graphically demonstrated, selection and drift are capable of producing very rapid short-term evolutionary changes within local populations. Though I have discussed (above) why the *net effect* of such change nonetheless sums to zero—why stasis is the rule despite the potential and indeed the proven reality of rapid evolution within demes—I also agree that just why patterns of variation and selection-mediated change are not routinely accumulated species-wide through time needs further analysis.

Yet it is empirically the case that species overwhelmingly remain in stasis throughout their often-long histories. Natural selection may be strongly directional locally, but the net effect is "stabilizing" summed over an entire species, especially in the truly short term.

LEVEL 2

Short-to-longer term, highly localized to subregional ecosystemic disturbance. Hubbell and colleagues [34] (see also Tilman [57]) have recently examined a spectrum of colonization events that took place over a span of 13 years in a tropical forest in Panama. Though their aim was to evaluate conflicting hypotheses pertaining to the maintenance of diversity levels within species-rich habitats, their results are pertinent here as well. The spectrum of disturbance they examined spanned the gamut from the space created by the loss of a few individual trees (25–49 square meters), to clearings greater than 400 square meters in areal size. Hubbell et al. [34] conclude that most recruitment comes from immediately adjacent plant species—rather than from individuals or demes of pioneer species recruited from farther afield. At the spectrum of spatiotemporal levels they investigated, they concluded that succession—involving especially recruitment of pioneer species from afar—was far less important than direct recruitment of whatever types of plants (regardless of their placement within the normal successional sequence) happened to be nearest to the new clearing.

It is noteworthy, however, that, progressively, the larger clearings in the study reported by Hubbell et al. do show higher proportions of pioneer species as the successful colonists [34, table 1, p. 556]—somewhat vitiating their conclusions, as the larger in area the disturbance, the greater the proportion of pioneers relative especially to shade-tolerant trees of more mature forest. Thus, though Hubbell et al. point out that there are no records of major disturbances, such as hurricanes, striking Barro Colorado island (where the study was carried out), it is as well to consider that disturbances of somewhat larger spatial dimensions (such as the aftermath of fires or storm blowdowns)—events that typically happen at lower frequencies, such that decades, centuries, or even millennia are required to elapse before events of such magnitude may be expected to occur—would in fact trigger true ecological succession, and contribute as well to maintenance of long-term diversity levels—an effect not considered in the smaller spatiotemporal scales of the Hubbell et al. study.

What, in an evolutionary sense, might be expected to happen during full-blown succession? Most well-studied examples of succession emphasize the patchiness of the process: where degraded habitat might sit directly beside mature, undamaged ecosystems. (See Johnson [39] for a particularly graphic example involving the benthic invertebrate communities of Tomales Bay, California.) What seems to happen is that outlying demes offer recruits in more-or-less orderly succession—meaning simply that existing adaptations of all species—from the pioneers up through the dominant species of climax communities—are put into play. (Here is a prime example of genealogical systems—in this case, demes of conspecifics—recruited to rebuild parts of local ecosystems whose avatars have been eliminated through physical disturbance.) Selection might change those adaptations locally—but succession is a fact of ecological structural organization—and stabilizing selection seems to be the norm under these circumstances—which embrace, as well, the features of Level 1.

LEVEL 3

Larger-scale physical disturbance. The upper level of Level 2 grades into Level 3—on regional and sometimes larger scales and in temporal dimensions ranging from decades through hundreds of thousands of years. This is the level of true habitat tracking (discussed above under Pattern 1), where ecosystems are systematically replaced by others (for example, when continental glaciers move south, displacing biomes before them), but at a pace that fails to reach threshold levels. Species, by and large, track habitat successfully and remain stable. Normal processes of ecological succession, as well as within- and among-deme selection and drift, of the constituent lower levels, also take place. Some extinction and speciation (Pattern 2) may occur, but persistence (and stasis) of species prevails. Once again, the net effect is stability, even in the face of major environmental change.

LEVEL 4

Threshold regional and global disturbance. When disturbances characteristic of Level 3 cause changes of ecosystem composition too rapidly—so that habitat tracking is no longer a viable option for many species—Pattern 3 results. Many species in unrelated lineages become extinct and speciation is triggered in some, generally in conjunction with the extinction of ancestral species. Habitat tracking, ecological succession, survival, and stasis of some species—characteristic features of lower levels—also obtain.

It is at this threshold level that most adaptive change in evolution seems to occur—in conjunction with the origin of new species. Whether speciation rates are higher—or whether survival of fledgling species is simply higher—at Level 4, is not fully understood.

LEVEL 5

Global disturbances/mass extinction. Of the five global mass extinction events that have struck earth's biota in the past half-billion years, I mention two, to characterize in brief the effects such large-scale events have had on the history of life.

The largest mass extinction event so far known to have struck the earth's biota occurred some 245 million years ago—at the end of the Permian Period—effectively ending the Paleozoic Era and simultaneously ringing in the Mesozoic. Raup [47] has estimated that somewhere between 70% and 96% of the earth's species may have gone extinct during this event—the causes of which remain incompletely understood.

As an example of the evolutionary consequences of such epochal extinctions, consider the "corals." There were two major groups (classified generally as "orders") of calcified Anthozoa in the Paleozoic, the Rugosa (also known as "tetracorals"), and the Tabulata. Both disappeared at the end of the Permian, after some 400 million years or so of life (including extensive reef-building) in Paleozoic seas. After a lag of some 5 to 7 million years, a new group of calcified Anthozoans appears in the Lower Triassic—members in good standing of the modern calcified Anthozoan group, Order Scleractinia.

Paleontologists had struggled for years trying to understand how the Rugosa—with their calyces of calcite and their internal four-fold symmetries—could have given rise to the Scleractinia—with their aragonitic calyces and six-fold internal symmetry. It turns out that the sister-group of Scleractinia are actually sea anemones, which are known to have been present in the Paleozoic. The best interpretation, then, of the transition from Rugosa and Tabulata of the Paleozoic to the Scleractinia of the Mesozoic and Cenozoic is that it is no such thing at all: extinction claimed the former two Orders and corals were, in effect, "reinvented" by evolution when members of the anemone clade gained the ability to secrete calcified calyces.

Mammals and dinosaurs provide a similar example. As is by now well known, mammals and dinosaurs both evolved in the Upper Triassic. But for reasons no one claims to understand, it was the dinosaurs, rather than the mammals, that radiated into a variety of sizes, shapes, and ecological roles. Mammals remained the "rats of the Mesozoic," little differentiated for some 150 million years. Only after the bolide impacts disrupted the world's terrestrial and marine ecosystems so severely 65 million years ago—and only when the last of the terrestrial (i.e., "non-bird") dinosaurs became extinct, did the mammals (again, after a characteristic lag) evolve into the variety of sizes, shapes, and ecological roles formerly displayed in dinosaur (and collateral reptilian kin) diversity.

Hence the *sloshing bucket:* Think about these five levels in reference to the hierarchies depicted in figure 3. A Dawkins model would see the structure of both sides of the diagram dictated by competition for reproductive success. I suggest an alternative picture, where evolution is variably constrained or forced by the effects of physical environmental events on ecosystems. At the lowest levels, minor such events cause little or minor disturbance and the status quo remains. The same is true even when larger-scale environmental events disturb entire regional ecosystems, but at rates sufficiently gradual enough for habitat tracking to dominate recovery patterns. Only when the threshold is reached and extinction and speciation (correlated, but as yet without complete causal understanding) become dominant elements of response, do we encounter actual evolutionary change to any significant degree. Finally, the rarest but most devastating and encompassing of such events, affecting the entire world's biotas, drive larger-scale taxa extinct. Subsequent radiations (i.e., not origins) of clades tend to produce taxa that are, retrospectively, given coordinate taxonomic ranking (e.g., Orders Rugosa and Tabulata vs. Order Scleractinia; Class Dinosauria vs. Class Mammalia). The bucket is "sloshing," because the greater the disturbance, the higher level the taxa that become extinct, hence the higher level the taxa that subsequently evolve to play analogous ecological roles, hence the more different the subsequently rebuilt ecosystems will be: minor disturbance leads to recreation of ecosystems with the same players. Disturbance passing the extinction threshold leads to development of similar ecosystems, but with different species. Mass extinction leads to much more radically altered biotas—leading our predecessors, for example, to distinguish eras of "Ancient," "Middle," and "Modern" life.

One final note to the question, raised at the outset—i.e., Is evolution highly contingent or rather is it strongly deterministic? The answer has to be "yes": the accidents that took out the dinosaurs, triggering, sooner rather than later, the radiation of Mammalia, were just that—accidents, not "failures" of their adaptations. Score one for contingency. But the radiation of mammals, in effect recreating the ecological guilds (if not the exact physiognomies) of the now-departed dinosaurs, does show that most evolution is about reinventing the ecological wheel. There is a strong determinism in adaptation through natural selection. But the conclusion seems inevitable to me that nothing much in the way of adaptive evolutionary change takes place unless and until physical disturbance impacts ecosystems and that the amount of evolution is proportional to the amount of disturbance and extinction that affects ecological and genealogical systems.

REFERENCES

[1] Brett, C. E., and G. Baird. "Coordinated Stasis and Evolutionary Ecology of Silurian to Middle Devonian Faunas in the Appalachian Basin." In *New Approaches to Speciation in the Fossil Record*, edited by D. H. Erwin and R. L. Anstey, 285–315. New York: Columbia University Press, 1955.

[2] Carruthers, R. G. "On the Evolution of *Zaphrentis delanouei* in Lower Carboniferous Times." *Quart. J. Geol. Soc. Lond.* **66** (1910): 523–538.

[3] Cheetham, A. H. "Tempo of Evolution in a Neogene Bryozoan: Rates of Morphologic Change within and across Species Boundaries." *Paleobiology* **12** (1986): 190–202.

[4] Cheetham, A. H., and J. B. C. Jackson. "Process from Pattern: Tests for Selection Versus Random Change in Punctuated Bryozoan Speciation." In *New Approaches to Speciation in the Fossil Record*, edited by D. H. Erwin and R. L. Anstey, 184–207. New York: Columbia University Press, 1995.

[5] Cheetham, A. H., and J. B. C. Jackson. "Speciation, Extinction, and the Decline of Arborescent Growth in Neogene and Quarternary Cheilostomes Bryozoa of Tropical America." In *Evolution and Environment in Tropical America*, edited by J. B. C. Jackson, A. F. Budd, and A. G. Coates, 205–233. Chicago, IL: University of Chicago Press, 1996.

[6] Coope, G. R. "Late Cenozoic Fossil Coleoptera: Evolution, Biogeography and Ecology." *Ann. Rev. Ecol. Syst.* **10** (1979): 247–267.

[7] Damuth, J. "Selection among 'Species': A Formulation in Terms of Natural Functional Units." *Evolution* **39** (1985): 1132–1146.

[8] Darwin, C. *On the Origin of Species.* London: John Murray, 1859.

[9] Darwin, C. *The Descent of Man, and Selection in Relation to Sex.* London: John Murray, 1871.

[10] Davis, M. "Quarternary History of Deciduous Forests of Eastern North America and Europe." *Ann. Missouri Bot. Gard.* **20** (1983): 550–563.

[11] Dawkins, R. *The Selfish Gene.* New York and Oxford: Oxford University Press, 1976.

[12] Dawkins, R. *The Extended Phenotype. The Gene as the Unit of Selection.* Oxford and San Francisco: W. H. Freeman and Co., 1982.

[13] Dobzhansky, T. "A Critique of the Species Concept in Biology." *Phil. Sci.* **2** (1935): 344–355.

[14] Dobzhansky, T. *Genetics and the Origin of Species.* New York: Columbia University Press, 1937. Reprint, 1982.

[15] Dobzhansky, T. *Genetics and the Origin of Species.* 3d ed. New York: Columbia University Press, 1951.

[16] Eldredge, N. "The Allopatric Model and Phylogeny in Paleozoic Invertebrates." *Evolution* **25** (1971): 156–167.

[17] Eldredge, N. *Unfinished Synthesis. Biological Hierarchies and Modern Evolutionary Thought.* New York: Oxford University Press, 1985.

[18] Eldredge, N. "Information, Economics and Evolution." *Ann. Rev. Ecol. Sys.* **17** (1986): 351–369.

[19] Eldredge, N. *Macroevolutionary Dynamics. Species, Niches and Adaptive Peaks.* New York: McGraw-Hill, 1989.

[20] Eldredge, N. *Reinventing Darwin. The Great Debate at the High Table of Evolutionary Theory.* New York: John Wiley and Sons, 1995.

[21] Eldredge, N. *The Pattern of Evolution.* New York: W. H. Freeman, 1999.

[22] Eldredge, N., and J. Cracraft. *Phylogenetic Patterns and the Evolutionary Process. Method and Theory in Comparative Biology.* New York: Columbia University Press, 1980.

[23] Eldredge, N., and S. J. Gould. "Punctuated Equilibria: An Alternative to Phyletic Gradualism." In *Models in Paleobiology*, edited by T. J. M. Schopf, 82–115. San Francisco: Freeman, Cooper, 1972.

[24] Eldredge, N., and S. J. Gould. "Reply to Hecht." *Evol. Biol.* **7** (1974): 303–308.

[25] Eldredge, N., and S. J. Gould. "On Punctuated Equilibria. (Letter)." *Science* **276** (1997): 338–339.

[26] Eldredge, N., and M. Grene. *Interactions. The Biological Context of Social Systems.* New York: Columbia University Press, 1992.

[27] Eldredge, N., and S. N. Salthe. "Hierarchy and Evolution." *Oxford Survs. Evol. Biol.* **1** (1984): 182–206.

[28] Eldredge, N., and I. Tattersall. "Evolutionary Models, Phylogenetic Reconstruction, and Another Look at Hominid Phylogeny." In *Contributions to Primatology*, edited by F. S. Szalay, vol. 5, 218–242. Basel: S. Karger, 1975.

[29] Futuyma, D. J. "On the Role of Species in Anagenesis." *Amer. Natur.* **130** (1987): 465–473.

[30] Gingerich, P. D. "Stratigraphic Record of Early Eocene *Hyopsodus* and the Geometry of Mammalian Phylogeny." *Nature* **248** (1974): 107–109.

[31] Gingerich, P. D. "Paleontology and Phylogeny: Patterns of Evolution at the Species Level in Early Tertiary Mammals." *Amer. J. Sci.* **276** (1976): 1–28.

[32] Gould, S. J. "G. G. Simpson, Paleontology, and the Modern Synthesis." *The Evolutionary Synthesis: Perspectives on the Unification of Biology*, edited by E. Mayr and W. B. Provine, 153–172. Cambridge: Harvard University Press, 1980.

[33] Gould, S. J. *Wonderful Life. The Burgess Shale and the Nature of History.* New York: W. W. Norton, 1989.

[34] Hubbell, S. P., R. B. Foster, S. T. O'Brien, K. E. Harms, R. Condit, B. Wechsler, S. J. Wright, and S. Loo de Lao. "Light-Gap Disturbances, Recruitment Limitation, and Tree Diversity in a Neotropical Forest." *Science* **283** (1999): 554–557.

[35] Hull, D. L. *Darwin and His Critics.* Cambridge: Harvard University Press, 1973.

[36] Ivany, L. C. and K. M. Schopf, eds. "New Perspectives on Faunal Stability in the Fossil Record." *Palaeogeography, Palaeoclimatology, Palaeoecology* **127** (1996): vii–361.

[37] Jackson, J. B. C., and A. H. Cheetham. "Phylogeny Reconstruction and the Tempo of Speciation in Cheilostome Bryozoa." *Paleobiology* **20** (1994): 407–423.

[38] Jackson, J. B. C., and A. H. Cheetham. "Tempo and Mode of Speciation in the Sea." *Trends Ecol. & Evol.* **14** (1999): 72–77.

[39] Johnson, R. G. "Conceptual Models of Benthic Marine Communities." In *Models in Paleobiology*, edited by T. J. M. Schopf, 148–159. San Francisco: Freeman, Cooper, 1972.

[40] Lieberman, B. S., C. E. Brett, and N. Eldredge. "Patterns and Processes of Stasis in Two Species Lineages of Brachiopods from the Middle Devonian of New York State." *Amer. Museum Novitates* **3114** (1994): 1–23.

[41] Lieberman, B. S., C. E. Brett, and N. Eldredge. "A Study of Stasis and Change in Two Species Lineages from the Middle Devonian of New York State." *Paleobiology* **21** (1995): 15–27.

[42] Maynard Smith, J. "Palaeontology at the High Table." *Nature* **309** (1984): 401–402.

[43] Mayr, E. *Systematics and the Origin of Species.* New York: Columbia University Press, 1942. Reprints, 1982.

[44] Patterson, H. E. H. "The Recognition Concept of Species." *Species and Speciation. Transvaal Mus. Monogr.* **4** (1985): 21–29.

[45] Pictet, F. J. "On the Origin of Species by Charles Darwin." *Archives des Sciences de la Bibliotheque Universelle* **3** (1860): 231–255.

[46] Prothero, D. "Punctuated Equilibrium at Twenty: A Paleontological Perspective." *Skeptic* **1** (1992): 38–47.

[47] Raup, D. M. "Biological Extinction in Earth History." *Science* **231** (1986): 1528–1533.

[48] Rowe, A. W. "An Analysis of the Genus *Micraster*, as Determined by Rigid Zonal Collecting from the Zone of *Rhynchonella Cuvieri* to that of *Micraster cor-anguinum*." *Quart. J. Geol. Soc. Lond.* **55** (1899): 494–547.

[49] Schaeffer, B., M. K. Hecht, and N. Eldredge. "Phylogeny and Paleontology." *Evol. Biol.* **6** (1972): 31–46.

[50] Schankler, D. M. "Local Extinction and Ecological Re-entry of Early Eocene Mammals." *Nature* **293** (1981): 135–138.

[51] Sheldon, P. R. "Parallel Gradualistic Evolution of *Ordovician trilobites.*" *Nature* **330** (1987): 561–563.

[52] Simpson, G. G. *Tempo and Mode in Evolution.* New York: Columbia University Press, 1944.

[53] Simpson, G. G. *The Major Features of Evolution.* New York: Columbia University Press, 1953.

[54] Simpson, G. G. *Principles of Animal Taxonomy.* New York: Columbia University Press, 1961.

[55] Tattersall, I. *The Fossil Trail. How We Know What We Think We Know about Human Evolution.* New York and Oxford: Oxford University Press, 1995.

[56] Thompson, J. N. "Rapid Evolution as an Ecological Process." *Trends Ecol. & Evol.* **13** (1998): 329–332.

[57] Tilman, D. "Diversity by Default." *Science* **283** (1999): 495–496.

[58] Vrba, E. S. "Evolution, Species and Fossils: How Does Life Evolve?" *So. Afr. J. Sci.* **76** (1980): 61–84.

[59] Vrba, E. S. "Evolutionary Pattern and Process in the Sister-Group Alcelaphini-Aepycerotini (Mammalia:Bovidae)." In *Living Fossils*, edited by N. Eldredge and S. M. Stanley, 62–79. New York: Springer-Verlag, 1984.

[60] Vrba, E. S. "Environment and Evolution: Alternative Causes of the Temporal Distribution of Evolutionary Events." *So. Afr. J. Sci.* **81** (1985): 229–236.

[61] Wade, M. J., and C. J. Goodnight. "Perspective: The Theories of Fisher and Wright in the Context of Metapopulations: When Nature Does Many Small Experiments." *Evolution* **52** (1998): 1537–1553.

[62] Williams, G. C. *Adaptation and Natural Selection. A Critique of Some Current Evolutionary Thought.* Princeton: Princeton University Press, 1966.

[63] Williams, G. C. *Natural Selection. Domains, Levels, and Applications.* New York: Oxford University Press, 1992.

[64] Wright, S. "Evolution in Mendelian Populations." *Genetics* **16** (1931): 97–159.

[65] Wright, S. "The Roles of Mutation, Inbreeding, Crossbreeding, and Selection in Evolution." *Proc. Sixth Int. Congr. Genetics* **1** (1932): 356–366.

[66] Wright, S. "Tempo and Mode in Evolution: A Critical Review." *Ecology* **26** (1945): 415–419.

Developmental Morphospaces and Evolution

Gunther J. Eble

1 INTRODUCTION

The existence of evolutionary dynamics is conditioned on the preexistence of structured organismal entities through which deterministic and random evolutionary processes operate. Such entities, in turn, are products of evolution and therefore of a preexisting evolutionary dynamics. This dialectic between dynamics and organization is logically inescapable, and yet the evolutionary paradigm has historically concentrated on the dynamics of entities taken as a given, with the explanatory focus placed on the variation and evolution of individual traits and particulate inheritance, not of whole organisms. This approach has been very successful and has formed the stepping stone of evolutionary theory for decades, in the framework of population genetics, molecular evolution, and behavioral ecology. However, it has led to a theory of the evolution of genes as particles of information, in which the dynamics arise from the filtering of genetic inputs into reproductive output. Phenotypes have stood as epiphenomena of gene products. Form, cast in terms of lower-level explanations, became identical to the study of adaptation and design of malleable organismal black boxes of arbitrary decomposability.

Not surprisingly, this alienation of form produced lasting insights in microevolution, where the spatial and temporal scales of form are less important than those of population dynamics. Yet form, and more generally the phenotype, has always remained a fundamental problem demanding explanation in evolutionary biology. Darwin spoke of varieties and species as different forms, the theory of natural selection being, at its inception, a theory of the evolution of organic form and diversity. The diversity of form is a major signature of macroevolution, and its manifestations in space and time specify unique domains for paleobiological, systematic, ecological, and developmental studies. While the structured entities of evolutionary dynamics certainly include genes, structure is not equivalent to genetic information; genes are necessary but not sufficient for the emergence of form. The realization of form is tied to the structure of the genotype-phenotype map, and phenotypic variation is expressed in hierarchical fashion, through the nonlinearities and discontinuities characteristic of developmental pathways. The hierarchical nature of development thus invites consideration of intervening levels of variation between genotype and (adult) phenotype. Although such levels are causally related, their articulation in ontogenetic time leads to the successive creation of novel organizations, the character of which is only partly reducible to preceding levels. The unfolding of development produces heritable organizations which are encoded but not fully specified by genes. Because such organizations are heritable, they are subject to evolution along with the genes that they contain. And while their inheritance is primarily effected by genes, they do not exist in genotypic space, such that their evolutionary dynamics is subject to a different, if overlapping, set of constraints and opportunities. Such constraints and opportunities specify, at each level of organization, an ontologically distinct, semidiscrete, phenotypic space. The genotype-phenotype map has several pages, with genotypic space succeeded by several phenotypic spaces where form unfolds in ontogenetic time.

This is the current challenge of evolutionary theory: to recuperate the theoretical relevance of development and organismal form, at various levels of organization, and to make them an integral part of a richer evolutionary discourse, where objects of explanation are also themselves endowed with explanatory value, and where the differential organization of entities across the biological hierarchy interacts with, molds, is molded by, and justifies ontologically semi-independent dynamics of differential survival and reproduction.

Two recent disciplinary movements, with slightly different origins in time, suggest the desirability of an expanded evolutionary picture and a more diverse phenomenology:

1. Macroevolution, the study of evolution at and above the species level, has become an established discipline, articulating paleobiology and systematics and moving beyond the subordinate role that such disciplines had in the modern

synthesis. Over the last three decades, a plethora of macroevolutionary phenomena, expressed and studied in morphological terms, have been validated as unique to particular temporal scales and hierarchical levels. They include stasis and punctuation [32, 52]; large-scale trends in biodiversity change and its origination and extinction components [26, 90, 92, 102, 103, 108]; mass extinctions [59, 60, 93]; evolutionary radiations [20, 35, 62, 109]; species-level sorting and selection [30, 31, 51, 106, 115]; inhomogeneous morphospace occupation [27, 41, 44, 46, 47, 119]; discordant temporal patterns of diversity and disparity [27, 41]; and large-scale trends, or the absence thereof, in lineage evolution [78, 79]. Paleobiology has incorporated rigorous techniques for the mathematical description, measurement, and modeling of form across species, as well as sophisticated statistical methods for hypothesis testing, some of them never before applied to evolutionary problems. In systematics, the cladistic revolution has brought logical and numerical principles to bear on the analysis of morphological data, leading to increased objectivity and to the production of testable phylogenetic diagrams where novel macroevolutionary phenomena could be revealed. Macroevolution is today a well-grounded and expanding discipline [61], with quantitative morphological data used routinely to generate robust inferences, and with molecules and morphology often contributing together to the unraveling of patterns and processes of evolution.

2. Evolutionary developmental biology [54] has emerged as the domain where the mechanistic production of form and the genotype-phenotype map can be understood in evolutionary terms. Embryology and the study of morphogenesis was viewed as central to evolution in the nineteenth century. Its insights into organismic evolution, however, were eclipsed for most of the twentieth century, in large part because of the successes of genetics and molecular biology, and its role in the modern synthesis was nondescript. Over the last two decades, however, developmental biology built connections with paleontology and systematics, and resurrected interest in how rules of development at various stages in ontogeny may constrain and channel phenotypic evolution [3, 6, 8, 54, 86]. Heterochrony was restored to prominence [77], the logic of development was recast in terms of novel notions such as developmental constraints [70], evolvability [118], and modularity [86, 117], and the application of molecular techniques has revealed that underneath the bewildering diversity of form lie fundamental commonalities in developmental machinery [110, 111]. The overarching goal of evolutionary developmental biology is to understand macroevolution, and the evolution of form, in terms of both mechanistic (proximal) and evolutionary (ultimate) principles that relate to, but expand on, standard population genetic theory [34].

The companionship of macroevolution and evolutionary developmental biology is still poorly articulated, but its progressive realization is having profound effects on evolutionary theory. The successes of macroevolution have validated a hierarchical approach to evolution, with entities at different organizational and genealogical levels proving to be semi-independent and expressing relatively autonomous temporal dynamics. Evolutionary developmental biology, in turn, has led to the realization that it is possible to explicitly incorporate organization into dynamics, by casting descriptions of evolution not only in terms of genes, but also in terms of the ontogenetic combinatorics of genes and their products together in the context of spatially bounded developing forms.

In genotypic space, redundancy is very high, neutrality and selection are prominent, and intrinsic constraints, while present [19], may affect but do not appreciably prevent evolvability [64]. Entities in genotypic space are simple and the abundance of variation biases the investigator toward an appreciation of dynamics over organization. Form and its manifestations in phenotypic spaces, however, reflect state spaces of much lower redundancy, with fewer (though still very important) opportunities for drift and selection to operate, and with multiple constraints capable of significantly limiting and channeling variation and evolvability. Some of these constraints arise from the physics of pattern formation, others from the interplay of cells, tissues, and organs in development, and still others from the sheer imprint of the flow of history through underlying parameter space. As in a random walk, historical constraint is much more binding in the low-turnover phenotypic landscape than in its high-turnover genotypic counterpart. In phenotypic space, variation is still the raw material for evolution, but it is significantly structured by constraints and the complexity of form, inviting more proportionate consideration of dynamics and organization.

An expanded evolutionary theory that is capable of directly dealing with hierarchically expressed phenomena and processes, in ontogenetic and evolutionary time, brings the need for representation schemes, statistical and mathematical tools, and conceptual frameworks partially distinct from those that characterize the study of evolution in genotypic space. The similarity arises mostly from the hierarchical manifestation of evolutionary processes well characterized in population genetics (e.g., mutation, selection, drift), in other words, of ingredients of evolutionary dynamics. The distinctness arises from the need to emphasize, more than at the level of the genotype, the underlying state space and the relationship between actual and possible, as these will more directly color the temporal dynamics. Concomitantly, the analysis of constraints implies a concern for the randomness of variation [25] and for the robustness of organizations to change.

These issues can be approached with the rigorous portrayal of phenotypes against a set of possibilities, the state space. The notion of state space is well established in other fields and is amenable to sophisticated statistical and mathematical analysis. In terms of form, it is expressed in the notion of morphospace. In terms of development, it can be articulated as developmental morphospaces. The concept of morphospace promises to solidify the growing ties between the fields of macroevolution and evolutionary developmental biology and can serve as a tool for the joint study of organization and dynamics. In this chapter, characterizations of morphospaces reflecting the spread, spacing, location, and neighborhood relations of forms are presented and their bearing on the relationship between organizational constraints and evolutionary dynamics is discussed. The notion of disparity will be emphasized as an appropriate, and theoretically meaningful, large-scale statistic. After a discussion of the diversity of morphospace approaches in evolutionary biology, the notions of developmental morphospaces and developmental disparity will be introduced and illustrated as means of approaching issues such as vectors of change in ontogenetic and evolutionary time, general changes in rate and timing of development, testing of developmental "laws," and links between phylogenetic and ontogenetic trends. Emphasis will be placed on the inference of structural constraints to the evolutionary process. Sea urchin evolution is presented as a case study. Finally, it will be argued that developmental morphospaces, by providing a way of generating further insights into the dialectic between organization and dynamics, deserve to be more explicitly incorporated into theories of evolutionary dynamics.

2 WHY MORPHOSPACES OF DEVELOPMENT?

As statements about pattern, morphospace analyses are usually accompanied by inferences about possible processes involved, either through the comparison of morphological distributions and associated descriptors with ancillary data of taxonomic, environmental, or functional nature [41, 66, 67, 97], or through the recasting of the morphospace description itself in terms of such ancillary data [56, 80]. It is noticeable, however, that developmental data are only occasionally considered in morphospace studies [6, 18, 33, 125]. To the extent that morphological evolution expresses the interplay of underlying intrinsic (developmental) and extrinsic (ecological) factors, it is of interest to evaluate the relative importance of such factors in explaining patterns of morphospace occupation. While functional interpretations of morphospaces have a long tradition in ecomorphological and performance studies, functional studies are not always feasible when taxonomic breadth is wide (certain taxa may not be available for experimental work) or temporal coverage is extended (with inclusion of fossils). In addition, given the dominance of adaptational thinking in evolutionary biology, exclusive focus on function often masks a plethora of evolutionarily relevant questions such as heterochrony, heterotopy, and asymmetry of amount of variation in developmental time. If ontogenetic data are available, developmentally meaningful morphospaces can be constructed instead, thus complementing deficiencies (or, for some groups, absence) of functional studies and suggesting novel interpretations of morphological patterns. Logically, developmental morphology has the same status and the same potential utility in understanding evolution as functional morphology.

There are multiple ways to study the role of development in evolution (see Eble [20]). Although a phylogenetic framework is always beneficial and mapping of developmental information onto cladograms can be invaluable in ascertaining developmental trends, a morphospace approach to development and evolution can be advantageous for several reasons: (1) The precept that the order recovered through phylogeny is a mirror of ontogenetic order [83, 85], while true in many cases, has been disputed on theoretical and empirical [5, 57, 58] grounds—a morphospace, in turn, allows unambiguous representation of both ontogenetic and temporal order and "disorder," whether it is recoverable by phylogenetic algorithms or not. (2) Because of homoplasy, there will often be multiple equally or near-equally parsimonious trees, generating uncertainty in the choice of a reference phylogeny (Foote [46]; this is compounded by controversies over combinability of different kinds of data)—a morphospace, however dependent on choice of analytical technique, is a unique representation of the statistical structure of both homologous and homoplastic characters across taxa. (3) Phylogenies are statements about sameness and the hierarchical distribution of homology, and when homoplasies are considered, phylogenies can at most take into account total differences—morphospaces allow both sameness and difference to be economically considered together, and thus can capture the net differences (including plesiomorphies) that underlie the developmental hierarchy of levels of distinctness, or disparity [20]. Clearly, a morphospace approach to the study of the relationship between development and evolution has intrinsic value and is not subordinate to knowledge of phylogenetic relationships. Phylogenetic and morphospace approaches are best viewed as complementary [45].

3 DEVELOPMENTAL AND NONDEVELOPMENTAL MORPHOSPACES

Developmental morphospaces are here defined as morphospaces built in a manner that reflects development directly by reason of either the variables or the entities depicted in such morphospace carrying meaningful developmental information. Although nondevelopmental morphospaces need not be devoid of developmental meaning, their information content is often removed from the axis of ontogenetic action as a result of the descriptive representation of adult forms only. As will be seen below, the distinction between developmental and nondevelopmental morphospaces is not absolute; they can and will often overlap, but the former will tend to encapsulate implications of process (much in the same way as with functional or design spaces), while the latter will be imbued with a more pattern-oriented quality. Much of the field of morphospace studies is concerned with temporal, spatial, or taxonomic partitioning of nondevelopmental morphospaces. As inferences move from pattern to process, intrinsic ontogenetic biases stand as one possible explanation for observed regularities.

The idea of developmental morphospaces is by no means new, for it can be seen as a logical morphological extension of Waddington's metaphor of the "epigenetic landscape" [116], Goodwin's notion of "epigenetic space" [48], and Alberch's rendition of "parameter space" [6]. While such developmental spaces are suitable for probing the genotype-phenotype map [118], developmental morphospaces are more useful for the inference (though by no means ultimate explanation) of more focally phenotypic phenomena such as heterochrony, heterotopy, and developmental constraints. In addition, they allow coverage of patterns of developmental variation in the fossil record.

Although developmental morphospaces can be studied through clade/ clade or clade/subclade comparisons, our best hopes of understanding how developmental variation structures the evolution of adult form should be tied to the comparison of developmental and nondevelopmental morphospaces. This is equivalent to the time-honored principle in statistical data analysis of using ancillary data to better understand a given pattern. If the spaces are commensurate, much insight can be gained by analyzing the statistical concordance in range and/or location of forms across spaces. Similar analyses can also be performed with comparisons of noncommensurate spaces (e.g., a comparison of larval and post-metamorphic morphospaces), but in this case the range of inferences is restricted to statements about the relative distance of forms in each space. Thus, although the commensurability requirement is not absolute, it immediately suggests the use of homologous features whenever possible.

To illustrate developmental and nondevelopmental morphospace comparisons diagrammatically, an iconography recently used by Lauder [66, 67] in the context of functional analysis is adapted for the present purposes, and shown in figure 1. Both concordances and discordances can reveal a potential role of developmental factors in structuring (adult) morphological distributions in nondevelopmental morphospace. The interpretation of distributions, however, is dependent on the specific kinds of developmental/nondevelopmental comparisons, a function of which data and which theoretical outlook is used. Three kinds of contrasts are immediately relevant: theoretical/empirical, abnormal/normal, and juvenile/adult morphospaces [20]. Each of these is discussed in detail below, both in terms of a reappraisal and reconceptualization of past work and of suggestions for future research. The discussion on juvenile/adult morphospaces is accompanied by an empirical example using echinoids.

4 THEORETICAL VS. EMPIRICAL MORPHOSPACE COMPARISONS

Phenomenologically, the distinction between theoretical and empirical descriptions has occupied a central position in morphospace studies. The distinction is sometimes made not so much in terms of means of production of representation spaces (by simulation as opposed to by measurement) but in terms of scope: theoretical spaces are able to represent what is possible, or occupiable, and empirical spaces are renditions of what has actually been occupied [8, 28, 72, 74]. Theoretical morphospaces are "developmental" due to their generative nature (explanans), to their ability to specify predictions of possible natural occurrence. Empirical morphospaces are, in this context (but not in others—see discussion

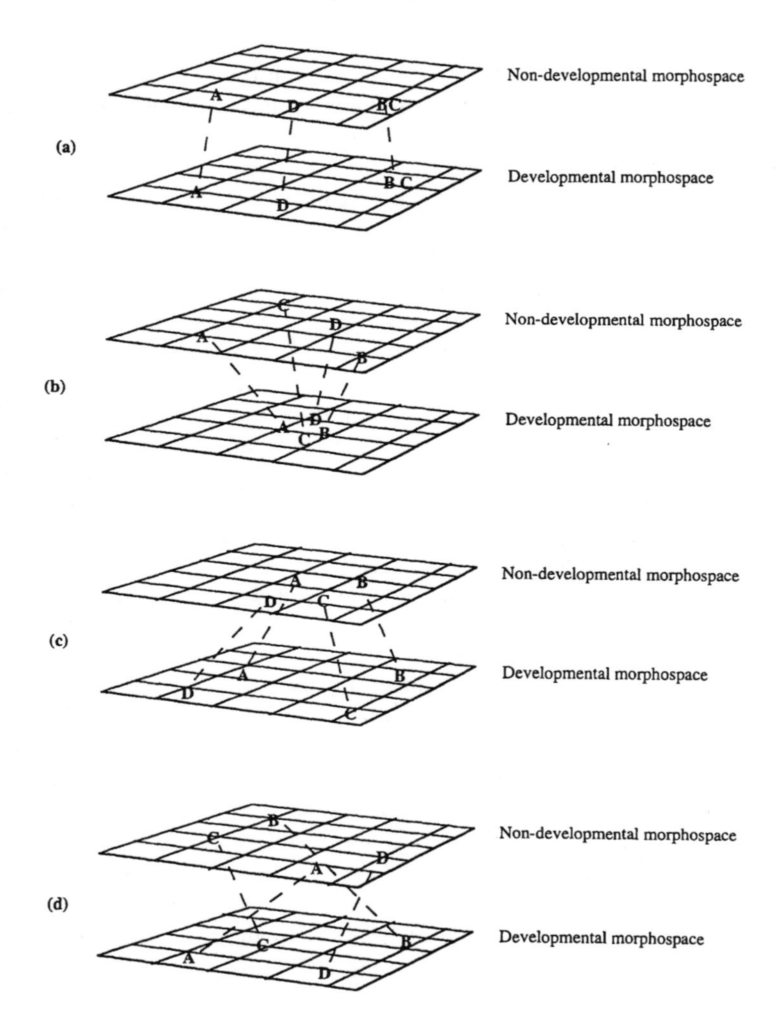

FIGURE 1 A typology of developmental versus nondevelopmental morphospace comparisons. (a) Clustering, location, and range are strictly predicted by developmental morphospace (this is an expectation in comparisons of theoretical and empirical distributions equally parametrized); for abnormal/normal and juvenile/adult comparisons this is consistent with a developmental control of adult distribution or with some other variable acting in both situations. (b) Location and range are not predicted by developmental morphospace because of deviation of forms across spaces. (c) Location and range are not predicted by developmental morphospace because of convergence of forms across spaces. (d) Extensive dissociation in ontogeny is suggested by lack of congruence in clustering and location of equivalent objects across both spaces. The clustering of forms from nondevelopmental and developmental morphospaces belonging to different species suggests much potential for developmental evolution. The similar variance and range implies developmental bounds to adult variation.

on juvenile morphospaces), nondevelopmental because they are built *in terms* of the theoretical generative parameters; they are the object of explanation (*explanandum*), and thus subordinate in this particular framework of deductive-nomological explanation [55].

The comparison of theoretical and empirical morphospaces allows, to different degrees as the case may be, the interpretation of empirical distributions in terms of boundary conditions specified by a relatively small number of parameters. The success of any theoretical model in portraying realistic possibilities, some or all of which are empirically realized, is already evidence not only for the potential simplicity of controls on morphological distributions (the machinery of variation), but also for the existence of a phenotypic logic of morphological transformations which is partly independent of direct gene control, implying a hierachical organization of development and a nonlinear genotype-phenotype map (see Alberch [3]; Wagner and Altenberg [118]; Webster and Goodwin [123]). Congruence of the kind portrayed in figure 1(a) should accrue.

Differential filling and density of actual occupation of different regions in theoretical morphospace, in turn, are separate issues, reflecting historical contingency, historical constraint, the distribution of biomechanical optima, and developmental constraints based on unconsidered generative parameters [91]. In effect, these issues suggest the desirability of complementing morphospace studies with approaches such as stochastic simulations and sensitivity analysis of morphospace colonization [37], mapping of phylogenies onto morphospaces [10, 18], assessment of functionality of different forms [56, 66, 71, 73, 88, 107], and modeling based on more refined developmental parameters, capable of codifying inequiprobable renditions of form [6, 50, 95, 104, 123].

Three points of contention are particularly relevant in theoretical/empirical morphospace comparisons and affect how much information about developmental controls on evolution can be extracted from theoretical morphospace constructs: (1) the extent to which anisometry is taken into account; (2) the biological meaning of different underlying models (geometric, physical, morphogenetic); and (3) the meaning of modeling in general and what it can tell us about the workings of development and its potential role in evolution.

1. Often, occupied empirical morphospaces have been contrasted with occupiable theoretical counterparts based on sampling of adults alone (e.g., Raup [87, 88]; Ward [121]; McGhee [71, 73]; but see Raup [88, 89]). But if inferences about constraints are to be sought in comparisons of the occupied and the occupiable, more attention should be payed to ontogenetic variation—regions inaccessible to adult morphologies may well be occupied at different stages in ontogeny. For example, adult cassiduloid echinoids do not have lanterns, but juveniles do. Atelostomate echinoids, having an oval or heart-shaped outline, actually start out after metamorphosis very round. Whenever allometric growth is present, the relationship between occupied and occupiable is likely to change through ontogeny. By the same token, explanations for differential

filling, such as those relying on biomechanical considerations and optimality criteria, must whenever possible incorporate the precept that different stages in life-history may be subject to different living requirements. In short, different morphological distributions may hold at different stages in ontogeny, such that the realized morphospace may often be broader than an adult-oriented representation may grant, with consequences for causal interpretation.

2. It has been argued that a desideratum of theoretical morphospace models should be their possession of biological meaning, often equated with the presence of morphogenetic parameters [95, 99]. Certainly, in a literal sense, a purely geometric morphospace model says nothing about underlying developmental controls. But meaning is what an effective translation preserves, literally or not [9], and it thus can be argued that the "translation" of form generation in terms of geometric (as opposed to morphogenetic) parameters still effectively permits exploration of the developmentally relevant issues of the relationship between possible and actual, the unevenness of filling of morphospace, and the suggestion that unoccupied regions may reflect developmental prohibitions. "Nonmorphogenetic" theoretical morphospaces are thus not devoid of developmental meaning, at least for the majority of questions raised by theoretical/empirical morphospace comparisons.

   In addition, it can be argued that the use of actual biological parameters is not the only way to study morphogenetically meaningful morphospaces. Physical models of form have been shown to be very suggestive of actual principles involved: some echinoid plate arrangements, including angles between units, resemble soap bubble arrangements and seem to obey purely physical rules of packing [89]; limpet-like, bivalve-like and brachiopod-like morphologies can be simulated by wax accretion at the interface of air and water, with temperature appearing to be an important variable [101]; and spatial aggregating patterns of slime mold amoebas mimic with remarkable fidelity the self-organizing spatial patterns of chemical waves produced by the Beloussov-Zhabotinsky reaction [49]. Many other examples are given in the work of D'Arcy Thompson [12]. To the extent that such physical principles are involved in development, they become part and parcel of the developmental matrix that is potentially involved in structuring evolution in morphospace. Theoretical morphospaces, however produced, are *always* bound to reflect principles of development, because they are invariably generative in producing expectations of natural behavior based on intrinsic principles of organization. Precise biological meaning is a goal, but this does not detract from the heuristic power that different degrees of approximation to the logic of development can afford.

3. What is the ultimate purpose of a theoretical model? Realism or simplicity? Realism is the ultimate judge of the worthiness of any model, but simplicity is the sine qua non. Mathematical models are not explanations [82], and thus the requirement of being realistic has to be a relative one. Overly complex models have the potential to be intractable, insoluble, and unintelligible [68],

thus deceiving the very purpose of modeling. Biologically meaningful insights can arise under varying approximations of reality. In terms of development, computer models that succeed in simulating morphology imply that the "actual biological system (...) need not be more complicated than that used in the computer simulation" [89]. Thus, even very simple models can be useful, by suggesting potentially simple underlying morphogenetic controls. While the goal of theoretical modeling is to achieve a realistic representation of an underlying generative logic, aspects of this logic may well be quite simple. This should encourage a more complete appreciation for the complementarity of models with different degrees of elaboration [95].

A final comment pertains to the problem of commensurability of empirical and theoretical morphospaces. Form is ultimately n-dimensional, and many empirical studies have quantified morphospace so as to take into account a broad range of features (see Foote [46]), regardless of whether a generative logic can be found for them through theoretical modeling. Indeed, for many discrete characters, theoretical morphospace construction is a rather intractable problem, although attempts in such direction have been made [107]. A possibility not yet explored, but which bridges the gulf of incommensurability, is the partitioning of n-dimensional empirical spaces into m- and p-dimensional components, with p representing parameters derived from a given theoretical model of form generation, and m corresponding to all additional variables. With appropriate scaling, p- and m-dimensional spaces can be compared in terms of relative spread and spacing of forms, behavior of extremes, differential filling, etc. The extent to which patterns suggested solely by generative theoretical parameters (p) are concordant with those derived from additional descriptive parameters (m) can thus be assessed, in effect achieving an evaluation of how predictive of patterns of morphological disparity (e.g., Foote [44, 46]; Roy [97]; Wagner [119]; Eble [27]) biological versatility (sensu Vermeij [114]) can be.

5 ABNORMAL VS. NORMAL MORPHOSPACE COMPARISONS

Alberch's work [2, 4, 6] resurrected the elegance and the goal of the work of nineteenth-century teratologists such as Isidore Geoffroy Saint-Hilaire. By studying the "logic of monsters" [6], the goal was to learn about the logic of morphological transformations in normal systems. Abnormalities become natural experiments in developmental transformation, and stand as manifestations of the underlying structure of developmental space. The importance of teratology to evolution is tied to the working hypothesis that the nature of normal forms in an evolutionary sense (and in a nondevelopmental space) is not different from the nature of abnormalities in a developmental sense (in a developmental space). In other words, discontinuous variation would be behind both phenomena, the respective phenotypic gaps would be isomorphic, and the kinds of developmental (epigenetic) transitions that occur in the generation of abnormalities could actually take place in evolution and contribute raw material to evolutionary change.

If abnormalities can ever be defined in a more or less objective way (see below), and if sizeable samples are available (not a trivial problem), interesting lines of investigation arise. Fundamentally, they hinge on comparisons of normal and abnormal forms. Such comparisons may involve:

1. Abnormal intraspecific versus normal interspecific variation. If certain classes of clear abnormalities appear again and again within species despite the assumption of negative selection, hints on the constraints imposed by development and the rationality of developmental transformations can be gained [6]. If they happen to mimic interspecific variation, a strong claim can be made to the effect that evolutionary transformation itself may ultimately hinge on radical developmental change. Most comparisons of intra- and interspecific variation to date have been qualitative (see Oster et al. [84]; Alberch [6]). A more rigorous approach would involve phenetic (or even phylogenetic) analyses of normal and abnormal forms.
2. Cladewide interspecific comparisons of morphospace occupation between normal and abnormal forms. If morphological evolution proceeds by selection acting on small variations, pushing morphologies toward adaptive optima, we should expect that the pattern of morphospace occupation for normal forms would be considerably different from that of abnormal forms. On the other hand, if development can play a role in evolution by providing not only constraints on variation but also raw material and creativity in terms of discontinuous variants, one should expect a coincidence in the pattern of total morphospace occupation of normal and abnormal forms. If coincidence of occupation holds, the claim that macroevolution is at least partially controlled by intrinsic developmental potentials is supported, and vice versa. An illustration of this is given by the work of Dafni [13, 14] (see also Seilacher [100]).

The approaches outlined above rely on the assumption that abnormalities can be identified unequivocally. But how to define an abnormality? Clear teratologies are obvious abnormalities, and an extensive descriptive literature exists for a number of model systems. The definition of a teratology is always somewhat subjective, however, and is often restricted to discrete variation, which allows all-or-none categorizations. It is more productive, thus, to focus on the notion of abnormality per se. An abnormality is, almost by definition, anything that extends beyond normality. At a minimum, this implies variants subject to strong negative, but not necessarily lethal, selection; at the limit, abnormalities may be so sweeping as to allow the epithet "teratological."

If the abnormal is to be identified objectively, rigorous comparison with the normal is necessary. This is a statistically tractable problem. For any measurable character, means and a standard deviations can be characterized for ensembles of specimens. Standard models of stabilizing selection naturally lead to the expectation that the higher the deviation from the mean, the lower the selective value. Since standard quantitative genetic models can be viewed at face value as a representation of the present null hypothesis (i.e., evolutionary change as a result of selection acting on small variants), one can refer to population and quantitative genetics for estimates of pronounced lack of fitness. Given uni- and multivariate normality, cutoffs usually involve two or three standard deviations. This approach is defensible because it stems from the null hypothesis and it is used in the testing of the alternative hypothesis. It also has the advantage of allowing consideration of both meristic and metric variation in a rigorous framework. This framework for the quantitative comparative study of normal and abnormal forms has never been implemented. Given an operational definition of abnormality as outlined above, studies of this kind should be tractable.

6 JUVENILE VS. ADULT MORPHOSPACE COMPARISONS

An extensive allometric and heterochrony literature exists with detailed descriptions of and comparisons between ontogenetic trajectories of related species [75, 77]. This taxon-by-taxon approach has yielded a wealth of information and delineated current theories about the relationship between development and evolution [7, 77]. From the perspective of morphospace studies, however, an ensemble approach focused on cladewide comparisons can allow a more synthetic account of general issues like clade shape in ontogenetic time, trends in changes of rate and timing, testing of developmental "laws," and parallels between phylogenetic and ontogenetic trends. Few studies have explored the approach to date [15, 16, 18, 21, 22].

Perhaps more so than in the case of theoretical morphospaces and abnormality morphospaces, the inferences from which must rely on consistency arguments, juvenile morphospaces are closer to the actual dynamics of evolutionary change in development. The intergradation of developmental and nondevelopmental morphospaces is more clear here as well, since at least in commensurate spaces (e.g., after metamorphosis) the total set of features under investigation can be easily modeled with allometric functions (in the case of abnormalities, outliers make such an exercise potentially more difficult). In this vein, juvenile morphospaces can be said to be an intermediate level within the genotype-phenotype map [118]. Purely phenotypic juvenile-adult comparisons can be very useful in understanding the large-scale structure of such mapping: because development is hierarchical and genes behave more as parameter values in epigenetic feedback loops than as a central directing agency [122], phenotypic variation is as useful a target of study in understanding the relationship between development and evolution (and of genotype and phenotype) as genetic variation. Even in the absence of innovation (see Rice [95]), the range of juvenile allometries can at least potentially suggest many pathways of change in adult form [15, 16, 17, 18, 124].

7 DEVELOPMENTAL DISPARITY

The concept of disparity grew out of phenetics and the notion of distance in state space [11, 37, 105, 112, 113]. Rather than a summarization of morphological distributions through ordination, the goal is to produce state space summaries that represent the average spread and spacing of forms in morphospace. Many morphospace studies nowadays thus incorporate the quantification of disparity as a major goal [20, 25, 38, 39, 41, 46, 97, 98, 119, 120]. A body of theory regarding expectations for the dynamics of disparity vis-à-vis diversity is also beginning to grow [41, 44].

A number of disparity measures are possible, by reference to an extensive literature on taxonomic distances [105]. It must be noted that measures of multivariate distance are most useful, and intuitive, when the variables are equally weighted [112], such that transformations of the data may be necessary. Because different disparity metrics are meant to reflect the same underlying morphospace, one can often find equivalence relations among different metrics [43, 112], with general patterns emerging regardless of metric (e.g., Foote [43]). Still, at its most basic level, two fundamentally different aspects of morphospace occupation, and thus different disparity measures, should be recognized when estimates of dispersion are sought: the variance and the range. The variance captures the average dissimilarity among forms in morphospace; the range reflects the amount of morphospace occupied [37]. In multivariate terms, these measures can be generalized as the total variance (the sum of the univariate variances) and the total range (the sum of the univariate ranges) (see Van Valen [112, 113]). Although empirically the total variance and the total range tend to correlate [40], the relationship is not always proportional or monotonic, since the total range is very sensitive to sample size [37, 42]. When sample size differs among samples under comparison, procedures such as morphological rarefaction [40] should be used. Other multivariate statistics of variation exist [36, 112, 113], but in general the total variance and the total range can effectively summarize dispersion in morphospace.

Developmental disparity can be formalized as the disparity among objects in any subregion of a developmental morphospace. Developmental disparity can be quantified either across taxa or within taxa in developmental time, depending on the purpose. Instead of changes in amount of variation in evolutionary time, the focus is shifted to changes in amount of variation in ontogenetic time.

At the limit, evolutionary and developmental renditions of morphospaces and disparity can be brought together, for one might be interested in tracking how changes in disparity and morphospace occupation through ontogeny themselves change in evolutionary time. While for sufficiently well sampled paleontological time series this can be determined directly against a time scale, in many cases the indirect route of mapping on a phylogeny can be used. An emphasis on developmental morphospaces and developmental disparity, especially from a multivariate perspective, expands the playing field of evolutionary developmental biology and further approximates it to macroevolution. [20].

8 AN EXAMPLE: DEVELOPMENTAL MORPHOSPACES AND DEVELOPMENTAL DISPARITY IN SEA URCHINS

An empirical study of juvenile vs. adult morphospace and disparity comparisons is presented next, using echinoids of the order Spatangoida as a model system.

Spatangoids. Spatangoids constitute a monophyletic group of heart urchins that appears in the early Cretaceous (145 M.Y. ago) and ranges to the Recent. They experienced substantial losses at both the end-Cretaceous and late Eocene extinction events. As heart urchins, they display bilateral asymmetry superimposed over radial symmetry, an anteriorly displaced mouth and a posteriorly displaced anus. In addition, ambulacral areas often have enlarged pores, producing a petaloid appearance; a frontal furrow of variable depth is often present; very enlarged plates form a plastron on the oral side; and diverse profiles of the test with different implications for burrowing [63] have evolved. Spatangoids are the most diverse group of sea urchins—12 families, circa 200 genera, and hundreds of species together display a number of themes of echinoid evolution (see Kier [65]).

Developmentally, however, little is known about this group. Interestingly, in at least some species it is known that the very youngest stages after metamorphosis display virtually no bilateral symmetry, resembling instead the more primitive regular echinoids in general outline [17, 53]. This suggests that von Baerian recapitulation [69, 83] might be expected for other features as well.

Data. Although spatangoids have a rich fossil record, and variation in size of the order of 4 can be commonly sampled, it is difficult to ascribe precise ages to different specimens beyond the categorization of sexual maturity in terms of the opening or not of the genital pores in the apical system. Growth is for the most part indeterminate, and although the episodic nature of plate growth registers discrete bands, these are very difficult to reveal even in Recent specimens. In addition, it has been suggested that allometric nonlinearities (implying raw material for extensive developmental variation) in heart urchins are particularly prominent in very small specimens [15, 16, 17, 18]. Thus, although juvenile-adult comparisons of fossil forms are amenable to study, I chose to focus on sampling of Recent species in long-ranging genera, and on representation of all the main families of spatangoids, so as to guarantee representation of the broad features of phenetic and phylogenetic differentiation in the group. A survey of dedicated collections at museums in Paris, London, Copenhagen, Washington, San Francisco, and Berkeley, as well as of illustrations in the literature dating back to the work of Agassiz [1] and Mortensen [81], allowed sampling of the smallest postmetamorphic specimens reliably identified in the past. The data used in this study comprise 14 juvenile-adult sequences in 14 species, 14 genera, and 8 families of spatangoids.

Methods. The primary data for the present morphospace and disparity analyses consists of (mostly) three-dimensional Cartesian coordinates for 18 landmarks

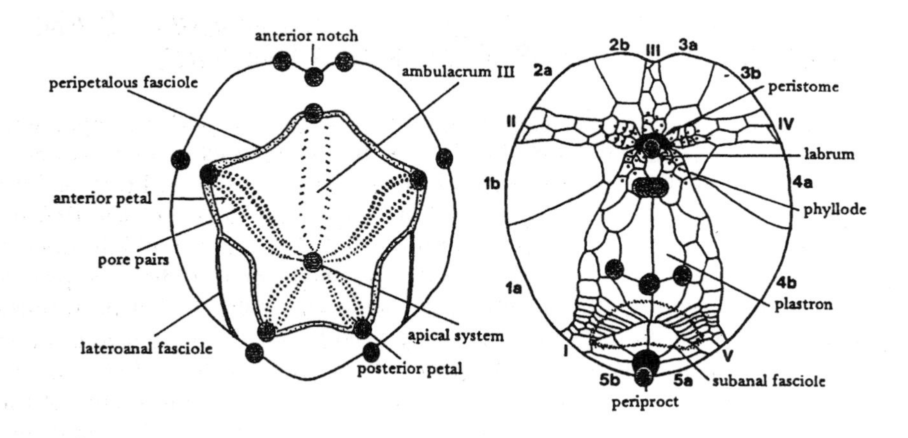

FIGURE 2 Morphometric scheme used in the study. Points represent landmarks from which x, y, z coordinates were collected for all of the specimens considered in this study (after Eble [27]).

that comprehensively capture test architecture (see Eble [24, 27]). The morphometric scheme is illustrated in figure 2. This scheme, with 18 repeatable and relevant landmarks (sensu Roth [96]), nonredundantly codifies several standard morphometric parameters (height, length, width, etc.) and accounts for important test features such as general profile, relative positions and relative size of aboral ambulacral regions and their modifications (petals), extent of frontal furrow, geometry and relative size of plastron plates, and relative positions of the apical system, peristome, and periproct. Because plastron landmarks vary little in the z-direction, and because such variation might be confounded by measurement and orientation error, they were constrained to lie in a single plane. Also, redundant landmark information was averaged and noninformative (invariant) or unreliable (due to difficulties in measurement for very small specimens) coordinates were removed. From 18 landmarks, the final number of variables was then 36. Landmark data collection consisted of image acquisition, digitalization, and analysis using a video camera and BioScan OPTIMAS® image analysis software.

Landmark coordinates for all specimens were scaled to a common size using a reference baseline (the y-coordinate for the tip of the periproct, with the origin of the coordinate system centered between the frontal genital plates of the apical disc), and row normalization was performed for each specimen, rendering the sum of squares of variates for each object equal to one. Row normalization retains the proportionality of variables within objects, and destroys differences in magnitude between objects [94]. Further, standardization was carried out to guarantee equal weighting of the variables; although they are all on the same scale, the range of plastron coordinates tended to be an order of magnitude smaller than that of other coordinates, thus making standardization necessary. The final data matrix used in the analyses consisted of standardized variates.

To produce an ordination space, the resulting correlation matrix was used as input in a principal components analysis. The first six ordered PCs were retained for further analysis, encompassing 83% of the total variance. Based on natural breaks in the decay of eigenvalues, the first 4 PCs (68% of the total variance) were chosen for graphic portrayal of dispersion and location of forms in morphospace. Morphological disparity per se was measured as the total variance in the original, size-normalized, and standardized morphospace. The total variance is equivalent to the mean squared Euclidean distance between each point and the centroid; it is also equal to the sum of the eigenvalues [112, 113]. In this study, the term "disparity" will be used in the sense of variance.

Results and Discussion. Figure 3 presents a comparison of total disparity between pooled adults and pooled juveniles. Three different criteria of partitioning of juveniles from adults were used to assess the sensitivity of the results. In figure 3(a), a strict maturational partitioning was used; i.e., specimens were considered juveniles if they were not sexually mature (genital pores were not open). In figure 3(b), a looser criterion was used, based on size differences: juveniles in a given species were defined as such if they were smaller than the largest specimens by at least an order of 2. Finally, in figure 3(c), an extreme size-based criterion was used, with only the smallest juvenile and the largest adult in each ontogenetic series being considered.

As is clear from figure 3, in all cases the distribution of bootstrapped disparities for adults is indistinguishable from that of juveniles. This is remarkable, given the common assumption that, especially after the phylotypic stage (in this case, the urchin rudiment upon metamorphosis), von Baerian deviation should be the rule. Interspecific adult variation would seem to occur within a developmentally constrained morphospace which is not statistically exceeded.

This, of course, is a general description of the large-scale structure of morphospace, in terms of the average spacing among forms. One might expect different variables to show different ontogenetic topologies of variation. Figure 4 presents principal component plots for several composite variables (maturational partitioning was used here; one outlier removed). The juvenile (developmental) morphospace is superimposed over the adult (nondevelopmental) morphospace (since the spaces are commensurate, one is in effect considering a single "developmental morphospace"). There is a lot of overlap, suggesting much fluidity in variation, but the variance behaves differently in different dimensions. The variance of adults and juveniles along each composite dimension is indicated

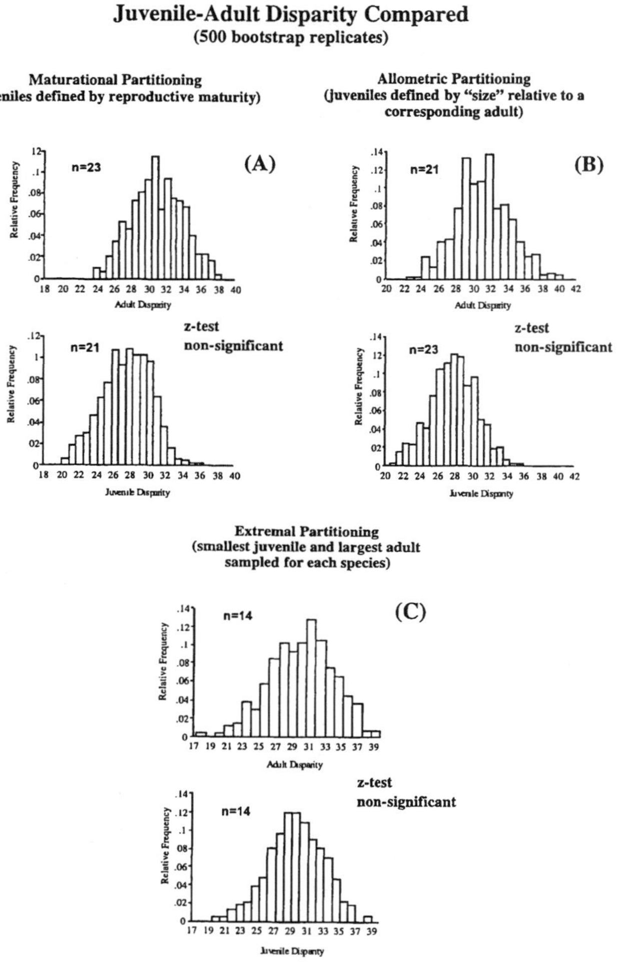

FIGURE 3 Comparison of bootstrapped total disparity between pooled adults and pooled juveniles across 14 genera of spatangoids, under three different criteria for partitioning juveniles. (a) Maturational partitioning, based on sexual maturity. (b) Allometric partitioning, based on size differences of the order of at least 2. (c) Extremal partitioning, with consideration of the smallest juvenile and the largest adult sampled only. In all three cases, no statistically significant difference in levels of disparity is detectable.

alongside each plot. PC III and PC IV display no difference (a bootstrap test was used) between the disparity of juveniles and that of adults, which supports the general inference of pervasive developmental constraint. PC I, however, has juveniles displaying significantly less disparity than adults, in conformity with von Baer's second law. In contrast, PC II has adults showing significantly less disparity than juveniles, thus contra von Baer's second law. Therefore, of the four main composite variables in the system, only one supports a major prediction of developmental evolutionary biology. It is possible to further examine this result by focusing on nearest-neighbor distance analyses (cluster analyses) in light of a phylogeny. A phylogeny for all the taxa involved (the first spatangoid species, appearing 145 million years ago, was used as an outgroup) is presented in figure 5. The tree depicted is the single most parsimonious tree, produced by a posteriori weighting by the consistency index. Except for the node that follows *Hemiaster* and the *Abatus-Pericosmus* node (51% bootstrap support), all other nodes had a bootstrap support of 100%.

The spatangoid phylogeny of figure 5 was used to assess the relative primitiveness or derivedness of species in the nearest-neighbor distance analyses. Samples of the overall cluster analysis are presented in figure 6. It is clear that the overlap in juvenile and adult distributions in morphospace results from taxonomically heterogeneous adult-juvenile clusters. Much dissociation of ontogenetic sequences in morphospace is apparent, and this by itself suggests that changes in rate and timing of development are taking place. In figure 6(a) there is association of a derived juvenile with more primitive adults, thus suggesting overall peramorphosis leading to derived adults. Figure 6(b) shows an association of a primitive juvenile with a derived adult, suggesting overall paedomorphosis leading to derived adults. In figure 6(c), overall peramorphosis is again suggested, with derived juveniles clustering with a more primitive adult. Finally, in figure 6(d), one finds evidence in support of von Baer's second law, with association of juveniles from different taxa (this is also apparent from figure 6(c)). Although here we find some evidence for von Baer's second law, it is by no means pervasive.

von Baer's second law would also predict that over phylogenetic/stratigraphic time, adults from different species would be more and more divergent in comparison with juveniles. This prediction is tested in figures 7 and 8. In figure 7, the disparity implied by adults and by juveniles in each subclade is shown. In figure 8, the respective clade ranks are plotted against the levels of disparity of adults and juveniles. A Wilcoxon rank test indicates that within subclades, juveniles tend to be more variable than adults, which runs against the prediction of von Baer's second law.

As we see, although total juvenile disparity falls within the range of bootstrapped adult disparities (in effect suggesting that overall variation is similarly constrained), a multiplicity of patterns is encountered. In a few cases the patterns are in agreement with von Baer's second law, but at least in spatangoids the evidence is overwhelmingly against it. This is in conflict with what is generally predicted from vertebrate model systems. The multiplicity of results pinpoints

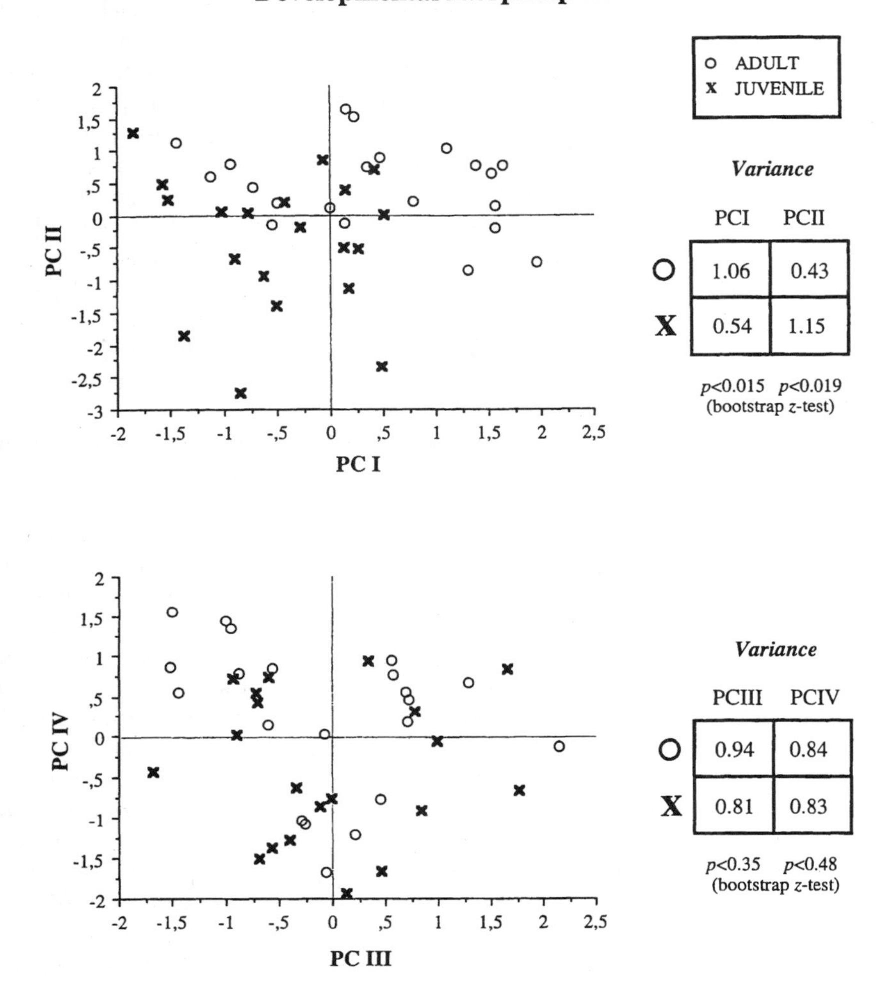

FIGURE 4 Juvenile and adult morphospaces superimposed into a single "developmental morphospace." PC I is plotted against PC II, and PC III is plotted against PC IV. The variance of juveniles is significantly smaller than the variance of adults along PC I, in agreement with von Baer's second law; the variance of juveniles is significantly larger than the variance of adults along PC II, contra von Baer's second law; and no difference in variance between juveniles and adults in apparent for PC III and PC IV. A bootstrap z-test (1000 replicates) was used in each case.

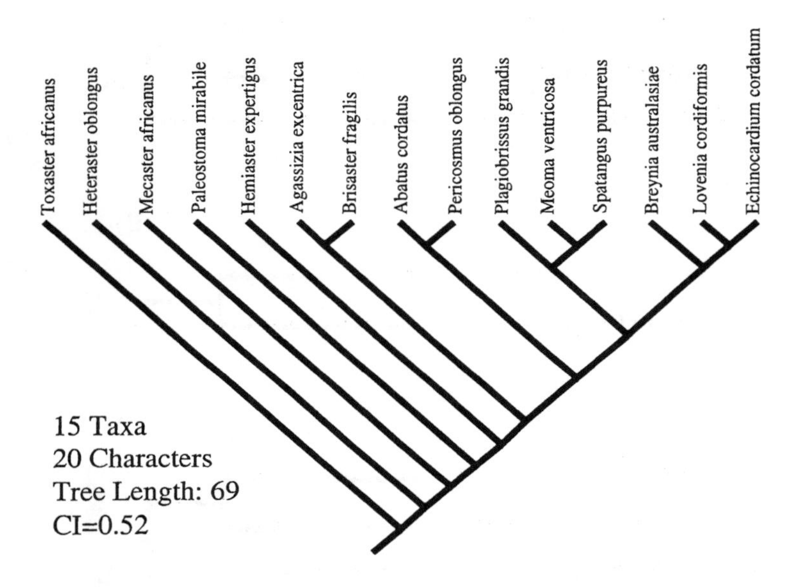

FIGURE 5 A phylogeny of spatangoids. All taxa included in the juvenile-adult study were incorporated. The tree is the single most parsimonious topology generated by successive reweighting by the consistency index, with the first spatangoid in the fossil record used as the outgroup. High bootstrap support is found for the majority of nodes. Data matrix available from the author.

the complexity of the relationship between development and evolution. There is a lot of potential for developmental evolution to happen, which is explored by the group, and no single vector of change in disparity through ontogeny. Still, all such developmental action potentially driving evolution occurs within a constrained morphospace which is fully (and somewhat haphazardly) occupied but never truly statistically exceeded. While the generality of the present results should be explored in other groups (especially invertebrates), the approaches outlined do underscore the general utility of the comparative analysis of cladewide juvenile and adult morphospaces.

9 CONCLUSIONS

Morphospace and disparity studies have over the years suggested a number of large-scale inhomogeneities in morphological evolution, from discontinuities inherent in the construction of organisms to general asymmetries in disparity pro-

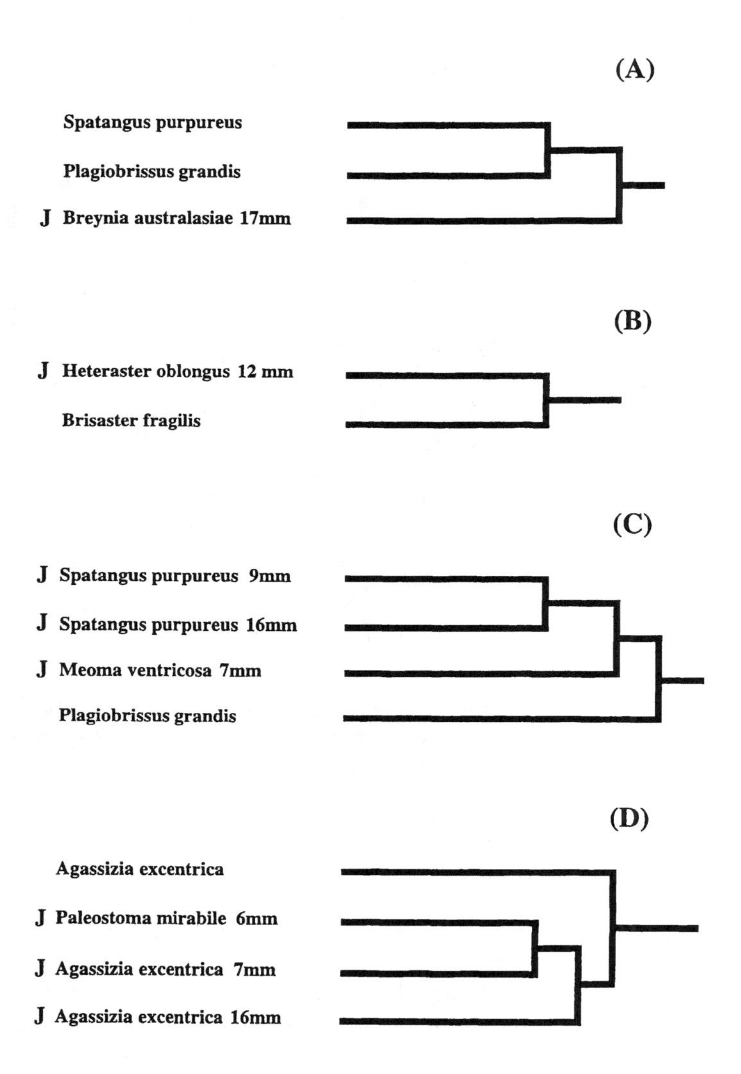

FIGURE 6 Samples of overall cluster analysis (using nearest-neighbor distances) of all spatangoid juveniles and adults considered in this study, so as to illustrate the multiplicity of phenomena encountered. Note the taxonomic heterogeneity of juvenile-adult clusters. In (a), a derived juvenile clusters with more primitive adults, suggesting overall peramorphosis. In (b), a primitive juvenile clusters with a derived adult, suggesting overall paedomorphosis. The same is suggested in (c). In (c) and (d), clustering of juveniles from different taxa supports von Baer's second law.

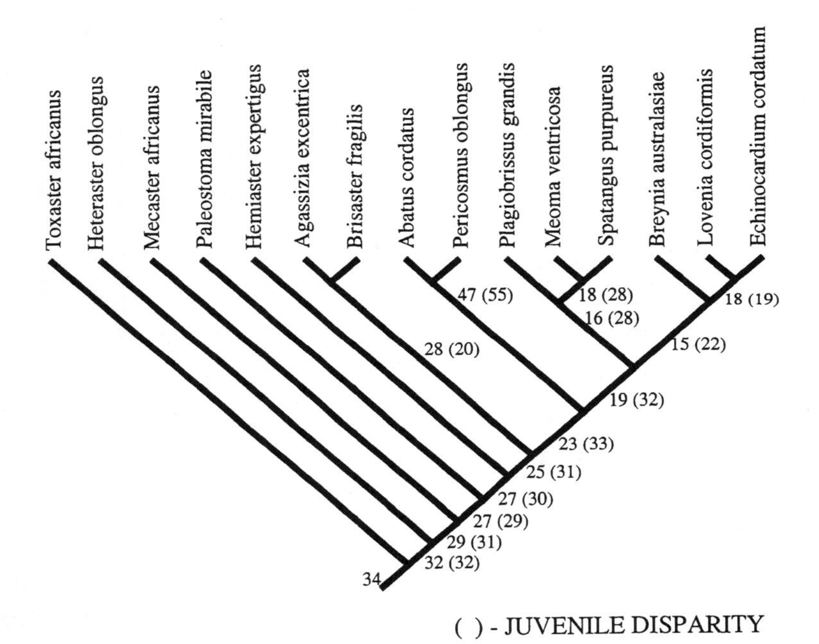

FIGURE 7 Mapping of the disparity implied by adults and juveniles in various subclades against the phylogeny illustrated in figure 5. Values in parentheses correspond to the disparity of juveniles. See text for discussion.

files through time. When interpretations of process are attempted, it is important to balance the usual focus on functional and ecological explanations with consideration of the potential structuring role of development itself. Intrinsic ontogenetic biases stand as one major framework of explanation in evolution and, to the extent that morphospaces can be partitioned into developmental and nondevelopmental components, the notions of developmental disparity and developmental morphospaces become powerful tools in the determination of the relative importance of development in evolution and of organization in dynamics. The morphospace contrasts outlined in this chapter (theoretical vs. empirical, abnormal vs. normal, and juvenile vs. adult) are not meant to be exhaustive and, as implied, the establishment of the developmental/nondevelopmental dichotomy is more heuristic than real, since development usually manifests itself along a continuum. It is hoped that further work can elaborate and refine the

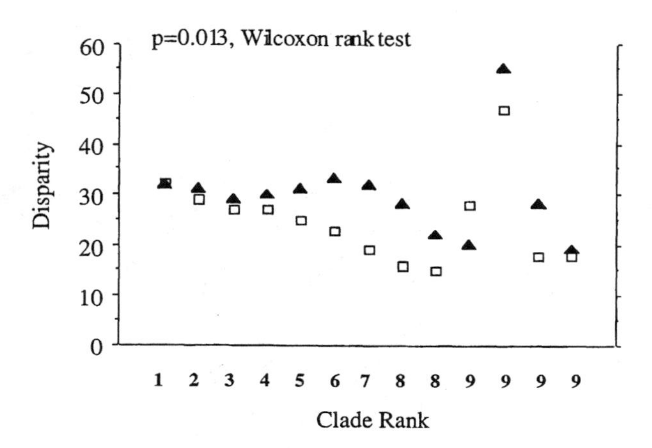

FIGURE 8 Plot of clade ranks versus the disparity of juveniles and adults for each subclade. A Wilcoxon rank test indicates significantly higher disparity of juveniles within subclades. This disagrees with von Baer's second law.

approaches here introduced. The construction of explicitly developmental morphospaces should become an important complement to other approaches to the study of evolution.

ACKNOWLEDGMENTS

I thank B. David, D. Erwin, W. Fontana, M. Foote, D. Jablonski, and M. Zelditch for comments and suggestions. Research was supported by a joint postdoctoral fellowship from the Santa Fe Institute and the Smithsonian Institution.

REFERENCES

[1] Agassiz, A. "Report on the Scientific Results of the Voyage of H. M. S. Challenger during the Years 1873–1876." Report on the Echinoidea. *Zoology* **III(9)**. London: Her Majesty Stationery Office, 1881.

[2] Alberch, P. "Ontogenesis and Morphological Diversification." *Amer. Zool.* **20** (1980): 653–667.

[3] Alberch, P. "Developmental Constraints in Evolutionary Processes." In *Evolution and Development*, edited J. T. Bonner, 313–332. Berlin and Heidelberg: Springer, 1982.

[4] Alberch, P. "Morphological Variation in the Neotropical Salamander Genus Bolitoglossa." *Evolution* **37** (1983): 906–919.

[5] Alberch, P. "Problems with the Interpretation of Developmental Sequences." *Syst. Zool.* **34** (1985): 46–58.

[6] Alberch, P. "The Logic of Monsters: Evidence for Internal Constraint in Development and Evolution." *Geobios* (mémoire spécial) **12** (1989): 21–57.

[7] Alberch, P., S. J. Gould, G. F. Oster, and D. B. Wake. "Size and Shape in Ontogeny and Phylogeny." *Paleobiology* **5** (1979): 296–317.

[8] Arthur, W. *The Origin of Animal Body Plans.* Cambridge, MA: Cambridge University Press, 1997.

[9] Audi, R., ed. *The Cambridge Dictionary of Philosophy.* Cambridge, MA: Cambridge University Press, 1995.

[10] Bookstein, F., B. Chernoff, R. Elder, J. Humphries, G. Smith, and R. Strauss. "Morphometrics in Evolutionary Biology." Special Publication 15, The Academy of Natural Sciences of Philadelphia, 1985.

[11] Cherry, L. M., S. M. Case, J. G. Kunkel, J. S. Wyles, and A. C. Wilson. "Body Shape Metrics and Organismal Evolution." *Evolution* **36** (1982): 914–933.

[12] D'Arcy Thompson, W. *On Growth and Form*, 2d ed. Cambridge, MA: Cambridge University Press, 1942.

[13] Dafni, J. "A Biomechanical Model for the Morphogenesis of Regular Echinoid Tests." *Paleobiology* **12** (1986): 143–160.

[14] Dafni, J. "A Biomechanical Approach to the Ontogeny and Phylogeny of Echinoids." In *Echinoderm Phylogeny and Evolutionary Biology*, edited by C. R. C. Paul and A. B. Smith, 175–188. Oxford: Clarendon Press, 1988.

[15] David, B. "Jeu en mosaïque des hétérochronies: Variation et diversité chez les Pourtalesiidae (échinides abissaux)." *Geobios* (mémoire spécial) **12** (1989): 115–131.

[16] David, B. "Mosaic Pattern of Heterochronies. Variation and Diversity in Pourtalesiidae (Deep-sea Echinoids)." *Evol. Biol.* **24** (1990): 297–327.

[17] David, B., and B. Laurin. "L'ontogenése complexe du spatangue Echinocardium cordatum: Un test des standards des trajectoires hétérochroniques." *Geobios* **24** (1991): 569–583.

[18] David, B., and B. Laurin. "Morphometrics and Cladistics: Measuring Phylogeny in the Sea Urchin Echinocardium." *Evolution* **50** (1996): 348–359.
[19] Dover, G. "A Molecular Drive through Evolution." *BioScience* **32** (1982): 526–533.
[20] Eble, G. J. "The Role of Development in Evolutionary Radiations." In *Biodiversity Dynamics: Turnover of Populations, Taxa, and Communities*, edited by M. L. McKinney and J. A. Drake, 132–161. New York: Columbia University Press, 1988.
[21] Eble, G. J. "Developmental and Non-developmental Morphospaces in Evolutionary Paleobiology." *Paleontol. Soc.* (Special Publication) **8** (1996): 111.
[22] Eble, G. J. "Disparity in Ontogeny and Paleontology." *J. Vertebrate Paleontol.* **17** (1997): 44A.
[23] Eble, G. J. "Developmental Disparity and Developmental Morphospaces." *Geol. Soc. Am. Mtg. Abs.* **30** (1998): A–327.
[24] Eble, G. J. "Diversification of Disasteroids, Holasteroids and Spatangoids in the Mesozoic." In *Echinoderms through Time*, edited by R. Mooi and M. Telford, 629–638. Rotterdam: Brookfield, A. A. Balkema Press, 1998.
[25] Eble, G. J. "On the Dual Nature of Chance in Evolutionary Biology and Paleobiology." *Paleobiology* **25** (1999): 75–87.
[26] Eble, G. J. "Originations: Land and Sea Compared." *Geobios* **32** (1999): 223–234.
[27] Eble, G. J. "Contrasting Evolutionary Flexibility in Sister Groups: Disparity and Diversity in Mesozoic Atelostomate Echinoids." *Paleobiology* **26** (2000): 56–79.
[28] Eble, G. J. "Theoretical Morphology: State of the Art." *Paleobiology* **26** (2000): 498–506.
[29] Eble, G. J. "Multivariate Approaches to Development and Evolution." In *Human Evolution through Developmental Change*, edited by K. McNamara and N. Minugh-Purvis. Baltimore: Johns Hopkins University Press, 2001.
[30] Eldredge, N. *Unfinished Synthesis: Biological Hierarchies and Modern Evolutionary Thought.* New York: Oxford University Press, 1985.
[31] Eldredge, N. *Macroevolutionary Dynamics.* New York: McGraw-Hill, 1989.
[32] Eldredge, N., and S. J. Gould. "Punctuated Equilibria: An Alternative to Phyletic Gradualism." In *Models in Paleobiology*, edited by T. J. M. Schopf, 82–115. San Francisco: Freeman, 1972.
[33] Ellers, O. "A Mechanical Model of Growth in Regular Sea Urchins: Predictions of Shape and a Developmental Morphospace." *Proc. Roy. Soc. Lond. B* **254** (1993): 123–129.
[34] Erwin, D. H. "Macroevolution is More than Repeated Rounds of Microevolution." *Evol. & Dev.* **2** (2000): 78–84.
[35] Erwin, D. H., J. W. Valentine, and J. J. Sepkoski, Jr. "A Comparative Study of Diversification Events: The Early Paleozoic versus the Mesozoic." *Evolution* **41** (1987): 1177–1186.

[36] Foote, M. "Nearest-Neighbor Analysis of Trilobite Morphospace." *Syst. Zool.* **39** (1990): 371–382.

[37] Foote, M. "Analysis of Morphological Data." In *Analytical Paleobiology*, edited by N. L. Gilinsky and P. W. Signor, 59–86. Knoxville, TN: The Paleontological Society, University of Tennessee, 1991.

[38] Foote, M. "Morphological and Taxonomic Diversity in a Clade's History: The Blastoid Record and Stochastic Simulations." *Contributions from the Museum of Paleontology, University of Michigan* **28** (1991): 101–140.

[39] Foote, M. "Paleozoic Record of Morphological Diversity in Blastozoan Echinoderms." *PNAS USA* **89** (1992): 7325–7329.

[40] Foote, M. "Rarefaction Analysis of Morphological and Taxonomic Diversity." *Paleobiology* **18** (1992): 1–16.

[41] Foote, M. "Discordance and Concordance between Morphological and Taxonomic Diversity." *Paleobiology* **19** (1993): 185–204.

[42] Foote, M. "Human Cranial Variability: A Methodological Comment." *Am. J. Phys. Anthropol.* **90** (1993): 377–379.

[43] Foote, M. "Morphological Diversification of Paleozoic Crinoids." *Paleobiology* **21** (1995): 273–299.

[44] Foote, M. "Models of Morphological Diversification." In *Evolutionary Paleobiology*, edited by D. Jablonski, D. Erwin, and J. Lipps, 62–86. Chicago: University of Chicago Press, 1996.

[45] Foote, M. "Perspective: Evolutionary Patterns in the Fossil Record." *Evolution* **50** (1996): 1–11.

[46] Foote, M. "The Evolution of Morphological Diversity." *Ann. Rev. Ecol. & Syst.* **28** (1997): 129–152.

[47] Foote, M. "Morphological Diversity in the Evolutionary Radiation of Paleozoic and Post-Paleozoic Crinoids." *Paleobiology* (Memoirs) **25(2)** (1999): 1–115.

[48] Goodwin, B. C. *Temporal Organization in Cells.* London: Academic Press, 1963.

[49] Goodwin, B. C. *How the Leopard Changed Its Spots.* New York: Charles Scribner's Sons, 1994.

[50] Goodwin, B. C., and L. E. H. Trainor. "The Ontogeny and Phylogeny of the Pentadactyl Limb." *Development and Evolution*, edited by B. C. Goodwin, N. Holder and C. C. Wylie, 75–98. Cambridge: Cambridge University Press, 1983.

[51] Gould, S. J. "Darwinism and the Expansion of Evolutionary Theory." *Science* **16** (1982): 380–387.

[52] Gould, S. J., and N. Eldredge. "Punctuated Equilibrium Comes of Age." *Nature* **366** (1993): 223–227.

[53] Gordon, I. "On the Development of the Calcareous Test of *Echinocardium cordatum*." *Phil. Trans. Roy. Soc. Lond. B* **215** (1926): 255–313.

[54] Hall, B. K. *Evolutionary Developmental Biology*, 2d ed. London: Chapman and Hall, 1998.

[55] Hempel, C. G. *Philosophy of Natural Science.* Englewood Cliffs, NJ: Prentice-Hall, 1966.

[56] Hickman, C. S. "Theoretical Design Space: A New Program for the Analysis of Structural Diversity." *Neues Jahrbuch für Geologie und Paläontologie Abhandlungen* **190** (1993): 183–190.

[57] Ho, M.-W. "An Exercise in Rational Taxonomy." *J. Theor. Biol.* **147** (1990): 43–57.

[58] Ho, M.-W. "Development, Rational Taxonomy and Systematics." *Rivista di Biologia-Biology Forum* **85** (1992): 193–211.

[59] Jablonski, D. "Background and Mass Extinctions: The Alternation of Macroevolutionary Regimes." *Science* **231** (1986): 129–133.

[60] Jablonski, D. "Extinctions in the Fossil Record." In *Extinction Rates,* edited by J. H. Lawton and R. M. May, 25–44. Oxford: Oxford University Press, 1995.

[61] Jablonski, D. "Micro- and Macroevolution: Scale and Hierarchy in Evolutionary Biology and Paleobiology." *Paleobiology* **26** (supplement) (2000): 15–52.

[62] Jablonski, D., and D. J. Bottjer. "Environmental Patterns in the Origin of Higher Taxa: The Post-Paleozoic Fossil Record." *Science* **252** (1991): 1831–1833.

[63] Kanazawa, K. "Adaptation of Test Shape for Burrowing and Locomotion in Spatangoid Echinoids." *Palaeontology* **35** (1992): 733–750.

[64] Kauffman, S. A. *The Origins of Order: Self-Organization and Selection in Evolution.* Oxford: Oxford University Press, 1993.

[65] Kier, P. M. "Evolutionary Trends and Their Functional Significance in the Post-Paleozoic Echinoids." *J. Paleontol.* (Memoir) **48** (1974): 1–95.

[66] Lauder, G. V. "On the Inference of Function from Structure." In *Functional Morphology in Vertebrate Paleontology,* edited by J. J. Thomason, 1–18. Cambridge: Cambridge University Press, 1995.

[67] Lauder, G. V. "The Argument from Design." In *Adaptation,* edited by M. R. Rose and G. V. Lauder, 55–91. San Diego: Academic Press, 1996.

[68] Levins, R. "The Strategy of Model Building in Population Biology." *Amer. Sci.* **54** (1966): 421–431.

[69] Loevtrup, S. "Recapitulation, Epigenesis and Heterochrony." *Geobios* (mémoire spécial) **12** (1989): 269–281.

[70] Maynard Smith, J., R. Burian, S. Kauffman, P. Alberch, J. Campbell, B. Goodwin, R. Lande, D. Raup, and L. Wolpert. "Developmental Constraints and Evolution." *Quart. Rev. Biol.* **60** (1985): 265–287.

[71] McGhee, G. R., Jr. "Shell Form in the Biconvex Articulate Brachiopoda: A Geometric Analysis." *Paleobiology* **6** (1980): 57–76.

[72] McGhee, G. R., Jr. "Theoretical Morphology: The Concept and Its Applications." In *Analytical Paleobiology,* edited by N. L. Gilinsky and P. W. Signor, 87–102. Knoxville, TN: The Paleontological Society, 1991.

[73] McGhee, G. R., Jr. "Geometry of Evolution in the Biconvex Brachiopoda: Morphological Effects of Mass Extinction." *Neues Jahrbuch für Geologie und Paläontologie Abhandlungen* **197** (1995): 357–382.

[74] McGhee, G. R., Jr. *Theoretical Morphology.* New York: Columbia University Press, 1999.

[75] McKinney, M. L., ed. *Heterochrony in Evolution: A Multidisciplinary Approach.* New York: Plenum Press, 1988.

[76] McKinney, M. L. "Classifying and Analyzing Evolutionary Trends." In *Evolutionary Trends*, edited by K. J. McNamara, 28–58. Tucson, AZ: The University of Arizona Press, 1990.

[77] McKinney, M. L., and K. J. McNamara. *Heterochrony: The Evolution of Ontogeny.* New York: Plenum Press, 1991.

[78] McNamara, K. J., ed. *Evolutionary Trends.* Tucson, AZ: The University of Arizona Press, 1990.

[79] McShea, D. W. "Mechanisms of Large-Scale Evolutionary Trends." *Evolution* **48** (1994): 1747–1763.

[80] Moore, A. M. F., and O. Ellers. "A Functional Morphospace, Based on Dimensionless Numbers, for a Circumferential, Calcite, Stabilizing Structure in Sand Dollars." *J. Theor. Biol.* **162** (1993): 253–266.

[81] Mortensen, T. *The Danish Ingolf Expedition*, vol. 4, pt. 2. Copenhagen: Bianco Luno, 1907.

[82] Murray, J. D. *Mathematical Biology.* Berlin, New York: Springer-Verlag, 1989.

[83] Nelson, G. J. "Ontogeny, Phylogeny, Paleontology, and the Biogenetic Law." *Syst. Zool.* **27** (1978): 324–345.

[84] Oster, G., N. Shubin, J. Murray, and P. Alberch. "Evolution and Morphogenetic Rules: The Shape of the Vertebrate Limb in Ontogeny and Phylogeny." *Evolution* **42** (1988): 862–884.

[85] Patterson, C. "How does Phylogeny Differ from Ontogeny?" In *Development and Evolution*, edited by B. C. Goodwin, N. Holder and C. C. Wylie, 1–31. Cambridge: Cambridge University Press, 1983.

[86] Raff, R. A. *The Shape of Life: Genes, Development, and the Evolution of Animal Form.* Chicago: The University of Chicago Press, 1996.

[87] Raup, D. M. "Geometric Analysis of Shell Coiling: General Problems." *J. Paleontol.* **40** (1966): 1178–1190.

[88] Raup, D. M. "Geometric Analysis of Shell Coiling: Coiling in Ammonoids." *J. Paleontol.* **41** (1967): 43–65.

[89] Raup, D. M. "Theoretical Morphology of Echinoid Growth." *J. Paleontol.* **42** (1968): 50–63.

[90] Raup, D. M. "Taxonomic Diversity during the Phanerozoic." *Science* **177** (1972): 1065–1071.

[91] Raup, D. M. "Neutral Models in Paleobiology." In *Neutral Models in Biology*, edited by M. H. Nitecki and A. Hoffman, 121–132. New York: Oxford University Press, 1987.

[92] Raup, D. M., and G. E. Boyajian. "Patterns of Generic Extinction in the Fossil Record." *Paleobiology* **14** (1988): 109–125.

[93] Raup, D. M., and Sepkoski. "Mass Extinctions in the Marine Fossil Record." *Science* **215** (1982): 1501–1503.

[94] Reyment, R. and K. G. Jöreskog. *Applied Factor Analysis in the Natural Sciences.* Cambridge: Cambridge University Press, 1993.

[95] Rice, S. H. "The Bio-geometry of Mollusc Shells." *Paleobiology* **24** (1998): 133–149.

[96] Roth, V. L. "On Three-Dimensional Morphometrics, and on the Identification of Landmark Points." In *Contributions to Morphometrics*, edited by L. F. Marcus, E. Bello and A. García-Valdecasas, 41–61. Madrid: Museo Nacional de Ciencias Naturales, 1993.

[97] Roy, K. "Effects of the Mesozoic Marine Revolution on the Taxonomic, Morphologic, and Biogeographic Evolution of a Group: Aporrhaid Gastropods during the Mesozoic." *Paleobiology* **20** (1994): 274–296.

[98] Roy, K. "The Roles of Mass Extinction and Biotic Interaction in Large-Scale Replacements: A Reexamination using the Fossil Record of Stromboidean Gastropods." *Paleobiology* **22** (1996): 436–452.

[99] Savazzi, E. "Theoretical Shell Morphology as a Tool in Constructional Morphology." *Neues Jahrbuch für Geologie und Paläontologie Abhandlungen* **195** (1995): 229–240.

[100] Seilacher, A. "Self-Organizing Mechanisms in Morphogenesis and Evolution." In *Constructional Morphology and Evolution*, edited by N. Schmidt-Kittler and K. Vogel, 251–271. Berlin, Heidelberg: Springer-Verlag, 1991.

[101] Seilacher, A. "Candle Wax Shells, Morphodynamics, and the Cambrian Explosion." *Acta Paleontologica Polonica* **38** (1994): 273–280.

[102] Sepkoski, J. J., Jr. "Biodiversity: Past, Present, and Future." *J. Paleontol.* **71** (1997): 533–539.

[103] Sepkoski, J. J., Jr. "Rates of Speciation in the Fossil Record." *Phil. Trans. Roy. Soc. Lond. B* **353** (1998): 315–326.

[104] Shubin, N., and P. Alberch. "A Morphogenetic Approach to the Origin and Basic Organization of the Tetrapod Limb." *Evol. Biol.* **20** (1986): 319–387.

[105] Sneath, P. H. A., and R. R. Sokal. *Numerical Taxonomy.* San Francisco: Freeman, 1973.

[106] Stanley, S. M. "A Theory of Evolution above the Species Level." *Proc. Nat. Acad. Sci. USA* **72** (1975): 646–650.

[107] Thomas, R. D. K., and W.-E. Reif. "The Skeleton Space: A Finite Set of Organic Designs." *Evolution* **47** (1993): 341–360.

[108] Valentine, J. W. "Patterns of Taxonomic and Ecological Structure of the Shelf Benthos during Phanerozoic Time." *Palaeontology* **12** (1969): 684–709.

[109] Valentine, J. W., S. M. Awramik, P. W. Signor, and P. M. Sadler. "The Biological Explosion at the Precambrian-Cambrian Boundary." *Evol. Biol.* **25** (1991): 279–356.

[110] Valentine, J. W., D. H. Erwin, and D. Jablonski. "Developmental Evolution of Metazoan Body Plans: The Fossil Evidence." *Devel. Biol.* **173** (1996): 373–381.
[111] Valentine, J. W., D. Jablonski, and D. H. Erwin. "Fossils, Molecules, and Embryos: New Perspectives on the Cambrian Explosion." *Development* **126** (1999): 851–859.
[112] Van Valen, L. "Multivariate Structural Statistics in Natural History." *J. Theor. Biol.* **45** (1974): 235–247.
[113] Van Valen, L. "The Statistics of Variation." *Evol. Theory* **4** (1978): 33–43.
[114] Vermeij, G. J. "Biological Versatility and Earth History." *PNAS USA* **70** (1973): 1936–1938.
[115] Vrba, E. S., and S. J. Gould. "The Hierarchical Expansion of Sorting and Selection: Sorting and Selection Cannot be Equated." *Paleobiology* **12** (1986): 217–228.
[116] Waddington, C. H. *The Strategy of the Genes.* London: Allen and Unwin, 1957.
[117] Wagner, G. P. "Homologues, Natural Kinds and the Evolution of Modularity." *Amer. Zool.* **36** (1996): 36–43.
[118] Wagner, G. P., and L. Altenberg. "Complex Adaptations and the Evolution of Evolvability." *Evolution* **50** (1996): 967–976.
[119] Wagner, P. J. "Testing Evolutionary Constraint Hypotheses with Early Paleozoic Gastropods." *Paleobiology* **21** (1995): 248–272.
[120] Wagner, P. J. "Patterns of Morphologic Diversification among the Rostroconchia." *Paleobiology* **23** (1997): 115–145.
[121] Ward, P. "Comparative Shell Shape Distribution in Jurassic-Cretaceous Ammonites and Jurassic-Tertiary Nautilids." *Paleobiology* **6** (1980): 32–43.
[122] Webster, G., and B. C. Goodwin. "The Origin of Species: A Structuralist Approach." *J. Soc. Biol. Struct.* **5** (1982): 15–47.
[123] Webster, G., and B. C. Goodwin. *Form and Transformation.* Cambridge, MA: Cambridge University Press, 1996.
[124] Zelditch, M. L., F. L. Bookstein, and B. L. Lundrigan. "Ontogeny of Integrated Skull Growth in the Cotton Rat *Sigmodon fulviventer.*" *Evolution* **46** (1992): 1164–1180.
[125] Zelditch, M. L., and W. L. Fink. "Heterochrony and Heterotopy: Stability and Innovation in the Evolution of Form." *Paleobiology* **22** (1996): 241–254

The Dynamics of Large Biological Systems: A Statistical Physics View of Macroevolution

Stefan Bornholdt

The biological evolution of species is difficult to model in its entirety. A wealth of knowledge exists about small subsystems of the biosphere as specific organisms or communities. However, any quantitative model of the whole biosphere's evolutionary dynamics is practically out of reach. Dynamical models using the perspective of statistical physics, sacrificing predictability in order to identify underlying mechanisms in complex systems, may provide a first step toward quantitative models of biological evolution. New sources of biological data available from paleontology and molecular biology provide the necessary foundation. Here, possible steps of such an approach are sketched with a focus on the question of the origin of punctuated equilibrium in the macroevolution of species as it is observed in the fossil record.

Evolutionary Dynamics, edited by
J. P. Crutchfield and P. Schuster. Oxford University Press.

1 INTRODUCTION: STATISTICAL PHYSICS AND BIOLOGICAL EVOLUTION

Erwin Schrödinger, in his famous lectures "What is life?" on a physicist's view of the biological research of his time [24], gave an impressive example of how methods of statistical physics can contribute to answer questions posed by biological research. At the time of his lectures in 1943, empirical data about the structure of the chromosome had only become available recently, for example, the structure of the chromosome from fluorescence microscopy, the point event nature of mutation from experiments with gamma rays on flies, and a possible molecular nature of genetic information storage [29]. These data eventually led to the identification of gene size and loci and helped to determine the process of point mutation. The statistical approach of Schrödinger demonstrated how to obtain an accurate upper bound for the size of a gene as well as the localization of mutations and quantified the enormous stability of genetic information storage. The emerging picture of the deterministic nature (the "clockwork mode") of information processing in genetic evolution as well as the suppression of fluctuations in this process are still valid today.

While gene loci and the genetic code have been deciphered to a large extent since that time, many open questions remain to be answered in the domains of genetic information processing and biological evolution. The exact gene functioning, in particular with respect to mutual genetic interaction, is still only partially understood. On a much larger scale, when the wealth of interaction between organisms and entire species leads to the unique dynamics of the biosphere called biological evolution, hard theoretical problems are left unanswered. Today, again, new sources of empirical data from biological research are available which might help to address these still open questions. Large numbers of genome sequences have emerged from current sequencing programs, gene expression data describing gene interaction have been mapped [28], and detailed databases of the fossil record on earth have been created recently [22, 25]. This situation of new availability of empirical data is reminiscent of Schrödinger's time. It is tempting to study recent data from the perspective of statistical physics again in order to ask questions about the mechanisms of biological evolution.

The statistical physics approach has been successful in describing natural phenomena where large numbers of interacting parts are involved (for example, atoms or molecules in liquids, gases, and solids). Although biological evolution is far from being a pure physical system, some of its functions and mechanisms may be accessible to methods of statistical physics developed to deal with large amounts of data or uncertainty. Typical questions are: What are relevant global variables describing the process of biological evolution? Is there a major fraction of irrelevant variables which we are allowed to neglect in the process of model building? When relevant global observables can be identified, a second step is the interesting possibility of building statistical models. Just as ecological models based on differential equations, for example, are different from predictive models, statistical models only try to capture properties of a few global variables. Ideally this may provide information about basic mechanisms at work in an otherwise extremely complex system. This is a possible way for models of biological evolution to include mathematically quantitative elements, even where truly predictive models remain out of reach. In the following, let us sketch such a scenario for a specific aspect of biological evolution. A prominent observational feature of evolution is punctuated equilibrium [6], describing the rapid appearance of new species in the fossil record after extended intervals of stasis. This phenomenon reminds us of similar dynamical behavior in some physical systems, as intermittency or metastable states, which in general point to an interesting underlying mechanism. It is therefore interesting to study whether punctuated equilibrium in evolution can be similarly traced back to an underlying mechanism. Possible observables relevant to this question may be found in statistical features of the fossil record. In the following, we will have a closer look at the statistics of species lifetimes as observed in the fossil record. Then it is described how a stochastic model of the evolution of regulatory genes may be related to this aspect of macroevolution. Finally, any scientific model has to be able to yield predictions or hypotheses in order to be falsifiable and testable. For statistical models as dealt with here, this is of particular interest.

2 PUNCTUATED EQUILIBRIUM: OBSERVATION AND THEORY

The picture of biological evolution that Charles Darwin drew more than a century ago is remarkably up to date [5]. Competition between species and selection of the fittest proves to be a useful hypothesis for explaining most of the observational picture of biological evolution. However, one prominent and highly nontrivial aspect of the evolutionary record is not part of this picture: The largely discontinuous nature of the emergence of new species observed in the fossil record, which Eldredge and Gould [6] coined "punctuated equilibrium." While the Darwinian picture assumes smooth adaptation and gradual speciation, the fossil record shows speciation mostly as discontinuous jumps in morphology space that cannot be further resolved on the stratigraphic time scale of the fossil record. An example is shown in figure 1. Hardly any gradual change is seen in the branching of many species, in contrast to the central postulate of gradual evolution in Darwin's original theory. For a long time this had been assigned to the incompleteness of the fossil record and a major critic of Darwinian evolution was the large number of "missing links" in this sense. However, today it is well established that speciation occurring in a noncontinuous fashion is the rule rather than the exception in the fossil record. This evidence has first been carefully documented by Schindewolf [23], but found wider acceptance only after

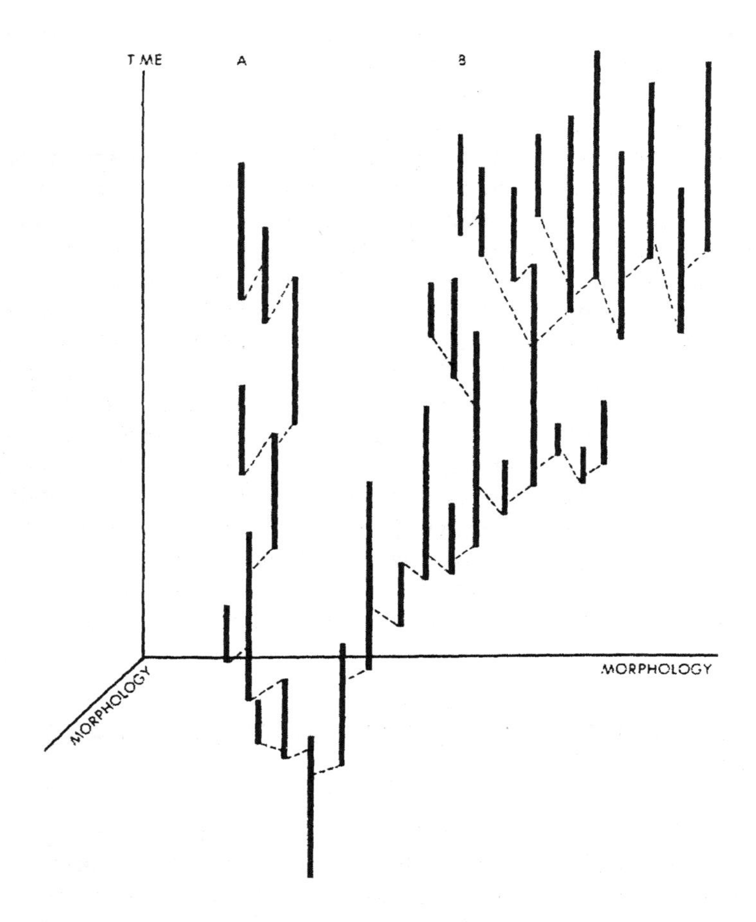

FIGURE 1 Biological evolution as punctuated equilibrium: morphological stasis within species is punctuated by morphological jumps at speciation.

further work by Eldredge and Gould [6] and several years of active discussion in the scientific community [10, 11].

What is the underlying reason for this remarkable dynamical feature of macroevolution? A convincing model for this behavior is still lacking. Ultimately, any valid theory of macroevolution should be able to provide a mechanism that explains this dynamics. In an early evolutionary theory Goldschmidt [9] postulates the clear distinction between gradual evolution within species and inter-species evolution which is then allowed to be discontinuous. However, the origin of what he calls "macromutations" that generate speciation events remain unresolved. Let us now turn to existing mathematical models of the emergence of new species and their possible accommodation of punctuated equilibrium.

Macroevolution focuses on the evolution of interacting species. As compared to intraspecies evolution, new types of evolutionary dynamics emerge, often summarized under the term coevolution: competition between different species, exploitation, and mutualism are examples of these dynamics. Most of all, when considering evolution of more than one species, the origin of species and the dynamics of speciation become legitimate questions within the theory.

Textbook examples for models of interacting species often consider a small number of species and the interaction of their phenotypes. A popular example is the interaction of two species in models of predator-prey interaction [17]. However, when studying true interspecies evolution, describing the birth and death of species, one soon has to go beyond this framework and allow for genetic modification and, eventually, branching of the considered species. With this addition, elementary processes of true coevolution can be modeled [8]. Still, existing models do not reach beyond a modest number of species, usually much smaller than the number of actively interacting species in any region of the biosphere where emergence and evolution of new species occur. Large-scale dynamical phenomena and statistical properties of macroevolution may therefore be still out of the reach of mathematical modeling. As for the observed punctuations in the dynamics of macroevolution, a conclusive explanation has yet to be given in the framework of these models.

As a first step toward understanding punctuated equilibrium, phenomenological models have been constructed that model this aspect of metastability in evolution [15, 19]. Their basic assumption is that species evolve in some potential landscape, similar to the evolution in a fitness landscape postulated by Wright at the level of populations [32]. For a species captured in a valley of this landscape, it may take some waiting time to exit to a neighboring, lower lying valley. This waiting time resembles the periods of stasis in punctuated equilibrium. One example is the model of Newman et al. [19] where a stochastic diffusion process in a double-well potential exhibits metastable states in one of the wells with considerable exit times to get to the neighboring well. The dynamics of such a transition is shown in figure 2. This waiting time for a transition from one well to another in the potential landscape is an analogue of the stasis in evolution. However, caveats are due here, in particular in questioning the applicability of these exact models to biological evolution. To what does the potential used here correspond in nature? Does the evolving biosphere stick to a potential function at all; i.e., is macroevolution an optimization process, locally in every single species? It is not clear whether any optimization process goes on at the species level as it does on the organism level.

While current evolution theory does not answer such questions, clearly new data is needed to address these hard problems. In particular, statistical data from new sources of biology and paleontology may open the way to alternative approaches to the above questions.

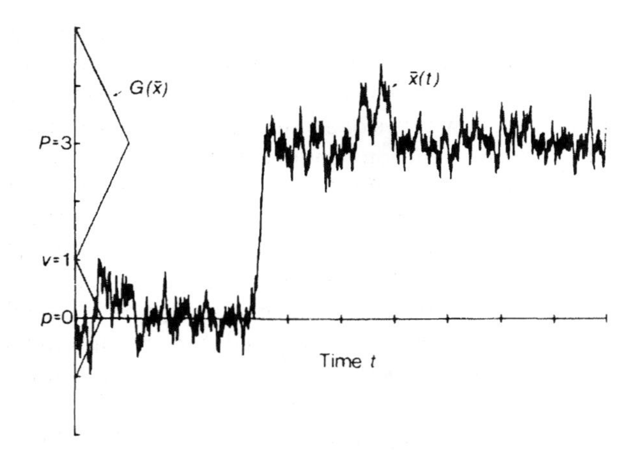

FIGURE 2 Macroevolutionary dynamics according to the model of C. M. Newman et al. [19]. Reprinted by permission of the authors.

3 EMPIRICAL DATA AND NEW MODELS OF EVOLUTIONARY DYNAMICS

Since the early days of building theories of evolution, new sources of empirical data have become available that may allow the testing of evolution theories on a more quantitative basis. Among those are sources of molecular biology supplying data on genetic sequences, gene expression patterns [28], and genetic regulatory domains [18]. Another valuable source is the paleontological fossil record as collected by paleontologists and its statistical properties, e.g., as compiled into a large database by Sepkoski [22, 25] containing the record of about 50,000 marine genera.

As questions of macroevolution often address variables that are connected to the biosphere as a whole, a statistical approach on the basis of such large databases seems promising in this regard. Furthermore, it may complement traditional approaches of biology and paleontology, studying evolutionary dynamics mostly on the basis of case studies on the levels of organisms, populations, or species. In order to approach biological evolution with statistical methods, suitable observables have to be found that can be derived from the data and contain valid information about the complex system under consideration. Regarding punctuated equilibrium in evolution, one such observable can be found in the example of a speciation tree (fig. 1): the observed lifetime of a species, from its first to its last occurrence in the stratigraphic record. It is a particu-

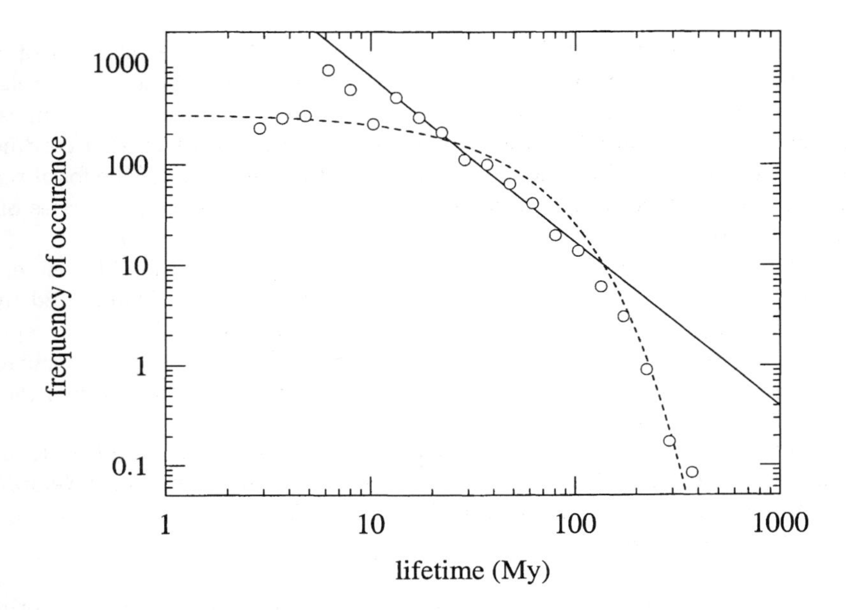

FIGURE 3 Genera lifetimes from the marine fossil record, from the database of Sepkoski [25], plotted after Newman and Palmer [21]. Reprinted with permission of the authors.

larly well-defined quantity in the fossil record, due to the punctuated nature of speciation, reducing ambiguity in estimating intervals among speciation events.

What do we naïvely expect for the distribution of species lifetimes, when considering the entire biosphere? If one does not know any details about the actual interaction among all species, the simplest assumption would be to treat the species independently. Furthermore, it might be reasonable to assume some average species lifetime (of, say, a few million years). Then the answer is readily calculated: The expected statistics of a large number of independent processes (here, species) is governed by the central limit theorem and obeys Poissonian statistics, with long lifetimes exponentially suppressed.

The corresponding distribution of species lifetimes extracted from Sepkoski's database [22, 25] of marine species is shown in figure 3, here in the form of the genera lifetimes. In the range of time intervals that can be determined from stratigraphic dating in geology, which lies in the range of about 7 Myr to at least 100 Myr, the observed lifetime distribution deviates from an exponential decline that would have been expected for independent species. Rather, the observed lifetime statistics in the above range is compatible with a power-law decay of exponent $\alpha = -1.7 \pm 0.3$.

What does this mean for the traditional picture of speciation? First of all, a nonrandom distribution of species lifetimes indicates that these are correlated, beyond a simple crowding around a common mean. Thus, species lifetimes are not randomly distributed in the biosphere, but rather organized according to some unknown underlying rule. The above statistical property of the fossil record thus provides a substantial experimental fact about large-scale properties of the evolving biosphere.

What does this result mean for existing models of speciation? Consider, for example, the above double-well model of Newman et al. [19]. Dominated by an average waiting time to pass to a lower lying valley in the imaginary species fitness landscape, this model predicts an exponential decrease of the occurrence of long species lifetimes. It is thereby falsified by the above data, which require stretched distributions to fit the data.

More generally, what are possible causes for the deviation from the statistics of independent events? Answers to this question may guide further development of models for speciation. Two immediate possibilities that could lead to the observed statistics are:

1. interactions between species result in an overall correlation of species lifetimes, and
2. species are correlated through their common biological origin.

The first possibility has been explored in recent years by stochastic models of evolution, with an emphasis on modeling the role of species interactions in extinction events. Among the mechanisms employed are self-organized criticality [1, 26], interaction networks [27], and external perturbation [20]. Most of these types of mathematical models also give predictions for the distribution of species lifetimes; some are compatible with the above data. For a detailed review of this family of models, see Newman and Palmer [21].

The second possibility has not been explored as extensively and will be the focus of the following discussion. The marine species considered in the dataset above are, indeed, closely related to each other, simply by their common evolutionary origin. This leaves room for a mechanism which organizes the abundance of species lifetimes in a nonrandom way. A prime candidate that may operate in such a way is the gene regulatory circuit of a species. It plays a key role in ontogeny of each multicellular organism, thus determining the morphology to a high degree. In a simplified picture, regulatory genes activate and deactivate each other, thereby forming a tight network of interaction. The resulting dynamics coordinates the details of an organism's ontogeny. Furthermore, with respect to sexual reproduction, the regulatory genes define a species on the genetic level; this requires compatibility of the regulatory circuits of the mating partners if they are to translate each other's genome into a functioning organism.

In recent years these developmental mechanisms have been studied in organisms [18], as in the regulatory genetic system of drosophila. In paleontology the importance of regulatory genes for the evolution of morphology has been discussed regarding punctuated equilibrium [10] and is currently a field of active research [7]. One prominent class of highly conserved regulatory genes are the homeobox genes which appear in virtually all multicellular organisms. This group of genes affects the organization of body axes. For example, the bilateral symmetry of vertebrates is affected by a small, albeit complicated contextually, set of genes, the same which, in echinoderms, generates a radial symmetry by means of a different interaction pattern [16]. With this power over morphological changes, regulatory genes are in a key position also in the process of speciation and, therefore, may influence species lifetimes. When considering common properties of coevolved species, as done above, one has to keep in mind the common origin of the gene regulation machinery. In particular, this applies to the marine species considered in the above data. Let us therefore, in the last part, consider a model of an interacting network of regulatory genes under evolutionary dynamics.

4 A GENETIC NETWORK MODEL OF SPECIATION

When building models of complex biological processes, the first problem one faces is to specify what aspect of reality is being modeled. Modeling a true genetic regulatory network in a predictive sense is clearly impossible: only a tiny fraction of the interactions between regulatory genes in eukaryotes is known today. However, modeling statistical properties of regulatory networks is still possible, e.g., by considering large numbers of different networks and studying their average properties. This is the approach of statistical physics. With respect to our present problem we might ask, what are the properties of the ensemble of genetic regulation circuits emerging from biological evolution?

Circuits of regulatory genes have come to operate in the "clockwork regime" of a highly deterministic mechanics, since any error in its regulation will propagate throughout ontogeny, with most probably devastating effects. This is presumably one of the reasons why regulatory genes in nature evolve on a very slow time scale, as can be seen from the high overlap in the regulatory genes of diverse species in the animal kingdom. However, once a regulatory network of genes has been mutated, it may form the genetic starting point of a speciation event. Let us take this as a motivation for modeling the evolution of genetic networks.

Stochastic models of genetic interaction networks have been proposed, the first by Kauffman [12, 13], to study statistical properties of such networks. The network is represented as a random Boolean network, consisting of binary switches (or nodes) that are asymmetrically connected to other nodes. Each node has a lookup table which determines its state in the next time step, given the incoming signals from other nodes. Connections, lookup tables, and the initial state of the network are chosen randomly and then the dynamics of the network is iterated. After a transient period the dynamics enters a periodic attractor.

There has been speculation about similarities to dynamical patterns in cell expression cycles. These discrete dynamical networks will be used here to study the generic behavior of such networks under evolutionary processes.

A model of the evolutionary dynamics of genetic networks on the basis of such Boolean networks has been proposed in Bornholdt and Sneppen [2, 4] and will be summarized in the following. To study evolution of Boolean networks, a random network is set up as described above (a mother network), then a copy is made (a daughter network) which is mutated by adding or removing a connection weight between two nodes. The dynamics of the two is compared, starting from an identical but otherwise random initial state. Often, the two end up in different dynamical attractors since any initial damage spreads fast through the network. Whenever this occurs, another mutation is tried. If the attractor dynamics of the two networks are identical, a neutral mutant of the mother network has been found. In the terms of this model, this is called a speciation event. (Even if the mutation occurs neutrally, without substantial phenotypic changes, it might surface at a later time, through a change in structural genes or other factors that interact with gene regulation.) Then, the mother network is replaced by the newly found mutant and the algorithm is iterated further. This requirement of consistency, that the daughter should be able to read the mother's structural genes, is the only selection criterion. No fitness function is assumed. The principle of this algorithm is depicted in figure 4.

The evolutionary dynamics of this model exhibits long waiting times for a starting point of new speciation. One might argue that this represents an internal evolutionary clock of the regulatory circuit, governing speciation rate for a given species. In macroevolutionary models where speciation and extinction are in a dynamical equilibrium as, for example, in Van Valen's Red Queen model [30], this can determine also species lifetimes. The waiting time distribution of the simulation is shown in figure 5 and exhibits a power-law decay with exponent $\alpha = -2$ and thus would be compatible with the lifetime fossil data.

What does this have to do with nature? We simulated a stochastic process in the computer, looking at the properties of purely random networks with random inputs, not even defining a fitness function—altogether this looks like a rather artificial process. However, this minimal stochastic model exhibits mechanisms, which in principle could be at work in very complex natural systems and which cannot be deduced from studying single parts of this natural system alone. Only a hard piece of scientific work remains to be done to, in principle, be able to identify such a mechanism in a natural system. A possible way is to find testable hypotheses made by the evolutionary process in the computer. As this is based on the consistency criterion of mother-daughter networks which end up in the same dynamical attractor, it is able to generate a chain of conditional probabilities. In other words, this minimal evolutionary process might generate networks with nonrandom properties. One such property which might be considered is the fraction of the nodes in the network that are dynamically frozen. Previously this so-called frozen component has been considered with respect to genetic networks

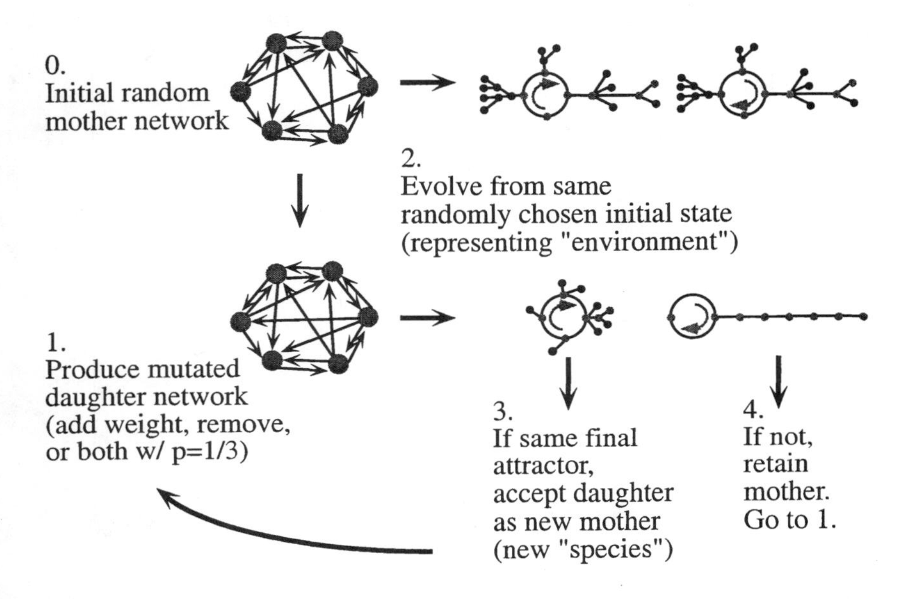

FIGURE 4 Evolution model of binary genetic networks. Networks evolve by mutation of links, but only those mutations are accepted that do not change the dynamics for a given environment, which this model considers to be an indication of a functional ontogeny.

by Kauffman [14]; he concludes that a large fraction of regulatory genes are kept in a constant state. In the simulation model considered here, evolved networks develop a frozen component which is larger than that of random networks with comparable number of nodes and connections [4]. An example is shown in figure 6. While being consistent with Kauffman's findings, a quantitative test of this hypothesis must wait. To measure this property in natural regulatory networks of eukaryotes a large fraction of the regulatory network has to be known. Further approaches to experimental tests may be provided by gene-knockout experiments or real-time evolution experiments as discussed in Bornholdt and Sneppen [4]. Also other properties of genetic regulatory networks may be suitable for future experimental tests as basic measures of their connectivities, which could be compared to predictions of network models [3]. Experimental programs are underway and the first small domains of regulatory networks have been mapped [31] which could provide data in this direction. Here is where current achievements in molecular biology may write exciting future chapters in evolution theory.

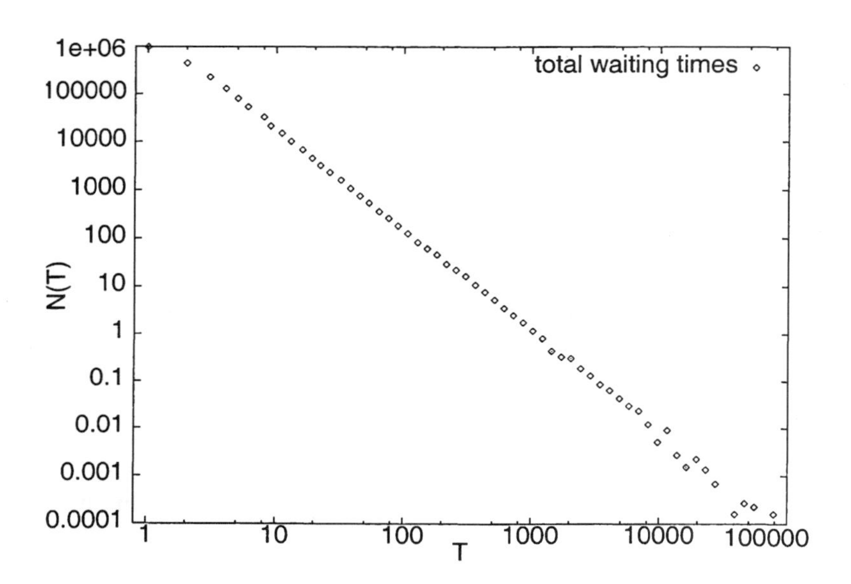

FIGURE 5 Lifetime distribution in the genetic evolution model, here for a Boolean network with 16 nodes. Bornholdt, S., and K. Sneppen [2]. Reprinted by permission of the authors.

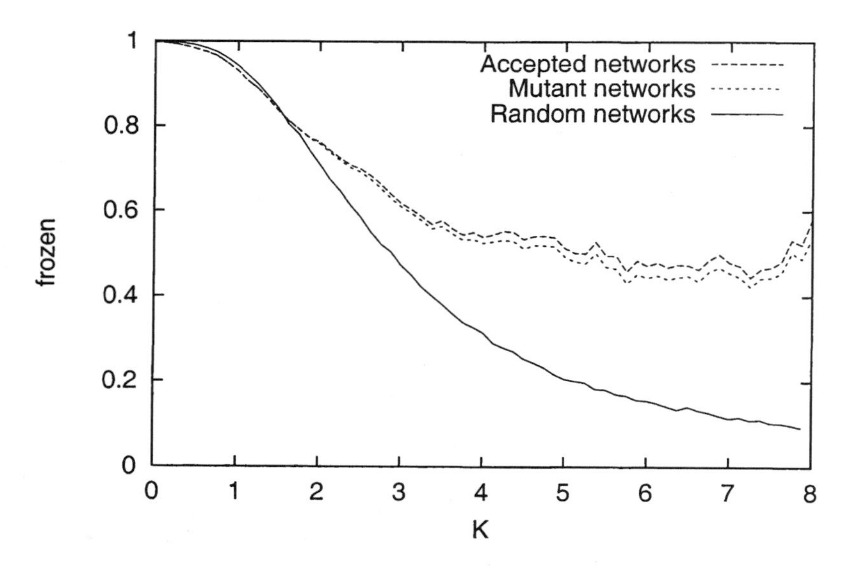

FIGURE 6 Large frozen components in networks during evolution. Bornholdt, S., and K. Sneppen [4]. Reprinted by permission of the authors.

REFERENCES

[1] Bak, P., and K. Sneppen. "Punctuated Equilibrium and Criticality in a Simple Model of Evolution." *Phys. Rev. Lett.* **71** (1993): 4083.

[2] Bornholdt, S., and K. Sneppen. "Neutral Mutations and Punctuated Equilibrium in Evolving Genetic Networks." *Phys. Rev. Lett.* **81** (1998): 236.

[3] Bornholdt, S., and T. Rohlf. "Topological Evolution of Dynamical Networks: Global Criticality from Local Dynamics." *Phys. Rev. Lett.* **84** (2000): 6114.

[4] Bornholdt, S., and K. Sneppen. "Robustness as an Evolutionary Principle." *Proc. Roy. Soc. Lond. B* **267** (2000): 2281–2286.

[5] Darwin, C. *The Origin of Species by Means of Natural Selection.* Harmondsworth: Penguin, 1968.

[6] Eldredge, N., and S. J. Gould. "Punctuated Equilibria: An Alternative to Phyletic Gradualism." In *Models in Paleobiology*, edited by T. J. M. Schopf, 82–115. San Francisco: Freeman Cooper, 1972.

[7] Erwin, D., J. Valentine, and D. Jablonsky. "The Origin of Animal Body Plans." *Amer. Sci.* **85** (1997): 126.

[8] Futuyama, D. J., and M. Slatkin, eds. *Coevolution.* Sunderland: Sinauer, 1983.

[9] Goldschmidt, R. B. *The Material Basis for Evolution.* 1940. Facsimile reprint. New Haven: Yale University Press, 1982.

[10] Gould, S. J., and N. Eldredge. "Punctuated Equilibria: The Tempo and Mode of Evolution Reconsidered." *Paleobiology* **3** (1977): 115–151.

[11] Gould, S. J., and N. Eldredge. "Punctuated Equilibrium Comes of Age." *Nature* **366** (1993): 223.

[12] Kauffman, S. A. "Metabolic Stability and Epigenesis in Randomly Connected Nets." *J. Theor. Biol.* **22** (1969): 437.

[13] Kauffman, S. A. "Emergent Properties in Random Complex Automata." *Physica D* **42** (1990): 135.

[14] Kauffman, S. A. "Whispers from Carnot: The Origins of Order and Principles of Adaptation in Complex Nonequilibrium Systems." In *Complexity: Metaphors, Models, and Reality*, edited by G. A. Cowan, David Pines, and David Meltzer, 83–160. Santa Fe Institute Studies in the Sciences of Complexity, Proc. Vol. XIX. Reading, MA: Addison Wesley, 1994.

[15] Lande, R. "Expected Time for Random Genetic Drift of a Population between Stable Phenotypic States." *Proc. Natl. Acad. Sci. USA* **82** (1985): 7641.

[16] Lowe, C. J., and G. A. Wray. "Radical Alterations in the Roles of Homeobox Genes during Echinoderm Evolution." *Nature* **389** (1997): 718.

[17] Maynard Smith, J. *Evolutionary Genetics*, 2d ed. New York: Oxford University Press, 1998.

[18] Melton, D. A. "Pattern Formation during Animal Development." *Science* **252** (1991): 234.

[19] Newman, C. M., J. E. Cohen, and C. Kipnis. "Neo-Darwinian Evolution Implies Punctuated Equilibria." *Nature* **315** (1985): 400.
[20] Newman, M. E. J. "A Model of Mass Extinction." *J. Theor. Biol.* **189** (1997): 235.
[21] Newman, M. E. J., and R. G. Palmer "Models of Extinction: A Review." *Paleobiology* (2001): submitted.
[22] Raup, D. *Bad Genes or Bad Luck?* New York: W. N. Norton, 1991.
[23] Schindewolf, O. H., Grundfragen der Paläontologie, E. Schweizerbart'sche Verlagsbuchhandlung, Erwin Nägele, Stuttgart, Germany; English language edition: Schindewolf, O. H. *Basic Questions in Paleontology* Chicago: The University of Chicago Press, 1993.
[24] Schrödinger, E. *What is Life? The Physical Aspect of the Living Cell.* Oxford, UK: Oxford University Press, 1946.
[25] Sepkoski, J. J. *A Compendium of Fossil Marine Animal Families*, 2d ed. Milwaukee Public Museum Contributions in Biology and Geology 83. Milwaukee: Milwaukee Public Museum, 1993.
[26] Sneppen, K., P. Bak, H. Flyvbjerg, M. H. Jensen. "Evolution as a Self-Organized Critical Phenomenon." *Proc. Natl. Acad. Sci. USA* **92** (1995): 5209–5213.
[27] Solé, R. V., and J. Bascompte. "Are Critical Phenomena Relevant to Large-Scale Evolution?" *Proc. Roy. Soc. Lond. B* **263** (1996): 161.
[28] Somogyi, R., and C. A. Sniegoski. "Modeling the Complexity of Genetic Networks: Understanding Multigenic and Pleiotropic Regulation." *Complexity* **1(6)** (1996): 45–63.
[29] Timoteef-Ressovsky, N. W., K. G. Zimmer, and M. Delbrück. "Über die Natur der Genmutation und Genstruktur." *Nachr. Ges. Wiss. Göttingen, Fachgr.* **1(13)** (1935): 190–245.
[30] Van Valen, L. "A New Evolutionary Law." *Evol. Theor.* **1** (1973): 1.
[31] Wen, X., S. Fuhrman, G. S. Michaels, D. B. Carr, S. Smith, J. L. Barker, and R. Somogyi. "Large-Scale Temporal Gene Expression Mapping of CNS Development." *Proc. Natl. Acad. Sci. USA* **95** (1998): 334–339.
[32] Wright, S. "Character Change, Speciation and the Higher Taxa." *Evolution* **36** (1982): 427.

Epochal Evolution

On the Population Genetics of Punctuation

Aviv Bergman
Marcus W. Feldman

Elementary considerations from Markov chain theory applied to a wide array of evolutionary dynamic models are used to explain patterns of stasis and punctuation. This analysis explains both gradual and punctuated evolution exhibited in a series of numerical simulations of finite, spatially distributed populations with mutation and genetic drift. Neutral, stabilizing, and rugged selection landscapes are studied. In particular, a multilocus study of rugged fitness landscapes reveals punctuated evolution in average phenotypic value and mean fitness, and that the punctuation in these may not be synchronized in time. Finally, we show that these findings hold independently of the evolution equations, and in particular are not limited to Mendelian inheritance.

1 INTRODUCTION

Recently, Elena et al. [6] reported on, what they called, "A clear and unambiguous case of punctuated evolution...." They studied the evolutionary trajectory of the cell size of *E. coli* population evolving for 3,000 generations. Their interpretation is that patterns of punctuated evolution [5, 11] are caused by natural selection of rare advantageous mutations that successively bring the population to fixation states.

In a subsequent letter, Coyne and Charlesworth [4] pointed out two limitations of Elena et al.'s [6] experimental design. First, the population of *E. coli* is asexual, and the results may not be extrapolated to sexual organisms. Second, the population began near fixation, where there was no genetic variation, so that the evolutionary trajectory could only progress through successive fixation of new advantageous mutations. They also pointed out that in a relatively constant environment, which is a more biologically relevant situation than a regime of strong artificial selection, stabilizing selection slows down morphological changes, thus creating the appearance of stasis. In response, Elena et al. [7] indicated that the high mutation rate, 10^6 new mutations per day, maintains sufficient genetic variation that stasis is the result of rare, highly advantageous mutations. They conclude their response by emphasizing that punctuated evolution can result from a mechanism that requires only the two basic components of population genetics—mutation and natural selection.

In the past, considerable attention was devoted to ascertaining which fitness landscapes would give rise to punctuated evolution. The connection between Wright's shifting balance process and observations of punctuated equilibrium [5, 11] has been stressed by Wright [22] and Charlesworth et al. [3]. For the case of weak Gaussian selection, Lande [17] found that between the transitions from one peak to another, there will be periods of stasis whose expected duration increases exponentially with population size. Lande's results relate to the claim by Charlesworth et al. [3] that "if prolonged morphological stasis exists in fossil populations, it must usually be caused by stabilizing selection, rather than gene flow or developmental constraints." A summary of selection models and other biological constraints that have been used to explain apparent punctuations in the fossil record is offered by Maynard Smith [18].

Newman et al. [19] also demonstrated that long periods of stasis with occasional rapid unidirectional jumps between fitness peaks of a morphological character may be triggered by random effects such as genetic drift. They also suggest that this pattern of punctuation is unlikely to be observed in highly structured or very large populations. Such punctuations were also demonstrated by Kirkpatrick [16] when the environmental variance was altered in a two-peak model of Gaussian stabilizing selection, or when the mutation rate changed.

Charlesworth et al. [3] stress the importance of what population genetics can tell us about patterns of morphological change over time. They point out that a more detailed understanding of the relevant population genetic theory is desirable before neo-Darwinian theory is discarded as an explanation of paleontological observations.

We begin the present study with short exposition of the analytic framework relevant to the study of evolutionary dynamics. A new interpretation of a well-studied body of knowledge in probability theory and its application to population genetics is then provided. This allows us to offer a plausible explanation, using classical population genetic theory, for the emergence of two, seemingly different, evolutionary dynamics.

Next we present a numerical investigation of the evolutionary trajectories of the mean phenotypic value of single and multilocus populations under neutral evolution, subject only to mutation and drift. We assume a simple additive mapping from genotype to phenotype to determine an individual's phenotype. We investigate the dynamics of the mean phenotype in a population under phenotypic stabilizing selection. The same additive mapping from genotype to phenotype is assumed, and individual fitness is proportional to $w(x) = e^{-(x-\xi)^2/2\sigma^2}$ where x is the phenotype, ξ is the optimal phenotype, and σ measures the strength of selection.

Bergman et al. [1] developed a diploid version of Kauffman's NK model of rugged fitness landscapes [12, 14] in order to investigate properties of Wright's shifting balance theory of evolution (reviewed in Wright [21]). The NK model is a relatively simple way to extend to the case of multiple fitness peaks many earlier analyses which discussed fitness surfaces with two peaks. As a final fitness model, we use our NK-diploid model to investigate the evolutionary trajectory of the mean genotypic fitness as well as the mean phenotypic value (with the same additive mapping from genotype to phenotype) for a given realization of a fitness landscape.

We shall see that in all fitness regimes tested here, neutral, stabilizing, and rugged landscape, a remarkable pattern of stasis and punctuation emerges. Furthermore, when evolving on rugged landscapes, the dynamics of the populations' mean phenotypic value exhibits stasis and punctuation whose pattern may or may not parallel that of the mean fitness. We conclude with an examination of the trajectories of haplotype frequencies which allows us to offer a plausible explanation for these observations, consistent with our interpretation of classical as well as cultural evolutionary process.

2 ANALYTIC FRAMEWORK

In this section we outline the theoretical framework that governs the process of evolution and causes the apparent punctuation in evolutionary trajectories.

Let $f_{ij}^{(n)}$ stand for the probability that in a process starting from a state E_i the first entry to E_j occurs at the nth step ($f_{ij}^{(0)} = 0$). E_i can stand for the state of a population, e.g., in a diallelic (A and a) population, E_i can represent the frequency q of allele A, and $f_{ij}^{(n)}$ represent the probability that after n generations the population is in state q' represented by E_j. Let $p_{ij}^{(n)}$ stand for the probability of transition from E_i to E_j in exactly n steps; i.e., $p_{ij}^{(n)}$ is the conditional probability of entering E_j at the nth step given the initial state E_i.

The probability, f_{ij}, that the process, starting from E_i will ever pass through state E_j is given by:

$$f_{ij} = \sum_{n=1}^{\infty} f_{ij}^{(n)} .$$

If $f_{ij} = 1$, then $\{f_{ij}^{(n)}\}$ is a proper probability distribution and it is called the *first-passage distribution.* The notation $\{f_{ii}^{(n)}\}$ represents the distribution of *the recurrence time for* E_i; i.e., the probability that the process starting at state E_i will return to that state at the nth step.

The *mean recurrence time for* E_i is given by:

$$\mu_i = \sum_{n=1}^{\infty} n f_{ii}^{(n)} .$$

If E_i is aperiodic, and $f_{ij} = 1$, $p_{ij}^{(n)}$ tends to the limit

$$p_{ij}^{(n)} \rightarrow \mu_j^{-1} ,$$

as $n \rightarrow \infty$. In an irreducible chain with only ergodic elements we have the limits

$$u_j = \lim_{n \rightarrow \infty} p_{ij}^{(n)} = 1/\mu_j ,$$

where u_j is the *stationary probability distribution* (for the given Markov chain); that is, the recurrence time is inversely proportional to the stationary probability distribution.

The evolution of a finite diallelic diploid population model with Mendelian segregation and bidirectional mutation can be studied as a Markov process with reflecting barriers. Furthermore, this model results in an aperiodic and irreducible Markov process; i.e., the population has a nonzero probability of moving from any allelic frequency (state) to any other, and it satisfies the condition that all of its states are ergodic. Thus, the recurrence time for each state is finite, and is, as shown above, inversely proportional to its stationary probability. (For further information see Feller [9, vol. 1, pp. 387–394].)

Wright [20] showed that the stationary allele frequency distribution for a one-locus diploid population under mutation and random genetic drift is the Beta distribution with density function

$$u_q = \frac{\Gamma\{4M\mu + 4M\nu\}}{\Gamma\{4M\mu\}\Gamma\{4M\nu\}} q^{4M\mu-1}(1-q)^{4M\nu-1} ,$$

where M is the population size, μ is the mutation rate from allele $A \rightarrow a$, ν is the mutation rate from allele $a \rightarrow A$, and q is the frequency of allele A in a diallelic population. When mutation rates μ, and ν, are sufficiently low; i.e., $4M\mu < 1$, and $4M\nu < 1$, u_q is a U-shaped distribution.

From the theory above, and the U-shaped distribution, one can conclude that for μ and ν small, since the recurrence time is shorter near fixation points, i.e., $q \approx 0$ or $q \approx 1$, where the stationary probability distribution has high values, the process will reside in those states for proportionally longer periods than in the intermediate states, where the stationary probabilities are low. Thus, given that a transition occurs from, say, $q \approx 0$ to $q \approx 1$, this transition will take a much shorter time than is spent in the neighborhood of these near-fixation points.

When mutation rates μ and/or ν are sufficiently high, i.e., $4M\mu > 1$, and/or $4M\nu > 1$, the stationary probability distribution, u_q, is unimodal or J-shaped. In these cases the population is most likely to evolve from its initial state to states where the stationary distribution, u_q, has a maximum. Transition from initial to final allelic frequencies can occur abruptly or gradually, depending of the exact shape of u_q, but for states other than those near the maximum of u_q, long residence times are not expected.

This analysis holds independently of the evolution equations, and in particular is not limited to Mendelian inheritance nor to neutral evolution. Any Markov process that satisfies the above conditions, and whose stationary distribution has a high density around two or more states, will exhibit successive rapid transitions among these states between long period of stasis near these states.

In what follows we present a series of results and interpretations of numerical simulations done under different fitness regimes, different degrees of population structure, and different inheritance schemes.

3 THE MODEL

Our simulation method follows that of Bergman et al. [1]. A population of M individuals is arranged as a one-dimensional array of cells forming a circle, each cell housing one individual. Goldstein and Holsinger [10] showed that there was little qualitative difference between one and two dimensions; only the scale on which genetic differentiation occurs is different (see also Kimura and Weiss [15]). We chose the one-dimensional array (to reduce computation time) with population sizes of M =100, 200, and 500 diploid individuals.

Each diploid individual consists of two 10-locus haplotypes with allele 1 or 0 at each locus. An offspring generation is constructed from the parental generation as follows. For each cell in the array, two parents are chosen at random from the set of locations that lie within $\pm d$ of that cell's position; d is called the maximal dispersal distance. Note that d incorporates effects of both migration and subpopulation size. The parents undergo Mendelian segregation with recombination. We simulate recombination by choosing a fixed probability r that a single crossover event occurs in each 10-locus parent and then selecting the location of the crossover uniformly across the 10 loci; multiple crossovers are not allowed. Two values, $r = 0.01$ and $r = 0.5$, are investigated. Mutation is symmetric; 1 mutates to 0 and 0 mutates to 1 each at rate $\mu = 10^{-4}$ per locus per generation.

Thus in all simulations presented here, $4M\mu < 1$. Once the genotype of the offspring is chosen in this way, the probability that it survives is calculated using the appropriate fitness scheme. If it survives, it occupies the originally chosen cell; if it does not survive, the process is repeated until that cell is filled. This is done for all cells, and once all are filled, they constitute the offspring generation that become the adults to produce the next generation, etc. This process is repeated T times. We choose a set of values up to $d = n/2$, which is equivalent to random mating. The populations are initially chosen either to have all chromosomes fixed on allele 0 at each locus, or to have 0 and 1 equally frequent but randomly associated across loci. We assume a simple additive mapping from genotype to phenotype to determine an individual's phenotype; an individual's phenotype is defined as the total number of 1s in its genotype.

3.1 CASE 1: NEUTRAL EVOLUTION

In the case of no selection, the procedure described in the previous section is followed except that the selection step is deleted.

Figure 1 demonstrates that punctuation occurs in the absence of selection. This figure shows a typical run for a 10-locus diploid population of 100 individuals with random mating in the absence of selection. In this simulation we record the mean and variance of the phenotype in the population. As can be seen from this figure, the population proceeds from one stasis period to the next during a relatively short (compared to the stasis periods) transition period. Stasis is observed as a period where the population's mean phenotype remains relatively constant. Note also that during the stasis period, the phenotypic variance remains low compared to its level during the transition periods.

3.2 CASE 2: STABILIZING SELECTION

Here we describe the dynamics of a 10-locus diploid population under phenotypic stabilizing selection where, again, the phenotype is just the total number of 1 alleles in the genotype. Individual fitness is proportional to

$$w(x) = e^{-(x-\xi)^2/2\sigma^2},$$

where x is the phenotype, ξ is the optimal phenotype, and σ measures the strength of selection. For an intermediate optimum, we set $\xi = 10$. When selection is strong, e.g., $\sigma = 1$, there is quasi fixation [8] on one of the genotypes that gives rise to the optimal phenotype, 10, and remains near this state throughout the entire 10,000 generations. Different initial conditions result in fixation on different genotypes. If selection is weaker, e.g., $\sigma = 7.5$, the evolutionary trajectories of the mean phenotype and mean fitness are flat, except for small random fluctuations. However, at the genotypic level, there are transitions between the different genotypes that give rise to the optimal phenotype. The evolutionary

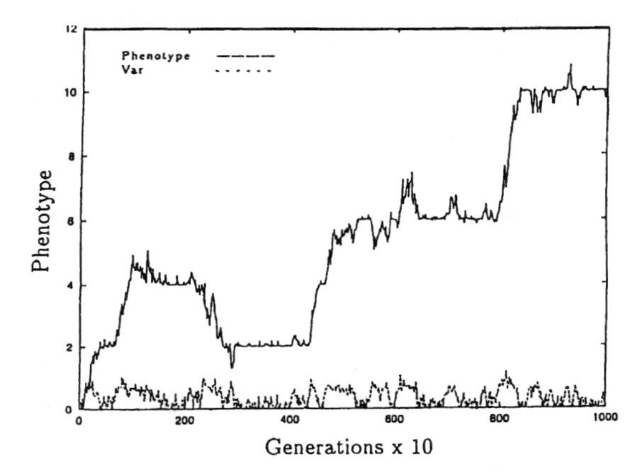

FIGURE 1 Evolutionary trajectory of the mean phenotypic value of a population of 100 diallelic 10-locus diploid individuals with random mating in the absence of selection.

trajectory of the genotypes exhibits rapid changes preceded and followed by long periods of stasis at the genotypic level, i.e, "genetic punctuated" equilibrium. These findings are illustrated in figure 2. Figure 2(a) shows the phenotypic trajectory of a population under stabilizing selection. No trend is apparent during the course of evolution. In figure 2(b) we monitor the average haplotype in the population, represented here as the integer from the binary representation of haplotypes.[1] There is a sharp transition between near-fixation states; this is punctuation in the genotype rather than in fitness or phenotype. The nature of the genotypic transition of a population under stabilizing selection as described above is such that at least two loci change their fixation states simultaneously, i.e., to maintain the phenotypic value constant, say 10, one gene has to change its near-fixation state from 0 to 1, while another has to change its near-fixation state from 1 to 0. When the mapping from genotype to phenotype, and/or the selection regime are more complicated, more than two loci can exhibit similar dynamics. Furthermore, we conjecture that if a set of genes which contributes to one phenotypic character under selection is linked to another set of genes which contributes to a second phenotypic character that may be under weaker or no selection, the dynamics of the latter set will be largely determined by the dynamics

[1]Note, in this binary representation the magnitude of the change is not a direct reflection of the number of genetic changes that occur. In figure 2(b), for example, the number of different alleles between the first and second, second and third, and between the first and third stasis periods, is the same, namely two. Large punctuations, e.g., between the first and second, or second and third stasis periods, are the result of a change at the high-significance bits, while the small difference between the first and third stasis periods are due to changes at the low-significance bits of the binary representation. This representation artificially inflates the height of the punctuation episodes.

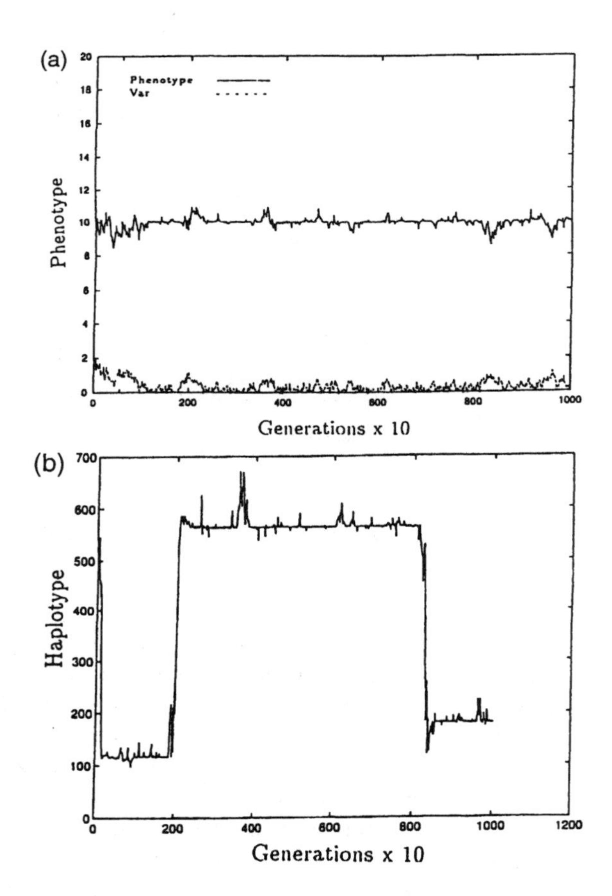

FIGURE 2 (a) Phenotypic trajectory of a population under stabilizing selection. (b) Average haplotype in the population, represented here as the integer from the binary representation of haplotypes.

of the former. Thus, observed phenotypic punctuation may be the result of drift and selection operating on a linked set of genes rather than those controlling the trait under study.

3.3 CASE 3: RUGGED FITNESS LANDSCAPE

3.3.1 Genotypic Selection. We use a diploid version of the NK selection scheme [12, 14] that was introduced in our work on the shifting balance theory [1]. Mutation occurs in the same way as in the previous simulation and again the population is distributed at integer points around a circle. In the NK model, N is the number of loci and K the number of other genes that affect the fitness contribution of any gene. To calculate the fitness contribution of the genotype at a specific locus, say, at position i, we choose K loci, independently of i, at random from among the other $N-1$. Then, for every possible genotype at these K loci, the genotype at locus i is assigned a fitness value chosen uniformly on $[0, 1]$ and this value is stored. This process is repeated until all genotypes at locus i have been assigned fitness values in all possible backgrounds. The fitness value of a given N-locus genotype is obtained by summing all of the stored values appropriate for each single locus genotype making up that N-locus genotype, using all appropriate K-locus backgrounds. Clearly $K = 0$ corresponds to additive fitnesses across loci, because the contribution of any locus to overall fitness does not depend on the genotype at any other locus. As K increases, the fitness surface becomes increasingly rugged [12]. Ruggedness may be described in a number of ways, but one obvious measure is the number of mutational steps at the genotypic level that separates local optima on the fitness surface. Local optima in this sense are points where all one-step mutation neighbors have a lower fitness value. The number of mutation steps separating such local optima seems to decrease as an approximately linear function of K [13], suggesting that these surfaces become more rugged in a regular way as K increases.

3.3.2 Qualitative Analysis. In most of our simulations, N was set at 10 loci, and for each value of $K \in \{0, 1, 2, 3, 4\}$ a realization of the fitness regime was chosen. In most cases, the population evolved for $T = 10,000$ generations, although for validation several runs were extended to one million generations. Each simulation was repeated 25 times, each with a different random seed, resulting in 25 evolutionary trajectories on the different fitness landscape. To validate our observations, a second set of experiments was conducted. For each parameter setting, we randomly selected a single realization of a fitness landscape on which 25 evolutionary trajectories, each with a different random seed, were followed.

During the course of the simulation we tracked the mean fitness, $\bar{w}$, in the population (except in the case without selection). Again, we use an additive mapping from genotype to phenotype; each individual's phenotype is measured by the total number of 1s in its genotype. We tracked the evolution of the mean phenotype, denoted by $\bar{x}$, over time in all cases. At crucial stages in this process, we also examined the distribution of haplotype frequencies in the form of a histogram.

Figure 3 illustrates the effect of population size (500 in fig. 3(a) and 100 in fig. 3(b)) on the evolutionary trajectory of a 10-locus diploid population with $K = 2$ and dispersal distance $d = 10$. First, as expected, the phenotypic standard deviation is larger when the population size is smaller. Second, the transition period is longer for the larger population size. This observation is in accord with the result of Lande [17] who found that the expected duration of the transition between two adaptive peaks under Gaussian stabilizing selection increases approximately logarithmically with the population size.

Figure 4 shows the effect of the number of loci, N, on the existence and nature of punctuation. Here $K = 0$ and $d = 1$. Figure 4(a) shows a typical evolutionary trajectory for a 5-locus diploid population of size 100. A punctuation event can be observed around generation 3,000. Note the increase in the phenotypic variation during the punctuation period relative to the stasis period (the dotted line). Figure 4(b) illustrates a typical trajectory for a 10-locus diploid population of size 100 with $d = 1$ in which there are no obvious punctuations. A closer study is required to further explore the interaction between the number of loci and population structure as it pertains to the emergence of punctuation events. In our examples, an extremely structured population, as in the case $d = 1$, very rarely produced punctuations.

The effect of epistasis among the loci on the nature of punctuation events is shown in figure 5, where we set $N = 10$ and $d = 50$. Figure 5(a) shows an evolutionary trajectory of a population on an additive fitness landscape, $K = 0$. Though a typical trajectory would not exhibit any punctuation, here we detect punctuation at generations 1,000 and 2,000. Note the high level of phenotypic variation during the stasis period, reducing the level PL of punctuation.[2] Figure 5(b) shows a typical trajectory for $K = 3$ with all other parameters as in figure 5(a). Punctuation at generation 5,000 is clearly observed. Note the low phenotypic variation throughout the trajectory, with a slight increase during the punctuation event, which results in a high punctuation level, $PL \approx 85$.

Figure 6 demonstrates the effect of dispersal, d. When dispersal distance is small, migration is infrequent and effective subpopulation size is low. Evolutionary trajectories with dispersal distances $d = 1, 3$, and 50 are reported in figures 6(a), (b), and (c), respectively, the last representing a panmictic population. Note first that the phenotypic standard deviation is highest for the lowest dispersal distance $d = 1$ throughout the evolutionary trajectory, even during the stasis periods. Observe also that the duration of punctuation, i.e., the sojourn time, decreases as the dispersal distance increases.

In most cases, punctuation in mean phenotype and punctuation in mean fitness are observed simultaneously. It is possible, however, for punctuation to occur at the phenotypic level but not in the mean fitness and vice versa. In figure 7 we show an evolutionary trajectory where some punctuation events are observed simultaneously for the mean phenotype (fig. 7(a)) and mean fitness (fig. 7(b)) (generation 3,000), while other punctuation events (generations 2,000 and 5,000) are seen only in the mean phenotype. Figures 7(c) and 7(d) show an

[2]To quantitatively evaluate punctuation events we use an adaptive algorithm to detect the location of punctuation and stasis periods (details are available from the authors). We then compute the punctuation level, PL, as the ratio between the average change per generation during the punctuation event, J_P, and the larger of the two average changes, preceding, J_{S_p}, and following, J_{S_f}, stasis periods; $PL = J_P / \max(J_{S_p}, J_{S_f})$. High values of PL indicate a short sojourn time preceded and followed by periods of low variation, relative to the level of change during the punctuation period ("clean" punctuation). A longer sojourn time, preceded and followed by periods of high variation, will result in low PL values ("noisy" punctuation).

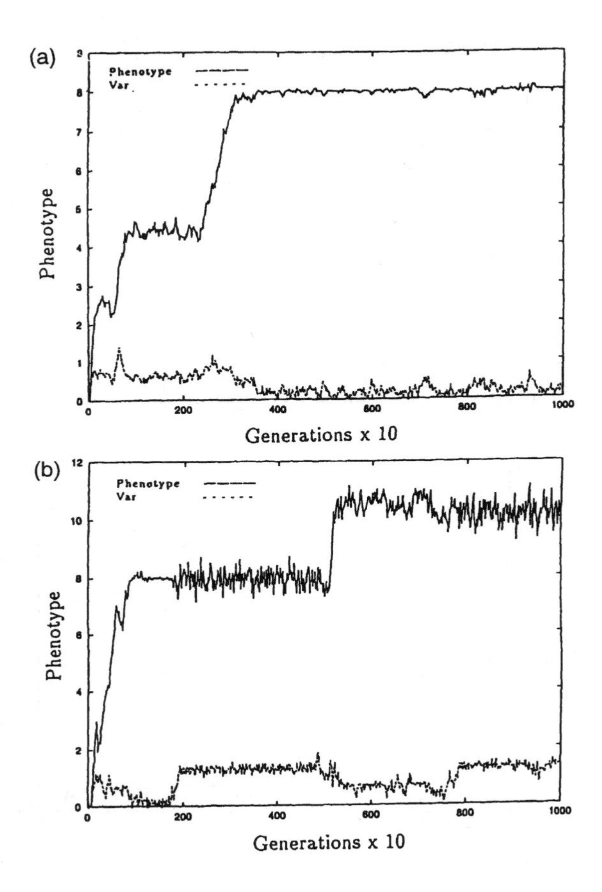

FIGURE 3 Effect of population size. Phenotypic trajectory of a population of 10-locus diallelic diploids evolving on a rugged landscape, $K = 2$ and $d = 10$. (a) Population size 500. (b) Population size 100.

example where punctuation is observed for the mean fitness (fig. 7(d)) but not for the mean phenotype (fig. 7(c)).

In order to better understand the dynamics of the multilocus genotypic system evolving on a rugged landscape, we constructed histograms of haplotypes. Our hypothesis is that populations move from near fixation on one haplotype to near fixation on another during the punctuation events. In figure 8 the haplotype histograms before and after the punctuation, generations 455 and 466 respectively, are presented, for a population of 100 diploid individuals with $N = 10$, $K = 3$, and $d = 50$. Figure 8(a), shows a closeup view of the phenotypic changes that occur during punctuation. Figure 8(b) shows the haplotype histogram just

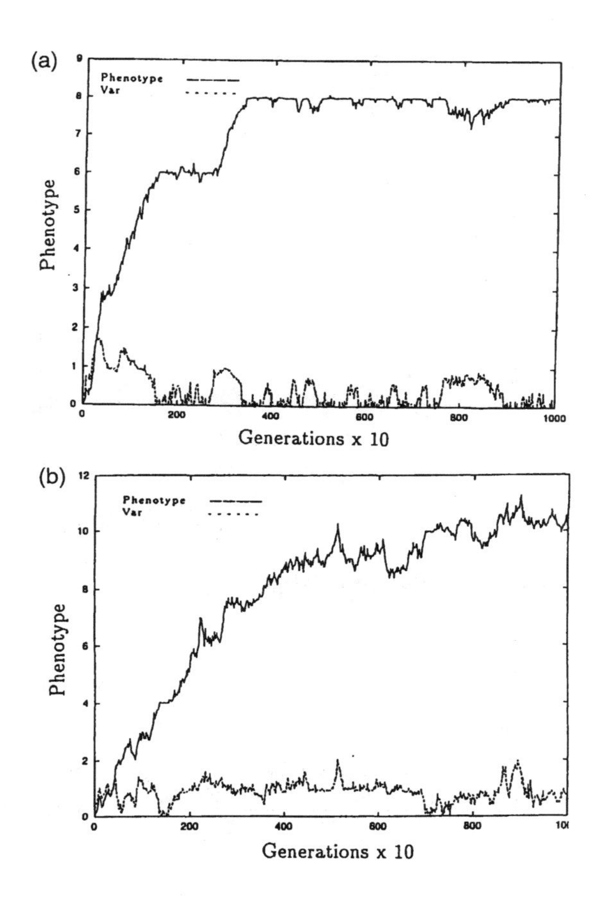

FIGURE 4 Effect of number of loci. Phenotypic trajectory of a population evolving on a rugged landscape, $K = 0$ and $d = 1$. (a) Evolutionary trajectory of a 5-locus diploid population of size 100. (b) Evolutionary trajectory of a 10-locus diploid population of size 100.

before the punctuation, where almost all haplotypes (195 out of 200) are of one type. Figure 8(c) shows the haplotype histogram just after punctuation where again, almost all (198 out of 200) are of the same haplotype, but different from that before punctuation occurs.

4 DISCUSSION

The analytic interpretation based on classical population genetic theory and the analysis of the haplotypic histograms show that punctuated evolution can

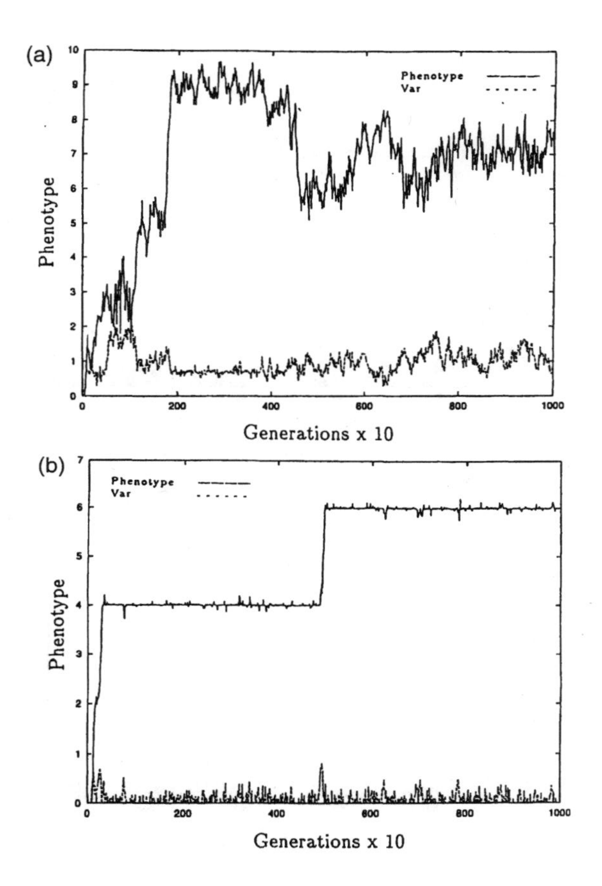

FIGURE 5 Effect of epistasis, $N = 10$ and $d = 50$. (a) Evolutionary trajectory of a population of a 10-locus diploid population of size 100 evolving on an additive fitness landscape, $K = 0$. (b) A typical trajectory of the same population with $K = 3$.

be explained as a succession of rapid transitions between long sojourns near fixation. We show that the process of genetic drift and mutation is sufficient to create punctuation in an evolutionary trajectory. Selection, tested here in the form of stabilizing, and multi-peak fitness landscapes, may enhance the effect and produce "clean" punctuations, i.e., high PL value. It may also provide the forces necessary for simultaneous transition of multiple genes.

Close attention is given to the dynamics of the haplotype distribution during the evolutionary process. Our numerical study clearly demonstrates the rapid transition between population states which differ in their genetic makeup, but lack phenotypic variation. During the rapid transition, i.e., punctuation, an in-

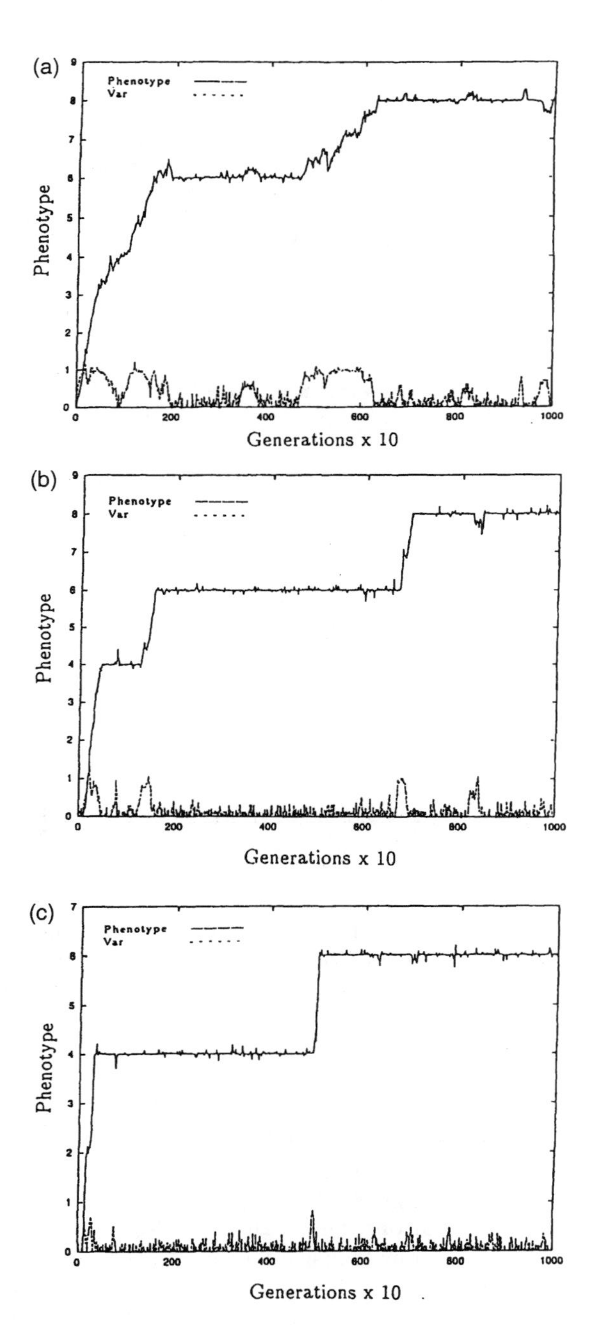

FIGURE 6 The effect of dispersal distance, d. Phenotypic trajectory of a 10-locus diploid population of size 100 evolving on a rugged landscape, $N = 10$ and $K = 3$. (a) $d = 1$, (b) $d = 3$, and (c) $d = 50$.

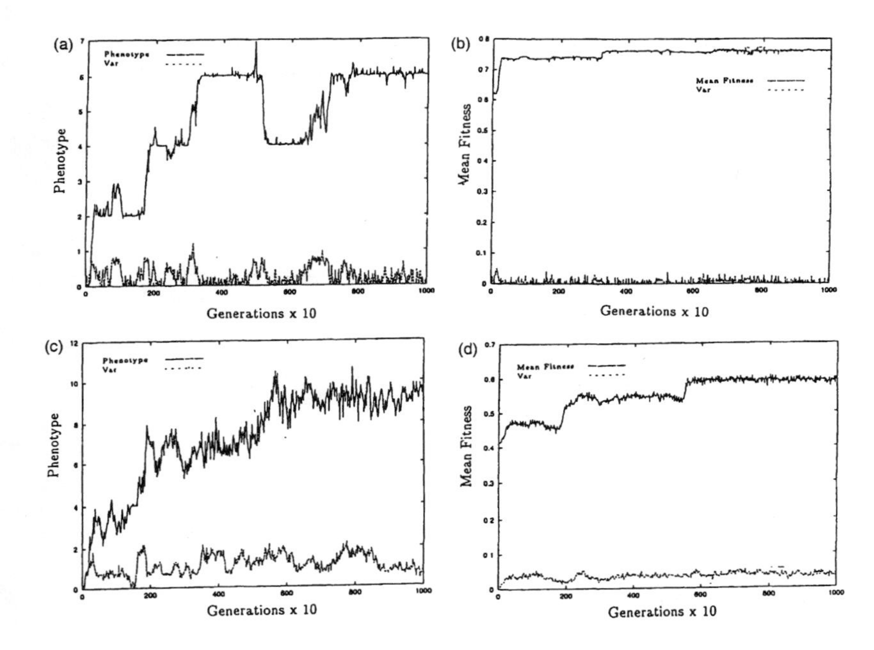

FIGURE 7 Discordance in punctuations. Evolutionary trajectory of a population evolving on a rugged landscape. (a) and (b) $N = 10$, $K = 1$, and $d = 10$. (c) and (d) $N = 10$, $K = 0$, and $d = 50$. (a) and (c) show trajectories of the mean phenotype. (b) and (d) show trajectories of the mean fitness.

crease in the variation is observed, preceded and followed by long periods of no haplotypic variation, i.e., stasis. The haplotypic dynamics seem to be rather independent of the selection regime; neutral evolution, stabilizing selection, and multi-peak fitness landscape, produced similar dynamics. Under neutral evolution, haplotypes went through a rapid change in only one locus at a time. Stasis and expected sojourn time have been studied in the past [8, p. 145]. With no selection, it is a rare event for more than one locus to undergo punctuation, although as seen in figure 6(a), it is possible. However, with phenotypic stabilizing selection, the selection forces are such that in order to maintain the phenotypic value constant, when one gene changes its near-fixation state from 0 to 1, another has to change its near-fixation state from 1 to 0. Punctuation events that involve more than one locus at a time have also been observed with rugged fitness landscapes; see figure 5(a).

Throughout this study we assume a simple additive mapping from genotype to phenotype. If the mapping from genotype to phenotype, and/or the selection regime, are more complicated than the ones studied here, we conjecture that phenotypic traits controlled by many genes, each with a limited contribution to

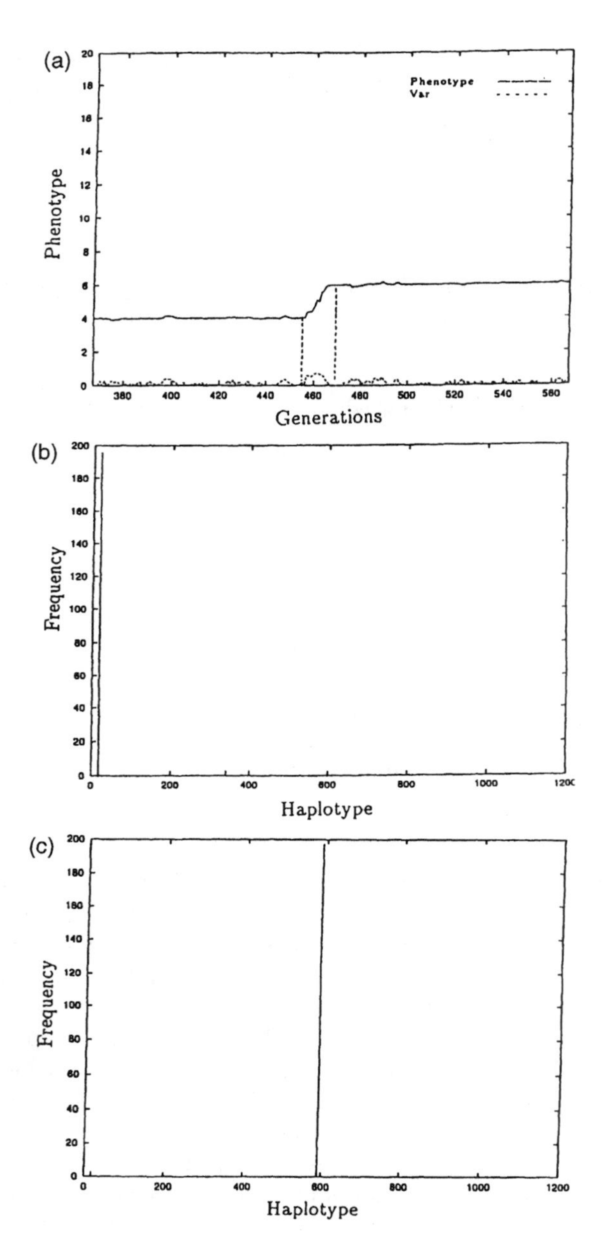

FIGURE 8 Haplotype histogram of a population of 100 diploid individuals evolving on a rugged landscape with $N = 10$, $K = 3$, and $d = 50$. (a) A closeup view of the phenotypic changes that occur during punctuation. (b) The haplotype histogram just before the punctuation, generation 455. (c) The haplotype histogram just after punctuation, generation 466.

the trait under study, can also undergo phenotypic punctuation. Furthermore, we conjecture that an observed phenotypic trait whose controlling genes are tightly linked to genes controlling a phenotypic trait under, say, stabilizing selection, can undergo punctuated evolution as a result of hitchhiking. Thus, observed phenotypic punctuation may not be the result of drift and selection operating on the trait under study, but rather on a linked trait.

This study also sheds some light on the role of population structure on the emergence of punctuation during the course of evolution. Here we controlled population structure using one parameter, the dispersal distance, d. This parameter incorporates effects of both migration and subpopulation size. Our results show that as the population becomes more panmictic, more and cleaner punctuations emerge. For a highly stuctured population, $d = 1$, punctuations are unlikely to occur. An interesting phenomenon which requires closer study is the interaction between the number of loci and population structure, as illustrated in figure 4.

In a preliminary study of the relationship between punctuation level, PL, and the degree of epistasis, K, we observed higher PL values (> 20) for intermediate epistasis ($K = 2$ and $K = 3$), while with low epistasis ($K = 0$ and $K = 1$), or high epistasis ($K = 4$), punctuations, if present, had low values, $PL < 5$. When selection is close to additive, there exist many small beneficial mutations that invade and spread throughout the population. When K is large, single-step mutations can cause such large fitness differences that drift cannot overcome selection and carry the population through adaptive valleys. The population, once at a local peak, will resist perturbations.

We conclude with a brief discussion of two systems; the first is a one-locus diallelic asexual finite population with symmetric mutation and alleles A and a. The following is the linear fractional evolutionary dynamic:

$$x_{t+1} = \frac{x_t(1+s)(1-\mu) + (1-x_t)\mu}{x_t(1+s) + (1-x_t)},$$

where x_t is the frequency of allele A, s is the strength of selection, and μ is the symmetric mutation rate. Binomial sampling is then performed using x_{t+1} as the probability with population size M.

Figure 9 shows a typical (for the chosen parameter set) evolutionary trajectory for a population of size $M = 100$ over 5,000 generations. Punctuations can be observed at generations 1,000 and 2,000. Note that even though selection favors A, the population remains near fixation on allele a for a substantial amount of time. As noted by Coyne and Charlesworth [4], it is not surprising that an asexual population of *E. coli* exhibits punctuation; they also suggested, however, that the results may not be extrapolated to sexual organisms. Our analysis indicates that the dynamical behavior of sexual and asexual organisms can both exhibit qualitatively indistinguishable punctuated evolution.

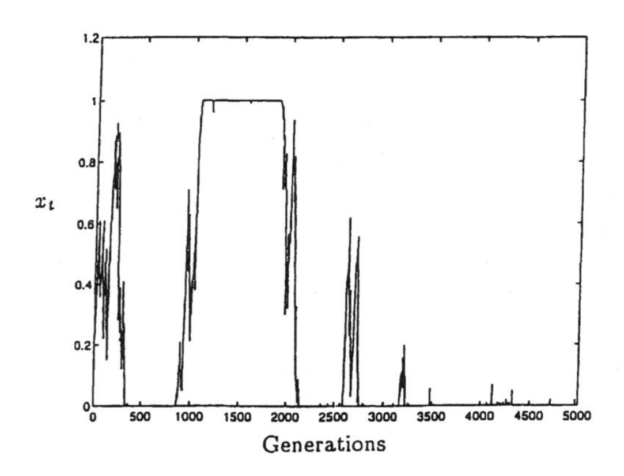

FIGURE 9 Asexual dynamics. Typical evolutionary trajectory of a population over 5,000 generations. 100 diallelic haploid individuals, $\mu = 10^{-4}$, $s = 10^{-4}$ (see text).

Finally, we also studied the dynamical behavior of a non-Mendelian transmission system, namely a cultural transmission model proposed by Cavalli-Sforza and Feldman [2]. In this model, a cultural trait can take one of two states, H and h, whose transmission probability depends on the parental cultural states. The probabilities that an H offspring results from maternal $\times$ paternal matings of type $h \times h$, $h \times H$, $H \times h$, $H \times H$, are denoted by b_0, b_1, b_2, and b_3, respectively. The frequency of trait H, u_t, as a function of time, t, is given by:

$$u_{t+1} = u_t^2 B + u_t C + b_0 \,,$$

where $B = b_3 + b_0 - b_1 - b_2$ and $C = b_2 + b_1 - 2b_0$ (see Cavalli-Sforza and Feldman [2, pp. 78–84]). Binomial sampling is then performed using u_{t+1} as the probability with population size M.

Figure 10 illustrates the evolutionary dynamics of a population of size $M = 100$ for the cultural trait, H, over 5,000 generations. Here too, punctuations are observed near generation 2,000 and 3,500.

The results obtained here strongly suggest that evolutionary dynamics can result in punctuated evolution whether the evolution is sexual, asexual, or non-Mendelian. Punctuation can occur in a neutral evolutionary system, under stabilizing selection, or with a rugged fitness landscape. The evolutionary behavior seems to depend solely on the properties of the Markov process governing the evolutionary process (see also Newman et al. [19]). Any Markov process that satisfies the conditions discussed in section 2, and whose stationary distribution has a high density around two or more states will exhibit successive rapid transitions among these states between long period of stasis.

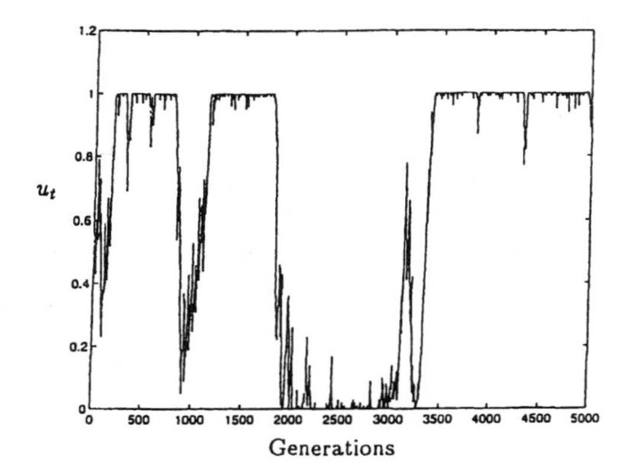

FIGURE 10 Evolutionary dynamics under cultural transmission, a typical trajectory of a population over 5,000 generations. 100 individuals with a trait that can take one of two values and is inherited with the following parameters $b_0 = 0.0005$, $b_1 = 0.5$, $b_2 = 0.5$, and $b_3 = 0.9995$ (see text).

ACKNOWLEDGMENT

The authors thank the Santa Fe Institute and the members of the Adaptive Computation Program at the Santa Fe Institute for many stimulating discussions. We are grateful to professor W. J. Ewens for his valuable comments on an earlier draft. Research supported in part by NIH grant GM28016.

REFERENCES

[1] Bergman A., D. B. Goldstein, K. E. Holsinger, and M. W. Feldman. "Population Structure, Fitness Surfaces, and Linkage in the Shifting Balance Process." *Genet. Res.* **66** (1995): 85–92.

[2] Cavalli-Sforza, L. L., and M. W. Feldman. *Cultural Transmission and Evolution: A Quantitative Approach.* Princeton, NJ: Princeton University Press, 1981.

[3] Charlesworth B., R. Lande, and M. Slatkin. "A Neo-Darwinian Commentary on Macroevolution." *Evolution* **36** (1982): 474–498.

[4] Coyne, J. A., and B. Charlesworth. "Mechanisms of Punctuated Evolution." *Science* **274** (1996): 1748–1749.

[5] Eldredge, N., and S. J. Gould. "Punctuated Equilibria: An Alternative to Phyletic Gradualism." In *Models in Paleobiology*, edited by T. J. M. Schopf, 82–115. San Francisco, CA: Freeman & Cooper, 1972.

[6] Elena, S. F., V. S Cooper, and R. E. Lenski. "Punctuated Evolution Caused by Selection of Rare Beneficial Mutations." *Science* **272** (1996): 1802–1084.

[7] Elena, S. F., V. S. Cooper, and R. E. Lenski. "Mechanisms of Punctuated Evolution." *Science* **274** (1996): 1749–1750.

[8] Ewens, W. J. *Mathematical Population Genetics.* Berlin: Springer-Verlag, 1979.

[9] Feller, W. *An Introduction to Probability Theory and Its Applications*, vol. 1. New York: John Wiley & Sons, 1968.

[10] Goldstein, D. B., and K. H. Holsinger. "Maintenance of Polygenic Variation in Spatially Structured Populations: Roles for Local Mating and Genetic Redundancy." *Evolution* **46(2)** (1992): 412–429.

[11] Gould, S. J., and N. Eldredge. "Punctuated Equilibria: The Tempo and Mode of Evolution Reconsidered." *Paleobiology* **3** (1977): 115–251.

[12] Kauffman, S. A. "Adaptation on Rugged Fitness Landscapes." In *Lectures in the Sciences of Complexity*, edited by D. L. Stein, 527–618. Santa Fe Institute Studies in the Sciences of Complexity, Lect. Notes Vol. I. Reading, MA: Addison-Wesley, 1989.

[13] Kauffman, S. A. *The Origin of Order.* New York: Oxford University Press, 1993.

[14] Kauffman, S. A., and S. Levin. "Towards a General Theory of Adaptive Walks on Rugged Landscapes." *J. Theor. Biol.* **128** (1987): 11.

[15] Kimura, M., and G. H. Weiss. "The Stepping Stone Model of Population Structure and the Decrease of Genetic Correlation with Distance." *Genetics* **49** (1964): 561–576.

[16] Kirkpatrick, M. "Quantum Evolution and Punctuated Equilibria in Continous Genetic Characters." *Amer. Natur.* **119** (1982): 833–848.

[17] Lande, R. "Expected Time for Random Genetic Drift of a Population between Stable Phenotypic States." *Proc. Natl. Acad. Sci. USA* **82** (1985): 7641–7645.

[18] Maynard Smith, J. "The Genetics of Stasis and Punctuation." *Ann. Rev. Genet.* **17** (1983): 11–25.

[19] Newman, C. M., J. E. Cohen, and C. Kipnis. "Neo-Darwinian Evolution Implies Punctuated Equilibria." *Nature* **315** (1985): 400–401.

[20] Wright, S. "Evolution in Mendelian Populations." *Genetics* **16** (1931): 97–156.

[21] Wright, S. *Evolution and the Genetics of Populations, Vol. 3. Experimental Results and Evolutionary Deductions.* Chicago, IL: University Chicago Press, 1977.

[22] Wright, S. "Character Change, Speciation, and the Higher Taxa." *Evolution* **36** (1982): 427–443.

When Evolution is Revolution—Origins of Innovation

James P. Crutchfield

A pictorial tour of the theories of epochal evolution and structural complexity is presented with a view toward the dynamical origins, stabilization, and content of evolutionary innovations. A number of alternative explanations for the occurrence of long periods of stasis that are interrupted by sudden change have been proposed since the first days of mathematical evolutionary theory. Here contrasts are drawn between the mechanisms underlying epochal evolution and those implicated in the classical theory of stochastic intermittency (drift) due to Fisher, Wright's adaptive landscapes, Kimura's neutral evolution, and Gould and Eldredge's notion of punctuated equilibria. The comparisons suggest what a synthetic theory of the evolution of complexity might look like, while at the same time emphasizing that it will remain incomplete without a theory of biological structure. The computational mechanics theory of structural complexity is offered as an approach to the latter.

Evolutionary Dynamics, edited by
J. P. Crutchfield and P. Schuster. Oxford University Press.

1 EPOCHAL EVOLUTION AND INNOVATION

The emergence of biological form and function through evolution is often considered to happen by a process of gradual adaptation: through a series of small changes, observable features and improved behaviors appear. When niches alter in character or when a species first moves into a existing niche, the context of prior diversity is changed and relative fitnesses in a population adjust. Selection then acts to reshape the cloud of diverse individuals in directions appropriate to the new environment. The diversity of individuals is the expression of genetic variation—variations whose origins are not correlated with individual fitness. Darwin's analysis of Galapagos finches is the paradigmatic case: The diversity in beak length and shape was seen as reflecting incremental adaptations to small geographic variations in type of food source [14].

This has given rise to a view of evolutionary dynamics as an optimization process: The environment provides constraints and species either go extinct or are able to incrementally change in ways that take advantage of or mitigate the constraints. When there is this kind of tight coupling between organism and environment and when the time scales of change allow adaptation, the form and behavior of the resulting organisms mirror niche structure and environmental constraint. Moreover, individual biological traits take on functional meaning, since they reflect the "solutions" to "problems" imposed by the environment. Finally, in this view the environment is a source of novelty and the instigator of change. Though it can react, Darwinian evolution cannot, in and of itself, produce novel biological structures and functions.

For well over a half century, however, it has been known that gradual adaptation is a substantially incomplete picture of evolutionary dynamics. Early mathematical analyses of stochastic processes, the rise of molecular genetics, investigations of the fossil record, the development of nonlinear population dynamics, and recent laboratory evolutionary experiments reveal that evolution need not be gradual, but can be episodic. Perhaps the most extreme examples are seen in *evolutionary metastability*: Long periods of stasis are interrupted by rapidly emerging innovations. Importantly, evolutionary metastability loosens the coupling between individual diversity and adaptive response to the environment: There can be substantially more individual diversity than adaptation to the environment requires. Another consequence is that this loose coupling opens up the possibility that evolutionary dynamics can produce novel structures on its own, not only in lock-step response to environmental change. Unfortunately, current explanations of metastability do not define what biological structure is and so are not yet complete theories of the innovation of form and function.

One of the earliest recognitions of metastability in evolutionary dynamics is Fisher's analysis of *stochastic intermittency* in multi-allele drift processes [19]. He showed that in the absence of selection, when only drift was operating, there could be a transient fixation on one or another allele and that when a shift to a new allele occurred, the transition came quickly—compared to the time scale of

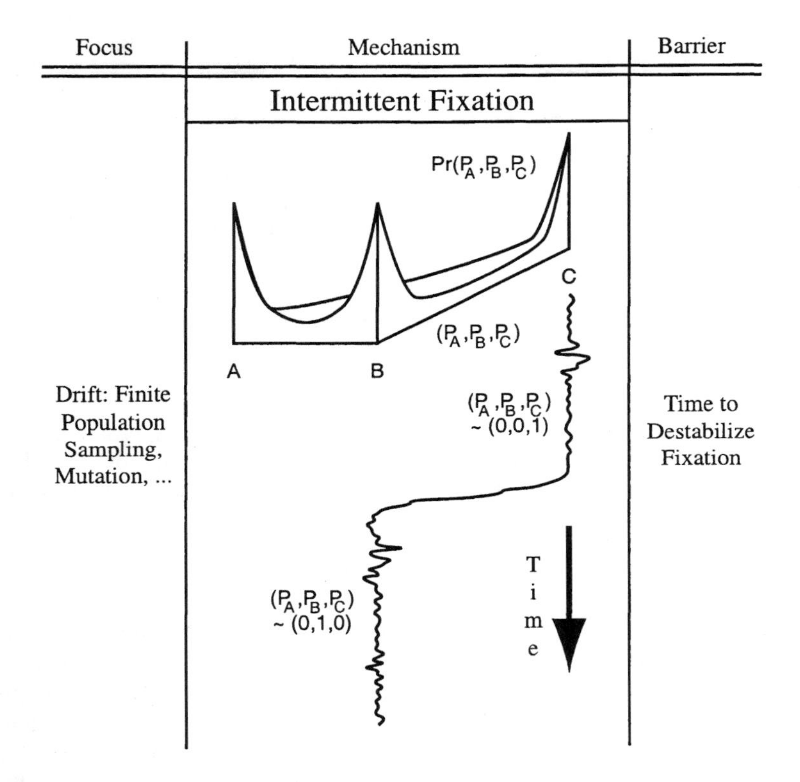

FIGURE 1 Metastability through intermittent fixation: A population is described by the proportion $(P_{\mathrm{A}}, P_{\mathrm{B}}, P_{\mathrm{C}})$ of individuals with one of three alleles A, B, and C. The probability $\Pr(P_{\mathrm{A}}, P_{\mathrm{B}}, P_{\mathrm{C}})$ of the population exhibiting proportion $(P_{\mathrm{A}}, P_{\mathrm{B}}, P_{\mathrm{C}})$ is highly peaked at the pure populations: $(1,0,0)$, $(0,1,0)$, and $(0,0,1)$. Nonetheless, there is some (low) probability of being in intermediate, "mixed" populations, and so transitions between the pure populations are possible. Interestingly, the transitions when they occur, occur rapidly.

allele fixation. (This is illustrated in figure 1.) Being a fundamental property of random finite-sample processes, stochastic intermittency can occur at a number of different levels in an evolutionary process: e.g., at both the genotype and phenotype (character) levels [4].

Another description of metastability is found in Wright's early attempt to explain the dynamics of evolutionary change. Wright introduced the notion of *adaptive landscapes* to describe the (local) stochastic adaptation of populations to environmental constraints [70]. This geographical metaphor has had a pervasive influence on theorizing about natural and artificial evolutionary processes. The basic picture is that of a gradient-following dynamics moving over a "land-

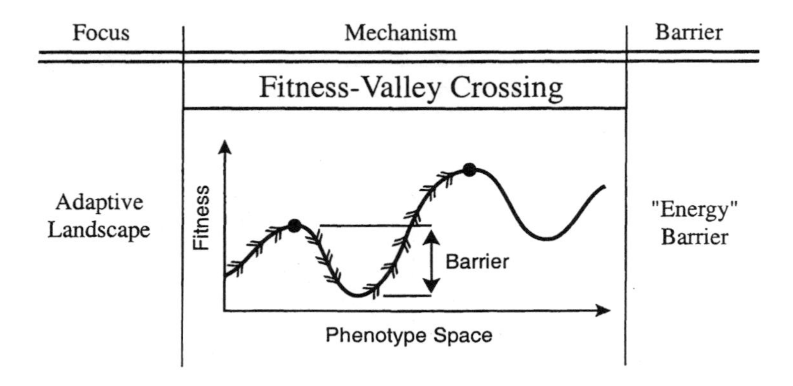

FIGURE 2 Metastability through fitness-valley crossing: A population residing at a local optimum in a fitness landscape must cross through a valley of lower fitness in order to reach another, possibly higher-fitness, peak. The landscape is defined over a space of phenotypes or traits.

scape" determined by a fitness "gravitational potential." Adaptive landscapes admit two kinds of (related) metastability. First, in *fitness-valley crossing* an evolving population stochastically crawls along a surface determined, perhaps dynamically, by the fitness of individuals, moving to peaks and very occasionally hopping across fitness "valleys" to nearby, and possibly higher fitness, peaks. (See fig. 2.) The *barriers* to innovation here are determined by the depth of the valley intervening between two peaks. Due to this, they are sometimes referred to as "energy" barriers, highlighting the physical metaphor. Second, in the *shifting balance* theory, periods of stasis correspond to times when populations are isolated at local optima in the landscape, as before. Innovations, however, correspond to populations adapting in response to changes in stability of landscape extrema—changes that are initiated by exogenous forces (e.g., environmental) and that alter the locations of peaks and valleys. (See fig. 3.) In the shifting balance theory the barriers to innovation are determined by the time scale of behaviors largely external to the evolutionary process.

More recently, extending Wright's notion of adaptive landscapes, it has been proposed that the processes underlying combinatorial optimization and biological evolution can be modeled as "rugged landscapes" [38, 44]. These are landscapes with wildly fluctuating fitnesses even at the smallest scales of single-point mutations. It is generally assumed that these "landscapes" possess a large number of local optima. With this picture in mind, the common interpretation of stasis and change in evolving populations is that of a population being "stuck" at a local peak, until a rare mutant crosses a valley of relatively low fitness to a higher peak—a picture more or less consistent with Wright's.

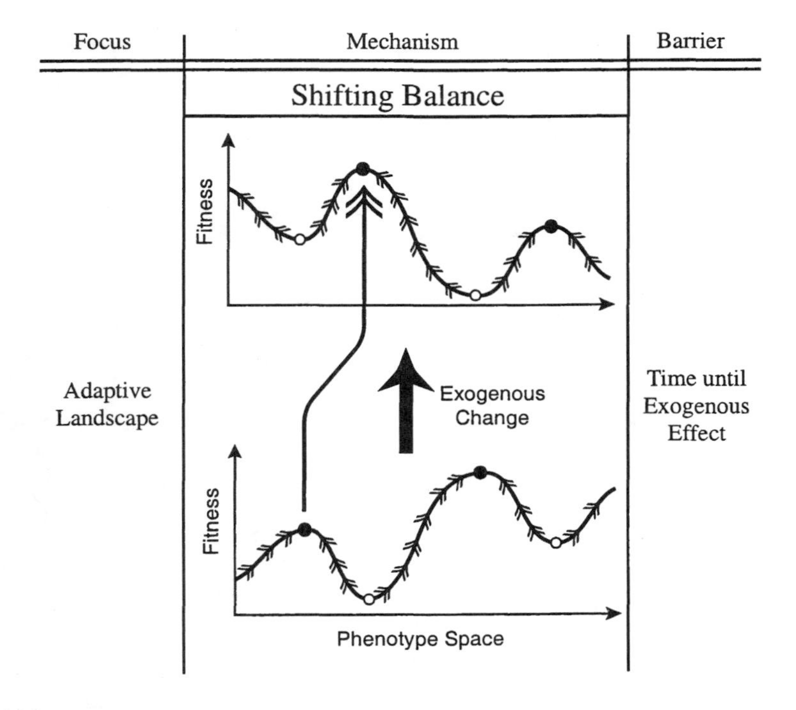

FIGURE 3 Metastability through shifting balance: A population resides at a local optimum in an adaptive landscape. The landscape is defined over a space of phenotypes or traits. An exogenous (e.g., environmental) change occurs that alters the shape of the adaptive landscape in such a way that the population's local optimality disappears and its stability is lost. The population then climbs to a neighboring peak.

Metastability also occurs when genetic variations do not produce changes in fitness. Since selection cannot act on those variations, there can be long periods of phenotypic constancy, despite the accumulation of substantial genomic change. The history of this idea—*neutral evolution*—goes back to Kimura [39, 53], who in the 1960s argued that on the genotypic level, most genetic variation occurring in evolution is adaptively neutral with respect to the phenotype. In this situation, many genotypes code for single phenotypes. Additionally, due to intrinsic or even exogenous variations (e.g., environmental fluctuations on relatively fast time scales), there simply may not exist a deterministic "fitness" value for each possible genotype. In this case, fluctuations induce variations in fitness such that genotypes with similar average fitness are not distinct at the level of selection. Differences in fitness are simply washed out and selection cannot act on them. Thus, metastability can be induced either by many genotypes coding for a given phenotype or by "noise" in the fitness evaluation of individuals. (See fig. 4.)

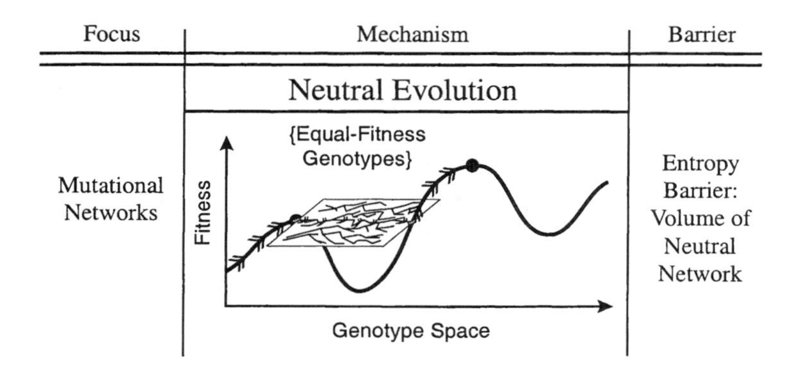

FIGURE 4 Metastability through neutral evolution: A population resides at a local optimum in a fitness landscape—here defined over the space of genotypes. Rather than passing through a valley of lower-fitness genotypes, driven by genetic variation (e.g., mutational) the population diffuses over a network of equal-fitness genotypes, until a higher-fitness genotype is found that leads to a new peak. While diffusing over the neutral network, the population's average fitness does not change. There is phenotypic stasis during a period of relatively rapid genotypic variation.

Today, the occurrence of neutral evolution is supported by a large and increasing body of evidence that there are substantial degeneracies (many-to-oneness) in genotype-to-phenotype and phenotype-to-fitness mappings. Neutrality has been implicated in the evolutionary optimization methods [9, 65] and the evolution of RNA structure [15, 22, 23, 24, 31, 32, 34, 35], protein structure [2, 36], and ribozymes [41, 69]. When degeneracies in the genotype-to-fitness map are operating, a large number of different genotypes in a population fall into a relatively small number of distinct fitness classes with approximately equal fitness, resulting in metastable evolution.

Probably the best known example of evolutionary metastability, though, is the *punctuated equilibria* behavior attributed to macroevolutionary processes by Gould and Eldredge [29]. They proposed punctuated equilibria to explain the observation in the fossil record of long periods of (morphological) constancy, which are interrupted by relatively short bursts of change, and so argued that gradual adaptation was inadequate. Although exact mechanisms supporting the metastable periods were not analyzed, the causes of punctuations were thought to originate typically in the environment, such as in planetwide climatic change [28].

The fossil record, however, is not amenable to experimental testing. Fortunately, new experimental techniques for the study of bacterial evolution have led to controllable laboratory model systems with sufficiently short replication times that evolution can now be observed in detail over many thousands of generations

[42]. These systems promise to yield the detailed and extensive data required for testing theories of evolutionary dynamics. In fact, recently Lenski and collaborators have reported punctuated-equilibrium-like behavior in the evolution of *E. coli* cell size—a proxy for fitness [16]. Even more recently, a genetic analysis of individuals taken from populations during the periods of stasis showed that there was substantial genetic variation—changes that were not phenotypically expressed [54]. Bacterial evolution appears to be a relatively clear and testable case of evolutionary metastability.

Metastability in artificial evolution has been observed in simulation studies of the population dynamics of machine-language programs [56]. In these studies, programs compete for memory and processing resources, replicate by copying themselves, and mutate when errors in copying occur. By directly observing changes in program structure and also by monitoring average replication rate—both of which are straightforward in simulation models, unlike biological experiment and the fossil record—periods of stasis and sudden change were observed over the course of many thousands of generations [1, 56].

There has also been a substantial amount of simulation and theoretical work recently on evolutionary search and optimization processes which exhibit metastability. One thread of this was directed at testing conjectures about evolution's ability to collect together functional "gene" groups by preferentially assembling *building blocks* or partial solutions [50]. In addition to concluding that building-block assembly was not responsible for the evolution of optimal solutions, it was discovered that the evolutionary search dynamics was not a gradual optimization process. Rather, it was dominated by periods of stasis and sudden change [65]. See figure 5 for an example run of a simple evolutionary algorithm that searches a space of binary strings for one with the largest number of functional gene groups.

Similar kinds of evolutionary metastability have been investigated in some detail in alternative models—"rugged landscapes" and others—by using discrete, rather than continuous, fitnesses in order to produce fitness plateaus over genotype space [3, 24, 25, 52].

Thus, it appears there is no shortage of examples, from basic theory and simulation to field and laboratory data, of metastable evolution—a behavior quite different from that implied by the view of evolution as gradual adaptation. Aside from the overt behavioral differences in the population dynamics, these cases indicate that evolution, on its own, can generate change and novelty. Except in the cases of shifting balance and punctuated equilibria, there are no appeals to environmental pressures that drive innovations. Moreover, in some of these cases, it has been reported that innovations do not lead to improved structures or functionality, calling into question functional ascriptions for evolutionary innovations.

These case studies of evolutionary metastability do not attempt to explain how novel form and function arise nor do the theories quantify form and functional change. One response to these concerns is found in early work on the arti-

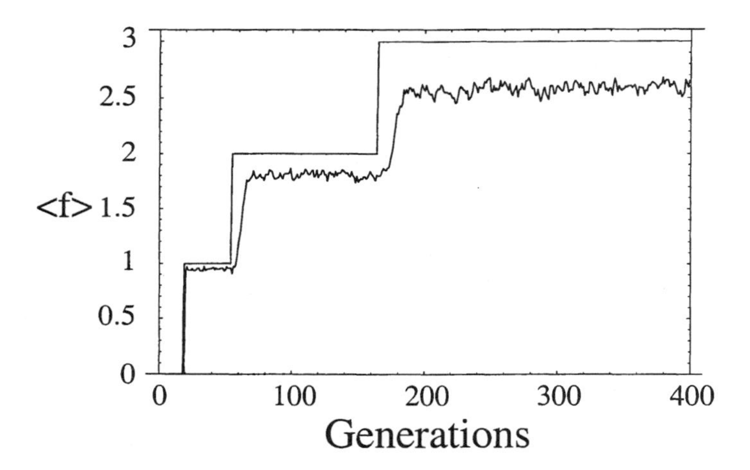

FIGURE 5 Macroscopic view of epochal evolution: survival dynamics, the level on which selection operates. Behavior of the average fitness $\langle f \rangle$ (lower curve) and best fitness (upper curve), for a population of individuals with 30 (binary) genes. They were evolved under a fitness function with three gene constellations, each consisting of 10 (binary) genes. *Constellations* are functional groups of genes that must all be properly set for a genotype to increase in fitness by one unit. The evolving population consisted of 250 individuals that at each generation were selected to replicate in proportion to their fitness and then mutated with probability 0.005 per gene. The fitness starts at 0 (no constellations set properly) and increases in a series of steps to a maximum fitness of 3 (all three constellations set properly). Reprinted with permission from the authors [66].

ficial evolution of computation. There the author introduced the phrase *epochal evolution* to describe the stepwise emergence of sophisticated strategies observed when evolving cellular automata to perform spatial computational tasks [49]. Why invent a new descriptor and not call these stages punctuated equilibria or not label them by one of the other alternatives? First, the theory of punctuated equilibria was introduced to describe the fossil record—a manifestly richer and more complex process than the artificial evolution of computational models. Second, it was clear that the evolutionary stages were due neither to Fisher's intermittent fixation nor to pinning at local optima. Third, more fundamentally it was important not to prejudice the analysis of the mechanisms driving the evolving cellular automata population dynamics. Now, however, based on the analyses of the cellular automata evolutionary dynamics, a theory of epochal evolution has been developed and the generality of the underlying mechanisms is better appreciated. Since there is little chance of confusion, we now refer to the examples of metastability given above as epochal evolution and ask which combinations of its constituent mechanisms produce the observed behaviors of stasis and rapid innovation in various cases.

The goals in the following discussion are twofold. The first motivation is to provide an accessible tour, augmented by illustrations, of recent theoretical results developed by Erik van Nimwegen and the author on the origins of metastability in evolutionary dynamics. The overview focuses on the central mechanisms underlying epochal evolution, leaving out the mathematical theory [11, 64, 65, 66]. (See van Nimwegen's thesis [62] for a detailed development.) The second motivation is to connect these ideas, which fall largely in the domain of mathematical population dynamics, with a parallel project on quantifying organization and structural complexity in natural systems, which falls largely in the domains of statistical physics and dynamical systems theory [7, 12, 58]. At the end, the discussion returns to compare epochal evolution to the various alternative mechanisms mentioned above and to the current doctrines of evolutionary theory. One conclusion drawn from the comparison is that at present the various evolutionary theories do not offer a mathematical basis on which to analyze the emergence of biological form and function. Thus, the ultimate goal, suggested by juxtaposing epochal evolution and structural complexity, is to knit the two threads of innovation and complexity together to build a predictive theory—an evolutionary mechanics [8]—of the emergence of novel structure.

2 STATISTICAL DYNAMICS OF EPOCHAL EVOLUTION

What, if any, are the common mechanisms that can explain the examples of epochal evolution given above? How are we to begin understanding the general process of epochal evolution? It turns out that answering these questions requires comparing, contrasting, and analyzing three different views of evolution: its appearance in genotype space, in phenotype space, and in a functional (fitness) space. (We have already seen an example of the latter in figure 5.) Comparing and contrasting these spaces is the burden of the following sections. The mathematical analyses that justify the approach and the results quoted in the following are found in the work just cited.

2.1 SUBBASINS AND PORTALS: MICROSCOPIC EVOLUTION

We think of *genotype space*—the collection of all genotypes—as a network whose nodes are genotypes and whose links connect genotypes that can be transformed into each other by simple genetic modifications, such as single-point mutations. Taking this and the biological facts of neutral evolution and the many-to-one structure of genotype-to-fitness maps into account, we see that genotype space decomposes into a set of neutral networks, or *subbasins* of approximately isofitness genotypes, which are entangled in a complicated fashion; see figure 6. As illustrated there, the space of genotypes is broken into strongly and weakly con-

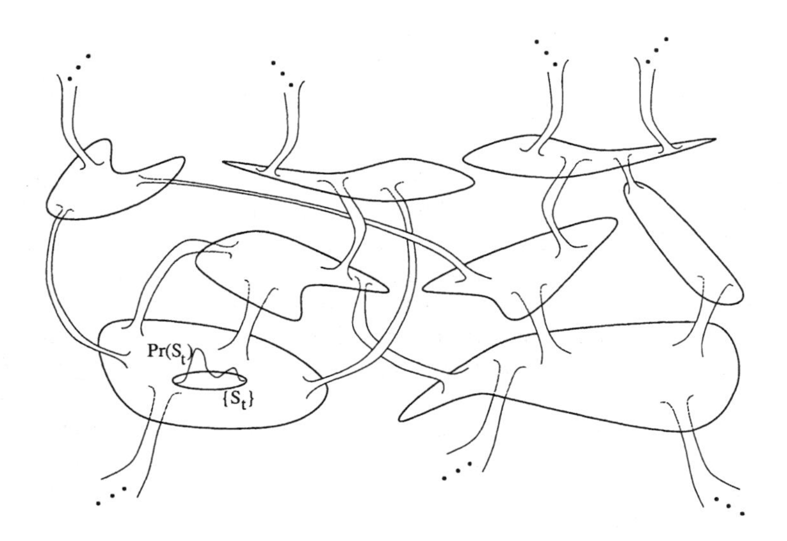

FIGURE 6 Subbasin and portal architecture in genotype space underlying epochal evolutionary dynamics. A population—a collection of individuals $\{S_t\}$ with distribution $Pr(S_t)$—diffuses in the subbasins (large sets) until a portal (tube) to a higher-fitness subbasin is found. Reprinted with permission from the authors [11].

nected sets with respect to paths generated by genetic modifications. Equal-fitness genotypes form one or several strongly connected neutral subbasins. The volume of each subbasin is determined by the number of genes that can vary without changing fitness: the more *wildcard* genes within a genotype, the larger the volume. In fact, subbasin volume grows exponentially with the number of wildcard genes. Moreover, subbasins of high fitness are generally much smaller than subbasins of low fitness, since higher-fitness genotypes (typically) tend to require more *fixed* genes to maintain their fitness. One consequence is that a subbasin tends to be only weakly connected to subbasins of higher fitness. This is depicted by the tubelike portals in figure 6.

The genotype space for the epochal evolution example of figure 5 consists of all genotypes of 30 (binary) genes, a set of 2^{30} ($\sim 10^9$) binary strings. There are three functional gene-constellations, the ten genes of which must be properly set to obtain a unit of fitness. Due to this, genotype space contains four subbasins of fitnesses 0, 1, 2, and 3, respectively. There is only one genotype with fitness 3; 3069 genotypes have fitness 2; $\sim 3 \times 10^6$ have fitness 1; and all others ($\sim 10^9$) have fitness 0. Thus, there are large degeneracies in the mapping from genotype to fitness.

Since the different genotypes within a subbasin are not distinguished by fitness selection, neutral evolution—driven by random sampling and genetic variation of individuals—dominates when the population resides in the subbasins. Selection stills acts to stabilize the population, of course, but only by culling low-fitness individuals, e.g., those in low-fitness subbasins. This leads to a rather different interpretation of the processes underlying stasis and change from that suggested by "landscape" models, for example. In landscape models a population stays pinned at a local optimum in genotype space, since all variation leads to decreased fitness. In epochal evolution, however, a population is free to diffuse randomly through subbasins of isofitness genotypes. A balance between selection increasing high-fitness individuals and deleterious mutations removing them leads to a (meta-) stable distribution of fitness (or of phenotype), while the population searches through spaces of neutral genotypic variants. During the neutral diffusion process the population of genotypes *accumulates* in the wildcard genes the history of the particular genetic variations that occurred. Even though there is no genotypic stasis during epochs, there is phenotypic stasis. As was first pointed out in the context of molecular evolution in Huynen et al. [35], through neutral mutations, the best individuals in the population diffuse over the neutral network of isofitness genotypes until one of them discovers a connection to a neutral network of higher fitness. The fraction of individuals on this network then grows rapidly, reaching a new equilibrium between selection and deleterious mutations, after which the new subset of most-fit individuals diffuses again over the newly discovered neutral network.

Note that in epochal dynamics, time scales are naturally separated. During an epoch, selection acts to establish an equilibrium in the proportions of individuals in the different subbasins, but it does not induce *adaptations* in the population. Adaptation occurs only in a short burst during an innovation (passage through a portal), after which equilibrium on the level of fitness is reestablished in the population. On a time scale much slower than that of innovations, members of the population diffuse through a subbasin of isofitness genotypes until a (typically rare) higher-fitness (portal) genotype is discovered. Thus, long periods of stasis occur because the population must diffuse over most of the subbasin before a portal to a higher-fitness subbasin is discovered. We refer to this as an *entropy barrier* to innovation, since the long duration of epochs is controlled by the *volume* of the most-fit neutral network on which the population resides.

In this way, we shift our view away from the geographic metaphor of evolutionary adaptation "crawling" along a "landscape," impeded by "energy" barriers (fitness-valleys), to the view of a diffusion process constrained by the subbasin-portal architecture. That architecture is induced, in turn, by degeneracies in the genotype-to-phenotype and phenotype-to-fitness mappings. This is not only a shift in architectural view, though, since it places a strong emphasis on the *dynamics* of populations as they move through subbasins, find portals, and so evolve increased fitness. It turns out that, while genotype-space architecture is a key component, it is not the only determinant of evolutionary population dynamics.

2.2 FINITE-POPULATION DYNAMICAL SYSTEMS: MESOSCOPIC EVOLUTION

From a microscopic point of view of genotype space, the exact state of an evolving population is only fully described when a list $\mathcal{S}$ of all genotypes with their frequencies of occurrence in the population is given. On the microscopic level, the evolutionary dynamics is implemented as a *Markov chain* with the conditional transition probabilities $\Pr(\mathcal{S}'|\mathcal{S})$ that the population at the next generation will be the collection $\mathcal{S}'$, given that the current population is $\mathcal{S}$. For any reasonable genetic representation, however, there is an enormous number of these microscopic states $\mathcal{S}$ and so too of their transition probabilities. The large number of parameters, $\mathcal{O}(2^L!)$ for L genes, makes it almost impossible to quantitatively study the dynamics at this microscopic level.

More practically, a full description of the dynamics on the level of microscopic states $\mathcal{S}$ is neither useful nor typically of interest. One is much more likely to be concerned with relatively coarse statistics of the dynamics, such as the evolution of the best and average fitness in the population or the waiting times for evolution to produce a genotype of a certain quality. The result is that quantitative mathematical analysis faces the task of finding a coarser description of the microscopic evolutionary dynamics that is simple enough to be tractable numerically or analytically and that, moreover, facilitates predicting the quantities of interest to an experimentalist. The key, and as yet unspecified, step in developing such a description of evolutionary processes is to find an appropriate set of intermediate-scale *mesoscopic* variables, or *mesostates*, with which to define the dynamics.

Fortunately, the very formulation of neo-Darwinian evolution suggests a natural decomposition of the microscopic population dynamics into a part that is guided by selection and a part that is driven by genetic diversification. Simply stated, selection is an ordering force that operates on the level of the phenotypic fitness in a population. In contrast, genetic diversification is a disordering and randomizing force that drives a population to an increased diversity of genotypes. Thus, it seems natural to choose as mesostates the proportions of genotypes in different *fitness classes* (subbasins). Additionally, one can assume that, due to random genetic diversification within each subbasin, the distribution of individuals within a subbasin is determined only by these proportions and is, otherwise, as random and unstructured as possible. (This is the *maximum entropy* assumption of statistical physics.)

Following this reasoning, we describe a population in terms of the proportions $P_0, P_1, \ldots, P_N$ of individuals located in each of the subbasins $B_0, B_1, \ldots, B_N$. The maximum entropy assumption entails that within subbasin B_i, individuals are equally likely to be any of the genotypes in B_i. (This is a

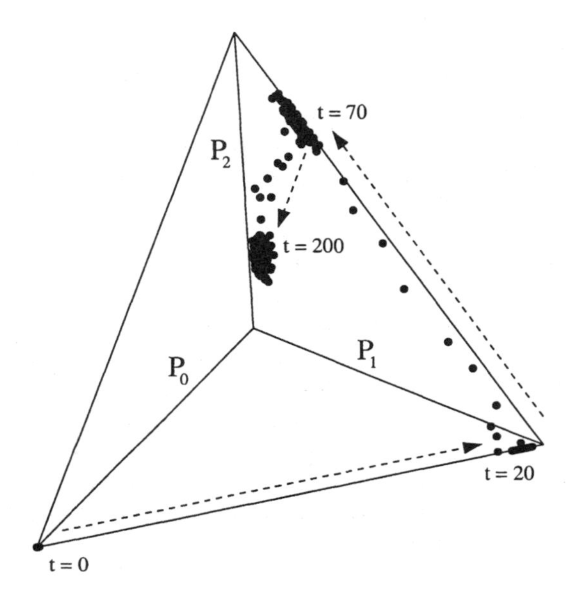

FIGURE 7 Mesoscopic view of epochal evolution—the level of population dynamics: the simplex (a solid tetrahedron) of allowed populations (dots) and the dynamic. The flow dynamics, including the clustering and regions of stability, is induced by selection and genetic variation. The stochasticity seen is the result of finite-population sampling. In this example, the fitness distribution $\vec{P} = (P_0, P_1, P_2, P_3)$ is shown for a population evolving under the fitness function of figure 5 which gives genotypes three levels of fitness: $0, 1, 2, 3$. The location of the fitness distribution at each generation is shown by a dot. The dashed lines indicate the direction in which the fitness distribution moves from metastable to metastable cluster. (The population data used here comes from the same run as in figure 5.) The times at which the different metastable states are first reached are indicated as well. These should be compared to the innovation times of figure 5. Reprinted with permission of the authors [66].

rather strong assumption that works surprisingly well in predicting observed population dynamics.) In other words, we assume that all wildcard genes are equally likely to be set in any possible way, as long as this does not lead to a portal configuration that changes fitness. Thus, we use the coarser *distribution of fitnesses*, rather than the much more unwieldy genotype distribution, to describe a population. (For simplicity of description, we are assuming that the subbasins have distinct fitnesses.) The immediate benefit is that we work with a space of populations that is vastly smaller—its dimension is the number N of subbasins—than the exponentially large space of genotypes.

Figure 7 illustrates how epochal evolution appears in the intermediate-scale mesoscopic representation afforded by fitness distributions. The figure plots fitness distributions $\vec{P} = (P_0, P_1, P_2, P_3)$ from the run of figure 5. In the figure the P_0 axis indicates the proportion of fitness-0 genotypes in the population, P_1 the proportion of fitness-1 genotypes, and P_2 the proportion of fitness-2 genotypes. Of course, since $\vec{P}$ is a probability distribution, $P_3 = 1 - P_0 - P_1 - P_2$ is completely determined, and the space of possible fitness distributions forms a solid three-dimensional *simplex.*

We see that initially $P_0 = 1$ and the population is located exactly in the lower-left corner of the simplex. Later, between $t = 20$ and $t = 60$, the population is located at a metastable fixed point on the line $P_0 + P_1 = 1$ and is dominated by fitness-1 genotypes ($P_1 \gg P_0$). Some time around generation $t = 60$ a genotype with fitness 2 is discovered, and the population moves into the plane $P_0 + P_1 + P_2 = 1$—the front plane of the simplex. From generation $t = 70$ until generation $t = 170$, the population fluctuates around a metastable fixed point in the upper portion of this plane. Finally, a genotype of fitness 3 is discovered, and the population moves to the asymptotically stable fixed point in the interior of the simplex. It reaches this fixed point around $t = 200$ and remains there fluctuating around it for the rest of the evolution experiment.

2.3 SURVIVAL DYNAMICS: MACROSCOPIC EVOLUTION

Having described epochal evolution at the microscopic level of diffusion through subbasins and portals and the mesoscopic level of the population dynamics, we can return to the highest level of evolution: survival dynamics in the space of functionality—the space on which selection acts. In the simple example already shown in figure 5 we used fitness as a proxy for functionality. Recall that figure 5 showed the fitness dynamics of a population of 30-gene individuals evolving under a three-constellation fitness function.

At time $t = 0$ the population started out with 250 random genotypes. As can be inferred from figure 5, during the first few generations all individuals were located in the largest subbasin with fitness 0, since both average and best fitness are 0. The population randomly diffused through this subbasin until, around generation 20, a portal was discovered that led into the subbasin with fitness 1. The population was quickly taken over by genotypes of fitness 1, until a balance was established between selection and deleterious mutation: selection increasing the fraction of fitness-1 individuals and deleterious mutations (that go from fitness 1 to 0) decreasing their number. The individuals with fitness 1 continued to diffuse and replicate through the subbasin with fitness 1, until a portal was discovered connecting to the subbasin with fitness 2. This happened around generation $t = 60$ and by $t = 70$ a new selection-mutation equilibrium was established. Individuals with fitness 2 continued diffusing through their subbasin until the globally optimal genotype with fitness 3 was discovered some time around generation $t = 170$. Descendants of this genotype then spread through the population until around $t = 200$, when a final stable equilibrium was reached.

2.4 PORTRAIT OF AN INNOVATION: UNFOLDING AND STABILIZING NOVELTY

Putting together the views of evolution at the three different levels of genotype subbasins and portals, population dynamics, and survival dynamics, one sees that epochal evolution is a process of state-space unfolding (see fig. 8):

1. Initially, the population moves in (say) n mesoscopic dimensions of the population-dynamics space of fitness distributions.
2. It is attracted to a (noisy) fixed point—the metastable collection of populations observed during the epoch.
3. At the same time it diffuses neutrally in the very high dimensional microscopic space of genotypes. During epochs, many genotypic changes occur and accumulate, but do not alter the phenotype. This invariance of the phenotype is a *symmetry* of the fitness distribution with respect to microscopic change.
4. An innovation occurs when, having accumulated a certain combination of changed genes, a portal to increased fitness is discovered in the microscopic space.
5. This breaks the existing epoch symmetry, since genetic changes now affect fitness.
6. A new mesoscopic dimension becomes activated, fitness increases and selection begins to stabilize the innovated feature by removing lower-fitness genotypes (without the feature) and adding higher-fitness ones (with the feature).
7. The mesoscopic population dynamics now moves in an $(n + 1)$-dimensional space.

In the unfolding process microscopic variation is amplified through the innovations and becomes locked in due to the dynamics at the mesoscopic and macroscopic levels. Randomness serves to drive the diffusion in the microscopic dimensions and eventually leads to the discovery of portals to innovation. Selection acts to stabilize the structure of the mesoscopic spaces, once a new dimension has been activated. Complementing this flow of information from the microscopic to the macroscopic, there is also feedback from the macroscopic level that determines the constraints on the microscopic dynamics. That is, the macroscopic organization of possible individuals—e.g., attainable fitness levels—is reflected in the subbasin-portal architecture of the microscopic space. The mathematical analysis of these mechanisms and their interaction we call *statistical dynamics*.

It will be helpful at this point, having outlined the statistical dynamics of epochal evolution, on the one hand, and having earlier mentioned several alternative descriptions of the causes of metastability, on the other, to draw the contrasts more sharply between them. First, epochal evolution is not Fisher's intermittent fixation. Though aspects of drift due to sampling and mutation are components of epochal evolution, the epochs are stabilized by selection removing low-fitness genotypes (see step 6 above), which is not part of Fisher's model.

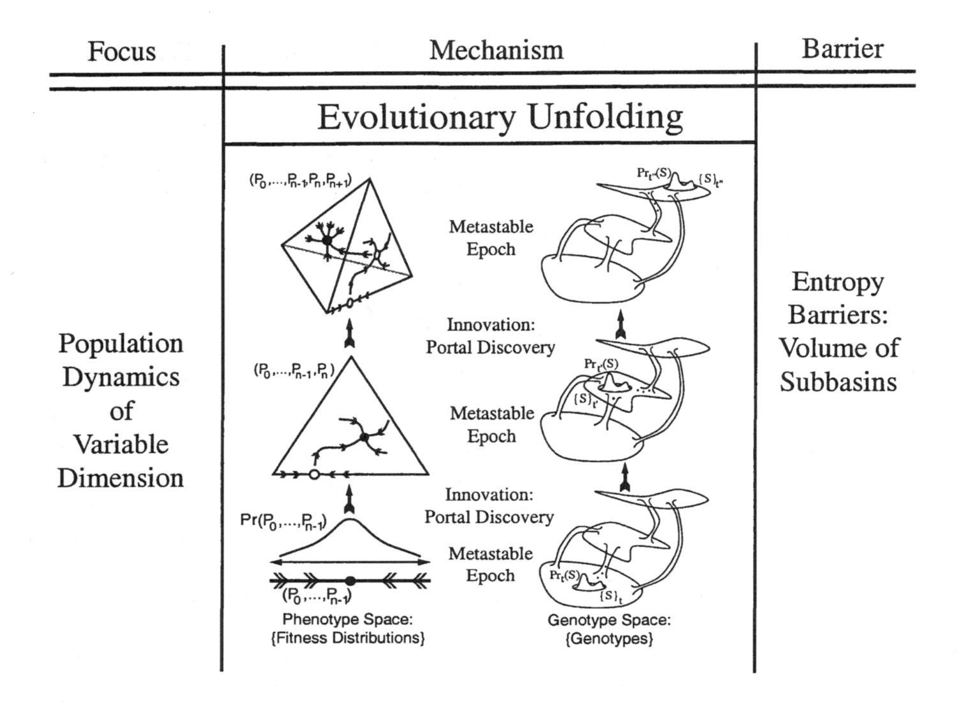

FIGURE 8 Portrait of an innovation: The mesoscopic (left) and microscopic (right) views of two innovations produced by epochal evolution. See text for description.

Second, epochal evolution is not Kimura's purely neutral evolutionary dynamic, since a key part of the former is an explicit mechanism for finding and then locking in structural innovations. In this sense, the theory of epochal evolution proposes an overall architecture for piecewise-neutral evolutionary processes. Third, epochal evolution is not Gould-Eldredge punctuated equilibria in that it is a predictive quantitative theory with specific mechanisms that, with new automated evolutionary experiments, will be laboratory testable. The theory of punctuated equilibria did not commit to much, if any, underlying mechanism, other than exogenous (environmental) causes of change. (The statistical dynamics of epochal evolution is a theory of endogenous change.) Punctuated equilibria served more as a descriptive summary of phenomenological aspects of the fossil record, as one naturally expects of paleontology. Perhaps at some future date, with more fossil data and a more elaborated theory of epochal evolution, it may be shown that punctuated equilibria in macroevolution is a kind of epochal evolution. At present, all one has is observational consistency, without the ability to positively identify underlying mechanisms from the fossil record.

One notable consequence of the statistical dynamics analysis is that epochal evolution is a kind of open-ended evolution. It is explicitly a dynamics by which a sequence of innovations can be discovered and then become the structural substrate for further evolution. Also, depending on which (randomly chosen) sequence of portals is realized, the course of macroscopic evolution can be very different and so reflect the accumulation of, what some call, *frozen accidents*. In these respects, the statistical dynamics of epochal evolution is a partial response to the criticism of population dynamical systems modeling of evolution as being evolutionarily closed and incapable of intrinsic novelty. The claim is that such models must at the outset "build in" the ultimate dimensionality of an evolutionary process which, in turn, caps evolutionary innovations [20]. Epochal evolution shows that this is not an intrinsic failing of population dynamical systems: they can be open-ended in the way epochal evolution unfolds and then stabilizes new state spaces. The main limit imposed on the continuing emergence of increasingly complex structures—assuming other parameters, such as population size and mutation rate, are compatible—comes from the structure of the space of individual function, not directly from population dynamics.

These observations on innovation processes in evolution lead immediately to questions about what one means by "structure" and "function"—largely open questions, as yet incompletely addressed by evolutionary theory. How can we ever say unambiguously, for example, that an evolutionary system evolved toward complexity, or that it was or was not open-ended, without a theory of structure and function that allows us to quantitatively monitor their change?

At its current stage of development, in the statistical dynamics of epochal evolution, "individuals" are simple and direct genotype-to-fitness maps. Beyond gene constellations that confer fitness when properly set, they are nearly structureless individuals. For example, they have no spatial structure and no temporal behavior. These aspects can play no role in determining individual fitness. Thus, there is no analog of development in the theory, except that which is implicit in the genotype-to-fitness maps. In contrast, the laboratory experiments and the simulations of evolving dynamical systems mentioned above do have structured individuals, often exhibiting complex structures and rich dynamical behavior. To develop a predictive theory of epochal evolution for these, one needs to be precise about how a given individual is structured, how it functions, and how its functionality confers fitness, in order to track increases (or decreases) in evolutionary and developmental complexity. In short, to quantify changes in structure and function one needs to define "complexity."

3 STRUCTURAL COMPLEXITY AND INNOVATION

Fortunately, recent progress has brought us to a level of understanding complexity that suggests we are close to defining it in ways that are germane to evolution. In particular, results on how complexity emerges in nonbiological systems give important insights into the structures that can emerge in evolutionary processes and also into the constraints that structural innovations must respect when they occur. Before describing these ideas, however, it will be helpful to set the historical context and to make several important distinctions.

As a label for natural systems that are difficult to model and analyze, over the last two decades "complexity" has served a useful role by ambiguously referring both to randomness and to organization. The study of *complex systems* has sometimes focused on simple (albeit, nonlinear) processes that appear random and are difficult to predict—e.g., deterministic chaos and fractal separatrices. The question there, to say it most directly, is, How does disorder emerge from simplicity? At other times, studies of complex systems have focused on large-scale processes consisting of many interconnected components—what one might call *complicated systems*. The question there has been, How is it that order arises despite so much possible disorganization? It is not surprising that two such opposite phenomena—disorder emerging from order and order from disorder—falling under the same rubric of complex systems would lead to confusion. Fortunately, the confusions and resulting debates about what "complexity" is led to a useful clarification. There are two basic and different categories of complexity: complication versus structure. As we now appreciate, although complication emerging from simplicity and organization emerging from disorder appear to be opposite kinds of phenomena, the complexities to which they refer are, in fact, complementary and not opposites.

On the one hand, we have complexity as varying degrees of randomness or complication in a system's behavior or in its architecture. The behavior of a dynamical system ranges, say as we change a control parameter to make it more nonlinear, from being regular, periodic, and predictable to chaotic and unpredictable. The organization of social systems ranges from the predictable delivery of vast amounts of food to major cities to the seeming turbulence and uncoordinated deal-making behavior of traders in a stock exchange. Thus, we think of natural complication as a spectrum of randomness: from pure order to utter disorder.

This spectrum is quite familiar to us. So much so that many fields have developed their own vocabularies for degrees of complicatedness. In physics, for example, one uses temperature (T) and thermodynamic entropy (S) to monitor where a system is in its spectrum of randomness: low temperature or entropy indicate an ordered system, high temperature or entropy a disordered one [55, 57, 71]. In the theory of communication, one uses Shannon's measure (H) of information: predictable messages are uninformative; unpredictable messages are highly informative [6, 59]. In the theory of computation, one uses Kolmogorov-Chaitin complexity (K) as an algorithmic measure of an object's randomness [5, 40, 43, 45], and so on. We can illustrate very simply, as done in figure 9, the relationship between these ways of measuring degrees of complicatedness: T, S, H, and K are all proportional to randomness.

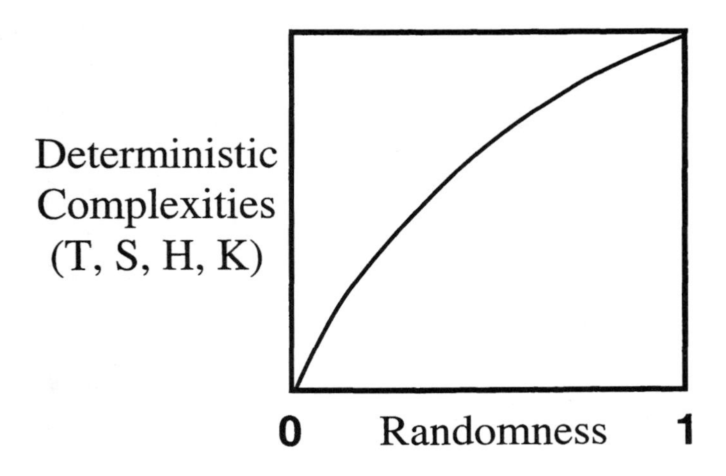

FIGURE 9 Measures of complicatedness—temperature T, thermodynamic entropy S, Shannon information H, and Kolmogorov-Chaitin complexity K—(vertical axis) are all proportional to the amount of randomness in a process (horizontal axis). In one way or another, these quantities assume that all of the randomness in a process must be described using deterministic models—such as a universal Turing machine on which the Kolmogorov-Chaitin complexity is defined. This is why they are referred to as *deterministic* complexities. Reprinted with permission from the author [7].

On the other hand, we have complexity as varying degrees of organization—or structure, regularity, symmetry, and intricacy—in a system's behavior or in its architecture. We say that a ferromagnet is more structured at the transition between its low-temperature ordered phase and its high-temperature disordered phase, since only there does it exhibit aligned-spin clusters of all sizes. The network of financial, technological, and industrial interdependencies that support the production of modern microprocessors is certainly neither a regular and fixed architecture, in which case it would be too rigid and unadaptive, nor one that is entirely unstructured, in which case it would simply be nonfunctional. The required institutional memory, flexibility, competition, and cooperation have led it to some state intermediate between these organizational extremes. In a way analogous to randomness, in the space of organizational architectures, we can think of a spectrum of structure: from simple symmetric architectures to sophisticated and hierarchical ones.

The corresponding measure of structuredness we call *structural complexity* (C). There have been many more or less specific proposals for structural complexity—some, it turned out, actually measure randomness. (See the review and especially the long list of citations in Feldman and Crutchfield [17, 18] and Shalizi and Crutchfield [58].) Nonetheless, we can summarize the basic idea be-

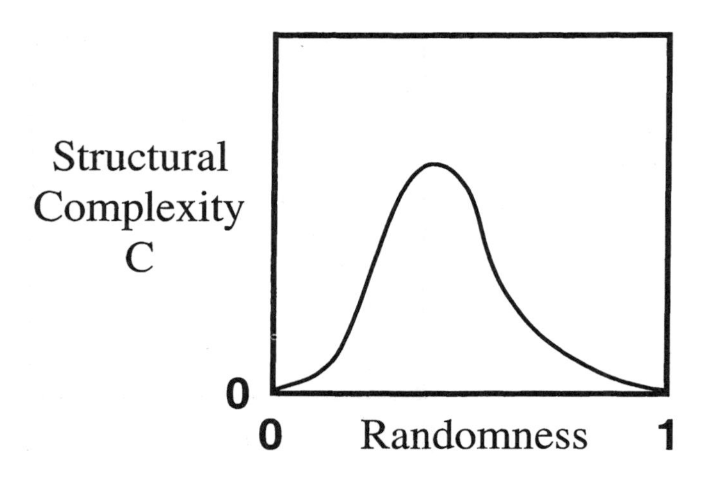

FIGURE 10 Structural complexity C (vertical axis) peaks in the intermediate region between the extremes of randomness (horizontal axis). Reprinted with permission from the author [7].

hind structural complexity by contrasting it with the spectrum of randomness. This is done schematically in the *complexity-entropy diagram* shown in figure 10. The peaked curve shows that (i) the extremes of randomness, highly predictable and highly random, are structurally simple (low C) and (ii) structural complexity is largest in the intermediate regime between the extremes.

We now consider structural complexity to be a complementary coordinate to degrees of randomness, as depicted in figure 10. It characterizes a different feature of a system—for temporal processes, the amount of historical *memory*—than randomness—which is the amount of information a system *produces*. That is to say, randomness and structural complexity are both necessary descriptors: the former captures surprise, the latter organization. In analyzing dynamical systems, for example, one uses complexity-entropy diagrams like that of figure 10 to display the spectrum of how a collection of systems generate and store information to varying degrees.

We refer to the theory of structural complexity as *computational mechanics* since it extends statistical mechanics—a theory of randomness—to include definitions of structure that capture computational architectures. The main questions asked when analyzing a system in this framework do not focus on the storage and transduction of energy. They ask instead how a system stores, transmits, and transforms *information*. Briefly, what are a system's *intrinsic computational* properties? Originally introduced over a decade ago [12], the mathematical foundations are now well developed. (See Upper [61] and, again, Feldman and Crutchfield [17] and Shalizi and Crutchfield [58], which also review alternative approaches to structural complexity.) Computational mechanics defines structural complexity (C) in terms of a decomposition of a system's behavior into its minimal causal architecture—a representation called an *ϵ-machine.* C is the amount of information, including spatial correlation and temporal memory, which this minimal architecture stores. In a well-defined sense C is the size of the set of minimal causal components embedded in a system. The procedure for identifying a system's minimal causal architecture is called *ϵ-machine reconstruction.*

Aside from providing a first-principles approach to extracting an ϵ-machine for a system and so measuring its structural complexity, one of the main results is that novel structures (forms of intrinsic computational architecture) emerge in pattern-forming systems that are at phase transitions. More generally, it is often observed that structural complexity emerges from the dynamical interplay of ordering and disordering forces—such as those operating when discovering portals in neutral networks. (These applications reify the rather coarse and schematic view captured in the complexity-entropy diagram of figure 10.)

Comparisons of how novel structural complexity emerges at different kinds of transition, such as phase transitions, and over time in cellular automata [33], for example, give some insight into the structural innovations that can emerge in evolutionary processes. (Investigations of evolving cellular automata give many examples of just this kind of structural innovation in an artificial evolutionary process—innovations that can be structurally analyzed in some detail using computational mechanics [9, 10].) First, the central way to detect that some new thing has emerged in an innovation is to monitor the causal architecture—either over time, if analyzing a temporal process such as the evolutionary population dynamics of cellular automata, or over a range of parameters, if it is a controlled process, such as a system undergoing a phase transition. Second, increased structural complexity can appear either smoothly, as shown in figure 10, or abruptly as in a critical phase transition; see, for example, [13]. In the latter case, there is a qualitative change in a system's causal architecture: a divergence in the number of causal components and a shift to a more powerful computational class. Innovations have been analyzed in systems that show a shift from a disorganized initial "heat bath" to patterned levels of coherent domains and particles, from finite-memory to infinite-memory processes, and from finite-state machine to pushdown-stack architectures. Such innovations have structural signatures and, using computational mechanics, there are now ways to detect them and quantify what novelty has emerged. Thus, the emergence of structural complexity from the interplay of a system's tendency to order and its tendency to disorder suggests where to look for innovations, how to detect them, and how to describe what has been created.

4 EVOLUTION TO COMPLEXITY

In the computational mechanics of structural complexity one sees the beginnings of a principled approach to form and function in evolutionary processes. First, identifying and then quantifying the kinds and amounts of structure embedded in natural systems are the first steps to making the concept of form precise and testable. To the extent that one considers biological form to include symmetry, regularity, hierarchy, pattern, modularity, and so on, structural complexity, as defined in computational mechanics, is an appropriate operational approach to it. Second, building on an unambiguous concept of form, one can view functionality as arising from the relationships between a system's intrinsic structures—their static architecture and their dynamical interaction—and intrinsic or externally determined evaluation of those structures.

In the computational mechanics view, then, evolutionary innovations are changes in the architecture of information processing. These changes can be reflected at the level of either a population or an individual. The novelty of an innovation is built out of structures on lower levels and occurs in "orthogonal" coordinates when something truly new emerges. Unlike the purely structural emergence observed in pattern formation processes—such as the appearance of spiral waves in a Belusov-Zhabotinsky chemical reaction-diffusion system—innovations can take on meaning and function *within* in an evolutionary process. Unlike the spiral waves, this *intrinsic emergence* does not require an outside observer to monitor the changes in structural complexity [8]. The meaning and function of intrinsically emerging organization derives from the fitness evaluation of individuals and the persistence of traits over time—features that are part and parcel of an evolutionary process. In a sense, selection performs the role of observer.

In computational mechanics, the process by which open-ended innovation can occur is called *hierarchical ϵ-machine reconstruction.* We think of hierarchical ϵ-machine reconstruction, or some dynamically instantiated version, as specifying the minimal requirements for open-ended evolution: successive innovation of levels of distinct structural classes that build on the lower levels' component structures. Figure 11 illustrates an open-ended series of evolutionary innovations: nested levels of information processing of increasing computational power. At each level, there is a spectrum of structures, some of which are more appropriate (e.g., useful or functional), since they balance both parsimonious resource use against minimal degrees of randomness. To store and use better structures, however, requires increased resources—they are in one or another sense larger than less optimal structures.

It is perhaps not surprising that finite resources drive the process of innovation. Why? If an individual at some level of organization had infinite resources, say to model its environment, then there would no benefit to restructure existing resources or to incorporate new ones. It would gain no predictive advantage,

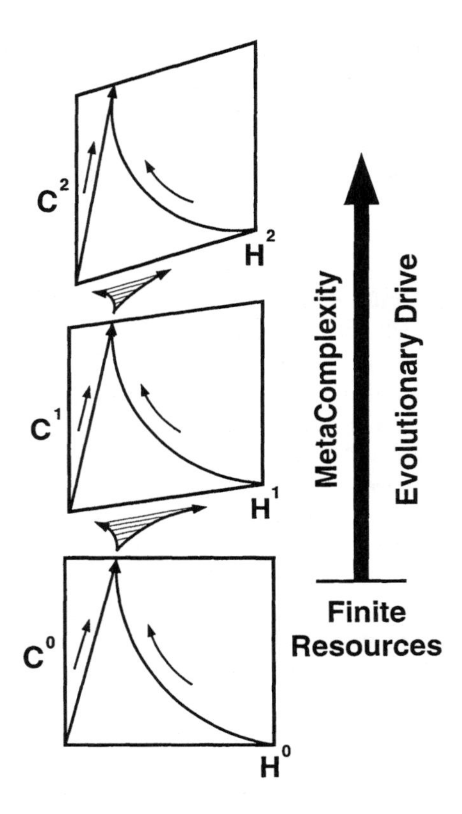

FIGURE 11 Computational mechanics view of open-ended evolution: a sequence of innovations leading up through a hierarchy of increasingly sophisticated classes of information processing. The distinct classes of structure are illustrated with a complexity-entropy diagram (H^i, C^i) that represents the tradeoffs between randomness and structural complexity appropriate at each organizational level. The arrows indicate an adaptive dynamics that leads within each level to increasingly sophisticated structures. When adaptation exhausts finite resources, there is an innovation of a new class of structure. Reprinted with permission from the author [7].

since the current model is as predictive as any alternative. When resources are limited there is an effective pressure to innovate a new class of organization—one that more efficiently uses the available resources while improving efficacy. In this way, computational mechanics describes an open-ended series of innovations as an unfolding hierarchy of recursively embedded structural classes. Though this view of open-ended evolution focuses on innovations in structural complexity, one sees the parallel between this process and how innovations arise during epochal evolution by a process of unfolding and then stablizing new state spaces. In computational mechanics one focuses on nested levels of information processing; in epochal evolution one focuses on activating and then stabilizing state spaces of increasing dimension.

5 ORIGINS OF FORM AND FUNCTION: SELECTION, ACCIDENT, OR MORPHOGENESIS?

It is often said that organisms today are more complex than in earlier times. But what (exactly) is this complexity and how did it emerge? Having reviewed the dynamics of epochal evolution and the theory of structural complexity and the roles they play in innovation, we can now contrast them more directly with views from within evolutionary biology on the emergence of form and function. According to Williams, three doctrinal bases have been used over the last century to address the evolution of complexity [68]:

1. *Natural selection*: "trial and error, as opposed to rational plan";
2. *Historicity*: "the role of historical contingency in determining the Earth's biota"; and
3. *Mechanism*: "only physico-chemical processes are at work in an organism."

Natural selection holds that structure in the biological world is due primarily to the fitness-based selection of individuals in populations whose diversity is maintained by genetic variation [46]. That is, genetic variation is a destabilizing force that provides the raw diversity of structure. Natural selection then is a stabilizing dynamic that acts on the expression of that variation, which is structural diversity. It "generates" organization by culling individuals based on their relative fitness, which is determined by their structure. This view identifies a source (genetic variation) of new structures and a mechanism (selection) for altering one form into another. Thus, the adaptiveness accumulated via selection is seen as the dominant mechanism driving the appearance of form and function.

The historicity doctrine acknowledges the Darwinian mechanisms of selection and variation, but emphasizes the accidental determinants of biological form [28, 51]. What distinguishes this position from the emphasis on natural selection is the claim that major changes in structure can be and have been nonadaptive.

While these changes have had the largest effect on the forms of present day life, at the time they occurred they conferred no survival advantage. Furthermore, today's existing structures need not be adaptive, reflecting instead a history of frozen accidents. One consequence is that a comparative study of parallel earths would reveal very different collections of life forms on each. Like the doctrine of natural selection, historicity accounts for the emergence of structure by a process of preferentially culling one or several structures within preexisting structural diversity. But it is a dynamics that is manifestly capricious or, at least, highly stochastic with few or no causal constraints. Due to this, the historicity doctrine is not a theory of the *origins* of diverse form and function.

In the mechanistic view of evolution the goal is to elucidate "principles of organization" that underlie the appearance of biological form. In this, it focuses directly on the question of what biological complexity might be. The doctrine contends that energetic, mechanical, biomolecular, and morphogenetic properties guide and limit the infinite range of possible biological form [20, 21, 26, 37, 60, 67]. The constraints result in a relatively small set of structure archetypes. In a sense, these play a role in morphology analogous to the Platonic solids in geometry: they preexist, before any evolution takes place. In the evolutionary emergence of complexity, then, natural selection chooses between these "structural attractors," possibly fine-tuning their adaptiveness. In this view, Darwinian evolution serves, at best, to fill the waiting attractors or not, depending on historical happenstance. It does not, however, create the structure of those attractors.

What is one to think of these conflicting theories of the evolution of complexity?

First, although natural selection's culling of genetic variation provides a theory of gradual structural transformation, it does not provide a theory of structure itself. For example, what is the average time under an evolutionary dynamic and under the appropriate environmental pressures for a fish fin to be transformed into a leg? If one knew the genetic trajectory—the required sequences of modified and innovated genes—in principle one could use the theory of population genetics to estimate how long the transformation would take. But this assumes and hides too much—How did those genes determine the functionality of fins and legs? To estimate from first principles the time to evolve a leg from a fin one needs a measure of the structures concerned and of the functionality they do or do not confer.

Second, historicity too provides a theory of transformation and not of structure. Moreover, for its highly stochastic transformational dynamics to be successful—or, at least, to not destroy all structures—and for there to be the requisite broad structural diversity on which it acts, historicity requires that the space of possible biological structures be populated with a high fraction that are functional. Whether this is true or not is simply unknown. Additionally, in emphasizing the dominance of historical accident, it advocates an extra-evolutionary theory for the origins of novel organization and form, side-stepping the issue of biology's role in actively producing them. An explanation that appeals to a meteor crashing into the earth simply falls outside the domain of evolutionary theory. Moreover, the occurrence of such events is unlikely ever to be explained by the principles of physical dynamics. The collision just happened: a consequence of particular initial conditions that occurred in a celestial dynamical system which is most likely chaotic and, if so, demonstrably unpredictable. Such accidents impose significant constraints; they do not constitute an explanation of the origins or biological form or function.

Finally, the mechanistic doctrine does not offer a theory of evolutionary transformation, though it focuses on morphogenesis which certainly interacts with evolutionary processes. Although it employs methods from pattern-formation and bifurcation theories, it too falls short in that it does not provide a theory of structure itself nor of the functions of evolved structures. In particular, the structural attractors are not quantitatively analyzed in terms of their internal architecture, nor in terms of system-referred functionality or fitness.

It would appear that the three doctrines rely on undefined concepts of form and function. What about the theory of neutral evolution? What does it say about form and function and their evolution? In natural selection's emphasis on gradual adaptation each and every biological thing, embodying the direct solutions to the survival problems posed by environmental constraints, has a function and so a "story." Neutrality, though, breaks the logic of functional ascription. The direct consequence of neutral evolution is the appearance of nonadaptive, nonfunctional, and nonfitness-conferring genotypes or phenotypes. Previously, Kimura argued that neutrality or near-neutrality is the rule in molecular evolution. One can also develop a different kind of argument, that one should expect neutrality to be common in the evolution of form and function, using the theory of structural complexity.

To say it most simply, whenever there is structure, there will be a many-to-one mapping from genotype to phenotype and to fitness. Why? To say there is "structure" is to say that the range of possible entities is constrained, not random. If the range of possible forms is not fully random (is not structureless), then the mapping of genotypes—an exponentially large number of long strings in a high-dimensional space—to structures is degenerate: many genotypes will code for individual structures, the former substantially out numbering the latter. The many-to-oneness derives most fundamentally, though, from a collapse of dimensionality in going from the microscopic realm of genotype space to the macroscopic realm of form and function. Even if genotypic coding consisted of continuous parameters rather than discrete genes so that the preceding (combinatorial) argument did not apply, any reduction in the dimension (from genotype space to phenotype space to function space) results in neutrality. Thus, the evolution of structurally complex organisms appears to implicate in a fundamental way neutral evolution and so inherently epochal population dynamics. Moreover, when properly calibrated against landscape-optimization processes, evolution along neutral pathways dominates since the time it takes to find innovations is markedly shorter than the time taken by fitness-valley crossing [63].

If neutral evolution is to be expected in the emergence of complex organisms, then there need not always be functional "stories" for each of their component structures, since some structures may have arisen during periods of stasis. Of course, they may become functional later on, say even contributing to an innovation. (Gould and Vrba call this recontextualizing originally nonfunctional traits *exaptation* [30].) Imagine examining a contemporary organism. Which of its structures emerged during periods of stasis and which not? The ambiguity here is only heightened when one realizes that, in many cases for which data is available, epochs of stasis are defined in terms of morphological, and not functional, constancy. Much of what we see in the biological world need not be there because it fulfills a purpose—not even for survival. Functionality, perhaps emerging through adaptive innovations, comes equally from the context of a given form—something much harder to detect than form itself. Thus, confronted with the possibility of metastable evolution, one comes to appreciate diversity of all kinds, even that which is not functional and which appears to serve no purpose. One may have to adopt a very long view. Present diversity may be highly determinant at some later time and in a different context.

6 NONE OF THE ABOVE

The impression the doctrinal debate leaves, though, is that there is a pressing need for both a qualitative dynamical theory of structural emergence [27] and a theory of biological structure itself [47, 48]. The main problem, at least to an outsider, does not reduce to showing that one or the other existing doctrine is correct. Each employs a compelling argument and often empirical data as a starting point. Rather, as a first step, the task facing us is to develop a synthetic theory that can balance the tensions between selection, neutrality, accident, and mechanism. The analysis of epochal evolution—how it unfolds and stabilizes novel mesoscopic spaces, when this occurs and when it is precluded—does suggest what this qualitative dynamical theory might look like. Foremost it would resolve the tension between the microscopic, mesoscopic, and macroscopic levels on which evolutionary processes act and give an architectural view of the microscopic and mesoscopic consequences of function.

If we ask about the *origin* of function, though, does it lie in selection, historical accident, morphogenesis, or some combination, the answer here has to be "none of the above." There is some basic thing missing in these three approaches. (Neutral evolution, as just noted, plays no direct role and mostly serves to complicate the question of function.) They do not directly address the question of functionality, nor are they equipped to do so. I argued that, on the way to addressing the origins of function, what is missing is a theory of form based on structural complexity and a theory of its emergence based on epochal evolution.

At this point, however, structural and functional constraints operating during epochal evolution are only reflected, and indirectly so, in the subbasin-portal architecture of genotype space. Like the three existing doctrines, the theory of epochal evolution is not a theory of structure, nor does it yet incorporate one. Thus, there is a second, much more difficult, step: to develop a quantitative theory of structure and then out of that, a theory of function. Without these, we appear to be in no position to explain the evolution of complexity. It would appear that if one stops here, when evolution is revolution, we simply cannot say what has been innovated.

The computational mechanics of nonlinear processes, however, is a theory of structure. Pattern and structure are articulated in terms of various types of causal architecture—what we called computational classes. The overall mandate there is to provide both a qualitative and a quantitative analysis of natural information-processing architectures. If computational mechanics is a theory of structure, then innovation via hierarchical ϵ-machine reconstruction is a computation-theoretic approach to the open-ended transformation of structure. It suggests one mechanism with which to study what drives (finite resources) and what constrains (intrinsic computation) the appearance of novelty.

The discussion has brought us to a possible next step toward an evolutionary dynamics of structural emergence. This would be to fold hierarchical ϵ-machine reconstruction into an evolutionary process, resulting in an intrinsic dynamics of innovation. In a rough way, something like this is observed in the evolution of cellular automata, mentioned earlier. The theoretical analysis there is incomplete. However, at least evolving cellular automata provides a concrete case, which appears to be tractable and which can be used to ferret out the many taxing definitional problems in the evolutionary dynamics of form and function.

There are two main points to draw from the parallel threads of epochal evolution and structural complexity. First, epochal evolution arises intrinsically: long periods of stasis and sudden change need not be driven by external forces. They are the product of the many-to-one mappings from genotype to phenotype and phenotype to fitness. Epochal evolution is to be expected and it occurs by an open-ended process of discovering and stabilizing novelty—novelty that becomes substrate for further evolution. Second, the emergence of structure can be monitored as an open-ended hierarchy of novel kinds of embedded computation and information processing. When these two threads are knitted together, one hopes that, when evolution is revolution, we will be able to say what novelty has been created.

ACKNOWLEDGMENTS

I thank Victoria Alexander, Dave Feldman, Erik van Nimwegen, and Cosma Shalizi for many stimulating and wide-ranging discussions. I also appreciate discussions with the participants of the workshop **Towards a Comprehensive Dynamics of Evolution—Exploring the Interplay of Selection, Neutrality, Accident, and Function**. Partial support for this work comes from

DARPA contract F30602-00-2-0583, NSF grant IRI-9705830, AFOSR via NSF grant PHY-9970158, and the Keck Foundation's Evolutionary Dynamics Program at the Santa Fe Institute.

REFERENCES

[1] Adami, C. "Self-Organized Criticality in Living Systems." *Phys. Lett. A* **203** (1995): 29–32.

[2] Babajide, A., I. L. Hofacker, M. J. Sippl, and P. F. Stadler. "Neutral Networks in Protein Space: A Computational Study Based on Knowledge-Based Potentials of Mean Force." *Folding Design* **2** (1997): 261–269.

[3] Barnett, L. "Tangled Webs: Evolutionary Dynamics on Fitness Landscapes with Neutrality." Master's thesis, School of Cognitive Sciences, University of East Sussex, Brighton, 1997. ⟨http://www.cogs.susx.ac.uk/lab/adapt/nnbib.html⟩.

[4] Bergman, A., and M. W. Feldman. "On the Population Genetics of Punctuation." This volume.

[5] Chaitin, G. "On the Length of Programs for Computing Finite Binary Sequences." *J. ACM* **13** (1966): 547–569.

[6] Cover, T. M., and J. A. Thomas. *Elements of Information Theory.* John Wiley & Sons, Inc., 1991.

[7] Crutchfield, J. P. "The Calculi of Emergence: Computation, Dynamics, and Induction." *Physica D* **75** (1994): 11–54.

[8] Crutchfield, J. P. "Is Anything Ever New? Considering Emergence." In *Complexity: Metaphors, Models, and Reality*, edited by G. Cowan, D. Pines, and D. Melzner, 479–497. Santa Fe Institute Studies in the Sciences of Complexity, Proc. Vol. XIX. Reading, MA: Addison-Wesley, 1994.

[9] Crutchfield, J. P., and M. Mitchell. "The Evolution of Emergent Computation." *Proc. Natl. Acad. Sci. USA* **92** (1995): 10742–10746.

[10] Crutchfield, J. P., M. Mitchell, and R. Das. "The Evolutionary Design of Collective Computation in Cellular Automata." This volume.

[11] Crutchfield, J. P., and E. van Nimwegen. "The Evolutionary Unfolding of Complexity." In *Evolution as Computation*, edited by L. F. Landweber, E. Winfree, R. Lipton, and S. Freeland. Lecture Notes in Computer Science. New York: Springer-Verlag, 2000.

[12] Crutchfield, J. P., and K. Young. "Inferring Statistical Complexity." *Phys. Rev. Lett.* **63** (1989): 105–108.

[13] Crutchfield, J. P., and K. Young. "Computation at the Onset of Chaos." In *Entropy, Complexity, and the Physics of Information*, edited by W. Zurek, 223–269. Santa Fe Institute Studies in the Sciences of Complexity, Proc. Vol. VIII. Reading, MA: Addison-Wesley, 1990.

[14] Darwin, C. *On the Origin of Species.* New York: Modern Library, 1993.

[15] Ekland, E. H., and D. P. Bartel. "RNA-Catalysed RNA Polymerization using Nucleoside Triphosphates." *Nature* **382** (1996): 373–376.

[16] Elena, S. F., V. S. Cooper, and R. E. Lenski. "Punctuated Evolution Caused by Selection of Rare Beneficial Mutations." *Science* **272** (1996): 1802–1804.

[17] Feldman, D. P., and J. P. Crutchfield. "Discovering Noncritical Organization: Statistical Mechanical, Information Theoretic, and Computational Views of Patterns in Simple One-Dimensional Spin Systems." *J. Stat. Phys.* (1997): submitted.

[18] Feldman, D. P., and J. P. Crutchfield. "Measures of Statistical Complexity: Why?" *Phys. Lett. A* **238** (1998): 244–252.

[19] Fisher, R. A. *The Genetical Theory of Natural Selection.* Oxford: The Clarendon Press, 1930.

[20] Fontana, W., and L. Buss. "'The Arrival of the Fittest': Toward a Theory of Biological Organization." *Bull. Math. Biol.* **56** (1994): 1–64.

[21] Fontana, W., and L. Buss. "What Would be Conserved if the Tape were Played Twice?" *Proc. Nat. Acad. Sci. USA* **91** (1994): 757–761.

[22] Fontana, W., and P. Schuster. "Continuity in Evolution: On the Nature of Transitions." *Science* **280** (1998): 1451–1455.

[23] Fontana, W., P. F. Stadler, E. G. Bornberg-Bauer, T. Griesmacher, I. L. Hofacker, M. Tacker, P. Tarazona, E. D. Weinberger, and P. Schuster. "RNA Folding and Combinatory Landscapes." *Phys. Rev. E* **47** (1992): 2083–2099.

[24] Forst, C. V., C. Reidys, and J. Weber. "Evolutionary Dynamics and Optimizations: Neutral Networks as Model Landscapes for RNA Secondary-Structure Folding Landscape." In *Advances in Artificial Life*, edited by F. Moran, A. Moreno, J. Merelo, and P. Chacon, 128–147. Lecture Notes in Artificial Intelligence, Vol. 929. Berlin: Springer, 1995.

[25] Gavrilets, S. "Evolution and Speciation in Hyperspace: The Roles of Neutrality, Selection, Mutation, and Random Drift." This volume.

[26] Goodwin, B. "Evolution and the Generative Order." In *Theoretical Biology: Epigenetic and Evolutionary Order from Complex Systems*, edited by B. Goodwin and P. Sanders, 89–100. Baltimore, MD: Johns Hopkins University Press, 1992.

[27] Goodwin, B., and P. Sanders, eds. *Theoretical Biology: Epigenetic and Evolutionary Order from Complex Systems*. Baltimore, MD: Johns Hopkins University Press, 1992.

[28] Gould, S. J. *Wonderful Life.* New York: Norton, 1989.

[29] Gould, S. J., and N. Eldredge. "Punctuated Equilibria: The Tempo and Mode of Evolution Reconsidered." *Paleobiology* **3** (1977): 115–251.

[30] Gould, S. J., and E. S. Vrba. "Exaptation: A Missing Term in the Science of Form." *Paleobiology* **8** (1982): 4–15.

[31] Grüner, W., R. Giegerich, D. Strothmann, C. M. Reidys, J. Weber, I. L. Hofacker, P. F. Stadler, and P. Schuster. "Analysis of RNA Sequence Structure Maps by Exhaustive Enumeration: I. Neutral Networks; II. Structure of

Neutral Networks and Shape Space Covering." *Monatsh. Chem.* **127** (1996): 355–389.

[32] Gutell, R., A. Power, G. Z. Hertz, E. J. Putz, and G. D. Stormo. "Identifying Constraints on the Higher-Order Structure of RNA: Continued Development and Application of Comparative Sequence Analysis Methods." *Nucleic Acids Res.* **20** (1993): 5785–5795.

[33] Hanson, J. E., and J. P. Crutchfield. "Computational Mechanics of Cellular Automata: An Example." *Physica D* **103** (1997): 169–189.

[34] Huynen, M. "Exploring Phenotype Space through Neutral Evolution." *J. Mol. Evol.* **43** (1996): 165–169.

[35] Huynen, M., P. F. Stadler, and W. Fontana. "Smoothness within Ruggedness: The Role of Neutrality in Adaptation." *Proc. Natl. Acad. Sci. USA* **93** (1996): 397–401.

[36] Huynen, M. A., T. Doerks, F. Eisenhaber, C. Orengo, S. Sunyaev, Y. Yuan, and P. Bork. "Homology-Based Fold Predictions for *Mycoplasma genitalium* Proteins." *J. Mol. Biol.* **280** (1998): 323–326.

[37] Kauffman, S. A. *Origins of Order: Self-Organization and Selection in Evolution.* New York: Oxford University Press, 1993.

[38] Kauffman, S. A., and S. Levin. "Towards a General Theory of Adaptive Walks in Rugged Fitness Landscapes." *J. Theor. Biol.* **128** (1987): 11–45.

[39] Kimura, M. *The Neutral Theory of Molecular Evolution.* Cambridge: Cambridge University Press, 1983.

[40] Kolmogorov, A. N. "Three Approaches to the Concept of the Amount of Information." *Prob. Info. Trans.* **1** (1965): 4–7.

[41] Landweber, L. F., and I. D. Pokrovskaya, "Emergence of a Dual-Catalytic RNA with Metal-Specific Cleavage and Ligase Activities: The Spandrels of RNA Evolution." *Proc. Natl. Acad. Sci. USA* **96** (1999): 173–178.

[42] Lenski, R. E. "Evolution in Experimental Populations of Bacteria." In *Population Genetics of Bacteria*, edited by S. Baumberg, J. P. W. Young, E. M. H. Wellington, and J. R. Saunders, 193–215. Cambridge, UK: Cambridge University Press, 1974.

[43] Li, M., and P. M. B. Vitanyi. *An Introduction to Kolmogorov Complexity and Its Applications.* New York: Springer-Verlag, 1993.

[44] Macken, C. A., and A. S. Perelson. "Protein Evolution in Rugged Fitness Landscapes." *Proc. Nat. Acad. Sci. USA* **86** (1989): 6191–6195.

[45] Martin-Lof, P. "The Definition of Random Sequences." *Info. Control* **9** (1966): 602–619.

[46] Maynard-Smith, J. *Evolutionary Genetics.* Oxford: Oxford University Press, 1989.

[47] McShea, D. "Complexity and Evolution: What Everybody Knows." *Biol. & Phil.* **6** (1991): 303–324.

[48] McShea, D. "Metazoan Complexity and Evolution: Is There a Trend?" *Evolution* **50** (1996): 477–492.

[49] Mitchell, M., J. P. Crutchfield, and P. T. Hraber. "Evolving Cellular Automata to Perform Computations: Mechanisms and Impediments." *Physica D* **75** (1994): 361–391.
[50] Mitchell, M., J. H. Holland, and S. Forrest. "When Will a Genetic Algorithm Outperform Hillclimbing?" In *Advances in Neural Information Processing Systems*, edited by J. D. Cowan, G. Tesauro, and J. Alspector, vol. 6, 51–56. San Mateo, CA: Morgan Kauffman, 1993.
[51] Monod, J. *Chance and Necessity: An Essay on the Natural Philosophy of Modern Biology.* New York: Vintage Books, 1971.
[52] Newman, M., and R. Engelhardt, "Effect of Neutral Selection on the Evolution of Molecular Species." *Proc. Roy. Soc. Lond. B.* **256** (1998): 1333–1338.
[53] Ohta, T., and J. Gillespie. "Development of Neutral and Nearly Neutral Theories." *Theor. Pop. Biol.* **49** (1996): 128–142.
[54] Papadopoulos, D., D. Schneider, J. Meier-Eiss, W. Arber, R. E. Lenski, and M. Blot. "Genomic Evolution during a 10,000-Generation Experiment with Bacteria." *Proc. Natl. Acad. Sci. USA* **96** (1999): 3807–3812.
[55] Plischke, M., and B. Bergensen. *Equilibrium Statistical Physics.* Englewood Cliffs, NJ: Prentice Hall, 1989.
[56] Ray, T. S. "An Approach to the Synthesis of Life." In *Artificial Life II*, edited by C. G. Langton, C. Taylor, J. D. Farmer, and S. Rasmussen, 371–408. Santa Fe Institute Studies in the Sciences of Complexity, Proc. Vol. X. Reading, MA: Addison-Wesley, 1992.
[57] Robertson, H. S. *Statistical Thermophysics.* Prentice Hall, 1993.
[58] Shalizi, C. R., and J. P. Crutchfield. "Computational Mechanics: Pattern and Prediction, Structure and Simplicity." *J. Stat. Phys.* (2001): in press.
[59] Shannon, C. E., and W. Weaver. *The Mathematical Theory of Communication.* Champaign-Urbana: University of Illinois Press, 1962.
[60] Thompson, D. W. *On Growth and Form.* Cambridge: Cambridge University Press, 1917.
[61] Upper, D. R. "Theory and Algorithms for Hidden Markov Models and Generalized Hidden Markov Models." Ph.D. Thesis, University of California, Berkeley, 1997. Published by University Microfilms Intl, Ann Arbor, Michigan.
[62] van Nimwegen, E. "Statistical Dynamics of Epochal Evolution." Ph.D. thesis, Universiteit Utrecht, The Netherlands, 1999.
[63] van Nimwegen, E., and J. P. Crutchfield. "Metastable Evolutionary Dynamics: Crossing Fitness Barriers or Escaping via Neutral Paths?" *Bull. Math. Biol.* **62** (2000): 799–848.
[64] van Nimwegen, E., and J. P. Crutchfield. "Optimizing Epochal Evolutionary Search: Population-Size Dependent Theory." *Machine Learning* (2001): in press.
[65] van Nimwegen, E., J. P. Crutchfield, and M. Mitchell. "Finite Populations Induce Metastability in Evolutionary Search." *Phys. Lett. A* **229** (1997): 144–150.

[66] van Nimwegen, E., J. P. Crutchfield, and M. Mitchell. "Statistical Dynamics of the Royal Road Genetic Algorithm." *Theor. Comp. Sci.* **229** (1999): 41–102. Special issue on Evolutionary Computation, A. Eiben and G. Rudolph, editors.

[67] Waddington, C. H. *The Strategy of the Genes.* London: Allen and Unwin, 1957.

[68] Williams, G. C. *Natural Selection: Domains, Levels, and Challenges.* New York: Oxford University Press, 1992.

[69] Wright, M. C., and G. F. Joyce. "Continuous *in vitro* Evolution of Catalytic Function." *Science* **276** (1997): 614–617.

[70] Wright, S. "Character Change, Speciation, and the Higher Taxa." *Evolution* **36** (1982): 427–443.

[71] Yeomans, J. M. *Statistical Mechanics of Phase Transitions.* Oxford: Clarendon Press, 1992.

Evolution and Speciation in a Hyperspace: The Roles of Neutrality, Selection, Mutation, and Random Drift

Sergey Gavrilets

The world as we perceive it is three dimensional. Physicists currently believe one needs on the order of a dozen dimensions to explain the physical world. However, biological evolution occurs in a space with millions of dimensions. Sewall Wright's powerful metaphor of rugged adaptive landscapes with its emphasis on adaptive peaks and valleys is based on analogies coming from our three-dimensional experience. Because the properties of multidimensional adaptive landscapes are very different from those of low dimension, for many biological questions Wright's metaphor is not useful or is even misleading. A new unifying framework that provides a plausible multidimensional alternative to the conventional view of rugged adaptive landscapes is emerging for deepening our understanding of evolution and speciation. This framework focuses on percolating (nearly) neutral networks of well-fit genotypes, which appear to be a common feature of genotype spaces of high dimensionality. A variety of important evolutionary questions have been approached using the new framework.

Evolutionary Dynamics, edited by
J. P. Crutchfield and P. Schuster. Oxford University Press.

1 THE PROBLEM OF SPECIATION

Between 1.4 and 1.8 million species have been described [124]. Current estimates are on the order of 10 million species with some estimates going as high as 100 million species [36, 60, 84]. There are 950,000 insect species of which 350,000 are beetles, 230,000 species of flowering plants, 69,000 fungal species, 25,000 bony fishes, 13,000 species of nematodes, 9,000 species of birds, about 4,200 mammal species, 1,814 species of rodents, 986 species of bats, hundreds of endemic species of Hawaiian Drosophila, 300 cichlid species in the Lake Victoria, and 250 species of gammarids in Lake Baikal [35, 129, 151]. It has been argued that the living species represent less than 1% of the number of extinct species [110]. This gives the "average rate of speciation" on the order of three new species per year [123]. (It is interesting that this number is very close to the rate of one new species per year estimated by Lyell in 1832; cited in Bennett [16].) Of course, any "average" rates of speciation are somewhat misleading for speciation takes place simultaneously in many different geographic locations, and its rates vary between different groups of organisms. Table 1 shows rates of genus origination (number of originations/standing diversity per one million years [M.Y.]) in marine animals in the fossil record for three major faunas [123]. As a rule of thumb, species origination rates are one order of magnitude higher than the corresponding genus origination rate. Thus, for example, an "average" species of Paleozoic crinoids produced a new species in 1 M.Y. with a probability of 60%.

The maximum speciation rates known are much higher. These are the rates of speciation in lakes and on islands. The following are some examples compiled by McCune [90]. The Hawaiian Islands have existed for about 5.6 M.Y. During this time, a large number of endemic species have originated there: 250 species of crickets, 860 species of drosophilids, 47 species of honeycreepers, 100 species of spiders, and 40 species of plant bugs. Fourteen endemic species of finches are known on Galapagos Islands, which have existed for 5–9 M.Y. Other examples are for fishes in lakes. Six species of semionotids originated in Lake P4, Newark Basin, in 5,000–8,000 years; five species of cyprinodontids in Lake Chichancanab in 8,000 years; and 22 species of cyprinodontids in Lake Titicaca in 20,000–150,000 years. Speciation of cichlids in Great African Lakes has been extremely rapid: five species in Lake Nabugabo in 4,000 years, 400 species in Lake Malawi in .7–2.0 M.Y., dozens of species in Lake Tanganyika in 1.2–12 M.Y., 11 species in Lake Borombi Mbo in 1.0–1.1 M.Y., and arguably the most spectacular speciation event known—speciation of 300 cichlid species in Lake Victoria in 12,000 years.

Speciation is a universal biological phenomenon that can be very rapid. Speciation has traditionally been considered to be one of the most important and intriguing processes of evolution. (Recall that Darwin's book [32] title begins with "The origin of species....") In spite of this consensus and significant advances in both experimental and theoretical studies of evolution, understanding speciation still remains a major challenge [27, 30, 62, 89, 126, 127]. A major reason for this situation is the ineffectiveness of direct experimental approaches

TABLE 1 Rates of genus origination (number of originations/standing diversity per 1 M.Y.) in marine animals in the fossil record (after Sepkoski [123]).

Fauna	Taxonomic class	Rate of origination
Cambrian	Trilobites	0.13
	"Monoplacophorans"	0.12
	Hyoliths	0.07
	"Inarticulates"	0.06
Paleozoic	Cephalopods	0.11
	Articulates	0.06
	Crinoids	0.06
	Corals	0.05
	Ostracodes	0.04
	Stenolaemates	0.03
Modern	Echinoids	0.03
	Crustaceans	0.03
	Foraminfera	0.03
	Gastropods	0.03
	Bivalves	0.03

because of the time scale involved. Experimental work necessarily concentrates on distinct parts of the process of speciation intensifying and simplifying the factors under study [116, 128]. In situations where direct experimental studies are difficult or impossible, mathematical modeling has proved to be indispensable for providing a unifying framework. Although numerous attempts to model parts of the process of speciation have been made, a quantitative theory of the dynamics of speciation is still missing. Currently, verbal theories of speciation are far more advanced than mathematical foundations.

2 RUGGED ADAPTIVE LANDSCAPES

Speciation is an extremely complex process influenced by a large number of genetical, ecological, environmental, developmental, and other factors. When one is trying to understand a very complex phenomenon, it is very helpful to have a simple model of this phenomenon. A minimal model for discussing evolution and speciation considers an organism as a sequence of genes that has some probability to survive to the age of reproduction. An individual's genes and the probability of survival are referred to as its genotype and fitness, respectively. The set of all possible genotypes is referred to as genotype space. (Wright referred to the "field of possible gene combinations.") Genotype space can be mathematically represented by the vertices of a (generalized) hypercube or an undirected graph [51, 111, 112]. It is useful to visualize each individual as a point in this genotype space. Accordingly, a population will be a cloud of points, and different populations (or species) will be represented by different clouds. Selection, mutation, recombination, random drift, and other factors change the size, location, and structure of these clouds.

The relationship between genotype and fitness is one of the most important factors in determining the evolutionary dynamics of populations. This relationship can be visualized using the metaphor of the adaptive landscape [147]. In what follows an adaptive landscape represents fitness as a function defined on the genotype space. To construct an adaptive landscape one assigns "fitness" to each genotype (or each pair of genotypes) in genotype space. Different forms of selection and reproductive isolation can be treated within this conceptual framework. For example, fitness can be a genotype's viability (in the case of viability selection); it can be fertility or the probability of successful mating between a pair of genotypes (in the case of fertility selection or premating isolation, respectively). Following Wright, adaptive landscapes are usually imagined as "rugged" surfaces having many local "adaptive peaks" of different height separated by "adaptive valleys" of different depth. Adaptive peaks are interpreted as different (potential) species, adaptive valleys between them are interpreted as unfit hybrids [9]; adaptive evolution is considered as local "hill climbing" [72], and speciation is imagined as a "peak shift" [147].

However, there are problems with this description and several of its implicit assumptions can be questioned. For instance, do different species really have different fitnesses (cf. Jongeling [70])? Are small differences in fitness important in speciation? (Note that Wright himself believed that "the principal evolutionary mechanism in the origin of species must...be an essentially nonadaptive one," [147, p. 364].) Are local peaks attainable given mutation and recombination, which destroy "good" combinations of genes, and finite population size, which results in the confinement of the population to an infinitesimal portion of the genotype space? Does formation of a new species always imply a (temporary) reduction in fitness while traveling through a valley? It does not seem that there are compelling reasons for positive answers to these questions. Finally, there is a fundamental problem, realized already by Wright himself: how can a population evolve from one local peak to another across an adaptive valley when selection opposes any changes away from the current adaptive peak?

A possibility of escaping a local peak that has received the most attention is provided by stochastic fluctuations in the genetic composition of the population (random genetic drift). Random genetic drift is always present if the population size is finite. The following two examples illustrate the difficulties of stochastic transitions in bistable systems arising in modeling evolution on a rugged landscape.

Fixation of an Underdominant Mutation. Let us consider a finite population of diploid organisms where a single diallelic locus controls fitness (viability). Let N be the population size and $w_{\mathbf{AA}} = 1, w_{\mathbf{Aa}} = 1 - s, w_{\mathbf{aa}} = 1$ be the fitnesses of genotypes **AA, Aa**, and **aa**, respectively (with $s > 0$). An adaptive landscape corresponding to this model has two "peaks" represented by the homozygous genotypes **AA** and **aa** which are separated by a "valley" represented by the heterozygous genotype **Aa**. Assume that initially all N organisms are homozygotes **AA**, and consider the fate of a single allele **a** introduced in the population by mutation. If the new allele is neutral, that is if $s = 0$, the probability of its eventual fixation is $1/(2N)$ [73]. Lande [75] has shown that if $s > 0$, the probability that allele **a** will be fixed in the population (and, thus, the population will shift to a new peak) is approximately

$$U = \frac{e^{-Ns}\sqrt{4Ns/\pi}}{erf(\sqrt{Ns})}$$

times smaller than the probability of fixation of a neutral allele. Here $erf(x)$ is the error function. Some numerical examples: if $Ns = 5$, then $U \sim 0.017$; if $Ns = 10$, then $U \sim 10^{-4}$; and if $Ns = 20$, then $U \sim 10^{-8}$. This shows that if the population is at least moderately large ($N > 200$) and the adaptive valley is at least moderately deep ($s > 0.1$), the probability of a stochastic transition across the valley is extremely small.

Shift between Stable Phenotypic States. In the model considered above there is a single major locus controlling fitness. However, the majority of traits affecting fitness are controlled by many loci with small effects. Such traits can be modeled using the standard framework of quantitative genetics. Let us consider a finite population of diploid organisms where a single additive quantitative trait z controls fitness. Assume that the distribution of the trait in the population is normal with a constant variance G. Let the fitness function $w(z)$ have two "peaks" at $z = a$ and $z = b$ with a "valley" between them at $x = \nu$. (For example, one can chose $w(z) = \exp(-(z-a)^2/V) + \exp(-(z-b)^2/V)$with a sufficiently small V.) Assuming that initially the population is at one peak, the expected time until the peak shift is approximately

$$T = \frac{2\pi}{G}\,(-c_a c_\nu)^{-1/2}\left[\frac{\overline{w}(a)}{\overline{w}(\nu)}\right]^{2N}$$

where $\overline{w}(x)$ and c_x are the average fitness of the population and the curvature of the fitness function at $z = x$, and N is the population size [10, 76]. For example [76], if the initial adaptive peak is 1.05 times higher than the valley (that is, $\overline{w}(a)/\overline{w}(\nu) = 1.05$), then, using realistic values of other parameters, if $N = 100$, then $T \sim 10^6$; whereas if $N = 200$, then $T \sim 10^{10} - 10^{11}$.

These two examples show that although stochastic transitions across very shallow valleys may sometimes occur, it is highly improbable that they can be a major mechanism of genetic divergence of populations on a large scale. This is especially so if the population size is larger than a few hundred individuals, and if the valley is sufficiently deep (that is, if the stochastic transitions are to result in significant reproductive isolation). Natural populations are usually much larger than few hundred individuals, and reproductive isolation, even between closely related species, tends to be the rule.

Shifting-Balance Theory. To solve the problem of stochastic transitions between different adaptive peaks, Wright [146, 148] proposed a shifting-balance theory. He considered populations to be subdivided into a large number of small subpopulations connected by migration. Because local subpopulations are small and there are many of them, there is a nonnegligible probability of a stochastic peak shift in at least some of them. Wright reasoned that having been established in a subpopulation, a new adaptive combination of genes can take over the whole system as a result of differential migration. Wright's argument was mainly verbal. Recent formal analyses of different versions of the shifting-balance theory [11, 17, 42, 75, 77] have lead to the conclusion that although the mechanisms underlying this theory can, in principle, work, the conditions are rather strict. The main problem is the third phase of the shifting-balance process—the spread of the new combination of genes from a local subpopulation throughout the whole system [31, 42, 59].

Founder-Effect Speciation. Another possibility of escape from a local adaptive peak is provided by founder-effect speciation [23, 87, 88, 125]. In this scenario, a few individuals found a new population that is geographically isolated from the ancestral species and that expands a new area. Here, a stochastic transition to a new peak happens during a short time interval when the size of the expanding population is still small. An advantageous feature of the founder-effect speciation scenario relative to the shifting balance process is that the new combination of genes does not have to compete with old combinations of genes which outnumber it. The proponents of this scenario proposed only verbal schemes without trying to formalize them. Later, formal analyses of founder-effect speciation on rugged adaptive landscapes using analytical models and numerical simulation have shown, however, that stochastic transitions between peaks after a founder event cannot result in a sufficiently high degree of reproductive isolation with a sufficiently high probability to be a reasonable explanation for speciation [9, 10, 26].

TABLE 2 Gene number (after Bird [19]).

Prokaryotes	1,000–8,000
Eukaryotes (except vertebrates)	7,000–15,000
Vertebrates	50,000–100,000

It appears that the crucial question about the mechanism of stochastic transitions between different adaptive peaks cannot be answered. Something must be wrong. At this moment we would do well to consider why we ask this question in the first place. The question about peak shifts appears to be a very natural one to ask within the realm of Wright's metaphor of rugged adaptive landscapes. However, one has to realize that the metaphors and simple models have defined the questions that we believe should be asked. Perhaps something is wrong with this metaphor. Even at a close inspection, the metaphor of rugged adaptive landscapes has seemed appropriate. Indeed, in our three-dimensional experience there is no way one can get from one peak to another without first descending to some kind of valley between the peaks. But are analogies coming from our three-dimensional experience appropriate for biological evolution?

3 NEARLY NEUTRAL NETWORKS AND HOLEY ADAPTIVE LANDSCAPES

The dimensionality of sequence space can be defined as the number of new sequences one can get from a sequence by changing its single elements. Even the simplest organisms known have on the order of a thousand genes (see table 2) and on the order of a million DNA base pairs. Each of the genes can be at several different states (known as alleles). Thus, the dimensionality of genotype space is at least on the order of thousands. It is on the order of millions if one considers DNA base pairs instead of genes. This results, on the one hand, in an astronomically large number of possible genotypes (or DNA sequences) which is much higher than the number of organisms present at any given time or even cumulatively since the origin of life. On the other hand, the number of different fitness values is limited. For example, if the smallest fitness difference that one can measure (or that is important biologically) is, say, 0.001, then only 1,000 different fitness classes are possible. Even if one wants to create an adaptive landscape for a set of binary sequences in a computer memory using double precision, then the assignment of different numerical fitness values to different sequences is only possible for sequences with the length $L < 64$.

There is an important consequence of this observation. Because of the redundancy in the genotype-fitness map, different genotypes are bound to have very similar (identical from any practical point of view) fitnesses. Unless there is a strongly "nonrandom" assignment of fitnesses (say, all well-fit genotypes are put together in a single "corner" of the genotype space), a possibility exists that well-fit genotypes might form connected clusters (or networks) that might extend to some degree throughout the genotype space. If this were so, populations might evolve along these clusters by single substitutions and diverge genetically without going through any adaptive valleys.

Another consequence of the extremely high dimensionality of the genotype space is the increased importance of chance and contingency in evolutionary dynamics. Because (a) mutation is random (which gene will be altered to which allele is unpredictable), (b) each specific mutation has a very small probability, and (c) the number of genes subject to mutation is very large, the genotypes present will be significantly affected by the random order in which mutations occur. Thus, mutational order represents a major source of stochasticity in evolution in the genotype hyperspace [82, 91, 92]. One should expect that even with identical initial conditions and environmental factors different populations will diverge genetically.

3.1 THE ORIGIN OF THE IDEA

These questions about transitions between adaptive peaks have been discussed in the literature many times. In particular, Dobzhansky [34] pointed out that if there are multiple genes producing isolation, then reproductive isolation between two species evolving from a common ancestor can arise as a by-product of fixing "complementary" genes, none of which has to be deleterious individually. To illustrate this he proposed a simple verbal model of a two-locus two-allele system in which well-fit genotypes formed a chain connecting two reproductively isolated genotypes. Dobzhansky noted that "this scheme may appear fanciful, but it is worth considering further since it is supported by some well-established facts and contradicted by none" (p. 282). Similar schemes were discussed by Bateson (1909, cited in Orr [106]), Muller [93], Maynard Smith [86], Nei [95], and Barton and Charlesworth [10]. Kondrashov and Mina [74] expressed this idea in terms of a "complex system of ridges in a genotype space" and illustrated it graphically (their fig. 2). The discussions of all these authors were restricted to the statement that if a specific kind of genetic architecture exists, then the problem of crossing adaptive valleys is solved. Maynard Smith [85] made one step further by concluding that this kind of architecture must be present: "It follows that if evolution by natural selection is to occur, functional proteins must form a continuous network which can be traversed by unit mutational steps without passing through nonfunctional intermediates" (p. 564).

Recently Maynard Smith's conjecture was put on firmer theoretical grounds. On the one hand, extensive "continuous networks" were discovered in numerical studies of RNA fitness landscapes [38, 57, 58, 68, 122] and also for protein fitness landscapes [3]. On the other hand, in analytical studies of different general classes of adaptive landscapes the existence of connected networks of well-fit genotypes has been shown to be inevitable under fairly general conditions [51, 111, 112, 113].

A few words about the terminology. In what follows, a *neutral network* is a contiguous set of sequences possessing the same fitness. This definition is in accord with that used in Barnett [5], Huynen et al. [68], Newman and Engelhardt [97], and is synonymous with "continuous network" used in Maynard Smith [85], "networks of neutral paths" used in Schuster et al. [122], "neutral nets" used in Huynen [66] and with "connected components" used in Gavrilets and Gravner [51]. This definition appears to be preferential to the broader definition of a neutral network that somewhat confusingly did *not* assume the connectivity [112, 114, 121]. A *nearly neutral network* is a contiguous set of sequences possessing approximately the same fitness. A *holey adaptive landscape* is an adaptive landscape where relatively infrequent well-fit (or as Wright put it, "harmonious") genotypes form a contiguous set that expands ("percolates") throughout the genotype space [51]. (An appropriate three-dimensional image of such an adaptive landscape is a flat surface with many holes representing genotypes that do not belong to the percolating set.)

3.2 SIMPLE MODELS

In this section I illustrate the origin of connected networks of well-fit genotypes in some simple models.

3.2.1 Russian Roulette Model. Let us assume that an individual's genotype can be completely specified by a binary sequence of length L. (Using the population genetics terminology, we consider haploid individuals different with respect to L diallelic loci.) In this case, the genotype space is equivalent to a binary hypercube. Let us consider a family of adaptive landscapes arising if genotype fitnesses are generated randomly and independently and are only equal to 1 (viable genotype) or zero (inviable genotype) with probabilities p and $1 - p$, respectively. (Here, one might think of the set of all possible genotypes playing one round of Russian roulette with p being the probability to get a blank.) The probability p can be interpreted as the probability to get a viable genotype after combining genes randomly. Thus, from biological considerations p is supposed to be rather small (definitely much smaller than that one in the nongenetic version of the Russian roulette). The number of loci, L, is very large. A counterintuitive feature of this model is that viable genotypes form neutral networks in the genotype space such that members of a neutral network can be connected by a chain of viable single-gene substitutions. Properties of these networks can be identified using methods from percolation theory and random graph theory [4, 51, 111, 112]. In general, there are two qualitatively different regimes: subcritical, which takes place when $p < p_c$, and supercritical, which takes place when $p > p_c$, where the critical value p_c, which is known as the percolation threshold, is approximately $1/L$. At the boundary of these two regimes, all properties of adaptive landscapes undergo dramatic changes, a physical analogy of which is a phase transition. In the subcritical regime there are many small networks; whereas, in the supercritical regime there is a single "giant component" that includes a significant part of all viable genotypes and "percolates" through the whole genotype space. The adaptive landscape corresponding to the supercritical regime is "holey." Biologically this means that there is a possibility for substantial evolution by fixing single mutations without crossing any adaptive valleys. In the subcritical regime, typical members of a network can be connected by a single sequence of viable genotypes. Thus, there is a single possible "evolutionary path." In contrast, in the supercritical regime, typical members of the percolating neutral network can be connected by many different evolutionary paths. It is very easy to see why the percolation threshold p_c in this model should be approximately equal to $1/L$. Indeed, if the number of loci L is very large and $p > 1/L$, then each viable genotype will have at least several viable neighbors (that is, genotypes different in a single gene) and the network of viable genotypes will "percolate." The percolation threshold decreases if there are more than two alleles. With k alleles at each locus, $P_c \approx 1/(L(k-1))$. If one allows for mutational events affecting more than one gene simultaneously, the percolation threshold decreases dramatically. For example, if a genotype cluster is defined as a set of genotypes that can be connected by a chain of viable single- or two-gene substitutions, the percolation threshold becomes equal to $p_c = 2/L^2$. In general, in high-dimensional genotype spaces (that is, when $L * (k-1)$ is large), even very small values of p will result in the existence of a percolating neutral network.

Different properties of a diploid version of this model are discussed in Gavrilets and Gravner [51]. Note that instead of assigning fitnesses 0 and 1, one can assign fitnesses 1 and $\sigma > 1$. In this case, the sequences with superior fitness σ will form a percolating neutral network if $p > 1/L$. Reidys et al. [114] studied the error threshold for a molecular quasi species evolving on a holey adaptive landscape arising in this model.

3.2.2 Uniformly Rugged Landscape. The assumption that fitness can only take two values might be viewed as a serious limitation. Here, I consider the same genotype space as in the previous section, but now I assume that genotype fitness, w, is a realization of a random variable having a uniform distribution between 0 and 1 [51]. The adaptive landscape arising in this model will be called a "uniformly rugged landscape." Let us introduce two threshold values, w_1 and w_2 such that $w_2 - w_1 = p > 0$, and let us say that a genotype belongs to a (w_1, w_2)-fitness band if its fitness w satisfies $w_1 < w \leq w_2$. According to the results from the previous section if $p > 1/L$, there is a percolating nearly neutral network of genotypes in a (w_1, w_2)-fitness band. The members of this network can be connected by a chain of single-gene substitutions resulting in genotypes that also belong to the network. If one chooses $w_2 = 1$ and $w_1 = 1 - p$, it follows that uniformly rugged landscapes have very high "ridges" (with genotype fitnesses between $1 - p$ and 1) that continuously extend throughout the genotype space. In a similar way, if one chooses $w_2 = p$ and $w_1 = 0$, it follows that uniformly rugged landscapes have very deep "gorges" (with genotype fitnesses between 0 and p) that also continuously extend throughout the genotype space. If p is small, the fitnesses of the genotypes in the (w_1, w_2)-fitness band will be very similar. Thus, with large L extensive evolutionary changes can occur in a nearly neutral fashion via single substitutions along the corresponding nearly neutral network of genotypes belonging to a percolating cluster. The maximum number of the nonoverlapping (w_1, w_2)-fitness bands is $1/p$, which with p just above the percolation threshold is about L. Thus, the maximum number of nonoverlapping percolating near-neutral networks of genotypes is L. In this model, there is a percolating network of well-fit genotypes and, thus, the corresponding adaptive landscape is holey.

3.2.3 Multiplicative Fitnesses. In a commonly used multiplicative fitness model, alternative alleles are interpreted as "advantageous" and "disadvantageous," and the fitness of an individual with k disadvantageous alleles is chosen to be $(1-s)^k$ with $s > 0$. Here, the fitness landscape has a single peak and $L+1$ different fitness values. Any two genotypes from the same fitness level can be connected by a chain of single substitutions leading not farther than the previous or the next fitness level. Thus, the number of distinct nearly neutral networks in this model is approximately $L/2$. These networks can be imagined as spherical shells in genotype space at a constant mean Hamming distance from the optimum genotype. In contrast to the previous model, in the multiplicative fitness model different networks have different sizes, with the diameter of the network (the maximum Hamming distance between its members) decreasing from L for genotypes with an equal number of advantageous and disadvantageous alleles to 0 for the most fit genotypes. Woodcock and Higgs [145] have studied this model in detail and shown that a finite population subject to mutation reaches a state of stochastic equilibrium staying close to the fitness level corresponding to U/s disadvantageous alleles. Here U is the rate of mutation per sequence. The whole of the population is clustered together in a particular region of genotype space wandering randomly through the corresponding nearly neutral network. Gavrilets and Gravner [51] have considered a diploid version of this model in that each genotype is assigned fitness 1 with a probability proportional to $p = (1-s)^k$ where k is the number of heterozygous loci.

3.2.4 Stabilizing Selection on an Additive Trait. A common model in evolutionary quantitative genetics is that of stabilizing selection on a trait z determined by the sum of effects of L diallelic loci, $z = \sum \alpha_i l_i$ where α_i is the contribution of the ith gene to the trait and $l_i = 0$ or 1 for $i = 1, \ldots, L$. The term "stabilizing selection" means that individual fitness $w(z)$ decreases with the deviation of the trait value from some optimum value θ. Assuming for simplicity that $\alpha_i = 1$ for all i and that optimum θ is at the mid value of the trait range ($\theta = L/2$) results in a single-peak fitness landscape with $L/2$ different fitness values. As in the multiplicative fitness model, any two genotypes at the same fitness level can be connected by a chain of single substitutions leading not farther than the previous or the next fitness level. Thus, the number of distinct nearly neutral networks in this model is approximately $L/4$. In contrast to the multiplicative model where the most-fit nearly neutral network has the smallest diameter, in the present model this network percolates (and has the largest diameter L) and, thus, the corresponding adaptive landscape is holey. This means that extensive nearly neutral divergence is possible under stabilizing selection. Barton [8] and Mani and Clarke [82] studied the divergence in this model in detail.

3.2.5 NK Model. Stabilizing (or any other nonlinear) selection on an additive trait z results in epistatic interactions between effects of different loci on fitness. The order of these interactions depends on the degree of nonlinearity of the fitness function $w(x)$ but each locus epistatically interacts with all other $L-1$ loci. For example, under quadratic stabilizing selection (that is, with $w(z) = 1 - sz^2$), there are pairwise additive-by-additive epistatic interactions in fitness between all $L(L-1)/2$ pairs of loci [50].

A structurally different class of epistatic model is a family of the so-called NK models [71] where each locus interacts only with a specified number K of other loci in such a way that interactions of all possible orders (from the second through the Kth order) are present. The existence of neutral and nearly neutral networks percolating through the genotype space in the NK models was demonstrated and considered in detail [5, 97]. Barnett [5] and Ohta [103, 104] have numerically studied the patterns of population evolution in these models.

3.2.6 Conclusions from Models. The existence of chains of well-fit genotypes that connect reproductively isolated genotypes was postulated by Dobzhansky and other earlier workers. In contrast, the models just described show it to be inevitable under broad conditions. The existence of percolating nearly neutral networks of well-fit genotypes which allow for "nearly neutral" divergence appears to be a general property of adaptive landscapes with a very large number of dimensions. Do existing experimental data substantiate this theoretical claim?

3.3 EXPERIMENTAL EVIDENCE

Although none of the examples listed below can be viewed as irrefutable evidence by itself, viewed as a whole these examples provide substantial and credible evidence that the genetic architecture leading to the extended nearly neutral networks of well-fit genotypes is widespread.

(i) The most straightforward approach is to analyze the relationships between genotype and fitness [143]. Results of many studies of epistatic interactions in plants, *Drosophila*, mammals, and moths [105, 138, 150] imply the existence of chains of well-fit genotypes connecting genotypes that are reproductively isolated to some degree.

(ii) "Ring species" can probably be considered as one of the best manifestations of holey adaptive landscapes. A ring species is a chain of "races" (or subspecies) with gradual transitions and no reproductive isolation between adjacent geographic races but abrupt changes and reproductive isolation where the terminal races come into contact. Nine cases of ring species were described in Mayr [87] and more than a dozen additional cases were documented in

Mayr [88]. In ring species, chains of genotypes connecting reproductively isolated forms are recreated in a natural way.

(iii) Strong artificial selection in a specific direction usually results in the desired response, but as a consequence of the genetic changes brought about by selection, different components of fitness (such as viability or fertility) significantly decrease [65]. Moreover, after relaxing selection, natural selection usually tends to return the population to its original state. These observations stimulated Wright's view of species as occupying isolated "peaks" in an adaptive landscape [109]. However, the size of the experimental populations under selection is usually very small—on the order of few dozen individuals. Small populations are characterized by low levels of genetic variation and may not "find" ridges in the adaptive landscape that may exist in a large population. Weber [142] performed selection experiments using very large populations of *Drosophila melanogaster*, thousands of individuals, selecting for the ability to fly in a wind tunnel. Weber was able to change the selected trait (which is obviously nonneutral) by many standard deviation but did not observe any significant reduction in fitness components nor any tendencies to return to the original state after selection was relaxed. A straightforward interpretation of Weber's results is that the large population was able to find a ridge of well-fit genotypes in the adaptive landscape.

(iv) Extensive natural hybridization in animals and plants [2, 21, 117] represents something that is very difficult to reconcile with isolated peaks but is well expected if there are ridges in the genotype space.

(v) The existence of fit intermediates between radically different morphologies has been observed in the fossil record [24].

(vi) The analyses of RNA sequences and secondary structures have provided abundant evidence for the existence of neutral networks [38, 40, 57, 58, 66, 68, 111, 112, 120, 122].

(v) Empirical evidence for extensive functional neutrality in protein space is presented by Martinez with co-workers [83].

(vi) Some additional evidence coming from the properties of hybrid zones and patterns of molecular evolution will be considered below.

3.4 A METAPHOR OF HOLEY ADAPTIVE LANDSCAPES

Wright's metaphor of rugged adaptive landscapes puts special emphasis on adaptive peaks and valleys. This metaphor is very useful for thinking about adaptation and optimization. However, its utility for understanding perpetual genetic divergence and speciation is questionable. Though I overstate the case to make the point, peaks are largely irrelevant because populations are never able to climb them (cf. Ohta [103]), and valleys are largely irrelevant because selection quickly moves populations out of them. A finite population subject to mutation is likely to stay within a fitness band determined by the balance of mutation, selection, and random drift. Under very general conditions, genotypes with fitnesses within this band form a connected network. Better understanding of the processes of genetic divergence and speciation can be achieved by focusing on these nearly neutral networks of well-fit genotypes, which are expected to extend throughout genotype space under fairly general conditions. A simplified view of adaptive landscapes that puts special emphasis on these networks is provided by the metaphor of holey adaptive landscapes [43, 51]. This metaphor disregards fitness differences between different genotypes belonging to the network of well-fit genotypes and treats all other genotypes as "holes." The justification for the latter is a belief that selection and recombination will be effective in moving the population away from these areas of genotype space on a time scale that is much faster than the time scale for speciation. Accordingly, microevolution and local adaptation can be viewed as the climbing of the population out of a "hole" in to the holey adaptive landscape; whereas, macroevolution can be viewed as a movement of the population along the holey landscape with speciation taking place when the diverging populations come to be on opposite sides of a "hole" in the adaptive landscape. In this scenario, there is no need to cross any "adaptive valleys"; reproductive isolation between populations evolves as an inevitable side effect of accumulating different mutations. As Charlesworth [25] put it, "the loss in fitness to species hybrids is no more surprising than the fact that a carburetor from a car manufactured in the USA does not function in an engine made in Japan" (p. 103).

4 APPLICATIONS

Simple models and metaphors train our intuition about complex phenomena, provide a framework for studying such phenomena, and help identify key components in complex systems. Next, I briefly review some important biological problems and processes that have been (or can be) studied using approaches focusing on (nearly) neutral networks and holey adaptive landscapes.

4.1 GENETIC DIVERGENCE AND MOLECULAR EVOLUTION

One of the consequences of the existence of the percolating nearly neutral networks of well-fit genotypes is the expectation that biological populations will evolve (and diverge) staying mainly within these networks. The metaphor of "holey" adaptive landscapes neglects the fitness differences between genotypes in the network but these differences are expected to exist and be apparent on a finer scale. If one applies a finer resolution, the movement along the network will be accompanied by slight changes in fitness. Evolution will proceed by fixation of weakly selected alleles, which can be advantageous, deleterious, over- and underdominant, or apparently neutral, depending on the specific area of genotype space the population passes through. Smaller populations will pass relatively quickly through the areas of genotype space corresponding to fixation of slightly deleterious mutations; whereas, larger populations will pass relatively quickly through the areas corresponding to fixation of (compensatory) slightly advantageous mutations. These patterns of molecular evolution and genetic divergence, as expected from the general properties of multidimensional adaptive landscapes, are similar to the patterns revealed by the methods of experimental molecular biology [80], which form the empirical basis for the nearly neutral theory of molecular evolution (Ohta [99, 102]; see Gillespie [55] for an alternative interpretation of the data).

4.2 SPECIATION

A classical view of speciation is that reproductive isolation arises as a by-product of genetic divergence [34, 147]. Models incorporating holey adaptive landscapes provide a way to evaluate whether the mechanisms implied by this view may result in (rapid) speciation. These models also train our intuitions of the speciation process.

Nei [95], Wills [144], and Bengtsson and Christiansen [15] initiated formal analyses of the Dobzhansky model. Nei and co-authors [96] studied one- and two-locus multiallele models with stepwise mutations and considered both postmating and premating reproductive isolation. Genotypes were reproductively isolated if they were different by more than one or two mutational steps. In their model, speciation was very slow. They conjectured, however, that increasing the number of loci may significantly increase the rate of speciation. Wagner and co-authors [138] considered a two-locus, two-allele model of stabilizing selection acting on an epistatic character. For a specific set of parameters, the interaction of epistasis in the trait and stabilizing selection on the trait resulted in a fitness "ridge." The existence of this ridge simplified stochastic transitions between alternative equilibria. Gavrilets and Hastings [53] formulated a series of two- and three-locus Dobzhansky-type viability selection models as well as models for selection on polygenic characters. They studied these models in the context of founder-effect speciation and noticed that the existence of ridges in the adaptive landscape made stochastic divergence much more plausible. In these models, the resulting reproductive isolation can be very high and can evolve with a high probability on the time scale of dozens or hundreds of generations. For appropriate parameter values, Gavrilets and Hastings' results have demonstrated that founder-effect speciation is plausible. Similar conclusions were reached by Gavrilets and Boake [48], who studied the effects of premating reproductive isolation on the plausibility of founder-effect speciation. The adaptive landscape considered by Gavrilets and Boake was defined for *pairs* of genotypes. They have demonstrated that after a founder event a new adaptive combination of genes may rise to high frequencies in the presence of an old combination of genes that is sympatrical (cf. Wu [149]).

The models just discussed were formulated for a small number of loci (or quantitative traits). Higgs and Derrida [63, 64] proposed a model with an infinitely large number of unlinked and highly mutable loci. In their model the probability of mating between two haploid individuals is a decreasing function of the proportion of loci at which they are different. Here, any two sufficiently different genotypes can be visualized as sitting on opposite sides of a hole in a holey adaptive landscape, which is defined for pairs of individuals. Higgs and Derrida, as well as Manzo and Peliti [81], studied this model numerically, assuming that mating is preferential. In their models populations undergo a continuous process of splitting into reproductively isolated groups with subsequent extinction and/or hybridization and loss of reproductive isolation. Models where the probability of mating between a pair of haploid individuals are mathematically equivalent to models where fitness (viability) of a diploid individual depends on its heterozygosity (that is, the number of heterozygous loci). Orr [105, 107] studied possibilities for allopatric speciation in a series of such models for the diploid case. Gavrilets with co-workers [54] have performed individual-based simulations to evaluate whether rapid parapatric speciation is possible if the only sources of genetic divergence are mutation and random genetic drift. Distinctive features of their simulations are the consideration of the complete process of speciation (from initiation until completion), and of a large number of loci, which was only one order of magnitude smaller than that of bacteria. To reflect the idea that reproductive isolation arises simultaneously with genetic divergence, it was posited that an encounter of two haploid individuals can result in mating and viable and fecund offspring only if the individuals are different in no more than K loci (cf., Higgs and Derrida [63, 64]). Otherwise the individuals do not mate (premating reproductive isolation) or their offspring is inviable or sterile (postmating reproductive isolation). In contrast to Higgs and Derrida approach, the encounters of individuals were random and mutation rates were more realistic (much lower). As a consequence, speciation was irreversible. These numerical results demonstrated that rapid speciation on the time scale of hundreds of generations is plausible without the need for extreme founder events or complete geographic isolation. Selection for local adaptation is not necessary for speciation (cf., Rice [115] and Schluter [118]). The plausibility of speciation is enhanced by population subdivision. Simultaneous emergence of more than two new species from a subdivided population is highly probable. Gavrilets [47] developed some analytical approximations for the dynamics observed in the numerical simulations.

4.3 ADAPTATION

Extended (nearly) neutral networks are important in adaptation for they can be "used" by a population to find areas in genotype space with higher fitness values [121]. Numerical and analytical results for RNA fitness landscapes [39, 40, 66, 68], for a single peak fitness landscape arising in the "Royal Road" genetic algorithm [130, 131, 132, 133], and for multipeak fitness landscapes of the NK model [97] show that evolutionary dynamics of adaptation proceed in a steplike fashion where short periods of jumps to a higher fitness level are interrupted by extended periods during which populations diffuse along neutral networks. Similar behavior has been observed in experiments [79] with bacteria populations adapting to a new environment. Thus, understanding evolution along nearly neutral networks (and on holey adaptive landscapes) may increase our understanding of local adaptation and microevolution.

4.4 HYBRID ZONES

Hybrid zone is a geographic region where genetically distinct populations meet and interbreed to some extent, resulting in some individuals of mixed ancestry. Analysis of hybrid zones provides insights into the nature of species, the strength and mode of natural selection, the genetic architecture of species differences, and the dynamics of the speciation process [11, 12, 61]. Many hybrid zones exhibit a gradual change ("cline") in a character or in allele frequency along a geographic transect. Theoretical studies of hybrid zones concentrate on the form of clines and the ability of genes to penetrate hybrid zones [6, 14, 45, 49].

Many hybrid zones are thought to be formed following a secondary contact of different populations, and to be maintained by a balance between selection against hybrids and recombinant phenotypes and dispersal [12]. Gavrilets [44] used the Dobzhansky model to contrast the properties of hybrid zones formed when adaptive peaks are isolated with those formed when adaptive peaks are connected by a chain of well-fit intermediates. A major difference between the two types of hybrid zones is expected to be in the distribution and fitnesses of genotypes in the center of the hybrid zone. If adaptive peaks are isolated, in the center of the hybrid zones besides the high-fitness parental forms, one should observe mainly low-fitness hybrids. Moreover, one expects concordant clines in neutral allele frequencies [7, 98]. With strong Dobzhansky-type epistatic selection, and low rates of migration reproductive isolation between allopatric populations on opposite sides of the hybrid zone will increase with distance between these populations. F_1 hybrids between individuals from allopatric populations on opposite sides of the hybrid zone will have low fitness. These F_1 hybrids will have genotypes that differ from hybrid genotypes common in the center of the hybrid zone, which will have high fitness. In general, clines in the frequencies of neutral marker alleles linked to selected loci will be disjoint and unsymmetric. Concordant clines are expected for neutral alleles unlinked to selected loci.

Hybrid zones with apparently discordant clines and apparently well-fit recombinant genotypes present are known for house mouse, grasshopper, common shrew, burney moth, and field vole [18, 61, 69, 98, 134]. A grasshopper hybrid zone studied by Virdee and Hewitt [134] is especially interesting in this regard. Here, crosses between the two pure taxa (*Chorthippus parallelus parallelus* and *Chorthippus parallelus erythropus*) result in sterile male offspring whereas no such dysfunction has been detected in hybrid males collected through the center of the hybrid zone. Crosses have revealed noncoincident clines for dysfunction near the center of the hybrid zone.

4.5 RNA AND PROTEINS

For RNA sequences neutral networks are defined as contiguous sets of sequences that fold into the same secondary structure. Different biological implications of the existence of neutral networks in both RNA sequence space and protein space are explored in much detail elsewhere [38, 40, 57, 58, 66, 68, 83, 111, 112, 120, 122].

4.6 GENE AND GENOME DUPLICATION

Conrad [29] puts forward an idea of an "extra-dimensional bypass" on adaptive landscapes. According to Conrad, an increase in the dimensionality of an adaptive landscape is expected to transform isolated peaks into saddle points that can be easily escaped resulting in continuing evolution. A straightforward mechanism for increasing the dimensionality of the adaptive landscape is an increase in the size of genome by gene or genome duplication. Similar ideas are discussed by Gordon [56]. The results on the existence of percolating nearly neutral networks of well-fit genotypes reviewed above provide a formal justification of the idea of an "extra-dimensional bypass" (see Gavrilets and Gravner [51]). In general, the percolation threshold decreases with increasing the dimensionality of genotype space which to a large degree is controlled by the genome size. Increasing the latter by gene or genome duplication [101, 100, 141] is expected to result in increasing the connectivity of the networks of well-fit genotypes which in turn will increase the possibilities for evolutionary change. Interestingly, it has already been argued that Cambrian explosion was a result of an increase in the gene number [94]. Increasing the genome size increases the redundancy in proteins and DNA [28, 137] which may facilitate evolution and result in increased canalization of development.

4.7 CANALIZATION OF DEVELOPMENT

There is a general belief that biological systems ought to evolve to a state of greater stability [119, 135, 140]. One usually distinguishes between genetic canalization (insensitivity to mutations) and environmental canalization (insensitivity to environmental variation). Evolution of environmental canalization has been considered elsewhere [52, 140]. The metaphor of holey adaptive landscapes is useful for thinking about the evolution of genetic canalization. From general considerations, one should not expect complete symmetry of "real" adaptive landscapes, which are supposed to have areas varying with respect to the width and concentration of ridges of well-fit genotypes. Numerical simulations show that populations tend to spend more time in areas of high concentration of well-fit genotypes [37, 67, 108]. One of the biological manifestations of this effect will be apparent reduction in the probability of harmful mutations, that is, evolution of genetic canalization [136]. Another manifestation will be a change in the ability of random genetic variation to produce phenotypic changes, that is, evolution of evolvability [1, 139]. There is a controversy regarding effects of recombination and sex on the ability of populations to find areas of high concentration of well-fit genotypes. Peliti and Bastolla [108] and Finjord [37] results suggest that only sexual populations tend to find these areas and hang there. On the other hand, Huynen and Hogeweg [67] observed the same effect in modeling asexual populations.

4.8 MORPHOLOGICAL MACROEVOLUTION

So far the discussion has been limited to genotype space where individuals were represented as sequences of genes or DNA and RNA base pairs. However, the results on the existence of nearly neutral networks and holey adaptive landscapes should be valid for any sequence space of high dimensionality. In particular, instead of sequences of genes one can consider sequences of discrete morphological characters and study morphological evolution.

Empirical studies of long-term morphological evolution are typically based on a large number of discrete characters. A common null model in interpreting patterns of morphological changes observed in the fossil record is random diffusion in morphospace [20]. An implicit assumption of this model is that all possible directions for evolution are equally probable. In terms of adaptive landscapes, this corresponds to a flat landscape of neutral evolution [33, 73]. In general, because of genetic, developmental, or ecological constraints some of the possible character combinations can be prohibited. In this case, the morphospace will be mathematically equivalent to a hypercube with "holes" (with "holes" representing prohibited character combinations) and the corresponding adaptive landscape will be "holey" rather than "flat." If the proportion of holes is not extremely high, "harmonious" character combinations will form a nearly neutral network extending throughout the whole morphospace. A characteristic signature of a random walk on a holey hypercube appears to be a stretched exponential dependence of the overlap between the current and initial positions of the walker on time (e.g., Campbell et al. [22] and Lemke and Campbell [78]). Gavrilets [46] has developed a model describing the dynamics of clade diversification on a morphological hypercube and applied this model to Foote's data [41] on the diversification of blastozoans. The fitting of the stretched exponential curve to blastozoan data has led to inconclusive results: although the fit is good, it is not better than the fit of a simple exponential curve expected for flat landscapes. More detailed data sets and theoretical results on random walks on hypercubes with holes are needed for more precise conclusions.

5 CONCLUSION

Although most attempts to use the approaches discussed here are relatively recent, the list of applications is already impressive. Still we have only started and this list will definitely grow in the near future. Among applications that appear to be especially important are adaptations of the approach for the case of continuous (morphological) hyperspace, incorporation of changes in the adaptive landscapes brought about by biotic and abiotic factors, development of a dynamical theory of random walks on neutral networks, analyses of cluster formation in hyperspaces with emphasis on the origin of hierarchies, and bridging results on the dynamics on neutral networks with the methods for reconstructing phylogenies.

Currently, the mathematics of high-dimensional spaces is very abstract and lacks any real applications. Genotype space arising in evolutionary biology is an example of a hyperspace that is both real and very important. The discovery of complex behaviors of simple ecological models a quarter of a century ago has stimulated the impressive development of the theory of low-dimensional dynamical systems. One can hope that recent advances in evolutionary biology reviewed here will have a similar effect on the development of mathematical theories of the structure of hyperspace and the dynamics it supposes.

ACKNOWLEDGMENTS

I am grateful to Jim Crutchfield and Peter Schuster for inviting me to participate in this workshop and to Mitch Cruzan for helpful comments on the manuscript. Partially supported by National Institutes of Health grant GM56693 and Santa Fe Institute.

REFERENCES

[1] Altenberg, L. "Genome Growth and the Evolution of the Genotype-Phenotype Map." In *Evolution and Biocomputation: Computational Models of Evolution*, edited by W. Banzhaf and F. H. Eeckman, vol. 899, 205–259. Lecture Notes in Computer Sciences. Berlin: Springer-Verlag, 1995.

[2] Arnold, M. L. *Natural Hybridization and Evolution.* Oxford: Oxford University Press, 1997.

[3] Babajide, A., I. L. Hofacker, M. J. Sippl, and P. F. Stadler. "Neutral Networks in Protein Space: A Computational Study Based on Knowledge-Based Potentials of Mean Force." *Fold. & Des.* **2** (1997): 261–269.

[4] Balobás, B. *Random Graphs.* New York: Academic Press, 1985.

[5] Barnett, L. "Tangled Webs: Evolutionary Dynamics on Fitness Landscapes with Neutrality" M.Sc. Thesis, University of Sussex, 1997.

[6] Barton, N. H. "Gene Flow Past a Cline." *Heredity* **43** (1979): 333–339.

[7] Barton, N. H. "Multilocus Clines." *Evolution* **37** (1983): 454–471.

[8] Barton, N. H. "The Divergence of a Polygenic System Subject to Stabilizing Selection, Mutation and Drift." *Genet. Res.* **54** (1989): 59–77.

[9] Barton, N. H. "Founder Effect Speciation." In *Speciation and Its Consequences*, edited by D. Otte and J. A. Endler, 229–256. Sunderland, MA: Sinauer, 1989.

[10] Barton, N. H., and B. Charlesworth. "Genetic Revolutions, Founder Effects, and Speciation." *Ann. Rev. Ecol. & Sys.* **15** (1984): 133–164.

[11] Barton, N. H., and K. S. Gale. "Genetic Analysis of Hybrid Zones." In *Hybrid Zones and the Evolutionary Process*, edited by R. G. Harrison, 13–45. Oxford: Oxford University Press, 1993.

[12] Barton, N. H., and G. M. Hewitt. "Adaptation, Speciation and Hybrid Zones." *Nature* **341** (1989): 497–503.

[13] Barton, N. H., and S. Rouhani. "Adaptation and the Shifting Balance." *Genet. Res.* **61** (1993): 57–74.

[14] Bengtsson, B. O. "The Flow of Genes through a Genetic Barrier." In *Evolution Essays in Honor of John Maynard Smith*, edited by J. J. Greenwood, P. H. Harvey, and M. Slatkin, 31–42. Cambridge: Cambridge University Press, 1985.

[15] Bengtsson, B. O., and F. B. Christiansen. "A Two-Locus Mutation Selection Model and Some of Its Evolutionary Implications." *Theor. Pop. Biol.* **24** (1983): 59–77.

[16] Bennett, K. D. *Evolution and Ecology: The Pace of Life.* Cambridge: Cambridge University Press, 1997.

[17] Bergman, A., D. B. Goldstein, K. E. Holsinger, and M. W. Feldman. "Population Structure, Fitness Surfaces, and Linkage in the Shifting Balance Process." *Genet. Res.* **66** (1995): 85–92.

[18] Bert, T. M., and W. S. Arnold. "An Empirical Test of Predictions of Two Competing Models for the Maintenance and Fate of Hybrid Zones: Both Models are Supported in a Hard-Clam Hybrid Zone." *Evolution* **49** (1995): 276–289.

[19] Bird, A. P. "Gene Number, Noise Reduction and Biological Complexity." *Trends Genet.* **11** (1995): 94–100.

[20] Bookstein, F. L. "Random Walks and the Biometrics of Morphological Characters." *Evol. Biol.* **9** (1988): 369–398.

[21] Bullini, L. "Origin and Evolution of Animal Hybrid Species." *Trends Ecol. & Evol.* **9** (1994): 422–426.

[22] Campbell, I. A., J. M. Flessels, R. Jullien, and R. Botet. "Random Walks on a Hypercube and Spin Glass Relaxation." *J. Phys. C* **20** (1987): L47–L51.

[23] Carson, H. L. "The Population Flush and Its Genetic Consequences." In *Population Biology and Evolution*, edited by R. C. Lewontin, 123–137. Syracuse, NY: Syracuse University Press, 1968.

[24] Carroll, R. L. *Vertebrate Paleontology and Evolution.* New York: Freeman, 1988.

[25] Charlesworth, B. "Speciation." In *Palaeobiology: A Synthesis*, edited by D. E. G. Briggs and P. R Crowther, 100–106. Oxford: Blackwell Scientific Publications, 1990.

[26] Charlesworth, B. "Is Founder-Flush Speciation Defensible?" *Amer. Natur.* **149** (1997): 600–603.

[27] Claridge, M. F., H. A. Dawah, and M. R. Wilson, eds. *Species. The Units of Biodiversity.* London: Chapman and Hall, 1997.

[28] Conrad, M. "The Mutation Buffering Concept of Biomolecular Structure." *J. Biosci.* (Supp.) **8** (1985): 669–679.

[29] Conrad, M. "The Geometry of Evolution." *BioSystems* **24** (1990): 61–81.

[30] Coyne, J. A. "Genetics and Speciation." *Nature* **355** (1992): 511–515.

[31] Coyne, J. A., N. H. Barton, and M. Turelli. "A Critique of Sewall Wright's Shifting Balance Theory of Evolution." *Evolution* **51** (1997): 643–671.

[32] Darwin, C. *The Origin of Species by Means of Natural Selection, or the Preservation of Favoured Races in the Struggle for Life.* New York: Modern Library, 1859.

[33] Derrida, B., and L. Peliti. "Evolution in a Flat Fitness Landscape." *Bull. Math. Biol.* **53** (1991): 355–382.

[34] Dobzhansky, T. H. *Genetics and the Origin of Species.* New York: Columbia University Press, 1937.

[35] Eldredge, N. *Life in the Balance: Humanity and the Biodiversity Crisis.* Princeton: Princeton University Press, 1998.

[36] Ehrlich, P. R., and E. O. Wilson. "Biodiversity Studies: Science and Policy." *Science* **253** (1991): 758–562.

[37] Finjord, J. "Sex and Self-Organization on Rugged Landscapes." *Intl. J. Mod. Phys. C* **7** (1996): 705–715.

[38] Fontana, W., and P. Schuster. "A Computer Model of Evolutionary Optimization." *Biophys. Chem.* **26** (1987) 123–147.

[39] Fontana, W., and P. Schuster. "Continuity in Evolution: On the Nature of Transitions." *Science* **280** (1998): 1451–1455.

[40] Fontana, W., and P. Schuster. "Shaping Space: The Possible and the Attainable in RNA Genotype-Phenotype Mapping." *J. Theor. Biol.* **194** (1998): 491–515.

[41] Foote, M. "Paleozoic Record of Morphological Diversity in Blastozoan Echinoderms." *Proc. Natl. Acad. Sci. USA* **89** (1992): 7325–7329.

[42] Gavrilets, S. "On Phase Three of the Shifting-Balance Theory." *Evolution* **50** (1996): 1034–1041.

[43] Gavrilets, S. "Evolution and Speciation on Holey Adaptive Landscapes." *Trends Ecol. & Evol.* **12** (1997): 307–312.

[44] Gavrilets, S. "Hybrid Zones with Dobzhansky-Type Epistatic Selection." *Evolution* **51** (1997): 1027–1035.

[45] Gavrilets, S. "Single Locus Clines." *Evolution* **51** (1997): 979–983.

[46] Gavrilets, S. "Dynamics of Clade Diversification on the Morphological Hypercube." **266** (1999): 817–824.

[47] Gavrilets, S. "A Dynamical Theory of Speciation on Holey Adaptive Landscapes." *Amer. Natur.* **154** (1999): 1–22.

[48] Gavrilets, S., and C. R. B. Boake. "On the Evolution of Premating Isolation after a Founder Event." *Amer. Natur.* **152** (1998): 706–716.

[49] Gavrilets, S., and M. B. Cruzan. "Neutral Gene Flow across Single Locus Clines." *Evolution* **52** (1998): 1277–1284.

[50] Gavrilets, S., and G. de Jong. "Pleiotropic Models of Polygenic Variation, Stabilizing Selection and Epistasis." *Genetics* **134** (1993): 609–625.

[51] Gavrilets, S., and J. Gravner. "Percolation on the Fitness Hypercube and the Evolution of Reproductive Isolation." *J. Theor. Biol.* **184** (1997): 51–64.

[52] Gavrilets, S., and A. Hastings. "A Quantitative Genetic Model for Selection on Developmental Noise." *Evolution* **48** (1994): 1478–1486.

[53] Gavrilets, S., and A. Hastings. "Founder Effect Speciation: A Theoretical Reassessment." *Amer. Natur.* **147** (1996): 466–491.

[54] Gavrilets, S., H. Li, and M. D. Vose. "Rapid Speciation on Holey Adaptive Landscapes." *Proc. Roy. Soc. Lond. B* **265** (1998): 1483–1489.

[55] Gillespie, J. H. *The Causes of Molecular Evolution.* Oxford: Oxford University Press, 1991.

[56] Gordon, R. "Evolution Escapes Rugged Fitness Landscapes by Gene or Genome Doubling: The Blessing of Higher Dimensionality." *Comp. Chem.* **18** (1994): 325–331.

[57] Grüner, W., R. Giegerich, D. Strothmann, C. Reidys, J. Weber, I. L. Hofacker, P. F. Stadler, and P. Schuster. "Analysis of RNA Sequence Structure Maps by Exhaustive Enumeration: Neutral Networks." *Monatshefte für Chemie* **127** (1996): 355–374.

[58] Grüner, W., R. Giegerich, D. Strothmann, C. Reidys, J. Weber, I. L. Hofacker, P. F. Stadler, and P. Schuster. "Analysis of RNA Sequence Structure Maps by Exhaustive Enumeration: Structure of Neutral Networks and Shape Space Covering." *Monatshefte für Chemie* **127** (1996): 375–389.

[59] Haldane, J. B. S. "Natural Selection." In *Darwin's Biological Work. Some Aspects Reconsidered*, edited by P. R. Bell, 101–149. New York: Wiley, 1959.

[60] Hammond, P. M. "The Current Magnitude of Biodiversity." In *Global Biodiversity Assessment*, edited by V. H. Heywood, 113–138. Cambridge: Cambridge University Press, 1995.

[61] Harrison, R. G. "Hybrid Zones: Windows on Evolutionary Process." In *Oxford Surveys in Evolutionary Biology*, edited by D. Futuyma, and J. Antonovics, vol. 7, 69–128. Oxford: Oxford University Press, 1990.

[62] Harrison, R. G. "Molecular Changes at Speciation." *Ann. Rev. Ecol. & Sys.* **22** (1994): 281–308.

[63] Higgs, P. G., and B. Derrida. "Stochastic Model for Species Formation in Evolving Populations." *J. Phys. A: Math. & Gen.* **24** (1991): L985–L991.

[64] Higgs, P. G., and B. Derrida. "Genetic Distance and Species Formation in Evolving Populations." *J. Mol. Evol.* **35** (1992): 454–465.

[65] Hill, W. G., and A. Caballero. "Artificial Selection Experiments." *Ann. Rev. Ecol. & Sys.* **23** (1992): 287–310.

[66] Huynen, M. A. "Exploring Phenotype Space through Neutral Evolution." *J. Mol. Evol.* **43** (1996): 165–169.

[67] Huynen, M. A., and P. Hogeweg. "Pattern Generation in Molecular Evolution—Exploitation of the Variation in RNA Landscapes." *J. Mol. Evol.* **39** (1994): 71–79.

[68] Huynen, M., P. F. Stadler, and W. Fontana. "Smoothness within Ruggedness: The Role of Neutrality in Adaptation." *Proc. Natl. Acad. Sci. USA* **93** (1996): 397–401.

[69] Jaarola, M., H. Tegelström, and K. Fredga. "A Contact Zone with Noncoincident Clines for Sex-Specific Markers in the Field Vole (*Microtus agrestis*)." *Evolution* **51** (1997): 241–249.

[70] Jongeling, T. B. "Self-Organization and Competition in Evolution: A Conceptual Problem in the Use of Fitness Landscapes." *J. Theor. Biol.* **178** (1996): 369–373.

[71] Kauffman, S. A. *The Origins of Order.* Oxford: Oxford University Press, 1993.

[72] Kauffman, S. A., and S. Levin. "Towards a General Theory of Adaptive Walks on Rugged Landscapes." *J. Theor. Biol.* **128** (1987): 11–45.

[73] Kimura, M. *The Neutral Theory of Molecular Evolution.* Cambridge: Cambridge University Press, 1983.

[74] Kondrashov, A. S., and S. I. Mina. "Sympatric Speciation: When Is It Possible?" *Biol. J. Linn. Soc.* **27** (1986): 201–223.

[75] Lande, R. "Effective Deme Sizes during Long-Term Evolution Estimated from Rates of Chromosomal Rearrangements." *Evolution* **33** (1979): 234–251.

[76] Lande, R. "Expected Time for Random Genetic Drift of a Population between Stable Phenotypic States." *Proc. Natl. Acad. Sci. USA* **82** (1985): 7641–7645.

[77] Lande, R. "The Fixation of Chromosomal Rearrangements in a Subdivided Population with Local Extinction and Colonization." *Heredity* **54** (1985): 323–332.

[78] Lemke, N., and I. A. Campbell. "Random Walks in a Closed Space." *Physica A* **230** (1996): 554–562.

[79] Lenski, R. E., and M. Travisano. "Dynamics of Adaptation and Diversification: A 10,000-Generation Experiment with Bacterial Populations." *Proc. Natl. Acad. Sci. USA* **91** (1994): 6808–6814.

[80] Li, W.-H. *Molecular Evolution.* Sunderland, MA: Sinauer Associates, 1997.

[81] Manzo, F., and L. Peliti. "Geographic Speciation in the Derrida-Higgs Model of Species Formation." *J. Phys. A: Math. & Gen.* **27** (1994): 7079–7086.

[82] Mani, G. S., and B. C. C. Clarke. "Mutational Order: A Major Stochastic Process in Evolution." *Proc. Roy. Soc. Lond. B* **240** (1990): 29–37.

[83] Martinez, M. A., V. Pezo, P. Marlière, and S. Wain-Hobson. "Exploring the Functional Robustness of an Enzyme by *in vitro* Evolution." *EMBO J.* **15** (1996): 1203–1210.

[84] May, R. M. "How Many Species?" *Phil. Trans. Roy. Soc. Lond. B* **330** (1990): 293–304.

[85] Maynard Smith, J. "Natural Selection and the Concept of a Protein Space." *Nature* **225** (1970): 563–564.

[86] Maynard Smith, J. "The Genetics of Stasis and Punctuation." *Ann. Rev. Genet.* **17** (1983): 11–25.

[87] Mayr, E. *Systematics and the Origin of Species.* New York: Columbia University Press, 1942.

[88] Mayr, E. *Animal Species and Evolution.* Harvard: Belknap Press, 1963.

[89] Mayr, E. *The Growth of Biological Thought.* Cambridge, MA: Harvard University Press, 1982.

[90] McCune, A. R. "How Fast is Speciation? Molecular, Geological, and Phylogenetic Evidence from Adaptive Radiations of Fishes." In *Molecular Evolution and Adaptive Radiation*, edited by T. J. Givnish, and K. J. Sytsma, 585–610. Cambridge: Cambridge University Press, 1997.

[91] Muller, H. J. "Reversibility in Evolution Considered from the Standpoint of Genetics." *Biol. Rev.* **14** (1939): 261–280.

[92] Muller, H. J. "Bearing of the *Drosophila* Work on Systematics." In *The New Systematics*, edited by J. Huxley, 185–268. Oxford: Oxford University Press, 1940.

[93] Muller, H. J. "Isolating Mechanisms, Evolution and Temperature." *Biol. Symposia* **6** (1942): 71–125.

[94] Nash, J. M. "When Life Exploded." *Time* **146** (1995): 66–74.

[95] Nei, M. "Mathematical Models of Speciation and Genetic Distance." In *Population Genetics and Ecology*, edited by S. Karlin, and E. Nevo, 723–768. New York: Academic Press, 1976.

[96] Nei, M., T. Maruyma, and C-I. Wu. "Models of Evolution of Reproductive Isolation." *Genetics* **103** (1983): 557–579.

[97] Newman, M. E. J., and R. Engelhardt "Effects of Selective Neutrality on the Evolution of Molecular Species." *Proc. Roy. Soc. Lond. B* **265** (1998): 1333–1338.

[98] Nurnberger, B., N. H. Barton, C. McCallum, J. Gilchrist, and M. Appleby. "Natural Selection on Quantitative Traits in the *Bombina* Hybrid Zone." *Evolution* **49** (1995): 1224–1238.

[99] Ohta, T. "Slightly Deleterious Mutant Substitutions in Evolution." *Nature* **246** (1973): 96–98.

[100] Ohta, T. "Evolution by Gene Duplication and Compensatory Advantageous Mutations." *Genetics* **120** (1988): 841–847.

[101] Ohta, T. "Time to Acquire a New Gene by Duplication." *Proc. Natl. Acad. Sci. USA* **85** (1988): 3509–3512.
[102] Ohta, T. "The Nearly Neutral Theory of Molecular Evolution." *Ann. Rev. Ecol. & Sys.* **23** (1992): 263–286.
[103] Ohta, T. "The Meaning of Near-Neutrality at Coding and Non-coding Regions." *Gene* **205** (1997): 261–267.
[104] Ohta, T. "Evolution by Nearly-Neutral Mutations." *Genetics* **102/103** (1998): 83–90.
[105] Orr, H. A. "The Population Genetics of Speciation: The Evolution of Hybrid Incompatibilities." *Genetics* **139** (1995): 1803–1813.
[106] Orr, H. A. "Dobzhansky, Bateson, and the Genetics of Speciation." *Genetics* **144** (1997): 1331–1335.
[107] Orr, H. A., and L. H. Orr. "Waiting for Speciation: The Effect of Population Subdivision on the Waiting Time to Speciation." *Evolution* **50** (1996): 1742–1749.
[108] Peliti, L., and U. Bastolla. "Collective Adaptation in a Statistical Model of an Evolving Population." *Comptes Rendus de l'Academie des Sciences Series III—Sciences de la Vie—Life Sciences* **317** (1994): 371–374.
[109] Provine, W. B. *Sewall Wright and Evolutionary Biology.* Chicago: The University of Chicago Press, 1986.
[110] Raup, D. M. "A Kill Curve for Phanerozoic Marine Species." *Paleobiology* **17** (1991): 37–48.
[111] Reidys, C. M. "Random Induced Subgraphs of Generalized n-Cubes." *Adv. Appl. Math.* **19** (1997): 360–377.
[112] Reidys, C. M., P. F. Stadler, and P. Schuster. "Generic Properties of Combinatory Maps: Neutral Networks of RNA Secondary Structures." *Bull. Math. Biol.* **59** (1997): 339–397.
[113] Reidys, C. M., and P. F. Stadler. "Neutrality in Fitness Landscapes." *Appl. Math. Comput.* **117** (2001): 321–350.
[114] Reidys, C. M., C. V. Forst, and P. Schuster. "Replication and Mutation on Mental Networks." *Bull. Math. Biol.* **63** (2001): 57–94.
[115] Rice, W. R. "Disruptive Selection on Habitat Preferences and the Evolution of Reproductive Isolation: A Simulation Study." *Evolution* **38** (1984): 1251–1260.
[116] Rice, W. R., and E. E. Hostert. "Laboratory Experiments on Speciation: What Have We Learned in 40 Years?" *Evolution* **47**(1993): 1637–1653.
[117] Rieseberg, L. H. "The Role of Hybridization in Evolution: Old Wine in New Skins." *Am. J. Botany* **82** (1995): 944–953.
[118] Schluter, D. "Ecological Causes of Adaptive Radiation." *Amer. Natur.* **148** (1996): S40–S64.
[119] Schmalhausen, I. I. *Factors of Evolution. The Theory of Stabilizing Selection.* Chicago: The University of Chicago Press, 1949.
[120] Schuster, P. "How to Search for RNA Structures: Theoretical Concepts in Evolutionary Biotechnology." *J. Biotechnol.* **41** (1995): 239–257.

[121] Schuster, P. "Landscapes and Molecular Evolution." *Physica D* **107** (1997): 351–365.

[122] Schuster, P., W. Fontana, P. F. Stadler, and I. L. Hofacker. "From Sequences to Shapes and Back: A Case Study in RNA Secondary Structures." *Proc. Roy. Soc. Lond. B* **255** (1994), 279–284.

[123] Sepkoski, J. J. "Rates of Speciation in the Fossil Record." *Phil. Trans. Roy. Soc. B* **353** (1998): 315–326.

[124] Stork, N. E. "How Many Species Are There?" *Biodiversity and Conservation* **35** (1993): 321–337.

[125] Templeton, A. R. "The Theory of Speciation via the Founder Principle." *Genetics* **94** (1980): 1011–1038.

[126] Templeton, A. R. "The Meaning of Species and Speciation: A Genetic Perspective." In *Speciation and Its Consequences*, edited by D. Otte, and J. A. Endler, 3–276. Sunderland, MA: Sinauer, 1989.

[127] Templeton, A. R. "The Role of Molecular Genetics in Speciation Studies." In *Molecular Ecology and Evolution: Approaches and Applications*, edited by B. Schierwater, B. Streit, G. P. Wagner, and R. DeSalle, 455–477. Basel: Birkhaüser, 1994.

[128] Templeton, A. R. "Experimental Evidence for the Genetic Transilience Model of Speciation." *Evolution* **50** (1996): 909–915.

[129] Thompson, J. N. *The Coevolutionary Process.* Chicago: The University of Chicago Press, 1994.

[130] van Nimwegen, E., J. P. Crutchfield, and M. Mitchell. "Finite Populations Induce Metastability in Evolutionary Search." *Phys. Lett. A* **229** (1997): 144–150.

[131] van Nimwegen, E., J. P. Crutchfield, and M. Mitchell. "Statistical Dynamics of the Royal Road Genetic Algorithm." *Theor. Comp. Sci.* **229** (1999): 41–102.

[132] van Nimwegen, E., and J. P. Crutchfield. "Optimizing Epochal Evolutionary Search: Population-Size Independent Theory." *Comp. Method. Appl. Mech & Eng.* **186(2–4)** (2000): 171–194.

[133] van Nimwegen, E., and J. P. Crutchfield. "Optimizing Epochal Evolutionary Search: Population-Size Dependent Theory." *Mach. Learn. Theor* (1999): submitted.

[134] Virdee, S. R., and G. M. Hewitt. "Clines for Hybrid Dysfunction in a Grasshopper Hybrid Zone." *Evolution* **48** (1994): 392–407.

[135] Waddington, C. H. *The Strategy of Genes.* New York: MacMillan Co., 1957.

[136] Wagner A. "Does Evolutionary Plasticity Evolve?" *Evolution* **50** (1996): 1008–1023.

[137] Wagner, A. "Redundant Gene Functions and Natural Selection." *J. Evol. Biol.* **12** (1999): 1–16.

[138] Wagner A., G. P. Wagner, and P. Similion. "Epistasis can Facilitate the Evolution of Reproductive Isolation by Peak Shifts—A Two-Locus Two-Allele Model." *Genetics* **138** (1994): 533–545.

[139] Wagner, G. P., and L. Altenberg. "Complex Adaptation and the Evolution of Evolvability." *Evolution* **50** (1996): 967–976.

[140] Wagner, G. P., G. Booth, and H. Bagheri-Chaichian. "A Population Genetic Theory of Canalization." *Evolution* **51** (1997): 329–347.

[141] Walsh, J. B. "How Often Do Duplicated Genes Evolve New Functions?" *Genetics* **139** (1995): 421–428.

[142] Weber, K. E. "Large Genetic Change at Small Fitness Cost in Large Populations of *Drosophila melanogaster* Selected for Wind Tunnel Flight: Rethinking Fitness Surfaces." *Genetics* **144** (1996): 205–213.

[143] Whitlock, M. C., P. C. Phillips, F. B.-M. Moore, and S. J. Tonsor. "Multiple Fitness Peaks and Epistasis." *Ann. Rev. Ecol. & Sys.* **26** (1995): 601–629.

[144] Wills, C. J. "A Mechanism for Rapid Allopatric Speciation." *Amer. Natur.* **111** (1977): 603–605.

[145] Woodcock, G., and P. G. Higgs. "Population Evolution on a Multiplicative Single-Peak Fitness Landscape." *J. Theor. Biol.* **179** (1996): 61–73.

[146] Wright, S. "Evolution in Mendelian Populations." *Genetics* **16** (1931): 97–159.

[147] Wright, S. "The Roles of Mutation, Inbreeding, Crossbreeding and Selection in Evolution." *Proceedings of the Sixth International Congress on Genetics* **1** (1932): 356–366.

[148] Wright, S. "The Shifting Balance Theory and Macroevolution." *Ann. Rev. Genet.* **16** (1982): 1–19.

[149] Wu, C-I. "A Stochastic Simulation Study of Speciation by Sexual Selection." *Evolution* **39** (1985): 66–82.

[150] Wu, C-I., and M. F. Palopoli. "Genetics of Postmating Reproductive Isolation in Animals." *Ann. Rev. Genet.* **28** (1994): 283–308.

[151] Yampolsky, L. Y., R. M. Kamaltinov, D. Ebert, D. A. Filatov, and V. I. Chernykh. "Variation of Allozyme Loci in Endemic Gammarids of Lake Baikal." *Biol. J. Linn. Soc.* **53** (1994): 309–323.

Molecular Insights into Evolution of Phenotypes

Peter Schuster

The success and efficiency of Darwinian evolution is based on the dichotomy of genotype and phenotype: The former is the object under variation; whereas, the latter constitutes the target of selection. Genotype-phenotype relations are highly complex and hence variation and selection are uncorrelated. Population genetics visualizes evolutionary dynamics as a process among genotypes. Phenotypes are represented through empirical parameters only. The quasi-species concept introduces the molecular mechanism of mutation. Optimization is seen as a process in genotype space. Populations optimize through adaptive walks. Selective neutrality leads to random drift. Understanding evolution will always be incomplete unless phenotypes are considered explicitly. At the current state of the art, almost all genotype-phenotype mappings are too complex to be analyzed and modeled. Only the most simple case of an evolutionary process, the optimization of RNA molecules *in vitro*, where genotypes and phenotypes are RNA sequences and structures, respectively, can be treated successfully. We derive a model based on a stochastic process, which includes unfolding of genotypes to form phenotypes as well as phenotype evaluation. Relations between genotypes and phenotypes are handled as mappings from sequence space onto the space of molecular structures. Generic properties of this map are analyzed for RNA secondary structures. Optimization of molecular properties in populations is modeled *in silico* through replication and mutation in a flow reactor. The path from a selected structure is monitored and reconstructed in terms of an uninterrupted series of phenotypes from initial to final structure, called a relay series. We give a novel definition of continuity in evolution that identifies discontinuities as major changes in molecular phenotypes.

1 GENOTYPES AND PHENOTYPES

Evolutionary optimization in asexually multiplying populations follows Darwin's principle and is determined by the interplay of two processes that exert counteracting influences on genetic heterogeneity: (i) Mutations increase diversity of genotypes and (ii) selection decreases diversity of phenotypes.[1] Recombination occurring obligatorily in sexually reproducing populations is another process contributing to the maintainance of diversity. The genotypes of the offspring combine parts from both parental genotypes. Variation and selection operate on different manifestations of the individual, genotype and phenotype, respectively. At the first glance, decoupling of the targets of mutation (or recombination) and selection may seem to be a disadvantage. As a result of uncorrelatedness an advantageous mutation does not occur more frequently because it has a better chance to become selected. Considering nonbiological complex optimization problems, however, random variation is well known to be an efficient means of optimization [71]. Deterministic optimization techniques, gradient techniques, for example, are too easily caught in local extrema and can approach neither optimal nor near-optimal solutions on multipeak landscapes derived from sophisticated cost functions. Genotype-phenotype dichotomy in nature guarantees randomness of moves in Darwinian optimization.

Separation of genotype and phenotype is trivially fulfilled in higher forms of life where the phenotype is an adult multicellular organism created through development that unfolds a version of the genotype in a manner that reminds one of the execution of a computer program. Genotype and phenotype are different entities, and with the exception of few examples, it is currently impossible

[1]The genotype is understood as the polynucleotide sequence that carries the genetic information to build the organism. The polynucleotide is commonly DNA, or RNA in the case of several families of viruses and viroids. The phenotype is the entity that carries all properties required to enter the reproductive phase. For higher forms of life the phenotype is the adult organism, for prokaryotes it is the bacterial cell or the virus particle. The phenotype thus determines fitness, which is commonly understood in evolutionary biology as the number of fertile descendants transmitted into the next generation.

to infer changes in phenotypic properties from known modifications in the DNA sequence of the genotype. In addition, epigenetics exerts influence on the phenotype, which are by definition distinct from genetics. In case of unicellular organisms, procaryotic or eucaryotic, the phenotype comprises cellular metabolism in its full complexity. In today's reality, metabolism is too complex to be deduced from the genomic DNA sequence. Again genotype and phenotype are clearly distinct features of the organism. *In vitro* evolution deals also with Darwinian optimization in populations of molecules that are capable of replication. Here the distinction between genotype and phenotype is more subtle. Therefore we shall consider these experiments in more detail.

Sol Spiegelman and his group [102] pioneered experiments on evolution of RNA molecules in test tubes (fig. 1). A sample of RNA of the bacteriophage Q_β was transferred into a solution containing an RNA replicase and the activated monomers for RNA synthesis (ATP, UTP, GTP, and CTP). RNA replication sets in, and the material in the solution is consumed. After some time, the consumed material is replaced through transfer of a small sample into fresh stock solution. This procedure is repeated some fifty to one-hundred times. In such serial transfer experiments the rate of RNA synthesis increases by more than one order of magnitude. Spiegelman identified the nucleotide sequence of an RNA molecule as its genotype and the molecular structure as its phenotype. Genotype and phenotype thus are two different manifestations of the same molecule, known to the biochemist as sequence and spatial structure, respectively. Here, we cannot be sure *a priori* that genotype and phenotype are sufficiently distinct in order to lead to *de facto* uncorrelatedness of mutation and selection. Considering RNA folding in detail, however, we realize that structure formation is a highly complex process that does not (yet) generally allow to infer structural changes from mutations in the sequence. At best we must go through a sophisticated algorithm that predicts structure from known sequence (fig. 2). A characteristic of sequence-structure relations is that small changes in sequence may, but need not, have small consequences for structure, and at a (sufficiently) coarse-grained level, the sequence-structure map appears to be almost random (see section 3). All three examples discussed above do not allow for a direct feedback mechanism, which would translate possible consequences of mutation into the frequency of mutant formation. In the absence of such feedback, random movement is certainly an efficient means of optimization.

Different notions of structure imply different models for the molecular phenotype. Examples are: (i) the conformation of minimal free energy (mfe), which is formed after sufficiently long time and at sufficiently low temperature, (ii) the mfe structure together with Boltzmann weighted suboptimal conformations in the sense of a partition function, and (iii) kinetic structures or ensembles of structures, which take available folding times into account and acknowledge the fact that RNA is produced in the cell through transcription that forms the newly synthesized RNA strand from 5′-end to 3′-end. In thermodynamic language we call the mfe conformation the structure in the limit $T \to 0K$ and $t \to \infty$ (i). The

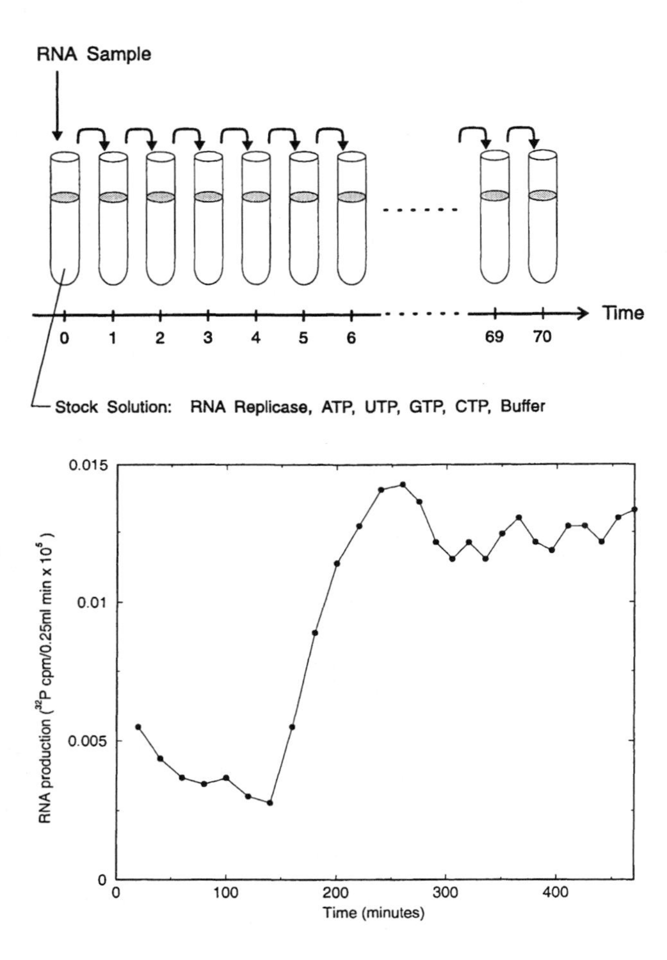

FIGURE 1 Stepwise increase in the rate of RNA production. The upper part shows the technique of serial transfer applied to evolution of RNA molecules in a test tube. The material consumed in the synthesis is replaced by transfer of a small sample into a new test tube with fresh stock solution. The stock solution contains an enzyme required for replication, Q_β-replicase, for example, and the activated monomers (ATP, UTP, GTP, and CTP), the building blocks for polynucleotide synthesis. The rate of RNA production (lower part) is measured through incorporation of radioactive GTP into the newly synthesized RNA molecules. This figure is redrawn from the data in Mills et al. [72].

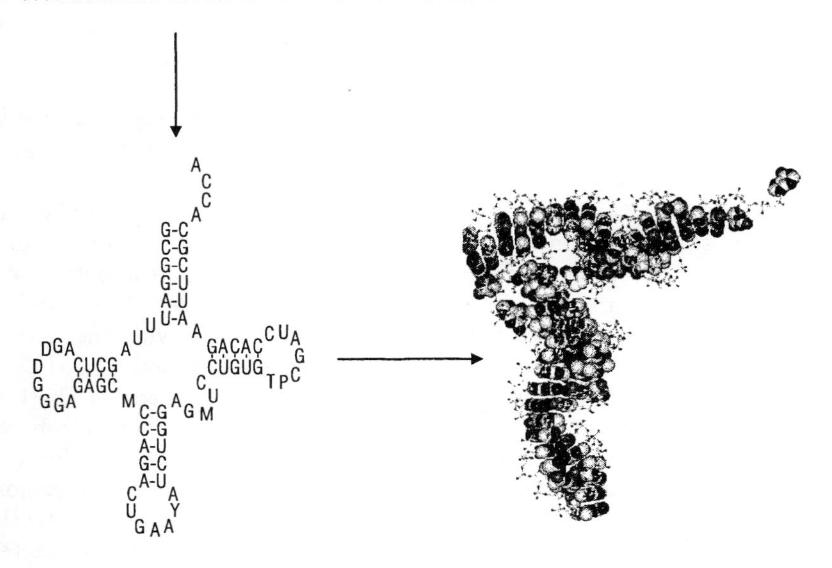

FIGURE 2 Folding of RNA sequences into structures. The folding is performed in two steps from the sequence to the secondary structure and from the secondary structure to the full spatial structure. The example shown is the transfer RNA molecule $tRNA^{phe}$. Both steps occur under the condition of minimal free energy (mfe). The secondary structure is commonly defined as a listing of base pairs, which is compatible with a planar graph without knots or pseudoknots.

partition function represents the ensemble of structures at finite temperature T and $t \to \infty$ (ii). A kinetic structure finally is defined for finite temperature and finite time (iii). It is commonly assumed that small RNA molecules form mfe structures on folding, but recent studies using a new algorithm, which resolves structure formation to formation and cleavage of single base pairs, have shown that this is not necessarily true. Kinetic structures, in the sense of metastable suboptimal conformations, may play a role also for rather small RNA molecules [29]. For longer RNA sequences the discrepancy between most stable and kinetically favored metastable structures is well established [77]. Refolding kinetics of RNA structures shows that only sufficiently low barriers between mfe structures and metastable conformations can be passed at room temperature. If high barriers separate different valleys of the conformational landscape, we observe only the subset of conformations that resides in the valley under consideration. These conformations are accessible within the (temperature dependent) time window of observations. Modified Boltzmann ensembles corresponding to such subsets are good candidates for an elaborate notion of biopolymer structure.

An interesting feature of Spiegelman's RNA evolution and other *in vitro* evolution experiments is stepwise increase in fitness or other quantities used to monitor optimization. Punctuation is observed even under controlled constant conditions (fig. 1). Epochal evolution [113] is not restricted to evolution of molecules in the test tube. It has been observed also with bacterial cultures under the constant conditions of precisely controlled serial transfer [27] as well as in evolution experiments *in silico* mimicking replication and mutation in a flow reactor [30, 34, 35, 51, 112]. A straightforward (but rather trivial) interpretation of the phenomenon says that during such quasi-stationary periods the population waits for some rare event. Basic questions concern the nature of the rare event and the means used by the population that allow for an infrequent incident. We shall try to give answers that are compatible with the now well established neutral [61] or nearly neutral [82] theory of evolution.

The mean generation time is the time unit of evolution. It decides whether or not evolution experiments are feasible. Higher organisms have generation times from several weeks to more than a decade. The time spans required for direct observation of evolutionary phenomena are at least hundreds to thousands of years and thus too long for experiments. At present study is confined to three experimental systems: polynucleotide molecules, viroids or viruses, and bacteria.

In order to set the stage for the development of a comprehensive theory of evolution, we describe a particularly illustrative experiment. Bacteria are well-suited objects for studies on evolution because generation times can be as short as 20 minutes under optimal conditions. The rate of mutation was determined for many DNA-based microbes and, interestingly, it was found to have a constant value of about 0.0033 per genome and generation independently of DNA chain length [18]. An elegant serial transfer experiment with *Escherichia coli* bacteria was carried out by Richard Lenski and coworkers [27, 63, 83]. Populations of 5×10^8 cells were diluted 1:100 every day and recorded for about three years leading to about 10,000 generations, which is tantamount to an average generation time of 3.6 hours.[2] Fitness, measured in terms of growth rate, increased by about 40% during a fast adaptive period over the first 2,000 generations. This increase occurs in steps and not continuously, as one might have expected [27]. After an adaptive initial period, the curve saturates in the remaining 8,000 generations and settles on a plateau at about 1.5 times the initial fitness. More recently, the rate of phenotypic evolution, as monitored via fitness or cell size, was complemented and compared with the rate of genomic evolution determined through DNA fingerprinting [83]: Phenotypic evolution is fast in the initial phase and slows down during saturation. Evolution of the genome, however, behaves differently: it does not slow down and seems even to speed up in the saturation phase. Although the values from experiments on two *E. coli* variants differ substantially, it is certain that the rate of genotypic change does not decrease during saturation

[2]More precisely the bacteria in the solution multiplied substantially faster at the beginning of a transfer period and slowed down later when the nutrient fluid became exhausted.

as phenotypic evolution does. This finding asks indeed for an explanation since a comprehensive theory of evolution should be in a position to make correct predictions on relative rates of genomic and phenotypic evolution.

2 A WORLD OF GENOTYPES

The theory of population genetics was conceived and built by the three famous scholars, Ronald Fisher, John Haldane, and Sewall Wright, and united the previously conflicting issues of Darwinian evolution and Mendelian genetics in an elegant and straightforward way. Evolution is considered as an optimization process at the level of populations, and the relevant variables are the frequencies of genes. The properties of phenotypes enter the model equations as parameters. Such parameters are, among others, lifetimes, litter sizes, survival probabilities of descendants, and rates of reproduction, all of them contributing to fitness values. The notion of "optimization process" implies the definition of a direction: Every change or "move" along the defined direction is accepted; every

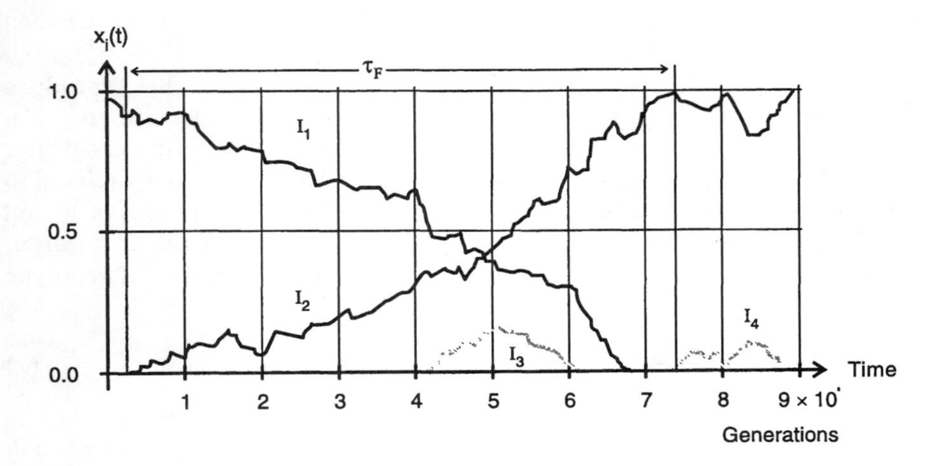

FIGURE 3 Evolution of genotype frequencies in asexual populations. The sketch shows typical solutions curves representing relative concentrations or frequencies $x_i(t)$ of a population modeled by equation (1). New variants are formed by rare mutation events. Depending on replication rates relative to the mean value the frequencies will increase $(a_i > \bar{a})$, decrease $(a_i < \bar{a})$ or drift randomly in the neutral case $(a_i \approx \bar{a})$. Stochastic theory shows that fixation of mutants occurs also in neutral evolution: According to Kimura's theory [61] the mean time from the appearence of a mutant, $x_m(0) = 1/N$, until its fixation in the population, $x_m(\tau_F) \approx 1$, is $\langle \tau_F \rangle = 2N$.

move in opposite direction is rejected.[3] The reduction of the phenotype to a set of input parameters for the differential equations of population dynamics is the basis of success and, at the same time, the most serious limitation of conventional population genetics. Proper choice of parameter values allows one to model and analyze typical idealized situations and to study the influences of quantities like, for example, relative fitness, mutation rate, recombination rate, or population size on the spreading of genes in populations. Problems arise when it becomes necessary to assign realistic values to the parameters, which are commonly very hard to determine experimentally, or when one aims at studies that deal with phenotypes explicitly. In the latter case, we require knowledge about the relation between genotypes and phenotypes in order to be able to derive or model the consequences of changes in the genomic nucleotide sequence for the phenotype. Genotype-phenotype maps are highly complex and hard to investigate, even in the most simple cases. We shall discuss a particularly simple example of phenotypes in test-tube evolution being represented by the spatial structures of RNA molecules in the section 3.

2.1 SELECTION EQUATION

It is straightforward to model selection in populations with asexual replication. Since there is little or no recombination, the appropriate variables are numbers (N_k), concentrations ($[I_k]$), or frequencies (x_k) of genotypes I_k rather than genes. The definitions are: $N_k = \#(I_k)$, the population size $N = \sum_{j=1}^{n} N_j$, $[I_k] = \#(I_k) \;/\; (V \cdot N_L)$ with V being the reaction volume and N_L Avogadro's number, and $x_k = [I_k]/\sum_{j=1}^{n}[I_j]$. Suitable conditions for selection are provided, for example, by serial transfer experiments or by a flow reactor, as discussed later on (fig. 6). A constraint known as *constant organization* [23] is closely related to that of a flow reactor and leads to constant population size. The normalization condition for the frequencies of genotypes, $\sum_{j=1}^{n} x_j = 1$, is then readily incorporated into the differential equation describing the time dependence of genotype distributions in the limit of infinite population size:

$$\frac{dx_k}{dt} = x_k \left(e_k - \Phi(t)\right) \,,\; k = 1, \ldots, n \,. \tag{1}$$

Herein $e_k = a_k - d_k$ is the net production rate constant of genotype I_k, which is obtained as the difference between replication rate constant a_k and degradation rate constant d_k, and $\Phi(t) = \bar{e}(t) = \sum_{j=1}^{n} e_j x_j(t)$ is the mean net production, which is tantamount to the excess reproduction rate of the population. Accordingly, we have $\sum_{j=1}^{n} dx_j/dt = 0$ leading to constant population size. Population geneticists measure progeny in terms of fitness. In the rare mutation case, fitness is identical to net production: $f_k = e_k$ and $\Phi(t) = \bar{f}(t) = \sum_{j=1}^{n} f_j x_j(t)$. For

[3] Acception and rejection may be bounded by predefined probability limits. Nothing is said so far about moves that are neither associated with progress nor with regression. Such "neutral" moves will be the subject of forthcoming sections.

equal degradation rates or lifetimes, $\tau_k = \ln 2/d_k$, the contributions of degradation rates to e_k and $\Phi(t)$ compensate each other exactly, and the fitness values are equal to the rate constants of replication, $f_k = a_k$. Then, the input parameters of the equations of populations genetics (1), i.e., the parameters mentioned above, are simply the replication rate constants a_k of the molecules, viruses, or microorganisms. They are determined, in essence, by the corresponding phenotypes, molecular structures, viral life cycles, or cellular metabolism, respectively.

Mutations are assumed to be rare events, and they are not considered explicitly in the differential equations. At finite population size, fluctuations become important. In addition, every mutant has to start from a single copy. Hence it is jeopardized by random elimination. In order to account for random events, selection in finite populations has been modeled by means of a master equation [21, 58, 64, 65, 106]. It turns out, however, that the convenient constraint of constant organization, as applied in equation (1), leads to instability in the sense that fluctuations in population size increase in time without limit. Two different modifications were applied that stabilize the stochastic selection equation:

(i) every random replication event is strongly combined to a random dilution event, which is tantamount to two-component elementary steps leaving the population size N strictly constant [76], and
(ii) the assumption of a dilution flux $\Phi_0(t) = \sum_{j=1} a_j N_j / N_0$ with constant N_0. This dilution flux is consistent with a population size fluctuating around the fixed value N_0 in the stationary state [58].

The first case (i) is known as the Moran model [76] and yields a strictly constant population size. The second approach (ii) corresponds to a population size $N(t)$ with fluctuations proportional to $\sqrt{N}$. In the limit of long times, $N(t)$ approaches the constant N_0. Van Kampen's expansion [111] was applied to derive stationary solutions of the stochastic selection problem [65]. Typical solution curves are shown in figure 3. Genotypes replace each other in the course of evolution. A typical snapshot will not show more than two genotypes at nonmarginal frequencies. Here we restrict ourselves to a brief presentation of results, which were derived from Motoo Kimura's stochastic theory of *neutral evolution* [61, 60]. This notion was coined because Kimura's concept allows one to investigate the neutral case, $a_1 = \ldots = a_k = \bar{a}$.

In Kimura's model the mean rate of evolution, $\langle k \rangle$, is measured as the number of mutant substitutions per generation time and can be expressed by

$$\langle k \rangle \;=\; N \cdot v \cdot u(N, s, p) \;=\; N \cdot v \cdot \frac{1 - \exp(-2Ns\,p)}{1 - \exp(-2Ns)} \,, \qquad (2)$$

where N is again the population size, v the mutation rate per genome and generation, and $u(N, s, p)$ represents the probability of fixation. This probability is a function of s, the selective advantage,[4] and p, the initial frequency of the mutant. Since every mutant starts inevitably from a single copy, we may put $p = 1/N$ and find

$$\langle k \rangle = N \cdot v \cdot \frac{1 - \exp(-2s)}{1 - \exp(-2Ns)} \ .$$

In the neutral case the rate of evolution is computed to be $\langle k \rangle = v$, and the mean time of the replacement for a given genotype by the next one is $\langle \tau_R \rangle = 1/\langle k \rangle = v^{-1}$ generations. For substantial selective advantage, $s > 1/(2N)$, we find $\langle k \rangle = 2Ns \cdot v$ since $u \approx 2s$: The rate of evolution increases linearly with the selective advantage s and the population size N. On the average, a genotype will be replaced by the next one after $\langle \tau_R \rangle = (2Ns \cdot v)^{-1}$ generations (which is always shorter than in the neutral case because $s > 1/(2N)$ holds). At still higher values of the selective advantage s, the probability of fixation, converges to $\lim_{s \to \infty} u(s) = 1$. Thus, we find in the limit of infinite advatage: $\lim_{s \to \infty} \langle k \rangle = N \cdot v$. Accordingly, the mean rate of evolution for neutral, weakly and strongly advantageous variants is confined by $v \leq \langle k \rangle \leq N \cdot v$. We can use this expressions for rough estimates on the time required for the observation of evolutionary phenomena. We should keep in mind, nevertheless, that the upper limit of $\langle k \rangle$ is highly unrealistic because it requires the maintainance of large increases in fitness through mutation, which do not occur over a sequence of many consecutive mutants under normal conditions (see, however, the initial period of *in silico* evolution of RNA molecules in section 4).

It is also worth considering the mean time for fixation of neutral and advantageous variants, τ_F. The solution of the deterministic selection equation for two genotypes,[5] $x_2 = x$, $x_1 = 1 - x$ and $f_0 = 1$, is readily obtained in analytical form:

$$x(t) \ = \ \frac{x_0}{x_0 \ + \ (1 - x_0) \cdot \exp(-s\, t)} \ .$$

From this equation we compute τ_F as the time it takes for a mutant to grow from a single copy, $x(0) = 1$, to population size, $x\,(\tau_F) \geq N - 1$ and find $\tau_F \approx 2 \ln N/s$. For sufficiently large selective advantage, $s > 1/(2N)$, the mean time of fixation is substantially shorter than in the neutral case, where $\tau_F = 2N$ (see fig. 3). In other words, the inverse of the mean fixation time, ${\tau_F}^{-1}$, decreases linearly with s in the deterministic limit $s \to 0$, but the term from neutral evolution guarantees that the reciprocal time of fixation does not fall below the limit ${\tau_F}^{-1} = 1/2N$.

---

[4]The selective advantage s is measured relative to the neutral case: The fitness value is $f = f_0(1 + s)$ and thus neutrality implies $f = f_0$ or $s = 0$.

[5]Thereby we mean equation (1) for $n = 2$.

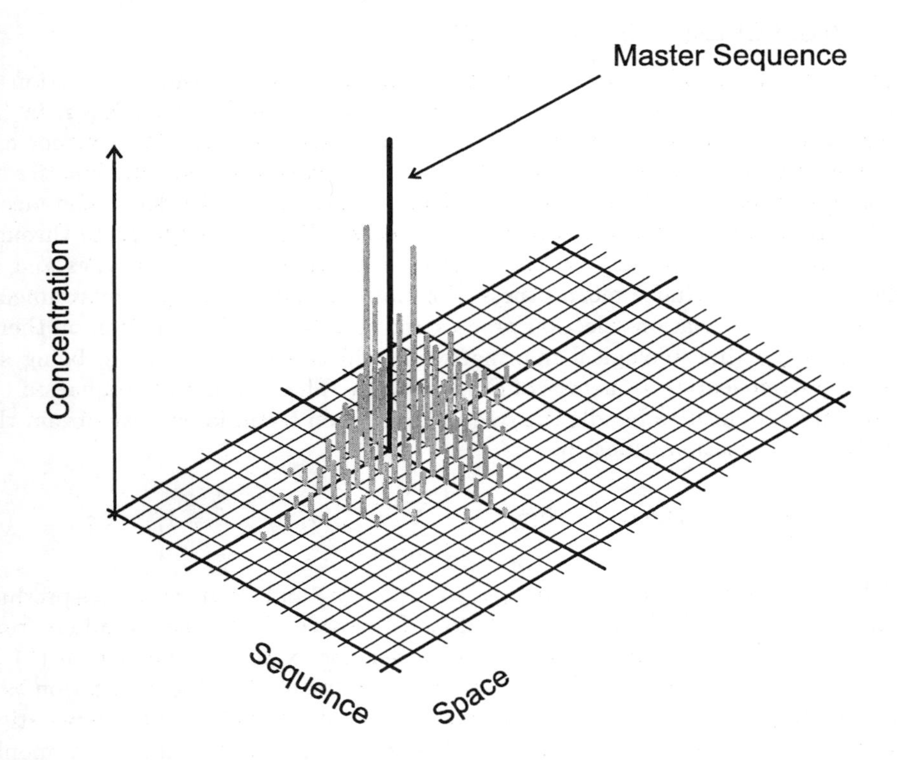

FIGURE 4 A quasi-species-type mutant distribution around a master sequence. The quasi species is an ordered distribution of polynucleotide sequences (RNA or DNA) in sequence space $\mathcal{I}_{\kappa}^{\ell}$. A fittest genotype or master sequence I_m, which is present at highest concentration, is surrounded in sequence space by a "cloud" of closely related sequences. Relatedness of sequences is expressed (in terms of error classes) by the number of mutations that are required to produce them as mutants of the master sequence. In case of point mutations, the distance between sequences is the Hamming distance.[6] In precise terms, the quasi species is defined as the stable stationary solution of equation (3) [22, 23], the mutant distribution described by the largest eigenvector of the matrix $W = \{W_{ij} = Q_{ij} \cdot a_j; i, j = 1, \ldots, n\}$ [57, 79, 105, 110]. (Its diagonal elements are approximations for fitness values, I_k: $f_k \approx W_{kk} = a_k \cdot Q_{kk}$.) In reality, such a stationary solution exists only if the error rate of replication lies below a maximal value called the error threshold. In this region, i.e., below the often sharply defined mutation rate of the error threshold, this eigenvector represents a structured population as shown in the figure. Above the critical error rate the largest eigenvector is (practically) identical with the uniform distribution. The uniform distribution, however, can never be realized in nature or *in vitro* since the number of possible nucleic acid sequences (4^{ℓ}) exceeds the number of individuals by many orders of magnitude even in the largest populations. The actual behavior is determined by incorrect replication leading to random drift: populations migrate through sequence space.

2.2 MOLECULAR QUASI SPECIES

An extension of conventional population genetics, which considers evolution as chemical reactions in genotype space, was proposed by Manfred Eigen in his seminal paper on the theory of the evolution of molecules [22]. His concept can be understood, in essence, as an application of chemical reaction kinetics to molecular evolution. A main issue of Eigen's approach was to derive the mechanism by which biological information is created. Populations migrate through sequence space as metastable but structured distributions of genotypes and at the same time optimize mean fitness. Populations diffuse through environments by means of a variation-selection process and thereby gain information on them. At the same time biological information is laid down in genotypes, being selected polynucleotide sequences of DNA or RNA. The deterministic equation (1) is readily extended to handle the frequent mutation scenario, and we obtain the replication-mutation equation:

$$\frac{dx_k}{dt} = x_k \Big(Q_{kk}a_k - \Phi(t)\Big) + \sum_{j=1, j\neq k}^{n} Q_{kj}a_j\, x_j \ , \quad k = 1, \dots , n \ , \tag{3}$$

with the mutation matrix $Q = \{Q_{ij}; i, j = 1, \dots , n\}$ and the mean excess production $\Phi(t) = \bar{a}(t) = \sum_{j=1}^{n} a_j\, x_j(t)$ as before. In essence, the ansatz (3) differs from conventional population genetics as expressed, for example, by equation (1) in the handling of mutations. The conventional treatment introduces mutation as a rare stochastic event in the environment that is not controlled by the replication mechanism. Mutation is thus characterized by a probability density, commonly by an expectation value and, eventually, a variance. The replication-mutation equation (3), however, deals explicitly with mutations and handles error-free and incorrect replication as parallel reactions. Relative frequencies of the corresponding reaction channels are given by the elements of the mutation matrix. In particular, Q_{kj} is the frequency at which the genotype I_k is synthesized as an error copy of I_j.[6] In general, equation (3) settles down to a stationary state corresponding to a stable or metastable[7] distribution of genotypes. Such distributions of genotypes, called *molecular quasi species* [23, 24], are ordered: In the center

[6]The symbol I_j is used here for genotypes or polynucleotide sequences (DNA, RNA) as well as $\mathcal{I}$ for the space of genotypes, called sequence space [45] in order to point out that the sequences are the carriers of biological information. In particular, $\mathcal{I}_\kappa^\ell$ is the sequence space of sequences of chain length ℓ over an alphabet of size κ (**GC**: $\kappa = 2$; **AUGC**: $\kappa = 4$). $\mathcal{I}_\kappa^\ell$ carries κ^ℓ different sequences. The Hamming distance [44] of two sequences I_i and I_j is denoted by d_{ij}^h and counts the number of positions in which two aligned strings differ. It induces a metric on the corresponding sequence space.

[7]There are two reasons why the infinite time solution of equation (3) may become unstable: (i) The steady state can never be reached because the population size is too small to approach the infinite population limit, and (ii) random drift in the sense of neutral evolution (which is not addressed by deterministic equations) may be followed by a new onset of selection after an epoch of stasis, thereby causing the population to switch from one (quasispecies-like) metastable state to another.

of the distribution we find a most frequent and commonly also fittest genotype called the *master sequence* (fig. 4). Quasi species represent the genetic reservoirs of asexually replicating species, for example, molecules, viroids, viruses, or bacteria.

Fitness landscapes are understood as distributions of fitness values over sequence space and can be represented by mappings from sequence space into the real numbers, $g : \{\mathcal{I}; d_{ij}^h\} \Rightarrow \mathbb{R}^1$. (For recent reviews, see Schuster [97] and Stadler [103].) Herein the sequence space space is denoted by $\mathcal{I}$ and the distance between two sequences by d_{ij}^h. Several classes of model landscapes were either invented, like the NK model [59], or adopted from physical models of spin glasses [109] in order to study replication and mutation with distributions of fitness values in sequence space, which are thought to be representative for real situations. Most landscapes sustain stable quasi species only for mutation rates below a certain critical value called the *error threshold* [23, 24]. At the critical point, an abrupt change in evolutionary dynamics is observed, which reminds one of a phase transition [66]. (Exceptions are classes of artificially smooth landscapes sometimes applied in population genetics, which show gradual transitions from quasi species to uniform distributions.) At error frequencies above threshold, populations migrate through sequence space in random walk manner and do not approach stationary states (fig. 4).

It is worth considering the parameter problem of population genetics once more, this time from the practical point of computer simulation. The numbers of possible RNA sequences or structures, the cardinalities of sequence space or shape space, are enormous, even for moderate chain length, and hence too large for any direct assignment of empirical parameters. Required are tractable models that allow one to compute all relevant phenotypic quantities, in particular the a_k- and Q_{kj}-values, from rules that make use of a few parameters only. As an example, we consider a useful and realistic model for the mutation matrix Q, which is based on the notion of sequence space $(\mathcal{I}_\kappa^\ell)$.[6] Restriction to point mutations and assumption of uniform error rates, implying that the probability of a mutation is independent of the nature of the base exchange and the position along the sequence, allow one to express all elements of Q in terms of only three parameters only, ℓ, q, and d_{kj}^h:

$$Q_{kj} = q^\ell \left(\frac{1-q}{q}\right)^{d_{kj}^h} . \tag{4}$$

Herein ℓ is the length of the polynucleotide chain. The single-digit accuracy of replication q implies a uniform error rate of $p = 1 - q$ per digit and replication event, which is independent of the position in the sequence, and d_{kj}^h, finally, is the Hamming distance between the two sequences to be interconverted by the mutation.

Within the frame of the uniform error rate model (4), it is straightforward to compute an approximate expression of the critical mutation rate [22]. First, mutational backflow (the sum in the right-hand part of equation 3) is neglected and, second, the stationary frequency of the master genotype is computed to $\bar{x}_m = (\sigma_m Q_{mm} - 1)/(\sigma_m - 1)$, wherein $\sigma_m = a_m/\bar{a}_{k \neq m}$ defines the superiority of the master sequence and $\bar{a}_{-m} = (\sum_{j=1, j\neq m}^{n} a_j x_j)/(1 - x_m)$ the mean replication rate of the population except the master. The expression for the error threshold is derived now by computing the error rate at which the concentration of the master sequence vanishes: $\bar{x}_m = 0 \rightarrow Q_{mm} = \sigma_m^{-1}$. Application of equation (4), $Q_{mm} = q^\ell$, yields two equations, one for the maximal mutation rate (or minimal replication accuracy, $p_{\max} = 1 - q_{\min}$) at constant chain length and one for the maximal chain length at constant mutation rate,

$$p_{\max} = 1 - \sigma_m^{-1/\ell} \quad \text{and} \quad \ell_{\max} = -\frac{\ln \sigma_m}{\ln q} \approx \frac{\ln \sigma_m}{(1-q)} = \frac{\ln \sigma_m}{p}, \tag{5}$$

which define the error threshold. Despite the simplifications made in the derivation of equation (5), the agreement between the exact curves $\bar{x}_m(q)$ and the approximation is surprisingly good [24]. For the sake of completeness we mention that the computation of an error threshold of replication and mutation has been extended to the diploid case [118] as well as to neutral evolution where stationarity refers to time-independent distributions of phenotypes rather than genotypes [89]. Quasi species formation has been studied also on a dynamical landscape [78].

Although all entries of the mutation matrix are now computable from a few input parameters, still more empirical data are required. As in the selection equation (1) the fitness values of phenotypes enter the kinetic differential equation (3) as parameters. Assignment of fitness values can be performed under model assumptions only. Considering, for example, a rather short RNA molecule of chain length $\ell = 100$ we are dealing with 1.6×10^{60} different genotypes, which may give rise to a smaller but still very large number of phenotypes. Commonly the problem is overcome by rather drastic simplifications. As an example we mention the single-peak fitness landscape: One replication rate $a_m = \sigma$ is assigned to the fittest genotype, I_m, and all other genotypes are assumed to have replication rate $a_{k\neq m} = 1$ [105]. This ansatz reminds one of the mean field approximation often used in physics. Since details of the distribution of fitness values are unknown, one replaces them by a mean value for all genotypes except the fittest one. In this sense, the single peak landscape has been used, for example, to derive analytical expressions for the threshold value of the error rate [22, 23, 24]. (For further work on replication and mutation on model landscapes see [1, 2, 8, 9, 69, 78, 94, 109].)

An analysis of stochastic effects in replication-mutation systems based on multitype branching processes [13] provides a mathematical interpretation of the error threshold in terms of first passage times: The probability of survival to infinite time of the master sequence is nonzero at error rates below threshold and becomes zero at the critical value. Above threshold, however, all genotypes have zero probability of survival to infinite time, which is tantamount to instability of stationary sequence distributions. A later approach modeled replication and mutation as a birth-and-death process [79] and resulted in an analytical expression for the error threshold in finite populations:

$$q_{\min}(N) = q_{\min}(\infty)\left(1 + \frac{2\sqrt{\sigma_m - 1}}{\ell\sqrt{N}} + \dots\right).$$

Other treatments of replication and mutation as stochastic processes were based on the corresponding master equation [21, 58, 64, 65, 106] and applied the same constraints as discussed for the selection equation, the Moran model [76] or the dilution flux $\Phi_0(t)$.

In summary, the quasi species concept (3) provides a solution for the mutation problem but does not yet deal explicitly with phenotypes and their properties. The problem in handling phenotypes is twofold. First, the mapping of genotypes into phenotypes is extremely complicated and generally very hard to model. Second, phenotypes have a large number of properties most of which contribute to fitness only in an indirect way. What is needed therefore is a realistic but sufficiently simple toy model that allows one to compute fitness values from a set of few (simple) rules.

Stationary mutant distributions were observed and analyzed in the case of evolution *in vitro* of RNA molecules [5, 90]. The data recorded in these studies reproduce well the predictions derived from equation (3). The threshold equation (5) can also be used to predict maximum genome lengths provided the population evolves at maximal mutation rate. This was indeed found to be the case with lytic RNA viruses [19] where the observed mutation rate p increases linearly with the reciprocal chain length ℓ^{-1}. A straightfoward interpretation of this finding says that these viruses mutate as fast as possible because they have evolved this way of coping with the powerful defense mechanisms of their hosts. A comparison of mutation rates in different groups of prokaryotes [18, 20] revealed a roughly constant mutation rate per genome length: $\ell \cdot p = \text{const}$. The constants are around 1 for lytic RNA viruses, roughly 0.1 for retroviruses and retrotransposons, and close to 1/300 for microbes with DNA-based genomes. (For an interpretation of these results on the basis of cost balance between reduction of deleterious mutants and precision of the replication machinery, see Drake et al. [20].) In addition, the quasi species concept turns out to be useful for the description of the evolution of RNA virus populations [15, 16] and provides hints for the development of novel antiviral strategies [17].

2.3 EXTENSION TO PHENOTYPES

The first explicit consideration of phenotypes in a model of molecular evolution was implemented *in silico* in order to simulate replication and mutation in a flow reactor (fig. 6) [30]. This simple model was already in a position to perform optimizations of RNA properties like thermodynamic stability or net productivity as expressed by the difference of replication and degradation rate constants ($e_k = a_k - d_k$). Later on, the relation between genotypes and phenotypes was made more precise and modeled as a mapping from sequence space into phenotype space. To this end we assume a metric phenotype space $\mathcal{S}$ with some (hypothetical) measure of distance between phenotypes, d^s_{ij}:

$$\psi : \{\mathcal{I}; d^h_{ij}\} \Rightarrow \{\mathcal{S}; d^s_{ij}\} . \qquad (6)$$

In other words, $S_k = \psi(I_k)$, implies that a phenotype S_k is uniquely assigned to the genotype I_k. The assignment expressed by equation (6) is tantamount to the formation of the phenotype S_k through unfolding of the genetic information stored in the genotype I_k. Fitness values are approximated by the product of replication rate constants and replication accuracy, $f_k \approx a_k \cdot Q_{kk}$, and can been seen as the result of a mapping f from the phenotypes into the nonnegative real numbers:

$$f : \{\mathcal{S}; d^s_{ij}\} \Rightarrow \mathbb{R}_+ . \qquad (7)$$

The map (7) evaluates the phenotype and returns its fitness value. In summary, we obtain fitness values from the genotype through the function: $f(S_k) = f(\psi(I_k)) = f_k \approx a_k \cdot Q_{kk}$. The mapping $\psi(.)$, in general, cannot be expressed in analytical terms. At best we have algorithms that allow to compute structures from sequences (see section 3). The situation is not less complex for the derivation of fitness values of phenotypes, but in this case the evaluation is often done by means of simple model functions. For example, $f(.)$ can be assumed to be a simple function of the distance between the phenotype and some target to be approached.

Now we are in a position to classify different mappings:

(i) $\psi(.)$ maps a discrete vector space, the sequence space $\mathcal{I}$, into another nonscalar discrete (or continuous) space $\mathcal{S}$. We call it a combinatory map [87] since the sequence space $\mathcal{I}$ is derived by a combinatory building principle (see also section 3).

(ii) $f(.)$ maps a discrete (or continuous) nonscalar space $\mathcal{S}$ into the nonnegative real numbers $\mathbb{R}_+$. It represents an example of a landscape or cost function. In particular, $f(.)$ is the fitness landscape assigning a fitness value f_k to a phenotype S_k.[8]

Finishing this section we consider environmental influences and indicate how one may generalize our approach to variable environments, $\mathcal{E}(t)$. Both the unfolding of the genotype as well as the evaluation of the phenotype depend on the environment $\mathcal{E}$. In other words, the same genotype, I_k develops different phenotypes, say S_k, S'_k, or S''_k, in different environments, $\mathcal{E}$, $\mathcal{E}'$, or $\mathcal{E}''$. The same

[8]The expression "landscape" is a generalization of the notion used in common-sense or geography for the representation of a three-dimensional relief on Earth as a mapping from two dimensions (longitude, latitude) into the real numbers (altitude).

phenotype may have different fitness values under different environmental conditions. In principle, the ansatz for evolutionary dynamics presented here can be readily extended to handle situations in variable environments by the introduction of time-dependent fitness values. Then we end up with the following equation which relates Darwinian fitness to genotypes:

$$f_k(t) = f\Big(S_k, \mathcal{E}(t)\Big) = f\Big(\psi(I_k, \mathcal{E}(t)), \mathcal{E}(t)\Big) . \qquad (8)$$

Incorporating time-dependent fitness values into equations (1) and (3) we obtain a differential equation which is the basis for the deterministic description of Darwinian evolution in asexually replicating populations:

$$\frac{dx_k}{dt} = x_k \Big(Q_{kk} f_k(t) - \Phi(t)\Big) + \sum_{j=1, j \neq k}^{n} Q_{kj} f_j(t)\, x_j \ , \quad k = 1, \ldots , n . \qquad (9)$$

Stochastic effects may be introduced into equation (9), for example, by means of a multidimensional master equation corresponding to a multivariate birth-and-death process with time-dependent birth and death rates. Alternatively, one may use Van Kampen's size expansion of the master equation [111] and finally end up with a stochastic differential equation. Stochastic effects are then incorporated into the deterministic differential equation through terms like $\eta_k(\mathbf{x}, t)\xi_k(t)$ which model fluctuations by a Wiener process whose amplitude depends on the variables of the deterministic solution. Separability of time scales is a prerequisite for the success of this approach: The environment-driven changes in the functions $f_k(t)$ must be slow compared to the progress of the evolutionary process in order to allow for decoupling of external and intrinsic dynamics.

Needless to say, the mappings (8) encapsulate a great deal of complexity, and there is no chance of finding simple solutions. They are, nevertheless, suitable for discussing special simplified cases, and they provide a proper reference for computer simulations. As said before, three experimentally accessible realizations of equation (9) are currently conceivable: (i) evolution of RNA molecules *in vitro*, (ii) life cycles and evolution of viral RNA (or DNA) in host cells, and (iii) metabolism and evolution of bacteria (under constant environmental conditions). In the following two sections we shall present a simplified model of (i) that allows one to simulate evolution according to equation (9) using a realistic algorithm to compute RNA structures from sequences. Virus evolution (ii) can be modeled in principle provided enough data are available on the influence of mutations on viral life cycles. Quantity and quality of these data are rapidly improving now, and we can expect a full understanding of virus evolution at the molecular level in the coming years. Although (iii) seems to be too complex by far for computer implementations, we may expect fast progress in the near future: Information on complete DNA sequences is already available in a few cases, and many more bacterial genomes will be sequenced soon. The current data are already used in the development of models for the genetic network controlling metabolism of prokaryotic cells [68]. Still, simulation of bacterial evolution based on such models will remain a great challenge for research in the future.

3 THE RNA MODEL

The phenotype in serial transfer or flow reactor experiments with RNA molecules is straightforwardly identified with the molecular structure of RNA [102]. Accordingly, the genotype-phenotype map relates RNA sequences with RNA structures. At the current state of the art our knowledge on RNA structures is far from complete, Hence prediction of RNA structures from known sequences is still a great challenge in bioinformatics and structural biology. The RNA case, however, is at least accessible by means of simplified but realistic models of sequence-structure mappings and thus contrasts the other, more complex phenotypes for which we have at best only pointwise genotype-phenotype information. The results of mathematical models and numerical computation on RNA optimization can be tested through comparison with the data from *in vitro* evolution experiments [5]. These data were complemented by results on RNA sequence-structure maps obtained from systematic studies based on site-directed mutagenesis in RNA sequences. (For example the work on tRNAs [86].) Additional, highly valuable information comes from SELEX (**s**ystematic **e**volution of **l**igands by **ex**ponential amplification) experiments with RNA molecules aiming at the production of aptamers [28] as well as from design of ribozymes, which are specific catalysts on RNA basis (poly**ribo**nucleotide en**zymes**), through selection methods [3]. (For an excellent and comprehensive review see Wilson and Szostak [119].) Aptamers are optimized RNA molecules that bind specifically to predefined targets. Successful selection of optimal binders to almost all classes of biomolecules have been reported. Ribozymes were even obtained for reactions that have no counterpart in nature [85]. A recent and particularly interesting paper [93] reports the design of an RNA sequence that adopts two conformations with different catalytic activities, specific RNA ligation and cleavage. Here we refrain from details, describe the current state of the art in the analysis of RNA sequence-structure maps, and mention only experimental results, which have direct implications for the theoretical concepts.

3.1 RNA PHENOTYPES

At present it is not yet possible to compute full three-dimensional RNA structures with sufficient reliability from sequences. A coarse-grained version of RNA structure, called secondary structure, however, is sufficiently simple in order to allow systematic investigations of genotype-phenotype maps (fig. 2). Secondary structures, in addition, are not only convenient theoretical constructs but also represent relevant and experimentally verified intermediates in the folding of RNA sequences into three-dimensional objects [4]. Moreover, secondary structures are conserved in nature, and they were used by biochemists for decades to interpret successfully the reactivities and other properties of RNA before three-dimensional structures became available. RNA secondary stuctures are understood best as a listing of Watson-Crick (**AU**,**GC**) and **GU** wobble base pairs, which are compatible with unknotted and pseudoknot-free two-dimensional graphs.[9]

The simplest notion of an RNA genotype-phenotype map, and the one we shall adopt here, assigns the minimum free-energy (mfe) structure, which can be obtained through application of a suitable folding algorithm (see section 3.2) to the sequence under consideration. This assignment, apparently, makes use of the thermodynamic concept of structure in the limit of 0 K. At nonzero temperatures we have to consider contributions from suboptimal foldings. The contributions of such configurations with energies higher than the mfe may be considered individually or handled collectively by choosing the partition functions rather than the single mfe structure as the phenotype. This choice leads to a temperature-dependent notion of phenotype. We are thus dealing with a concept that allows for a straightforward response of the phenotype to changes in an environmetal parameter, the temperature. As far as computational possibilities are concerned, both individual suboptimal foldings [121, 126] and partition functions [48, 70] are accessible by efficient algorithms based on dynamic programming.

The thermodynamic notion of structure, whether complemented through suboptimal conformations or not, supposes the existence of an observation window of infinite time for RNA folding. In reality, however, time is limited and accessible structures are restricted by the necessity to fold sufficiently fast. In this case, the conformations formed are often different from the thermodynamically most stable structures [77]. Such conformations are metastable states and commonly addressed as kinetically controlled structures. Kinetic folding has also been the subject of computations. Several computer programs were designed to determine kinetic structures [42, 43, 67, 73, 74, 75, 108, 104]. These algorithms are based on the concept of cooperative formation and melting of double helices. Hence, they treat whole stacks as the units of structure that form and open through all-or-none processes. A French group sucessfully intergrated RNA folding into the general concept of stochastic chemical kinetics [7, 53, 54]. In a more recent paper an attempt was made to drag folding down to the resolution of single base pair operations [29]. These operations include a closure and opening of base pairs as well as a shift move converting a base pair into another allowed pairing of nucleotides. Kinetic folding, in particular folding at single nucleotide

[9]RNA secondary structures can be represented by strings written in a shorthand notation using parentheses and dots. Parentheses correspond to bases combined in base pairs, dots represent single bases. The symbols for bases belonging together in pairs are interpreted unambiguously through reading them in the sense of mathematical notation, i.e., from outside to inside. For example, the string of a typical hairpin loop reads: ··((((····)))). A pseudoknot occurs when base pairs intercalate, for example in the secondary structure ··(((··[[···)))·]]·, where we need two classes of symbols, parentheses and square brackets, for an unambiguous grouping of bases into pairs.

resolution, introduces a new dimension into RNA phenotypes: Not only thermodynamic stability but also the probability of formation within a given time span determine the accessibilty of a phenotype. There is also a relevant third property, attainability through mutation [34, 35], which will be discussed in section 4.

How do the properties of RNA phenotypes change when the concept is extended from mfe structures to suboptimal conformations, partition functions, and kinetic folding? The answer in terms of biological concepts is straightforward: The mfe structure, regarded as a phenotype, is relatively independent of the environment, and its response to changes is very limited. In this case, adaptation is an almost exclusive property of populations, which as a whole can cope with variable environments through modifying and shifting genotype distributions in sequence space. Suboptimal conformations or the partition function introduce flexibility or "plasticity" in biological terms: The individual phenotype can adjust to the environment by changing the distribution of conformations. Considering RNA molecules in solution, variable environments may be visualized by changes in temperature, pH, or ionic strength. Alternatively, binding to other partners, for example small molecules, proteins, or nucleic acids, may also shift the conformational distribution, and hence flexible phenotypes can respond to the appearance of new molecules in their environments. Explicit consideration of folding kinetics brings the time coordinate on the stage. It matters whether a conformation can be adopted in sufficiently short time or not and whether a structure is formed with high or low probabilty. With respect to time, we see even an (admittedly vague) analogy between RNA folding and development: Embryonic pattern formation or morphogenesis is also bound to occur within a sufficiently short time span. Otherwise the phenotype could not compete successfully in evolution.

In the following sections we shall adopt the simplest possible notion of phenotype, the mfe structure. The more complex concepts discussed here can be incorporated straightforwardly into analysis of genotype-phenotype mappings and simulations of evolutionary optimization, although the higher computational efforts may be critical for the current possibilities.

3.2 SECONDARY STRUCTURES OF MINIMAL FREE ENERGIES

RNA secondary structures with minimum free energies are readily derived from sequences by means of fast algorithms based on dynamic programming [80, 124, 125, 48]. The mfe structures, sometimes called (RNA) shapes for short, were studied in order to explore the regularities of sequence to structure mappings by means of four strategies (table 1): (i) mathematical modeling based on random graph theory [87], (ii) folding all sequences belonging to sequence space $\mathcal{I}_{\kappa}^{\ell}$ and computation through exhaustive enumeration [40, 41], (iii) computation through statistics of properly chosen samples of sequences [32, 33, 95], and (iv) evaluation through evolutionary optimization [34, 35]. The following generic results were obtained.

TABLE 1 Various strategies applied to study sequence-structure maps of RNA.

	Method	Advantage	Disadvantage	Ref.
Mathematical model	Random graph theory	Analytical expressions	Limited validity of model assumptions	[87]
Exhaustive folding and enumeration	Folding algorithm and handling of large samples up to 10^{10} objects	Exact results	Limited to short chains: **GC**, $\ell \leq 30$ **AUGC**, $\ell \leq 16$	[40, 41]
Statistical evaluation	Inverse folding or random walks in sequence space	Applicability to longer sequences	Limited accuracy due to statistics	[32, 95]
Simulation of evolutionary dynamics	Chemical kinetics of replication and mutation	Evolutionary relevance	Restriction to small parts of sequence space	[34, 35] [51, 112]

(i) **More sequences than structures.** The numbers of acceptable secondary structures can be counted through combinatorial analysis of the assembly of conformations from elements by means of predefined rules [49, 115].[10] The calculations are done by means of the recursion shown in table 2. For large chain lengths ℓ the numbers $N_S(\ell)$ are well approximated by the expression:

$$N_S(\ell) \approx s(\ell) = 1.4848 \times \ell^{-3/2}(1.84892)^{\ell} .$$

$s(\ell)$ is an asymptotic upper limit for $N_S(\ell)$. (For $\ell = 30$, for example, the deviation is 20.8%; for $\ell = 100$ it is 6.0%; for $\ell = 300$ it is smaller than 2.0%; and for $\ell = 1000$ smaller than 0.65%.) Numbers computed from this expression (or the exact values) are many orders of magnitude smaller than 4^{ℓ}, and even orders of magnitude smaller than 2^{ℓ}, the cardinalities of sequence spaces built over four-letter and two-letter alphabets, respectively. Still we have only an

[10] A structure is considered acceptable if all hairpin loops contain three or more nucleotides and all stacks consist of at least two base pairs. Smaller loops are unstable because of high steric strain energies. Single base pairs are often unstable since the dominating stabilizing contribution comes from base pair stacking. Indeed, hairpin loops with one or two nucleotides are unknown in real structures, and single base pairs occur only rarely. Despite their high combinatorial probability, minimum free-energy conformations with single base pairs do not occur with **GC**-sequences at chain lengths $\ell \leq 13$. For longer sequences these conformations occur with increasing percentage: 5.7 % for $\ell = 14$; 12.8 % for $\ell = 20$; and 21.1 % for $\ell = 30$. The loopsize restriction is less powerful in reducing the number of shapes than the neglect of structures with isolated base pairs. For chain length $\ell = 30$ we find 2.15×10^{10} possible secondary structures, 2.41×10^8 structures with loopsize $n_{lp} \geq 3$, and only 760,983 structures with loopsize $n_{lp} \geq 3$ and stacksize $n_{st} \geq 2$.

TABLE 2 A recursion to calculate the numbers of acceptable RNA secondary structures, $N_S(\ell) = S_\ell^{(\min[n_{lp}],\min[n_{st}])}$ [49]. A structure is acceptable if all its hairpin loops contain three or more nucleotides (loopsize: $n_{lp} \geq 3$) and if it has no isolated base pairs (stacksize: $n_{st} \geq 2$). The recursion $m+1 \Longrightarrow m$ yields the desired results in the array Ψ_m and uses two auxiliary arrays with the elements Φ_m and Ξ_m, which represent the numbers of structures with or without a closing base pair $(1, m)$. One array, e.g., Φ_m, is dispensible, but then the formula contains a double sum that is harder to interpret.

Recursion formula:

$$\Xi_{m+1} = \Psi_m + \sum_{k=5}^{m-2} \Phi_k \cdot \Psi_{m-k-1}$$

$$\Phi_{m+1} = \sum_{k=1}^{\lfloor (m-2)/2 \rfloor} \Xi_{m-2k+1}$$

$$\Psi_{m+1} = \Xi_{m+1} + \Phi_{m-1}$$

$$\text{Recursion:} \quad m+1 \Longrightarrow m$$

Initial conditions:

$$\Psi_0 = \Psi_1 = \Psi_2 = \Psi_3 = \Psi_4 = \Psi_5 = \Psi_6 = 1$$

$$\Phi_0 = \Phi_1 = \Phi_2 = \Phi_3 = \Phi_4 = 0$$

$$\Xi_0 = \Xi_1 = \Xi_2 = \Xi_3 = \Xi_4 = \Xi_5 = \Xi_6 = \Xi_7 = 1$$

Solution: $S_\ell^{(3,2)} = \Psi_{m=\ell}$

upper limit for the number of shapes actually formed by folding all sequences of a given sequence space $\mathcal{I}_\kappa^\ell$, which evidently obeys $|\mathcal{S}_\kappa(\ell)| \leq N_S(\ell)$. The cardinality of shape space, $|\mathcal{S}_\kappa(\ell)|$, can be obtained only by exhaustive folding and enumeration of mfe structures.

As an example we consider binary **GC**-sequences of chain length $\ell = 30$. The number obtained from the recursion formula is $N_S(30) = 760,983$; whereas, exhaustive enumeration of the shapes formed by binary **GC**-sequences yields $|\mathcal{S}_{\mathbf{GC}}(30)| = 276,570$. (For the parameters used in these computations, see Walter et al [114].) As many as 58,252 structures contain single base pairs resulting in 218,318 conformations representing acceptable mfe structures and being comparable to the computed number $N_S(30) = 760,983$ (table 3).[11] This is only a fraction of 28.7% of all acceptable structures, $N_S(30)$. A com-

[11]The fraction of structures with isolated base pairs increases with the chain length ℓ. It starts in the **GC**-case with 5.7% at chain length $\ell = 14$, reaches 12.8% at $\ell = 20$ and 21.1% at $\ell = 30$.

TABLE 3 Comparison of exhaustively folded sequence spaces [38, 40, 41, 96, 100]. The values are derived through exhaustive folding of all sequences of chain length ℓ from a given alphabet. The parameters used were those currently implemented in the Vienna RNA Package [48, 114]. The numbers refer to actually occurring minimum free-energy structures without isolated base pairs, which are directly comparable to the total numbers of acceptable structures $N_S(\ell) = S_\ell^{(3,2)}$ (see table 2).

	Number of Sequences			Number of Structures			
ℓ	2^ℓ	4^ℓ	$S_\ell^{(3,2)}$	AUGC	UGC	GC	AU
7	128	1.64×10^4	2	1	1	1	1
8	256	6.55×10^4	4	3	3	3	1
10	1,024	1.05×10^6	14	13	13	13	1
12	4,096	1.68×10^7	37	36	35	35	1
15	3.28×10^4	1.07×10^9	174	152	145	130	14
16	6.55×10^4	4.29×10^9	304	257	245	214	25
20	1.05×10^6	1.10×10^{12}	2,741		2,112	1,599	128
25	3.36×10^7	1.13×10^{15}	44,695			18,400	1,471
30	1.07×10^9	1.15×10^{18}	760,983			218,318	13,726

parison of $|\mathcal{S}_{\mathbf{GC}}(30)|$ to the cardinality of sequence space, $|\mathcal{I}_{\mathbf{GC}}^{(30)}| = 1.07 \times 10^9$, shows that the ratio of these numbers is indeed very small, $|\mathcal{S}_{\mathbf{GC}}(30)|/|\mathcal{I}_{\mathbf{GC}}^{(30)}| = 2.04 \times 10^{-4}$. In other words, the mean number of sequences forming the same structure is 4901. In case of four-letter sequences the sequence to structure ratio would be much larger since we have $|\mathcal{I}_{\mathbf{AUGC}}^{(30)}| = 1.15 \times 10^{18}$ (see also table 4). Thus we are dealing with many more sequences than shapes. Clearly, the mapping from sequence space onto shape space is many to one and hence noninvertible.

(ii) **Few common and many rare shapes.** The distribution of the numbers of sequences forming the same shape, $|S_k|$, is rather broad and strongly biased toward the rare-shape end. Analysis through exhaustive folding [40, 41] yielded a clear result independently of chain lengths ℓ and size of alphabet (**AUGC**: $\kappa = 4$; **GC**: $\kappa = 2$). There are relatively few common shapes and many rare ones. In the above-mentioned example, **GC**-sequences of chain length $\ell = 30$ ($\mathbf{GC}_{30}$), more than 93% of all sequences fold into common shapes, which are made up of only 10.4% of all shapes. An increase or decrease in chain length causes these percentages to go up or down, respectively, and in the limit of long chains, almost all sequences fold into a vanishingly small fraction of all shapes. It is worth looking at the $\mathbf{GC}_{30}$ shape space more closely [97].[12] The most frequent structures are formed by more than 1.5 million sequences, which is about 0.15% of sequence space. The shape of rank 10 (the tenth common structure) has still a pre-image of more than 1.2 million sequences. A glance at the rare frequency end is also illuminating: 12,362 shapes are formed by a single sequence only, 41,487 shapes by five or less sequences. The average number of sequences forming the same shape is 4,906, but 124,187 shapes, which is more than 57%, are formed by ≤ 100 sequences.

(iii) **Shape space covering.** Sequences forming common shapes are distributed (almost) randomly in sequence space. Accordingly, one need not search the entire sequence space in order to find a sequence that folds into a given common shape. One can indeed show that is sufficient to screen a (high-dimensional) sphere around an arbitrarily chosen reference sequence in order to find (with probability one) at least one sequence for every common shape [95]. The radius of this shape space covering sphere, $r_{\text{cov}}(\ell)$, can be estimated straightforwardly [96, 97]:

$$r_{\text{cov}}(\ell) = \min\left\{ h = 1, 2, \ldots, \ell \,|\, B_h(\ell, \kappa) \geq \frac{\kappa^\ell}{N_S(\ell)} \right\} ,$$

where B_h is the number of sequences contained in a ball of radius h and can be easily obtained from the recursion

$$B_h(\ell, \kappa) = \sum_{i=1}^{h} b_i(\ell, \kappa) \, ; \; b_i = b_{i-1} \cdot \frac{(\kappa - 1)(\ell + 1 - i)}{i} \, ; \; b_0 = 1 \, .$$

The covering radius for common shapes is much smaller than the radius of sequence space ($\ell/2$). For example, it amounts to $r_{\text{cov}} = 15$ for **AUGC**-sequences of chain length $\ell = 100$, and thus one has to search only a fraction of sequence space $\mathcal{I}_{\mathbf{4}}^{(100)}$ that contains a 4.52×10^{37}-th of all sequences in order to find at least one sequence for each of the common shapes.

(iv) **Common structures form extended neutral networks.** The pre-image in sequence space of a given shape S_j is the set of sequences $M_j = \psi^{-1}(S_j) \doteq \{I_k | \psi(I_k) = S_j\}$. A set of sequences can be converted into a graph $\mathcal{G} = (v[\mathcal{G}], e[\mathcal{G}])$ in sequence space with $v[.]$ and $e[.]$ denoting the vertices and edges, respectively. The *neutral network* $\mathcal{M}_j$ of S_j is constructed by identifying the sequences in M_j with the nodes and drawing edges between all nearest neighbors in the sequence space $\mathcal{I}_\kappa^\ell$ (these are the pairs of sequences

[12] The numbers reported here are taken from the data of [40, 41], which were computed on the basis of a previous parameter set [55]. For comparison, the currently used parameters yield 161,756 shapes formed by ≤ 100 sequences and 16,211 shapes formed by a single sequence only.

TABLE 4 Frequency of common shapes formed by **AUGC**-, **GC**-, and **AU**-sequences of chain length $\ell = 15$ as minimal free-energy structures. Shapes are ranked according to their frequencies. The most frequent structure has rank no. 1, the next frequent one rank no. 2, etc. The open chain is not shown. It is the most frequent conformation in the **AU**- and **AUGC**-alphabet and has rank 44 in the **GC**-shape space.

Stucture	AU-Alphabet		GC-Alphabet		AUGC-Alphabet	
	Rank	No. of Sequences	Rank	No. of Sequences	Rank	No. of Sequences
(((((•••)))))••	2	748	5	980	61	4,695,813
•(((((•••)))))•	3	640	21	624	65	3,922,132
(((((••••)))))•	4	618	4	982	59	5,312,006
•((((•••••))))•	5	472	22	568	36	8,076,124
•(((((••••)))))	6	352	7	928	62	4,344,220
((((•••••))))••	7	264	10	837	19	9,506,300
((((((•••))))))	8	256	34	384	91	1,292,843
(((((•••••)))))	9	176	39	356	73	2,820,975
••(((((•••)))))	10	156	8	910	67	3,777,323
•((((••••))))••	11	80	11	789	37	8,062,983
••((((••••))))•	12	68	13	758	29	8,575,254
•((((•••))))•••	13	32	17	699	43	6,873,069
••((((•••))))••	14	16	16	702	47	6,616,063
•••((((•••))))•	15	16	14	719	41	7,336,950
((((••••))))•••			1	1,446	13	11,003,892
((((•••))))••••			2	1,365	22	9,379,688
•••((((••••))))			3	1,178	34	8,204,329
(((((••••)))))•	4	618	4	982	59	5,312,006
(((((•••)))))••	2	748	5	980	61	4,695,813
(((••••)))•••••			9	908	2	15,829,017
(((•••••)))••••			18	668	3	15,662,822
••••(((••••)))•			26	501	4	12,875,889
•••(((•••••)))•			35	380	5	12,416,824
(((••••••)))•••			25	505	6	12,302,038

with Hamming distance $d^h_{ij} = 1$):

$$\mathcal{M}_j = \Big(v[\mathcal{M}_j] = \{I_k \,|\, I_k \in M_j\}, \\ e[\mathcal{M}_j] = \{(\overline{I_k I_{k'}}) \,|\, I_k, I_{k'} \in M_j \;\text{ and }\; d^h_{k,k'} = 1\}\Big).$$

The question, how sequences belonging to a neutral network $\mathcal{M}_j$ are distributed in sequence space, was answered by means of random graph theory [87, 88]. The central quantity of this approach is the average degree of neutrality of a given network, $\bar{\lambda}(\mathcal{M}_j) = \bar{\lambda}_j$. It is, in other words, the mean fraction of neutral neighbors of sequences belonging to the network: $\bar{\lambda}_j = \sum_{I_k \in \mathcal{M}_j} \lambda_k / |M_j|$, where λ_k is the number of nearest-neighbor sequences of I_k, which form shape S_j divided by the total number of nearest neighbors, $\ell \cdot (\kappa - 1)$. Neutral networks show a kind of percolation phenomenon. They are connected and span entire sequence space if $\bar{\lambda}_j$ exceeds a critical threshold value; whereas, they are partitioned into components with one dominating giant part and many small "islands" when $\bar{\lambda}_j$ is below threshold:

$$\mathcal{M}_j \;\text{ is }\; \begin{cases} \text{connected:} & \bar{\lambda}_j > \left(\bar{\lambda}\right)_{\text{cr}} = 1 - \kappa^{-\frac{1}{\kappa-1}} \,, \\ \text{partitioned:} & \bar{\lambda}_j < \left(\bar{\lambda}\right)_{\text{cr}} = 1 - \kappa^{-\frac{1}{\kappa-1}} \,, \end{cases}$$

where κ is the number of digits in the alphabet of nucleotide bases (**AUGC**: $\kappa = 4$). Connected areas on neutral networks are important in evolution since they define regions in sequence space that are accessible to populations through random drift [51].

The predictions on sequence-structure mappings of RNA shapes made by random graph theory were tested through exhaustive folding of entire sequence spaces [38, 40, 41]. In some cases, we found deviations from generic behavior, and these deviations could be explained by or derived from specific molecular structures. One particularly relevant and illustrative example was observed in the partitioning of neutral networks in the $\mathbf{GC}_{30}$ case. Random graph theory predicts that networks are either connected or their partition contains one largest "giant component." Analysis of the sequence of components ordered with respect to sizes revealed, however, that there also networks with two or four dominant components of equal size, or with three components of size distributions 1:2:1. Most sequences of chain length $\ell = 30$ form shapes with one double-helical region. The four single-stranded chains coming out from the stack form a hairpin loop on one side and zero, one, or two free ends on the other side. Hairpin loops fall into two groups: (i) loops with $n_{lp} = 3, 4$ and (ii) loops with five or more single bases. Loops of the former group cannot be shortened by forming an additional base pair at the end of the stack since one- and two-membered hairpin loops do not occur in real structures. In contrast, five-membered loops can be converted into a base pair and a triloop, six-membered loops into a base pair and a tetraloop, etc. Similarly, we find at the other side of the stack: (i) no additional base pair can be formed if the number of free ends is zero or one, but (ii) shapes with two free ends allow for elongation of the stack, provided the corresponding sequence requirements are fulfilled. Combination of two elements at the two ends of the stack leads to three different classes of shapes (fig. 5), which form neutral networks with different sequence distributions in sequence space. Shapes with stacks containing two category (i) ends of stacks (class **0**) (tri- and tetraloops, as well as zero or one free ends) form generic neutral networks (connected or with one largest component); shapes with one category (i) and one category (ii) end (class **1**) form networks with two largest components; and shapes with two category (ii) ends (class **2**), eventually, form those with three or four largest components. Interpretation of this finding is straightforward: Generic neutral networks (class **0**) show a distribution of sequences in sequence space, which is close to the binomial distribution (being fulfilled by the distribution of all sequences of length ℓ over an alphabet of size κ in sequence space):

$$B(\ell, h, \kappa) \;=\; \binom{h}{\ell} (\kappa - 1)^h \Big/ \kappa^\ell ,$$

where h is the Hamming distance between the reference sequence and the error class under consideration (h). For binary sequences ($\kappa = 2$), the distribution $B(\ell, h, 2)$ is symmetric in sequence space and the majority of all sequences is found in the error class $\ell/2$ (or in the error classes $(\ell \pm 1)/2$, respectively). An arbitarily chosen most frequent binary sequence (50% **G**, 50% **C**) will form a shape of class **1** with a reduced probability because 50% of these sequences have complementary symbols (**G**,**C**) at the end of the stack and will spontaneously form the shape with the additional base pair. The largest probability to form a class **1** shape is therefore displaced by some percentage $\pm\delta$ from the zone of highest frequency. As a matter of fact, we find indeed two components lying symmetrically with respect to the center of sequence space, one with excess **G** ($+\delta$) and the other with excess **C** ($-\delta$). By the same token, we explain the occurrence of three or four largest components for shapes of class **2**: There are two labile category (ii) ends, and the excess percentages of **G** and **C** are superimposed independently yielding four components of equal size (displacements: $+2\delta, +\delta - \delta = 0, -\delta + \delta = 0, -2\delta$), two displaced ones and two in the middle. Three components of sizes 1:2:1 originate from a four-component system through a merging of the two central components.

In the following, we compare a few of the most common shapes formed by **AUGC**-, **GC**-, and **AU**-sequences of chain length $\ell = 15$ (table 4) [101]. The limitation to such short chains is dictated by the cardinality of the four-letter sequence space (table 3) since handling of more sequences than a few billions is difficult and extremely time consuming. Nevertheless, the chains are already sufficiently long to allow for a certain variety of shapes. We find simple hairpins, e.g., ••(((((•••))))), and hairpins with internal loops or bulges, e.g., ((•(((•••)))•)) or ((((••••))••))•. No shapes with two hairpins were found at $\ell =$

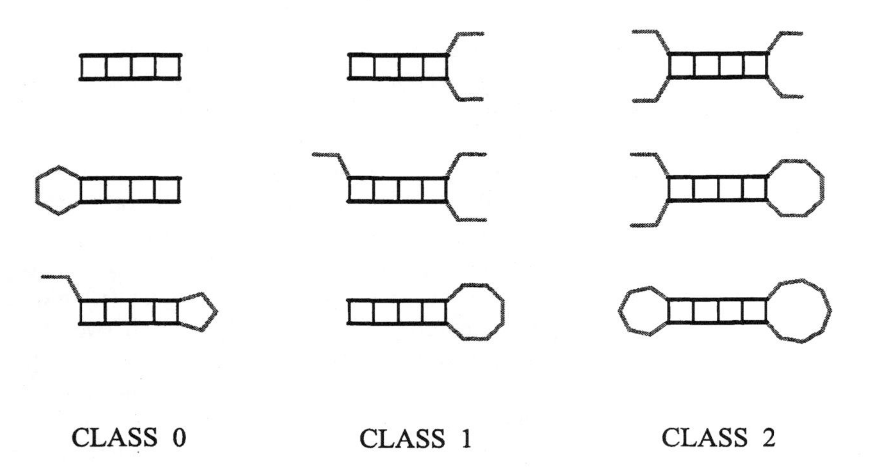

FIGURE 5 Classes of secondary structures with different distributions in sequence space. The three classes of structures sketched in the figure differ with respect to the ends of the stacking region. Class **0** structures contain a stack which cannot be extended by closing an additional base pair; class **1** structures can extend the stack on one end; and class **2** structures on both ends.

15. They occur only with longer chains, $\ell \geq 16$, e.g., ((•••)))•((•••))•. In table 4, we compare the most common shapes in the three alphabets by their ranks. All structures listed in the table are simple hairpin loops. More complex structures are less frequent. The frequency of the open chain conformation (not shown in the table) is a measure of the importance of end effects. In the sequence spaces $\mathcal{I}^{(15)}_{\mathbf{AUGC}}$ and $\mathcal{I}^{(15)}_{\mathbf{AU}}$ the open chain is the most frequent conformation, formed by 42.43% or 88.12% of all sequences, respectively, and hence, end effects dominate. In $\mathcal{I}^{(15)}_{\mathbf{GC}}$ the open chain has rank no. 44 and is formed by only 0.98%. Because of the weakness of the **AU** base pair relative to **GC**, mfe structures of **AU**-only sequences require long stacks in order to be sufficiently stable. Indeed, the fifteen structures of $\mathcal{S}_{\mathbf{AU}}(15)$ are tri-, tetra-, and pentaloops attached to stacks of four, five, and six base pairs. Almost always these conformations are less frequent in the other two sequence spaces where smaller numbers of base pairs are already sufficient for stable conformations. Comparing $\mathcal{I}^{(15)}_{\mathbf{AUGC}}$ and $\mathcal{I}^{(15)}_{\mathbf{GC}}$ we see that the **AUGC**-alphabet sustains about 15% more shapes than the **GC**-alphabet, 152 versus 130 (table 3). As expected, the **UGC**-case takes a position between **GC** and **AUGC**. Although the ranking of shapes shows substantial differences in the different alphabets, the most frequent shapes in one alphabet correspond to common shapes in the other, an exception being the **AU**-case, which does not sustain a sufficiently high number of stable conformations. The ranking of a shape according to its pre-image in sequence space is determined by two factors:

In order to be frequent a shape has to have (i) sufficient thermodynamic stability (sequences that form shapes with positive energies are counted for the open chain) and (ii) high combinatorial probability on the sequence level. Clearly, both factors depend on the size and the nature of the alphabet.

In order to close this section we compare the results presented here with experimental data from a recent investigation. Schultes and Bartel [93] chose two RNA-conformations of chain length $\ell = 88$. The two folds have no base pair in common, and the two reference sequences, LIG-P and HDV-P, represent two phylogenetically unrelated ribozymes with different catalytic activities, a synthetic specific RNA ligase [26] and a natural specific RNA cleaving ribozyme [84]. The authors then constructed one special RNA sequence, LIG-HDV, which is compatible with both secondary structures and thus can adopt both folds as stable or metastable conformations, respectively. The designed molecule did indeed show both catalytic activities, although the efficiencies were smaller by four orders of magnitude than those of the reference molecules. The chimeric ribozyme sequence, LIG-HDV, was then optimized independently for both catalytic functions by mutation and selection. This means two different optimization series were carried out, which led into different directions of sequence space. Only four- and two-point mutations were required to yield sequences, LIG-4 and HDV-2, which recovered the full catalytic efficiency of the "reference ribozymes," LIG-P and HDV-P, respectively. Still both optimized RNA molecules were more than Hamming distance 40 away from the reference sequences. For both RNA folds, the authors were able to track neutral paths of constant structure and full ribozymic activity from the mutant to the reference, i.e., from LIG-4 to LIG-P and from HDV-2 to HDV-P. This elegant work provides direct proof for the existence of extended neutral networks of both ribozymes. In addition, it shows that the two networks approach each other very closely in the surrounding of LIG-HDV, and thus can be understood as an experimental example of the intersection theorem [87]. Moreover, the sequence LIG-HDV is also an illustrative example of an RNA molecule which switches between two conformations [29].

4 DARWINIAN EVOLUTION IN SILICO

In this section the RNA model will be used as an example of a genotype-phenotype map in computer simulations of evolutionary optimization of RNA shapes or structure-related properties. At first, the simulation has to be embedded in a physically relevant environment, and we choose the flow reactor shown in figure 6 as an appropriate device. Such an experimental setup is known as "chemostat" in microbiology and used for continuous cultures of bacteria [62]. Apart from flow terms, the chemical reaction mechanism contains replication and mutation steps and is described by equations (3) or (9) with $\Phi(t) = \sum_j a_j x_j(t) V/N$. It is implemented as a stochastic process based on the underlying master equation. Individual trajectories are computed by means of an algorithm

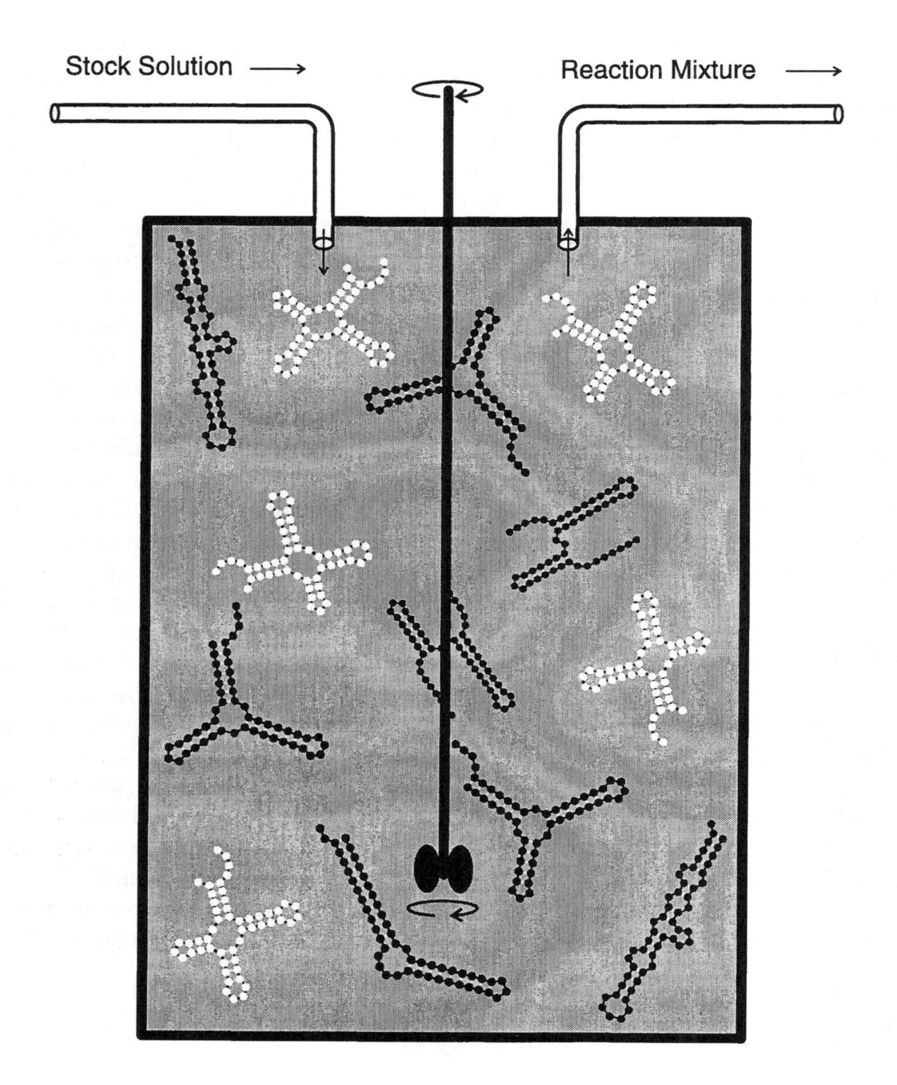

FIGURE 6 The flow reactor as a device for RNA structure optimization. RNA molecules with different shapes are produced through replication and mutation. New sequences obtained by mutation are folded into minimum free-energy secondary structures. Replication rate constants are computed from structures by means of predefined rules (see text). For example, the replication rate is a function of the distance to a selected stucture, which was chosen to be the clover-leaf-shaped tRNA shown above (white shape) in the reactor. Input parameters of an evolution experiment *in silico* are: the population size N, the chain length ℓ of the RNA molecules, and the mutation rate p.

conceived and analyzed by Daniel Gillespie [36, 37]. Under the constraints of the flow reactor, the population size fluctuates and has an expectation value of N and a standard deviation of $\sqrt{N}$. The replication rate constants are determined according to fitness criteria. In previous simulations [30, 31] the kinetic constants were derived from molecular structures by some predefined and biophysically motivated rules. Error-free replication and mutation are parallel reaction channels whose relative frequencies are given by equation (4). The single-digit accuracy of replication, q, corresponding to a mutation rate $p = 1 - q$ per site and generation, is an input parameter of the computations. Previous computer simulations confirmed three basic features of molecular evolution: (i) Population sizes of a few thousand molecules are sufficient for RNA optimization, (ii) stochastic effects dominate in the sense that the sequence of events recorded in one particular trajectory were never observed again in subsequent identical simulations,[13] and (iii) sharp error thresholds as predicted by the quasi species concept were observed in computer runs with different mutation rates.

More recently, computer simulations of replication and mutation in the flow reactor were used to show that evolution on the neutral network of a tRNA-structure corresponds to a diffusion process in sequence space, where the diffusion coefficient is proportional to the mutation rate [51]. In this simulation as well as in the computer experiments described below, replication rates were assumed to depend on the shape of the molecule independently of the sequence folding into it. Under this assumption, the neutrality condition for sequences folding into the same structure, $a_k = a(S_j) \, \forall \, I_k \in M_j$, is fulfilled. In particular, a function of the kind $a(S_j; S_\tau) = (\alpha + d^s_{j\tau}/\ell)^{-1}$ was used, where α is some constant, ℓ the chain length of the RNA, and $d^s_{j\tau}$ the distance between structure S_j and the target structure S_τ [35]. Many measures of distance between structures are conceivable [32], a particularly simple one is the Hamming distance between the shorthand "parentheses notations" (section 3) of the two shapes. For this goal shapes are understood as strings of the three symbols, '(', ')', and '•'.[9] Specific features like the efficiency of optimization and the time required to reach a particular goal are, of course, influenced by the model assumptions and parameters. Generic results concerning the course of evolution, however, were found to be largely independent of the specific choices of kinetic parameters, fitness functions, and distance measures.

Optimization of RNA structures was studied through simulations of the evolution of a population in the flow reactor [34, 35]. Every optimization run had an initial period during which almost every mutation led to an increase in fitness. After this first phase the approach toward the target structure, which happened to be a tRNA clover-leaf occurs in steps: Periods of fast decrease in

[13]By "two simulation experiments under identical conditions" we mean that everything was kept constant except the seeds for the random number generators.

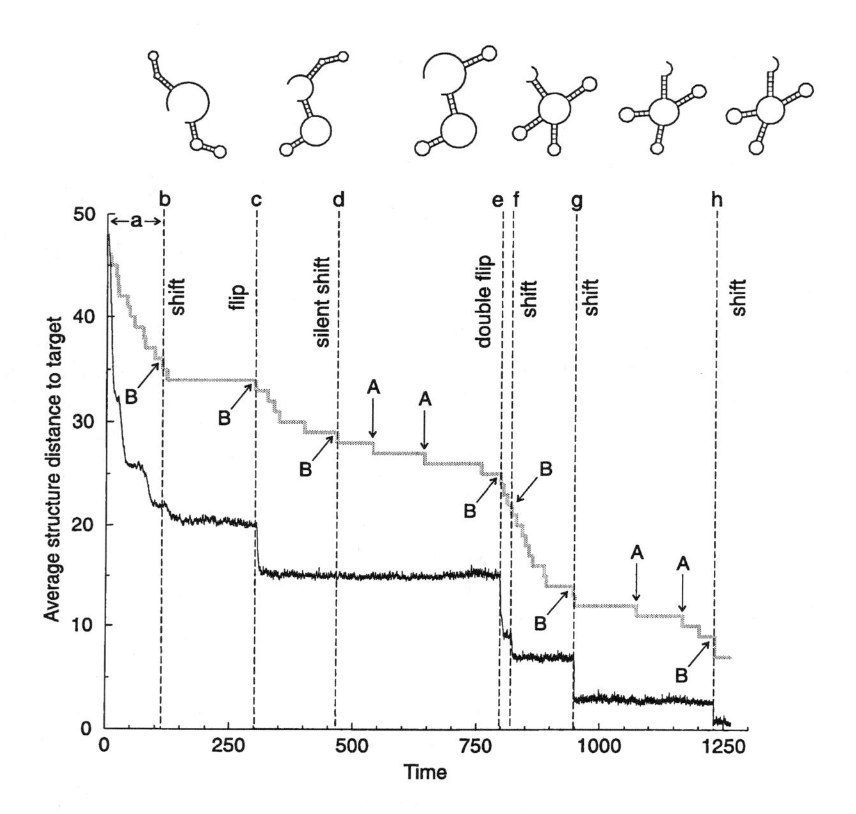

FIGURE 7 The recording of an RNA structure optimization experiment in the flow reactor. The computer experiment starts from a homogeneous initial population of 1000 RNA molecules with an arbitrarily chosen sequence folding into some initial shape. Fitness expressed as replication rate is computed as a function of the distance between current (S_j) and final structure (S_τ), $d^s_{j\tau}$. (For details, see text.) The final structure was chosen to be the clover leaf of tRNA^{phe} ($\ell = 76$). The mean distance to the final structure of the entire population, $\overline{d^s_\tau}(t)$ in equation (10) and plotted against time (black curve). The time scale represents the "real time" of the simulation experiment in arbitrary units. The whole simulation comprises about 1.1×10^7 replications. A mutation rate of $p = 0.001$ per site, and replication was applied. From this computer experiment a relay series of 42 shapes (or phenotypes) was reconstructed through backtracking the phenotypes that led to the final structure (see text and fig. 8). The six most important shapes are shown at the top of the figure. The relay series is indicated by the step function (grey), which assigns equal height to every shape. Transitions between phenotypes fall into two classes: (i) continuous (examples marked by A) and (ii) discontinuous (B). A more or less well-defined intial period of about one hundred time units is characterized by fast decrease in the distance to the final structure (a). The individual discontinous transitions are classified as "shifts," "flips," and "double flips," and marked by b to h. The "silent shift," at $t \approx 460$, is neutral with respect to distance to final structure. Discontinuous transitions lead to major changes in RNA shapes, which are followed by cascades of minor fitness-improving steps.

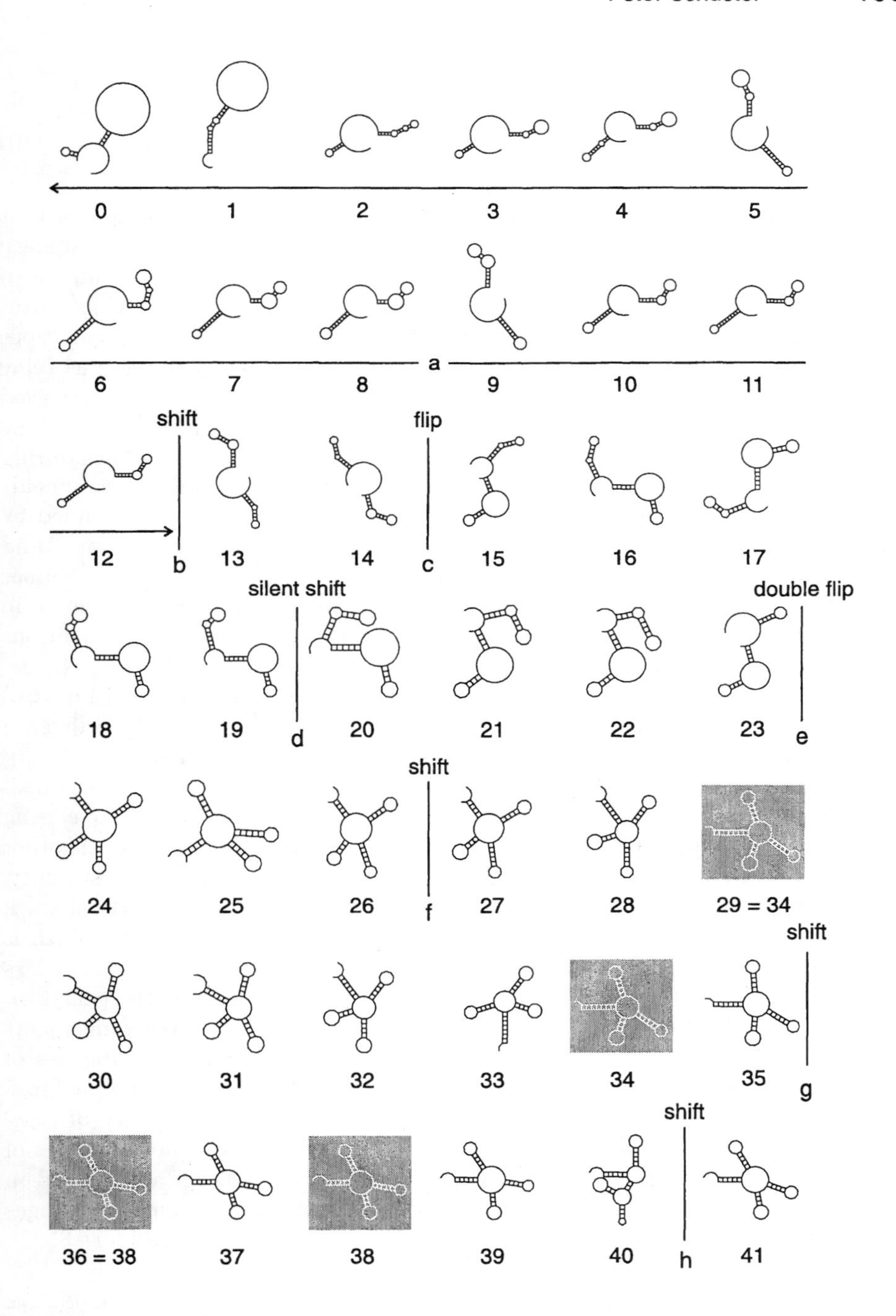

FIGURE 8 The relay series of the *in silico* optimization experiment described in figure 7. For details see text. It is worth noticing that a given shape may appear twice or more often in a relay series. Examples are the shapes $29 \equiv 34$ and $36 \equiv 38$.

the structure distance to final shape averaged over the whole population,

$$\overline{d_T^s}(t) = \sum_{j=1}^{n_s} p_j^s(t)\, d_{jT}^s \,, \tag{10}$$

are interrupted by long quasi-stationary phases or epochs of almost constant average fitness (fig. 7). In equation (10), n_s denotes the number of shapes; $p_j^s(t) = N_j^s(t)/N$ is the frequency of structure S_j; and $N_j^s(t)$ is the number of individual molecules with structure S_j. The course of the evolutionary optimization process was reconstructed through determination of a series of phenotypes leading from an intial shape to the final structure, called *relay series*. The relay series is a uniquely defined and uninterrupted sequence of shapes. It is retrieved through backtracking, that is, from the final structure to the initial shape. The procedure starts by highlighting the final structure and then traces it back during its uninterrupted presence in the flow reactor until the time of its first appearance. At this point, we search for the parent shape from which it descended by mutation. Then, we record time and structure, highlight the parent shape, and repeat the procedure. Recording farther back yields a series of shapes and times of first appearence, which ultimately ends in the initial population.[14] shown in figure 8. It contains 42 shapes produced through $n_{\text{rl}} = 41$ consecutive transitions (six characteristic structures along the series are shown on top of figure 7).

Transitions between two consecutive shapes in the relay series fall into two classes, A and B. Basis for this classification is the frequency of occurrence through mutations of the sequences from the reference neutral network (fig. 9), which manifests itself also in the underlying structural change. Class A transitions occur frequently on mutation and involve mostly minor changes like closing and opening of a base pair in the immediate neighborhood of a stack. Another example of a frequent transition is the opening of a stack of marginal stability, for example the terminal stack in the tRNA clover-leaf (the upper vertical stack in the secondary structure in figure 2). A mismatch in one base pair, which is readily produced by a single-point mutation is sufficient to open the stack. Class B transitions are rare events in the sense that they occur only with special sequences. They lead to major changes in structure. Such major rearrangements involve simultaneous displacement of several base pairs (Different subtypes of class B transitions were characterized as "shifts," "flips," and "double flips" depending on the details of the structural change [35]). The majority of rearrangements recorded in RNA optimization experiments with population sizes of a few thousand molecules are class A transitions. (Four of them are marked in figure 7.) Class B transitions are less frequent. For example, seven major changes are identified among the 41 transitions of the relay series shown in figure 8.

---

[14] It is important to stress two facts about relay series: (i) The same shape may appear two or more times in a given relay series. Then, it was extinct between two consecutive appearances. (ii) A relay series is not a genealogy which is the full recording of parent-offspring relations a time-ordered series of genotypes.

TABLE 5 Statistics of evolutionary trajectories. Different trajectories of *in silico* evolution toward a tRNA final structure were recorded for different values of the population size N [117]. All other parameters and conditions were chosen as described in figure 7. The length of the relay series is shown as the number of relay steps (n_{rl}). In addition, we show also the number of discontinuous or major transitions (n_{mt}) and the mean structure distance between the population at the end of the fast adaptive initial phase and the final shape: $\overline{d^s_{\mathrm{in},\mathcal{T}}} = \sum_{j=1}^{N(t_{\mathrm{in}})} d^s_{j,\mathcal{T}}$. Herein $d^s_{j,\mathcal{T}}$ is the structure distance between $S_j = \psi(I_j)$ and the final shape $S_{\mathcal{T}}$, and t_{in} the time at which the intital phase ends.

Population Size N	Number of Runs	Number of Relay Steps n_{rl}	Number of Transitions n_{mt}	Initial Phase $\overline{d^s_{\mathrm{in},\mathcal{T}}}$
1,000	10	120.1 ± 114.0	6.5 ± 1.7	17.6 ± 2.3
2,000	13	66.3 ± 25.8	6.5 ± 1.7	18.5 ± 2.3
3,000	12	41.9 ± 16.6	6.3 ± 2.2	17.6 ± 2.4
10,000	17	37.8 ± 11.8	5.7 ± 1.3	16.5 ± 1.0

Class A and class B transitions can be generalized in terms of neighborhood frequencies of neutral networks (fig. 9):

(i) **Continuous transitions** (A). They represent minor structural changes and lead to structures that are globally frequent in the neighborhood of the neutral network of the initial shape.

(ii) **Discontinuous transitions** (B). They involve major stuctural changes leading to globally rare and only locally frequent structures. Accordingly, discontinuous transitions require special sequences that allow major structural changes to occur on single-point mutations.

Typical simulations show an initial period ($0 \leq t \leq t_{\mathrm{in}}$; marked a in figure 7) of cascading discontinuous and continuous transitions followed by a stepwise optimization process with apparent regularities. Each epoch or quasi-stationary phase of evolution ends with a discontinuous transition. Discontinuous transitions (b to h), however, occur only rarely within quiescent periods.[15] Every discontinuous transition is followed by a cascade of continuous transitions, which are accompanied by fitness increase. Then, the population approaches the next plateau corresponding to an epoch of neutral evolution at approximately constant fitness. Along the plateau, the relay series shows neutral mutations with

[15]There is one case (d) at time $t \approx 460$ in the computer simulation of figure 7. A discontinuous transition is observed inside an epoch. We called it "silent" since it does not change the distance to target and is neutral with respect to fitness.

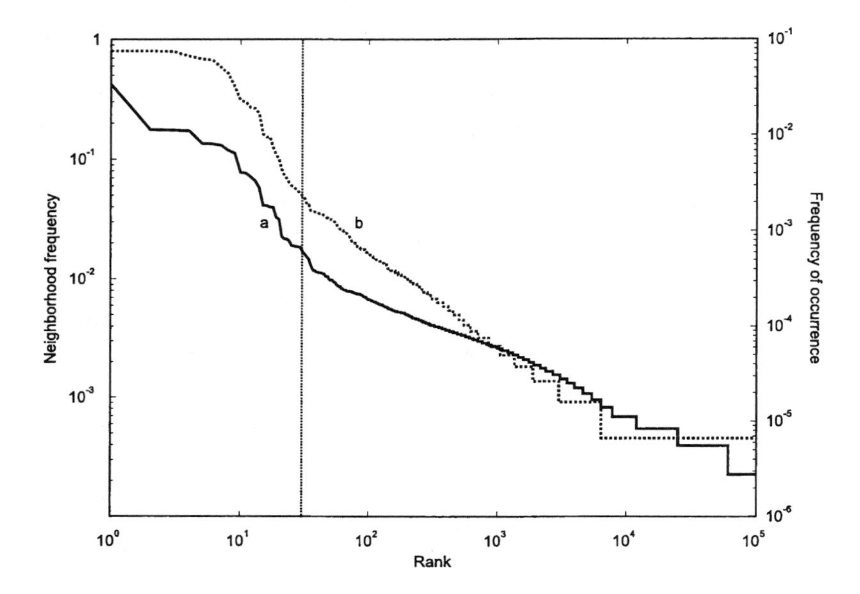

FIGURE 9 Statistics of shapes in the boundary of tRNA$^{\rm phe}$. The basis of the statistics are 2199 sequences folding into the clover-leaf structure shown in figure 2. All their one-error mutants, 501,372 in number, were folded. A fraction of 28% formed the same clover-leaf as the reference sequence and thus belonged to the neutral network. The remaining 358,525 sequences folded into 141,907 distinct shapes. Curve a is a log-log plot of the rank ordered frequency of occurrence (full line, right ordinate). The neighborhood frequency is plotted in curve b (dotted line, left ordinate). The dotted vertical line is meant to separate regions with different scaling: A region of frequent occurence (left) is distiguished from the power-law distribution (right), which is typical for scaling according to Zipf's law [122].

respect to structure or fitness neutral class A transitions until it reaches one of the special sequences from which a fitness-improving discontinuous transition is locally frequent and hence attainable with sufficiently high probability. Evolutionary optimization on landscapes with high degree of neutrality proceeds on two time scales: Fast periods containing cascades of adaptive changes are interrupted by long quasi-stationary epochs of neutral evolution during which populations drift randomly on neutral networks until they reach a neighborhood that is suitable for the next discontinuous transition.

The analysis of the computer simulation experiments led to a novel notion of evolutionary nearness between phenotypes, which is based on the concept of neutral networks [35]. In order to explain nearness we consider a shape S_j, its pre-image $M_j = \psi^{-1}(S_j)$, and the corresponding network $\mathcal{M}_j$. The

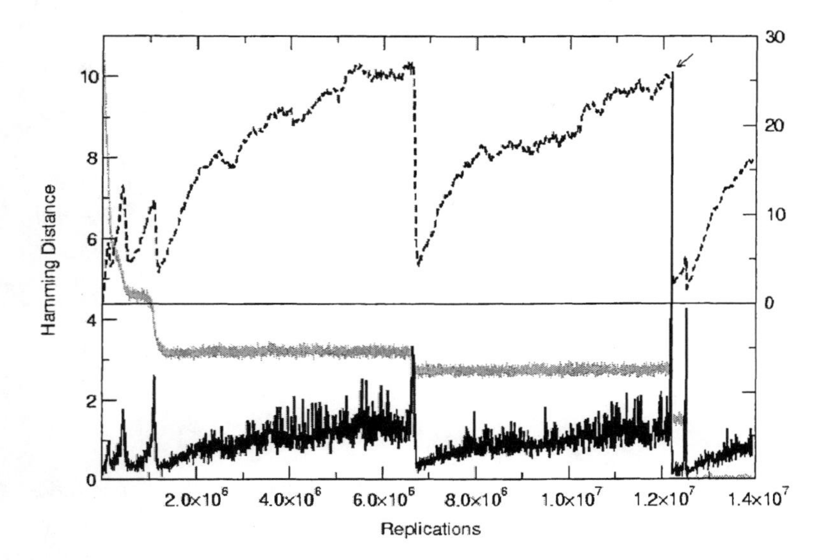

FIGURE 10 Variability in genotype space during punctuated evolution. Shown are the results of a simulation of RNA optimization between initial structure and a tRNA final structure (analogous to the run in figure 7) with population size $n = 3,000$ and mutation rate $p = 0.001$ per site and replication. The figure contains two plots with different measures of genetic diversity, $d_P(t, \Delta t)$ and $d_C(t, \Delta t)$ with $\Delta t = 8,000$ replications, against time, which is expressed as the total number of replications performed so far, and the trace (grey) of the underlying trajectory recording average distance from final structure. The upper plot contains the mean Hamming distance between the population (d_P; dotted line, right ordinate) at time t and time $t + \Delta t$ and the lower one shows the Hamming distance between the mean sequences at the same moments (d_C; full line, left ordinate). The arrow indicates a remarkably sharp peak of $d_C(t, 8,000)$ at the end of the second long plateau which reaches a Hamming distance of about 10. Every adaptive phase is accompanied by a drastic reduction in the genetic diversity while genetic variation increases during quasi-stationary epochs. The mutant cloud, whose average size is expressed by $d_P(t, \Delta t)$, expands quickly during neutral evolution and reaches diameters up to Hamming distance 25; whereas, the center of the cloud migrates only at a speed of Hamming distance 1 per 8,000 replications.

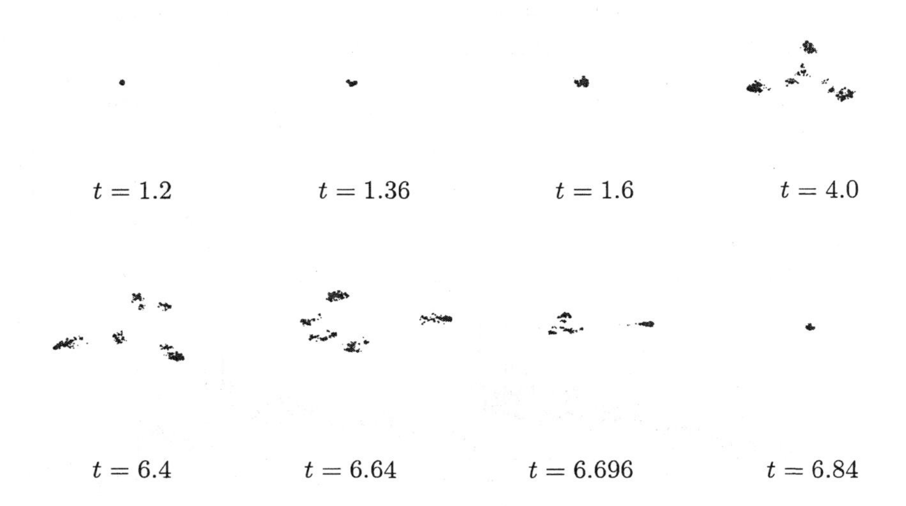

FIGURE 11 Spreading of a population in genotype space during a quasi-stationary epoch. The individual figures are snapshots of the genotype distribution at times corresponding to 1.2, 1.36, 1.6, 4.0, 6.4, 6.64, 6.696, and 6.84×10^6 replications. In order to visualize spreading, genotype distributions were transformed to principal axes and individual sequences were projected onto the plane spanned by the two largest eigenvectors. Along the series we observe an important and characteristic feature of population spreading in neutral evolution. The populations break up into smaller subclusters, which diffuse radially away from the center of the distribution (see also the model on neutral evolution discussed in [14, 51]). Whenever an innovation with increase in fitness happens in one of the subclusters, this subcluster takes over further development and all other subclusters die out.

boundary of the network, $B_j = \mathrm{bd}(M_j)$, is the set of sequences that can be reached from M_j by a single mutation event but do not belong to M_j. Folding the entire set into shapes yields a distribution of phenotypes, Σ_j, which is the image of B_j in shape space. The error rates applied in the computer simulations reported here (almost always) lead either to correct replication or single-point mutations. Hence, the boundary is the set of genotypes that are produced as one-error mutants of the genotypes belonging to the network.[16] A ranked frequency distribution of phenotypes in the boundary of a neutral network shows, in general, two clearly separable zones (fig. 9): A relatively small number of frequent phenotypes is contrasted by a large number of rare phenotypes. The most common shapes are of comparable frequency and usually closely related

[16]At higher error rates it might useful to define two-error, three-error or, in general n-error boundaries [35].

to the parent shape of the neutral network (S_j). The distribution of the rare phenotypes fulfills a power-law distribution, known as Zipf's law [122], which implies that the log(frequency)/ log(rank)-plot is a straight line. The results are essentially the same for the two different distributions presented in figure 9: (i) the frequency of occurrence, which counts the total number of sequences in the boundary that form the shape in question, and (ii) the neighborhood frequency, which counts the number of neighborhoods where the shape occurs. The transitions to high-frequency shapes in the boundary are the ones we called continuous (and, in other words, they occur readily on single-point mutations). The threshold between frequent and rare shapes in the boundary can be defined intuitively: It occurs near the ranks where the linear range of the log–logplot starts (see the straight line in fig. 9). Transitions to shapes of low frequency in the boundary do not occur readily because they have sufficiently high probability only at certain special positions on the network $\mathcal{M}_j$ (which, for example, have to be found by the population through random drift). Accordingly, we called them discontinuous.

The data on the distribution of shapes in the boundary of neutral networks suggest considering nearness in a statistical sense. A shape S_k is (statistically) "near" S_j when its frequency in the boundary B_j is above threshold. Let $\rho(S_k; S_j) = \gamma(S_k, S_j)/|B_j|$ be the frequency of occurrence of S_k in B_j where $\gamma(S_k, S_j)$ is the number of Hamming distance 1 contacts between the two neutral networks, and $|B_j|$ the cardinality of the boundary. Further, let ε be a properly defined threshold value for the frequency in the boundary, then the set $\Psi_\varepsilon(S_j) = \{S_k \in \Sigma_j | \rho(S_k; S_j) \geq \varepsilon\}$ defines the statistical neighbors of S_j which are accessible through continuous transitions. It is important to note that $\rho(S_k; S_j)$ does not fulfil the conditions of a metric. In general, it is neither symmetric, $\rho(S_k; S_j) \neq \rho(S_j; S_k)$ nor does it necessarily fulfill the triangle inequality (although, of course, $\gamma(S_k, S_j) = \gamma(S_j, S_k)$ is always true). In other words, the statement "S_a is statistically near S_b" does not imply that S_b is near S_a. This paradox, however, is readily solved when one recalls that two networks may have (very) different sizes in sequence space. The larger network can occupy a fairly high percentage of the positions in the boundary of the smaller network whereas, at the same time, the smaller one is present only at low frequency in the boundary of the larger network.[17] The proper mathematical context for accessibility

[17]For examples of pairs of neutral networks of RNA molecules lacking symmetry in the statistical neighborhood see [34]. We mention one representative case: a clover-leaf tRNA shape (S_{tRNA}) and a conformation with three hairpins, which originates from the clover-leaf through opening of the terminal stack (S_{3hp}; the terminal stack is the upper vertical stack of the secondary structure in figure 2). The latter, S_{3hp}, is near S_{tRNA} in a statistical sense, but the inverse is not true, S_{tRNA} is not statistically near S_{3hp}. The interpretation of this asymmetry is straighforward. Let $\gamma(S_{\text{tRNA}}, S_{\text{3hp}})$ be the number of Hamming distance 1 contacts between the two networks. The sizes of the two networks are very different: $|M_{\text{tRNA}}| \ll |M_{\text{3hp}}|$ and hence, the boundaries too, $|B_{\text{tRNA}}| \ll |B_{\text{3hp}}|$, which leads to

$$\rho(S_{\text{tRNA}}; S_{\text{3hp}}) = \gamma(S_{\text{tRNA}}, S_{\text{3hp}})/|B_{\text{3hp}}| \ll \rho(S_{\text{3hp}}; S_{\text{tRNA}}) = \gamma(S_{\text{3hp}}, S_{\text{tRNA}})/|B_{\text{tRNA}}|,$$

and thus the statistical neighborhood relation does not commute, q.e.d.

of phenotypes as well as continuity and discontinuity in evolution is still being developed [11, 12], but one can recognize already the usefulness of the concepts of statistical neighborhoods for simple as well as for general and highly complex genotype-phenotype mappings.

The time dependence of genetic diversity during evolution *in silico* is shown in figure 10. We apply two measures to visualize the diversity of genotypes in the population: (i) the mean Hamming distance within or between populations,

$$d_P(t,\Delta t) \;=\; \frac{\sum_{j=1}^{N(t)}\sum_{k=1}^{N(t+\Delta t)} d^h(I_j,I_k)}{N(t)\cdot N(t+\Delta t)}$$

and (ii) the Hamming distance between the mean nucleotide sequences at two different times t and $t+\Delta t$,

$$d_C(t,\Delta t) \;=\; \sum_{k=1}^{\ell}\sqrt{1-\sum_{j=\mathbf{A},\mathbf{U},\mathbf{G},\mathbf{C}} \pi_j^{(k)}(t)\,\pi_j^{(k)}(t+\Delta t)}\;.$$

The vector $\vec{\pi}^{(k)}(t)=\{\pi_{\mathbf{A}}^{(k)},\pi_{\mathbf{U}}^{(k)},\pi_{\mathbf{G}}^{(k)},\pi_{\mathbf{C}}^{(k)}\}$ is the square-normalized distribution of nucleotides at position k: $\pi_i^{(k)} \;=\; \alpha_i^{(k)} \;/\; \sqrt{\sum_{j=\mathbf{A},\mathbf{U},\mathbf{G},\mathbf{C}}\left(\alpha_j^{(k)}\right)^2}$ with $\sum_{j=\mathbf{A},\mathbf{U},\mathbf{G},\mathbf{C}}\alpha_j^{(k)}=1$. The former distance, $d_P(t,\Delta t)$, describes, in essence, the spreading of the population in sequence space; whereas, the latter, $d_C(t,\Delta t)$ is a measure for the migration of the center of the distribution. In figure 10, time is measured in terms of replications rather than in "real time" in order to come as close as possible to the number of generations. One generation corresponds to N replications on the average. We recognize an increase in genetic diversity during the quasi-stationary epochs of apparent constancy in shape space. The quantity $d_P(t,\Delta t)$ is an appropriate measure of the average diameter of the mutant cloud.[18] The size of the mutant cloud increases with time on the fitness plateaus and drops drastically when the population undergoes a discontinuous transition at the end of the epoch. The change in the mean nucleotide sequence of the population increases also during the quasi-stationary phase and saturates at values slightly above Hamming distance 1 per 8,000 replications. At the end of every epoch, we see a sharp or spike-like peak in $d_C(t,\Delta t)$, which indicates a bottleneck in genotype space through which the population passes during discontinuous transitions. In order to illustrate the spreading of populations, we recorded the image of the population in genotype space on the fitness plateau between 1.2 and 6.9×10^6 replications (fig. 11). After having passed the bottleneck of the previous discontinuous transition, the population starts

[18]The delay time Δt was chosen to be identical in both quantities, $d_P(t,\Delta t)$ and $d_C(t,\Delta t)$. We remark that $d_P(t,\Delta t)$ changes hardly within the time span considered: $d_P(t,8,000)$ is almost identical with $d_P(t,0)$.

instantaneously to expand. At a time corresponding to 2.8×10^6 replications, the population breaks up into subpopulations, which diverge further in sequence space. At 6.4×10^6 replications an advantageous mutant appears in one of the subclusters, this subcluster takes over, and all other subclusters die out. The corresponding fast shift in the consensus sequence of the population gives rise to the spike recorded in $d_C(t, \Delta t)$ (fig. 10). The observation of random drift and partitioning into subclusters is in agreement with previous results [14, 51] and shows that the replication-mutation mechanism cannot sustain diffuse mutant clouds because the positions of descendants in sequence space are inevitably close to those of their parents. Although the picture of genotypic versus phenotypic evolution obtained from *in silico* simulation is much more detailed than the results recorded with bacterial populations [83], we see general agreement in the fact that, compared to adaptive periods, genomic evolution is at least equally fast or even faster during the phases of phenotypic stasis in evolution.

In order to study the population size dependence of evolutionary optimization, we performed several computer runs and calculated the statistics of trajectories (table 5). The number of relay steps (n_{rl}) shows vast scatter but decreases considerably with increasing population size N. This finding is easily interpreted. At larger population sizes the relay series contains fewer structures because individual shapes are less likely to die out and thus stay longer in the population. Interestingly, the number of major transitions (n_{mt}) shows smaller scatter and stays fairly constant within the investigated variation of population sizes. Similarly, the mean distance from the shape at the end of the initial adaptive period (a in fig. 7) to the final structure, $\overline{d^s_{\text{in},\tau}}$, does not change significantly with population size. It is obvious to suggest that the number of major transitions and the distance from final structure after the initial phase are more or less set by the final shape itself. The path between a randomly chosen initial structure and a given final shape is determined by the size and the distribution of its neutral network in sequence space.

5 DARWINIAN EVOLUTION AND INFORMATION

The mechanism of Darwinian evolution makes use of the powerful interplay of chance and necessity being identified with variation and selection, respectively [99]. Both processes can be traced down to the molecular level and studied *in vitro* by means of physical and chemical techniques as well as *in silico* by computer simulation. In order to analyze evolutionary optimization in full detail, we introduce here a dynamical concept that starts out from chemical kinetics of replication and extends conventional population genetics (i) by visualizing mutation as a process in genotype or sequence space, and (ii) by introducing the phenotype as an integral part of the model. Central to this concept is a genotype-phenotype map, which is used to derive phenotypic properties from genomic sequences and, in particular, to evaluate fitness parameters that are incorporated into the rate equations for the genotypes. Because of the enormous complexity of ordinary phenotypes such maps are not available yet except for the most simple evolutionary scenario consisting of RNA evolution in test tubes where the phenotype is tantamount to the molecular structure and its properties. At present, the only tractable case of such a mapping is the sequence-secondary structure map of RNA molecules. An RNA model based on this admittedly simple but, nevertheless, realistic mapping was used, for example, in computer simulations of evolutionary optimization and yielded, in essence, four results that are readily generalized to other, more complex biological systems.

(i) Evolution may show punctuation even under precisely constant environmental conditions like those encountered in a flow reactor. In other words, no external triggers are required for a stepwise course of optimization processes. Evolution occurs on two time scales: Adaptive processes dominated by selection are comparatively fast, and random drift on neutral networks is slow. Both adaptive evolution and random drift contribute to the success of optimization processes. It was assumed for long time that random drift is a kind of unavoidable noise and has no positive effect on evolution. Diffusion on neutral networks, however, enables populations to escape from local fitness optima, which otherwise would act as evolutionary traps. This result supports the view on the role of neutral evolution elucidated by Emile Zuckerkandl [123] in the context of the development of genetic regulatory networks. He addresses neutral and nonneutral mutations as a "creative mix" in evolution.

(ii) Accessibility of phenotypes determines the progress of evolution. A notion of nearness between phenotypes has been developed that accounts for the existence of neutral networks [35]. Nearness is defined in a statistical sense with respect to the frequency of occurrence of phenotypes in the one-error neighborhood of neutral networks. A phenotype that is frequently found in this neighborhood can be reached from almost every sequence of the network, and the corresponding transitions were characterized as continuous because they occur almost instantaneously. Transitions to phenotypes that are infrequent in the neighborhood occur rarely and then only at special positions of the network. They were denoted as discontinuous. In other words, discontinuous transitions are locally frequent but globally improbable. Within the frame of the RNA model, the frequencies of transitions are readily interpreted in terms of probabilities of structural changes caused by single-point mutations. Discontinuous transitions and the major phenotypic changes they are associated with can be interpreted as the "real innovations" in Darwinian evolution.

(iii) Changes in genotypes may but need not be reflected by changes in the phenotypes, and thus the relative pace of genomic and organismic evolution is a central issue in biology. Computer simulations provide direct insight into this problem. The dispersion of genotypes in the population varies strongly and systematically during evolution. Strong increase of genomic diversity is observed during the diffusion on a neutral network. Simultaneously with the spread, the population is split into smaller clusters of sequences, which represent individual clones. At the same time, the center of the population in sequence space shows only small drift. A discontinuous transition is commonly manifested by a dramatic drop in the diversity of genotypes and a "jump" in the center of the population. Discontinuous transitions may be interpreted as "bottlenecks" in genomic evolution. The population becomes almost uniform during the passage and then spreads again on the next neutral network. The large shifts in genotype space observed with the population centers mean that the discontinuous transition starts out from one particular clone, which becomes dominant and then represents the new center on the population. It is straightforward to compare differences between evolution of genotypes and phenotypes in terms of changes per generation. Computer simulations show that genomic evolution speeds up through spreading of populations on neutral networks; whereas, phenotypic evolution measured in terms of fitness or distance to the final shape is slow or practically zero, as seen from the almost constant fitness plateaus. Major changes in phenotypes initiated by discontinuous transitions are accompanied by a drop in genetic diversity. Similar inverse relations between genomic and phenotypic change were found in the analysis of long time evolution experiments with bacteria [83].

(iv) How does the population size influence the evolutionary scenario described here? First of all, the larger the population size, the smaller is the number of individual steps in the relay series. Interestingly, the occurrence of major transitions as well as their numbers were found to be quite insensitive to variations of the population size. Punctuated evolution was found, after all, also with bacterial populations whose size was in the order of $5 \times 10^6 < N < 5 \times 10^8$ cells or in experimental setups with up to 10^{15} RNA molecules. Apparently, the transition from stochasticity to deterministic behavior occurs at much larger population sizes as far as these phenomena are concerned. Because of the enormously high numbers of genotypes evolution remains inherently stochastic even at the largest population sizes that can be realized. The same seems to be true for evolution on neutral networks. Clearly, larger populations would stay longer connected and break up in subclusters later (or not at all). The other qualitative features discussed here, in particular the time dependence of $d_P(t, \Delta t)$ and $d_C(t, \Delta t)$ shown in figure 10, would remain largely uneffected by a change in population size.

(v) The landscape concept, originally introduced by Sewall Wright [120] as a metaphor into evolutionary biology, was put on firm scientific grounds when applied to biopolymers, in particular proteins and RNA molecules. Molecular biophysics revealed the basis of neutrality and neutral evolution, which population geneticists could deduce only indirectly from comparisons of sequence data in contemporary organisms. The RNA model is currently based on shapes defined as minimum free-energy structures. The concept, however, is very flexible because additional structural features can be taken into account readily. Such features are, for example, the consideration of suboptimal conformations within reach from the ground state at room temperature or the kinetics of the folding process. In addition, consideration of tertiary interactions in RNA molecules will lead to truly three-dimensional structures. We can indeed expect a great variety of new phenomena that wait to be detected in such extended molecular models. These new regularities will help to illustrate and understand otherwise too-complex-to-analyze peculiarities of macroscopic evolution.

Finally, we address the question of genetic information and how it is created through evolution in Darwinian systems. The principle of variation and selection is a special case of self-organization and thus requires self-enhancement and nonequilibrium conditions. In biology and in assays that mimic biological evolution *in vitro*, self-enhancement is tantamount to multiplication. Furthermore, it is necessary to produce variants, on which selection can act, and to have a storage device, which keeps track of the past in the sense of inheritance. All these requirements are already fulfilled with RNA molecules replicating in test tubes and, indeed, the origin of genetic information can be visualized within the concept of the molecular quasi species [6, 22, 23, 25]. A population gains information in the course of the selection process since the selected molecules carry a kind of indirect "image" of their environment. The higher the fitness, the better the image, since it allows for better exploitation of the resources.

The ultimate basis of information in biology is interaction between molecules or molecular recognition. This recognition is improved during selection. Creation of information, however, requires more than just increase in strength and specificity of interactions, it requires the capacity of storage of past experience and retrieval in the sense of information processing. Although recognition is not restricted to a certain class of molecules, information processing is dependent on more stringent requirements that are fulfilled only by molecules that are both, sufficiently stable and able to act as templates in a copying process. Only a copying mechanism guarantees that replication errors, once they have occurred, can be transmitted to future generations and thus subjected to selection. One—and the only—class of molecules which are presently known to be suitable for this purpose are the nucleic acids. No wonder that evolution of molecules and its applications to biotechnology [28, 39, 107, 116] were more or less restricted to the use of RNA and DNA. Design of proteins or organic molecules by means of selection techniques requires either translation, which couples protein design to nucleic acid evolution, or direct intervention by the experimentalist through screening techniques.

How do neutral networks and explicit consideration of phenotypes modify or change the conventional view of the origin of genetic information? The answer is not straightforward. First, the widespread selectionist's view saying that (almost) "every adjustment to the environment is possible and can be achieved through an eventually very large number of infinitesimally small steps" is not true. The set of attainable conformations is usually substantially smaller than the set of all (possible) phenotypes, and steps need not be small. At the current state of the art, estimates on the accessible fraction of the "universe of possible phenotypes" are highly uncertain, and we have to wait for more information before we can give a decisive answer. However, it is certainly possible to find examples demonstrating lack of accessibility: So far all attempts failed to obtain a tRNA-like shape through evolutionary optimization of RNA molecules built over a two-letter **GC**-alphabet, although it is straightforward to show that such molecules do exist and are stable (see, for example, [35, p. 513]). Second, the sequences forming a neutral network for the mfe structure commonly differ with respect to suboptimal conformations or folding properties. Since these properties may contribute to fitness as well, the migration of the population on the neutral network corresponding to the ground state conformation may in fact follow a (small) fitness gradient. In the case of such flexible phenotypes, "neutral evolution" is not strictly neutral any more and may lead to structures that receive more complex properties through the appropriate suboptimal conformations and/or suitable folding patterns. Then, "guided drift" builds upon the properties that are not encoded in the conformation of the ground state, and thus diffusion on networks (which are strictly neutral only with respect to the mfe structure) may also generate genetic information. Further investigations on RNA evolution of and with adaptable phenotypes will shed more light on the underlying mechanisms.

ACKNOWLEDGMENTS

The work on the molecular quasi species is the results of a cooperation with Manfred Eigen and John S. McCaskill then at the Max-Planck-Institute of Biophysical Chemistry in Göttingen, Germany and Karl Sigmund from the Institute of Mathematics at the University of Vienna, Austria. The RNA model was developed during long-time research of our group at Vienna University. Many discussions with Manfred Eigen, Christoph Flamm, Walter Fontana, Ivo Hofacker, John McCaskill, Karl Sigmund, and Peter Stadler are gratefully acknowledged. Andreas Wernitznig kindly provided additional computer plots on *in silico* evolution experiments in the flow reactor. Financial support of the work presented here was provided by the Austrian *Fonds zur Förderung der wissenschaftlichen Forschung* (Projects P-13,093 and P-13,887), by the *Jubiläumsfonds der Österreichischen Nationalbank* (Project 7813), by the Commission of the European Union (Project PL-97,0189), and by the Santa Fe Institute.

REFERENCES

[1] Alves, Domingos, and Jose Fernando Fontanari. "A Population Genetic Approach to the Quasispecies Model." *Phys. Rev. E* **54** (1996): 4048–4053.

[2] Alves, Domingos, and Jose Fernando Fontanari. "Error Thresholds in Finite Populations." *Phys. Rev. E* **57** (1998): 7008–7013.

[3] Bartel, David P., and Jack W. Szostak. "Isolation of New Ribozymes from a Large Pool of Random Sequences." *Science* **261** (1993): 1411–1418.

[4] Batey, Robert T., Robert P. Rambo, and Jennifer A. Doudna. "Tertiary Motifs in Structure and Folding of RNA." *Angew. Chem. Int. Ed.* **38** (1999): 2326–2343.

[5] Biebricher, C. K., and W. C. Gardiner. "Molecular Evolution of RNA *in vitro*." *Biophys. Chem.* **66** (1997): 179–192.

[6] Brakmann, Susanne. "On the Generation of Information as Motive Power for Molecular Evolution." *Biophys. Chem.* **66** (1997): 133–143.

[7] Breton, Nicolas, Christine Jacob, and Patrick Daegelen. "Prediction of Sequentially Optimal RNA Secondary Structures." *J. Biomol. Struct. Dynam.* **14** (1997): 727–740.

[8] Campos, Paolo R. A., and Jose Fernando Fontanari. "Finite-Size Scaling of the Quasispecies Model." *Phys. Rev. E* **58** (1998): 2664–2667.

[9] Campos, Paolo R. A., and Jose Fernando Fontanari. "Finite-Size Scaling of the Quasispecies Model." *J. Phys. A: Math. Gen.* **32** (1999): L1–L7.

[10] Chao, Lin, and Edward C. Cox. "Competition between High and Low Mutation Strains of *Escherichia coli*." *Evolution* **37** (1983): 125–134.

[11] Cupal, Jan, Stephan Kopp, and Peter F. Stadler. "RNA Shape Space Topology." *Art. Life* **6** (2000): 3–23.

[12] Cupal, Jan, Peter Schuster, and Peter F. Stadler. "Topology in Phenoptype Space." In *Computer Science in Biology*, edited by R. Giegerich, R. Hofestädt, T. Lengauer, W. Mewes, D. Schomburg, M. Vingron, and E. Wingender, 9–15. GCB'99 Proceedings. Hannover, DE: University of Bielefeld, 1999.

[13] Demetrius, Lloyd, Peter Schuster and Karl Sigmund. "Polynucleotide Evolution and Branching Processes." *Bull. Math. Biol.* **47** (1985): 239–262.

[14] Derrida, Bernard, and Luca Peliti. "Evolution in a Flat Fitness Landscape." *Bull. Math. Biol.* **53** (1991): 355–382.

[15] Domingo, E. "Biological Significance of Viral Quasispecies." *Viral Hepatitis Rev.* **2** (1996): 247–261.

[16] Domingo, E., and J. J. Holland. "RNA Virus Mutations and Fitness for Survival." *Ann. Rev. Microbiol.* **51** (1997): 151–178.

[17] Domingo, E., L. Menéndez-Arias, M. E. Quinoñes-Mateu, A. Holguín, M. Gutierrez-Rivas, M. A. Martínez, J. Quer, and J. J. Holland. "Viral Quasispecies and the Problem of Vaccine-Escape and Drug-Resistance Mutants." *Prog. Drug. Res.* **48** (1997): 99–128.

[18] Drake, John W. "A Constant Rate of Spontaneous Mutation in DNA-Based Microbes." *Proc. Natl. Acad. Sci. USA* **88** (1991): 7160–7164.
[19] Drake, John W. "Rates of Spontaneous Mutation among RNA Viruses." *Proc. Natl. Acad. Sci. USA* **90** (1993): 4171–4175.
[20] Drake, John W., Brian Charlesworth, Deborah Charlesworth, and James F. Crow. "Rates of Spontaneous Mutation." *Genetics* **148** (1998): 1667–1686.
[21] Ebeling, Werner, and Reinhard Mahnke. "Kinetics of Molecular Replication and Selection." *Problems of Contemporary Biophysics (Zagadnienia Biofizyki Współczesnej)* **4** (1979): 119–128.
[22] Eigen, Manfred. "Self-Organization of Matter and the Evolution of Biological Macromolecules." *Naturwissenschaften* **58** (1971): 465–523.
[23] Eigen, Manfred, and Peter Schuster. "The Hypercycle. A Principle of Natural Self-Organization. Part A: Emergence of the Hypercycle." *Naturwissenschaften* **64** (1977): 541–565.
[24] Eigen, Manfred, John McCaskill, and Peter Schuster. "The Molecular Quasispecies." *Adv. Chem. Phys.* **75** (1989): 149–263.
[25] Eigen, Manfred. "The Origin of Genetic Information: Viruses as Models." *Gene* **135** (1993): 37–47.
[26] Ekland, E. H., Jack W. Szostak, and David P. Bartel. "Structurally Complex and Highly Active RNA Ligases Derived from Random RNA Sequences." *Science* **269** (1995): 364–370.
[27] Elena, S. F., V. S. Cooper, and R. E. Lenski. "Punctuated Evolution Caused by Selection of Rare Beneficial Mutants." *Science* **272** (1996): 1802–1804.
[28] Ellington, Andrew D. "RNA Selection. Aptamers Achieve the Desired Recognition." *Curr. Biol.* **4** (1994): 427–429.
[29] Flamm, Christoph, Walter Fontana, Ivo L. Hofacker, and Peter Schuster. "Elementary Step Dynamics of RNA Folding." *RNA* **6** (2000): 325–338.
[30] Fontana, W., and P. Schuster. "A Computer Model of Evolutionary Optimization." *Biophys. Chem.* **26** (1987): 123–147.
[31] Fontana, Walter, Wolfgang Schnabl, and Peter Schuster. "Physical Aspects of Evolutionary Optimization and Adaptation." *Phys. Rev. A* **40** (1989): 3301–3321.
[32] Fontana, Walter, Danielle A. M. Konings, Peter F. Stadler, and Peter Schuster. "Statistics of RNA Secondary Structures." *Biopolymers* **33** (1993): 1389–1404.
[33] Fontana, Walter, Peter F. Stadler, Erich G. Bornberg-Bauer, Thomas Griesmacher, Ivo L. Hofacker, Manfred Tacker, Pedro Tarazona, Edward D. Weinberger, and Peter Schuster. "RNA Folding and Combinatory Landscapes." *Phys. Rev. E* **47** (1993): 2083–2099.
[34] Fontana, Walter, and Peter Schuster. "Continuity in Evolution. On the Nature of Transitions." *Science* **280** (1998): 1451–1455.

[35] Fontana, Walter, and Peter Schuster. "Shaping Space. The Possible and the Attainable in RNA Genotype-Phenotype Mapping." *J. Theor. Biol.* **194** (1998): 491–515.

[36] Gillespie, D. T. "A General Method for Numerically Simulating the Stochastic Time Evolution of Coupled Chemical Reactions." *J. Comp. Phys.* **22** (1976): 403–434.

[37] Gillespie, D. T. "Exact Stochastic Simulation of Coupled Chemical Reactions." *J. Phys. Chem.* **81** (1977): 2340–2361.

[38] Göbel, Ulrike, Stephan Kopp, and Peter Schuster. "Complete Sequence-Secondary Structure Mapping of Oligo-Ribonucleotides of Chain Length $n = 16$." TBI-Preprint: No. pks-01-002, Univeristy of Vienna, Wien, AT, 2000

[39] Gold, Larry, Craig Tuerk, Pat Allen, Jon Binkley, David Brown, Louis Green, Sheela MacDougal, Dan Schneider, Diane Tasset, and Sean R. Eddy. "RNA: The Shape of Things to Come." In *The RNA World*, edited by Raymond F. Gesteland and John F. Atkins, 497–509. Plainview, NY: Cold Spring Harbor Laboratory Press, 1993.

[40] Grüner, Walter, Robert Giegerich, Dirk Strothmann, Christian Reidys, Jacqueline Weber, Ivo L. Hofacker, and Peter Schuster. "Analysis of RNA Sequence Structure Maps by Exhaustive Enumeration. I. Neutral Networks." *Mh. Chemie* **127** (1996): 355–374.

[41] Grüner, Walter, Robert Giegerich, Dirk Strothmann, Christian Reidys, Jacqueline Weber, Ivo L. Hofacker, and Peter Schuster. "Analysis of RNA Sequence Structure Maps by Exhaustive Enumeration. II. Structures of Neutral Networks and Shape Space Covering." *Mh. Chemie* **127** (1996): 375–389.

[42] Gultyaev, A. P., F. H. D. van Batenburg, and C. W. A. Pleij. "The Computer Simulation of RNA Folding Pathways using a Genetic Algorithm." *J. Mol. Biol.* **250** (1995): 37–51.

[43] Gultyaev, A. P., F. H. D. van Batenburg, and C. W. A. Pleij. "Dynamic Competition between Alternative Structures in Viroid RNAs Simulated by an RNA Folding Algorithm." *J. Mol. Biol.* **276** (1998): 43–55.

[44] Hamming, Richard W. "Error Detecting and Error Correcting Codes." *Bell Syst. Tech. J.* **29** (1950): 147–160.

[45] Hamming, Richard W. *Coding and Information Theory*, 2d ed. Englewood Cliffs, NJ: Prentice Hall, 1989.

[46] Higgs, Paul G. "Compensatory Neutral Mutations and the Evolution of RNA." *Genetica* **102/103** (1998): 91–101.

[47] Hofbauer, Josef, and Karl Sigmund. "Adaptive Dynamics and Evolutionary Stability." *Appl. Math. Lett.* **3** (1990): 75–79.

[48] Hofacker, Ivo L., Walter Fontana, Peter F. Stadler, L. Sebastian Bonhoeffer, Manfred Tacker, and Peter Schuster. "Fast Folding and Comparison of RNA Secondary Structures." *Mh. Chem.* **125** (1994): 167–188.

[49] Hofacker, Ivo L., Peter Schuster, and Peter F. Stadler. "Combinatorics of RNA Secondary Structures." *Discr. Appl. Math.* **88** (1998): 207–237.

[50] Huynen, Martijn A., Danielle A. M. Konings, and Pauline Hogeweg. "Equal G and C Content in Histone Genes Indicate Selection Pressures on Messenger RNA." *J. Mol. Evol.* **34** (1992): 280–291.

[51] Huynen, M. A., P. F. Stadler, and W. Fontana. "Smoothness within Ruggedness: The Role of Neutrality in Adaptation." *Proc. Natl. Acad. Sci. USA* **93** (1996): 397–401.

[52] Huynen, M. A. "Exploring Phenotype Space through Neutral Evolution." *J. Mol. Evol.* **43** (1996): 165–169.

[53] Jacob, Christine, Nicolas Breton, and Patrick Daegelen. "Stochastic Theories of the Activated Complex and the Activated Collision: The RNA Example." *J. Chem. Phys.* **107** (1997): 2903–2912.

[54] Jacob, Christine, Nicolas Breton, Patrick Daegelen, and Jean Peccoud. "Probability Distribution of the Chemical States of a Closed System and Thermodynamic Law of Mass Action from Kinetics: The RNA Example." *J. Chem. Phys.* **107** (1997): 2913–2919.

[55] Jaeger, John A., Douglas H. Turner, and Michael Zuker. "Improved Predictions of Secondary Structures for RNA." *Proc. Natl. Acad. Sci. USA* **86** (1989): 7706–7710.

[56] Jaeger, John A., Douglas H. Turner, and Michael Zuker, "Predicting Optimal and Suboptimal Secondary Structure for RNA." *Meth. Enzymol.* **183** (1990): 281–306.

[57] Jones, B. L., R. H. Enns, and S. S. Rangnekar. "On the Theory of Selection of Coupled Macromolecular Systems." *Bull. Math. Biol.* **38** (1976): 15–28.

[58] Jones, B. L., and H. K. Leung. "Stochastic Analysis of a Nonlinear Model for Selection of Biological Macromolecules." *Bull. Math. Biol.* **43** (1981): 665–680.

[59] Kauffman, Stuart A. *The Origins of Order. Self-Organization and Selection in Evolution.* Oxford, UK: Oxford University Press, 1993.

[60] Kimura, Motoo. "Evolutionary Rate at the Molecular Level." *Nature* **217** (1968): 624–626.

[61] Kimura, Motoo. *The Neutral Theory of Molecular Evolution.* Cambridge, UK: Cambridge University Press, 1983.

[62] Kubitschek, H. E. *Introduction to Research with Continuous Cultures.* Englewood Cliffs, NJ: Prentice Hall, 1970.

[63] Lenski, Richard E., and M. Travisano. "Dynamics of Adaptation and Diversification: A 10,000-Generation Experiment with Bacterial Populations." *Proc. Natl. Acad. Sci. USA* **91** (1994): 6808–6814.

[64] Leung, H. K. "Stability Analysis of a Stochastic Model for Biomolecular Selection." *Bull. Math. Biol.* **46** (1984): 399–406.

[65] Leung, H. K. "Expansion of the Master Equation for a Biomolecular Selection Model." *Bull. Math. Biol.* **47** (1985): 231–238.

[66] Leuthäusser, Ira. "Statistical Mechanics of Eigen's Evolution Model." *J. Stat. Phys.* **48** (1987): 343–360.

[67] Martinez, H. M. "An RNA Folding Rule." *Nucl. Acids Res.* **12** (1984): 323–334.

[68] McAdams, Herley H., and Adam Arkin. "Simulation of Prokaryotic Genetic Circuits." *Annu. Rev. Biophys. Biomol. Struct.* **27** (1998): 199–224.

[69] McCaskill, John S. "A Localization Threshold for Macromolecular Quasi-Species from Continuously Distributed Replication Rates." *J. Chem. Phys.* **80** (1984): 5194–5202.

[70] McCaskill, John S. "The Equilibrium Partition Function and Base Pair Binding Probabilities for RNA Secondary Structures." *Biopolymers* **29** (1990): 1105–1119.

[71] Metropolis, N., A. W. Rosenbluth, M. N. Rosenbluth, A. H. Teller, and E. Teller. "Equation of State Calculations by Fast Computing Machines." *J. Chem. Phys.* **21** (1953): 1087–1092.

[72] Mills, D. R., R. L. Peterson, and Sol Spiegelman. "An Extracellular Darwinian Experiment with a Self-Duplicating Nucleic Acid Molecule." *Proc. Natl. Acad. Sci. USA* **58** (1967): 217–224.

[73] Mironov, A., and A. Kister. "A Kinetic Approach to the Prediction of RNA Secondary Structures." *J. Biomol. Struct. Dyn.* **2** (1985): 953–962.

[74] Mironov, A., and A. Kister. "RNA Secondary Structure Formation during Transcription." *J. Biomol. Struct. Dyn.* **4** (1985): 1–9.

[75] Mironov, A., and V. F. Lebedev. "A Kinetic Model of RNA Folding." *BioSystems* **30** (1993): 49–56.

[76] Moran, P. A. P. "The Effect of Selection in Haploid Genetic Populations." *Proc. Camb. Phil. Soc.* **54** (1958): 463–474.

[77] Morgan, S. R., and P. G. Higgs. "Evidence for Kinetic Effects in the Folding of Large RNA Molecules." *J. Chem. Phys.* **105** (1996): 7152–7157.

[78] Nilsson, Martin, and Nigel Snoad. "Error Thresholds for Quasispecies on Dynamics Fitness Landscapes." *Phys. Rev. Lett.* **84** (2000): 191–194.

[79] Nowak, Martin, and Peter Schuster. "Error Thresholds of Replication in Finite Populations. Mutation Frequencies and the Onset of Muller's Ratchet." *J. Theor. Biol.* **137** (1989): 375–395.

[80] Nussinov, Ruth, and Ann B. Jacobson. "Fast Algorithm for Predicting the Secondary Structure of Single-Stranded RNA." *Proc. Natl. Acad. Sci. USA* **77** (1980): 6309–6313.

[81] Ochman, Howard. "Bacterial Evolution: Jittery Genomes." *Curr. Biol.* **9** (1999): R485–R486.

[82] Ohta, Tomoko. "The Nearly Neutral Theory of Molecular Evolution." *Ann. Rev. Ecol. Sys.* **23** (1992): 263–286.

[83] Papadopoulos, Dimitri, Dominique Schneider, Jessica Meier-Eiss, Werner Arber, Richard E. Lenski, and Michel Blot. "Genomic Evolution during a 10,000-Generation Experiment with Bacteria." *Proc. Natl. Acad. Sci. USA* **96** (1999): 3807–3812.

[84] Perrotta, A. T., and M. D. Been. "A Toggle Duplex in Hepatitis Delta Virus Self-Cleaving RNA that Stabilizes an Inactive and a Salt-Dependent Pro-Active Ribozyme Conformation." *J. Mol. Biol.* **279** (1998): 361–373.

[85] Prudent, J. R., T. Uno, and P. G. Schultz. "Expanding the Scope of RNA Catalysis." *Science* **264** (1994): 1924–1927.

[86] Pütz, Joern, J. D. Puglisi, Catherine Florentz, and Richard Giegé. "Identity Elements for Specific Aminoacylation of Yeast $tRNA^{asp}$ by Cognate Aspartyl tRNA Synthetase." *Science* **252** (1991): 1696–1699.

[87] Reidys, Christian, Peter F. Stadler, and Peter Schuster. "Generic Properties of Combinatory Maps. Neutral Networks of RNA Secondary Structure." *Bull. Math. Biol.* **59** (1997): 339–397.

[88] Reidys, Christian M. "Random Induced Subgraphs of Generalized n-Cubes." *Adv. Appl. Math.* **19** (1997): 360–377.

[89] Reidys, Christian, Christian Forst, and Peter Schuster. "Replication and Mutation on Neutral Networks." *Bull. Math. Biol.* **63** (2001): 57–94.

[90] Rohde, N., H. Daum, and C. K. Biebricher. "The Mutant Distribution of an RNA Species Replicated by $Q\beta$ Replicase." *J. Mol. Biol.* **249** (1995): 754–762.

[91] Savill, Nicholas J., D. C. Hoyle, and Paul G. Higgs. "RNA Sequence Evolution with Secondary Structure Constraints. Comparison of Substitution Rate Models using Maximum Likelihood Methods." *Genetics* **157** (2001): 399–411.

[92] Savill, Nicholas J., and Paul G. Higgs. "RNA Sequence Evolution with Secondary Structure Constraints. II. Comparison of Substitution Rate Models using Maximum Likelihood Methods. Preprint, 2000.

[93] Schultes, Erik A., and David P. Bartel. "One Sequence, Two Ribozymes: Implications for the Emergence of New Ribozyme Folds." *Science* **289** (2000): 448–452.

[94] Schuster, Peter, and Jörg Swetina. "Stationary Mutant Distribution and Evolutionary Optimization." *Bull. Math. Biol.* **50** (1988): 635–660.

[95] Schuster, Peter, Walter Fontana, Peter F. Stadler, and Ivo L. Hofacker. "From Sequences to Shapes and Back: A Case Study in RNA Secondary Structures." *Proc. Roy. Soc. Lond. B* **255** (1994): 279–284.

[96] Schuster, Peter. "How to Search for RNA Structures. Theoretical Concepts in Evolutionary Biotechnology." *J. Biotechnol.* **41** (1995): 239–257.

[97] Schuster, Peter. "Landscapes and Molecular Evolution." *Physica D* **107** (1997): 351–365.

[98] Schuster, Peter. "Genotypes with Phenotypes: Adventures in an RNA Toy World." *Biophys. Chem.* **66** (1997): 75–110.

[99] Schuster, Peter, and Walter Fontana. "Chance and Necessity in Evolution: Lessons from RNA." *Physica D* **133** (1999): 427–452.

[100] Schuster, Peter, and Peter F. Stadler. "Discrete Models of Biopolymers." In *Handbook of Computational Chemistry*, edited by M. J. C. Crabbe and M. Drew and A. Konopka. New York: Marcel Dekker, 2000.

[101] Schuster, Peter, M. Kospach, Christoph Flamm, and Ivo L. Hofacker. "RNA Structures over Different Nucleotide Alphabets: AUGC, AUG, UGC, AU, and GC." TBI-Preprint: No. pks-01-004, University of Vienna, Wien, AT, 2001.

[102] Spiegelman, Sol. "An Approach to the Experimental Analysis of Precellular Evolution." *Quart. Rev. Biophys.* **4** (1971): 213–253.

[103] Stadler, Peter F. "Fitness Landscapes Arising from the Sequence-Structure Maps of Biopolymers." *J. Mol. Struct. (Theochem)* **463** (1999): 7-19.

[104] Suvernev, A. A., and P. A. Frantsuzov. "Statistical Description of Nucleic Acid Secondary Structure Folding." *J. Biomol. Struct. Dynam.* **13** (1995): 135–144.

[105] Swetina, Jörg, and Peter Schuster. "Self-Replication with Errors—A Model for Polynucleotide Replication." *Biophys. Chem.* **16** (1982): 329–345.

[106] Swetina, Jörg. "First and Second Moments and the Mean Hamming Distance in a Stochastic Replication-Mutation Model for Biological Macromolecules." *J. Math. Biol.* **27** (1989): 463–483.

[107] Szostak, Jack W., and Andrew D. Ellington. "*In vitro* Selection of Functional RNA Sequences." In *The RNA World*, edited by Raymond F. Gesteland and John F. Atkins, 511–533. Plainview, NY: Cold Spring Harbor Laboratory Press, 1993.

[108] Tacker, Manfred, Walter Fontana, Peter F. Stadler, and Peter Schuster. "Statistics of RNA Melting Kinetics." *Eur. Biophys. J.* **23** (1994): 29–38.

[109] Tarazona, Pedro. "Error Threshold for Molecular Quasispecies as Phase Transitions: From Simple Landscapes to Spin-Glass Models." *Phys. Rev. A* **45** (1992): 6038–6050.

[110] Thompson, C. J. and J. L. McBride. "On Eigen's Theory of the Self-Organization of Matter and the Evolution of Biological Macromolecules." *Math. Biosci.* **21** (1974): 127–142.

[111] van Kampen, N. G. "The Expansion of the Master Equation." *Adv. Chem. Phys.* **34** (1976): 245–309.

[112] van Nimwegen, Erik, James P. Crutchfield, and Melanie Mitchell. "Finite Populations Induce Metastability in Evolutionary Search." *Phys. Lett. A* **229** (1997): 144–150.

[113] van Nimwegen, Erik. "The Statistical Dynamics of Epochal Evolution." Ph.D. Thesis, Utrecht, NL: Universiteit Utrecht, 1999.

[114] Walter, Amy E., Douglas H. Turner, James Kim, Matthew H. Lyttle, Peter Müller, David H. Mathews, and Michael Zuker. "Co-Axial Stacking of Helixes Enhances Binding of Oligoribonucleotides and Improves Predictions of RNA Folding." *Proc. Natl. Acad. Sci. USA* **91** (1994): 9218–9222.

[115] Waterman, M. S. "Secondary Structure of Single-Stranded Nucleic Acids." *Adv. Math. Suppl. Studies* **1** (1978): 167–212.

[116] Watts, Anthony, and Gerhard Schwarz, eds. "Evolutionary Biotechnology—From Theory to Experiment." In *Biophysical Chemistry*, vol. 66/2–3, 67–284. Amsterdam: Elesevier, 1997.

[117] Wernitznig, Andreas, Christoph Flamm, and Peter Schuster. "RNA Evolution *in silico*: The Flow Reactor." TBI-Preprint: No. pks-01-003, University of Vienna, Wien, AT, 2001.

[118] Wiehe, Thomas, Ellen Baake, and Peter Schuster. "Error Propagation in Reproduction of Diploid Organisms. A Case Study in Single Peaked Landscapes." *J. Theor. Biol.* **177** (1995): 1–15.

[119] Wilson, David S., and Jack W. Szostak. "*In vitro* Selection of Functional Nucleic Acids." *Ann. Rev. Biochem.* **68** (1999): 611–647.

[120] Wright, Sewall. "The Roles of Mutation, Inbreeding, Crossbreeding and Selection in Evolution." In *Intl. Proceedings of the Sixth International Congress on Genetics*, edited by D. F. Jones, vol. 1, 356–366. Ithaca, NY, 1932.

[121] Wuchty, Stefan, Walter Fontana, Ivo L. Hofacker, and Peter Schuster. "Complete Suboptimal Folding of RNA and the Stability of Secondary Structures." *Biopolymers* **49** (1999): 145–165.

[122] Zipf, G. K. *Human Behaviour and the Principle of Least Effort.* Reading, MA: Addison-Wesley, 1949.

[123] Zuckerkandl, Emile. "Neutral and Nonneutral Mutations: The Creative Mix—Evolution of Complexity in Gene Interaction Systems." *J. Mol. Evol.* **44** (Suppl. 1) (1997): S2–S8.

[124] Zuker, M., and P. Stiegler. "Optimal Computer Folding of Larger RNA Sequences using Thermodynamics and Auxiliary Information." *Nucl. Acids Res.* **9** (1981): 133–148.

[125] Zuker, M., and D. Sankoff. "RNA Secondary Structures and Their Prediction." *Bull. Math. Biol.* **46** (1984): 591–621.

[126] Zuker, M. "On Finding All Suboptimal Foldings of an RNA Molecule." *Science* **244** (1989): 48–52.

Population Genetics, Dynamics, and Optimization

The Nearly Neutral Theory with Special Reference to Interactions at the Molecular Level

Tomoko Ohta

The nearly neutral theory of molecular evolution is an extension of the neutral theory [17]. In the early 1970s, I argued that borderline mutations are important at the molecular level, when their behaviors are influenced by both random genetic drift and selection [26]. The rate of molecular evolution is highly dependent upon selective constraints of proteins or nucleic acids; highly constrained proteins like histone IV evolve very slowly, whereas weakly constrained ones like fibrinopeptides change rapidly. Under the neutral theory, it is assumed that a certain fraction of new mutations are free of constraint or are selectively neutral, while the rest have deleterious effects and are eliminated from the population. The nearly neutral theory regards the borderline mutations as most significant in molecular evolution and is directed toward understanding the interaction between random genetic drift and selection. Figure 1 depicts the comparison of how selection, neutral, and nearly neutral theories classify new mutations.

Evolutionary Dynamics, edited by
J. P. Crutchfield and P. Schuster. Oxford University Press.

The nearly neutral theory is summarized as follows. Random drift and selection both influence the behavior of very weakly selected mutations, with drift predominating in small populations and selection in large populations [26]. Most new mutations are deleterious, and most mutations with small effects are likely to be slightly deleterious. Such mutations are selected against in large populations but behave as if neutral in small populations. They are called nearly neutral mutations, and a negative correlation between evolutionary rate and population size is predicted. Quantitative treatment may be pursued in terms of the principle that the rate of gene substitution equals the number of new mutations multiplied by their fixation probability.

1 MEAN AND VARIANCE OF SEQUENCE DIVERGENCE

The nearly neutral theory predicts that evolution is rapid in small populations [26, 28]. Large organisms with long generation times tend to have small population sizes and vice versa [7]. Mutation rates probably depend on the number of cell generations, and the generation-time effect on mutation rate partially cancels the population-size effect for spreading of nearly neutral mutations [26]. This prediction has been examined by sequence analysis of mammalian genes.

Figure 2 shows the star phylogeny of the three mammalian orders with the numbers of synonymous and nonsynonymous substitutions per site beside each

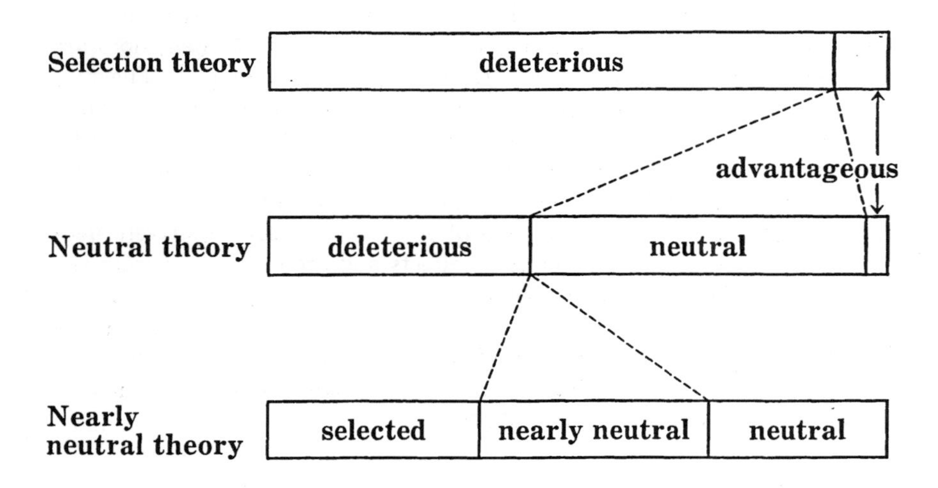

FIGURE 1 Diagram showing how new mutants are classified under selection, neutral, and nearly neutral theories.

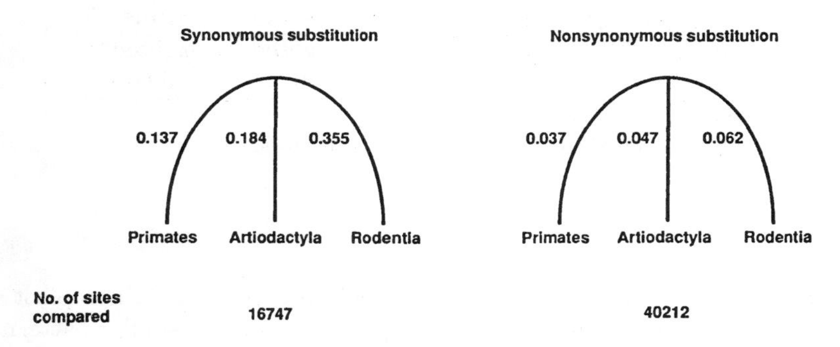

FIGURE 2 Star phylogenies of 49 mammalian genes. The numbers beside each branch are the estimated nucleotide substitutions per site (from Ohta [29]).

branch [29]. From the figure, it may be seen that the generation-time effect is more conspicuous for synonymous substitutions than for nonsynonymous substitutions. Note that rodentia tend to have larger population size and shorter generation length than primates. Therefore this result agrees with the prediction of the nearly neutral theory.

The variance of the evolutionary rate has also been studied using the same data. Following Gillespie [12], attention was paid to the dispersion index, R, that is the ratio of the variance to the mean number of substitutions. Using the same data as in figure 2, the dispersion indices were obtained [29]. Table 1 gives the results. For synonymous substitutions, the error variance is large [4], and one cannot conclude that the variance is larger than the value of the Poisson process. For nonsynonymous substitutions, the error variance is not large because of the short branch lengths [4], and R is significantly larger than unity. Therefore, the nonsynonymous substitution rate fluctuates, and protein evolution is likely to be episodic, as pointed out by Gillespie [12]. The above results have shown that the pattern of molecular evolution does not obey a simple Poisson process, and the nearly neutral theory should be studied in detail. Here interaction at various levels is thought to be responsible for the large R.

In the case of *Drosophila* species, the divergence pattern seems to be somewhat different from that of mammals. Zeng et al. [39] and Takano [36] found that the dispersion index is larger for synonymous substitutions than for nonsynonymous substitutions in *Drosophila.* In fact, R is significantly larger than unity for synonymous substitutions, but for nonsynonymous substitutions it is not so. These authors argue that weak selection for maintaining codon bias may be important and may elevate R in *Drosophila* but not in mammals, because the effective population size is larger in the former than in the latter. They also con-

TABLE 1 Average dispersion index.

Primate-Artiodactyl-Rodent, 49 Genes*			
Synonymous		Nonsynonymous	
unweighted	weighted**	unweighted	weighted**
25.01	5.89	8.46	5.60

*Same data set as in figure 2.

**Calculated after weighted by the lineage effect.

clude that nonsynonymous changes are not episodic. However, chloroplast genes of rice, tobacco, pine, and liverwort were reported to show a similar pattern with that of mammalian genes. Lineage effect on synonymous substitutions and the large variation of the rates of nonsynonymous substitutions were obtained [24].

2 THEORETICAL MODELS

Several models of weak selection or nearly neutral mutations have been analyzed. In the shift model, a certain distribution is assumed for selection coefficients of new mutations: an exponential function [27], a gamma function [18], and several discrete classes [20]. Whenever a mutant fixes in the population, the mean fitness shifts to the original value, so that the distribution for new mutations remains unchanged. In contrast, the fixed model, or the house-of-cards model, assumes that the population fitness depends upon the mutants' effects and moves. The mutants' effects are fixed here. By assuming a normal distribution for the selection coefficient of a mutation, simulation studies were performed [32, 34]. All these models predict that the rate of mutant substitution is negatively correlated with the species population size, although the magnitude of the correlation depends on the model. In the next section, another model incorporating interaction among amino acids within a protein is investigated.

In fact, genetic information may be regarded as interactive systems at various levels; the first level, those among amino acids or nucleotide sites (within-gene interaction); the second level, those among gene products (among-gene-products interaction); and the third level, those between regulatory regions and gene products (DNA-gene-products interaction). In order to understand the mechanisms that determine how natural selection works to keep or to refine interactive systems, we need to study such systems. However, most previous studies have been limited to the analyses of individual mutant substitutions that are assumed to be independent from the others. Wright's shifting balance theory [37, 38] is directed toward such an interaction system, but it is far from complete.

Here it is pertinent to connect the present discussion with the studies on the evolution of higher-order structure. Evolution of RNA secondary structure has been investigated in detail, and the properties of neutral networks have revealed some interesting patterns of punctuated equilibrium [9, 15]. The neutral network is, in the present context, equivalent to the nearly neutral class of sequences, of which the fitness values are not exactly the same but obey a certain distribution. The model of the holey landscape for analyzing speciation [10] has also some analogy with these studies, i.e., interactive loci results in neutral drift and speciation.

As a population genetics approach, the mutational landscape model was proposed for analyzing gene interaction [11, 12]. This model is based on strong selection, which is responsible for mutant substitutions. In other words, a population fitness climbs up to a local optimum, and environmental shift is needed to move the population away from the local peak. Kauffman [16] proposed a generalized model, the NK model. I adopt this model here, (as in the previous report [30]) for the analysis of the within-gene interaction.

The NK model assumes that each amino acid makes a fitness contribution that depends upon the amino acid and upon K other amino acids among the N that make the protein. In other words, this is a model of epistatic interaction among $K+1$ amino acids. According to Kauffman [16], the fitness landscape is very rugged if $K \geq 2$. In the original NK model, there are two states, 0 and 1 at each site. If $K+1$ sites interact, there are 2^{K+1} combinations for them. The fitness of each combination is assigned by drawing a uniform random number between 0.0 and 1.0. The fitness of an entire sequence is obtained by taking the average for all sites. The simple case, in which N sites are arranged in a circle and K sites are neighbors, was used here.

Newman and Engelhardt [25] introduced a tunable degree of neutrality into the NK model and found that the maximum fitness attainable increases with the increasing degree of neutrality. The degree of neutrality corresponds to the magnitude of random genetic drift, i.e., the reciprocal of the effective population size in the present analysis.

In my simulations, the original NK model [16] was modified to be slightly more realistic. First, there is assumed to be nine states at each site. Second, the average fitness of each site is assumed to decrease as the distance from the first site increases, by considering the presence of variable as well as conservative sites within an ordinary protein. The details are found in Ohta [30]. The average selection coefficient ($\overline{s}$) of random sequences may be chosen by a parameter in the simulation. The cases of $N = 24$ and 48, and $K = 0, 2$, and 4 with $\overline{s} = 0.25$ or 0.1 were simulated. Let v be the mutation rate per gene per generation. Each simulation was continued for the period of $132/v$ when $N = 24$, and of $264/v$ when $N = 48$. The entire period was divided into 11 terms. Except for the first term, various quantities of interest were calculated in each term. For a same set of parameters, the simulations were repeated 100 times. The effective population size is 50 to 400.

Quantities examined are the number of substitutions per each term (k), the mean fitness of the population (W), and the average number of substitutions at individual site. Also interesting are the variance of the number of substitutions and the pattern of polymorphism. The former was examined by the dispersion index, R, that is, the ratio of the variance to the mean of the number of substitutions per term (see Gillespie [12]). As for the latter, Tajima's D was studied [35]. Note that this is a statistic for testing selective neutrality of DNA polymorphisms. Under selective neutrality, $D = 0$, it becomes positive when some kind of balancing selection is at work, and it is negative when slightly deleterious selection is operating.

Through the analyses, important properties such as population-size effect on mutants' substitutions and the consequence of amino acid interaction are examined.

3 RESULTS OF SIMULATIONS

Results on the effect of population size are given here. Note that the population-size effect is essential for clarifying the role of random genetic drift. Figure 3 shows the average number of accumulated mutants (number of substitutions) per gene per term as a function of population size (haploid). The cases of $N = 24$ and $K = 0$, 2, and 4 are shown. Under selective neutrality, the expected number of substitutions is 12, and the number decreases as population size increases. This is because selection becomes more effective and mutant dynamics become slower as population size increases. This pattern clearly indicates the prevalence of slightly deleterious mutations at the level of occurrence. Also, the number decreases by increasing K, provided that the same condition holds on other parameters. This is thought to be caused by a larger number of interactive systems for larger K resulting in stronger constraints.

As it is expected, the average fitness of the population (W) increases by increasing population size. Figure 4 gives such a relationship, for $K = 0$, 2, and 4. It is also noted that W is generally larger for larger K, although there are considerable variations. This again indicates stronger selection for larger K.

Our next interest is the main problem, i.e., the variation of substitution rate. This is measured by the dispersion index, R. Table 2 gives the dispersion index for various parameter sets. Except for the case of $K = 0$, the value of R seems to be only slightly larger than unity. Therefore the model is not quite appropriate for explaining observed value of 2–7 for protein evolution [12, 29].

In what way should the model be modified so that it is compatible with the data? Araki and Tachida [3] have shown that when the population size changes cyclically, the fixed model (house of cards model) of nearly neutral mutations may explain the large values of the dispersion index. Simulations were performed with the NK model by incorporating the changing population size. In one set of simulations, population size was changed deterministically at the beginning of each term periodically, $N_1 \rightleftharpoons N_2$. In the other set, the change was done by using random numbers so that the mean length of one period becomes the same

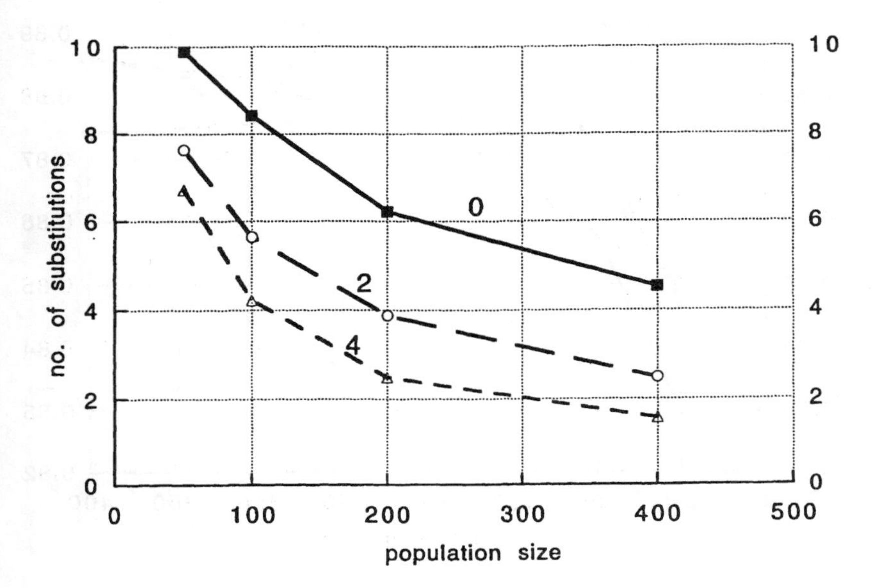

FIGURE 3 Results of simulations of the NK model (from Ohta [31]). The ordinate is the number of substitutions per term, for three values of K (0, 2, and 4 as shown). Other parameters are $N = 24$ and $\bar{s} = .25$ and the mutation rate $= 0.0001$ per site.

as the first set. Table 3 presents the results of simulations. As can be seen, the dispersion index is large.

4 PHENOTYPES AND THE NEARLY NEUTRAL THEORY

Muller's ratchet is the process in which deleterious mutations accumulate one after another in a population. It is related to the nearly neutral theory in its original form of slightly deleterious mutation theory. Muller's ratchet is based on phenotypic observations; whereas, the slightly deleterious theory is an extension of the neutral theory at the molecular level. Because of the different histories, the two theories have developed separately. Muller [23] noted that the accumulation of deleterious mutations is rapid in an asexual species (see also Felsenstein [8]). Chao [5] demonstrated Muller's ratchet by RNA virus; the fitness declined after several bottlenecks of one virus. He proposed that the Muller's ratchet and the slightly deleterious theory should be considered on the same ground [6]. At the gene level, accumulation of deleterious mutations was found in endosymbiotic bacteria by Moran [22] and Peek et al. [33], and in pathogenic bacteria by Andersson and Kurland [1]. These authors also note the common process between the Muller's ratchet and the nearly neutral theory. Andersson et al. [2]

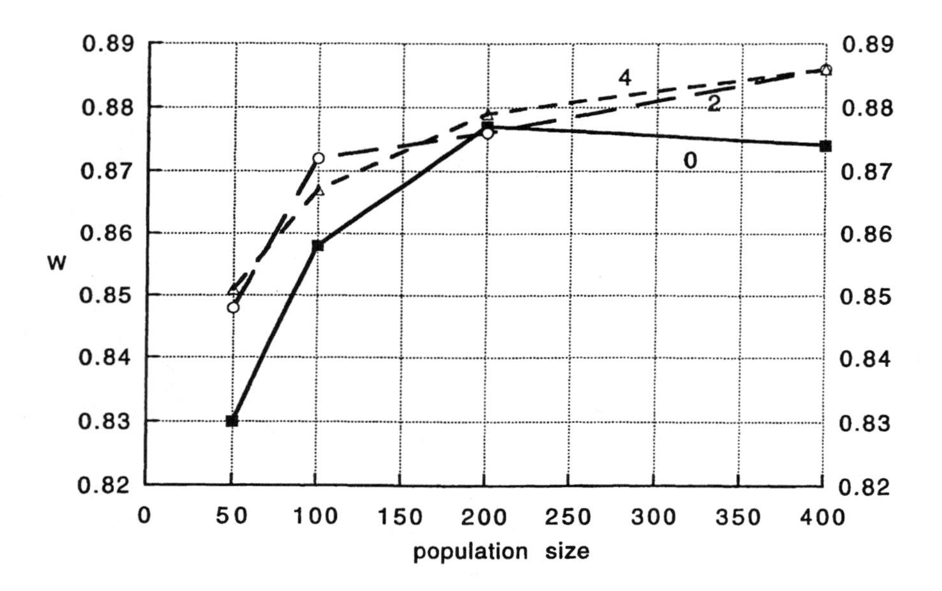

FIGURE 4 Results of simulations on average fitness, W, that is the mean of the population fitness for the period from the second to the last term (from Ohta [31]). Parameters are the same as in Figure 3.

reported the complete genome sequence of *Rickettsia prowazekü* and found that this genome contains a large fraction of noncoding regions. They suggest that such noncoding sequences are degraded remnants of genes accumulated in the course of evolution of the pathogenic bacteria.

Lambert and Moran [19] calculated the degraded amount of destabilization of the stem region of ribosomal RNA of endosymbiotic bacteria. They have shown that several independently derived insect endosymbionts have accumulated destabilizing base substitutions; i.e., the stability of a domain of 16S rRNA is 15%–25% lower in endosymbionts than in closely related free living bacteria. Lynch [21] did a similar analysis using tRNA of organelle of various species, with the conclusion that slightly deleterious mutations accumulated in genomes of organelle. In contrast, I have pointed out that the stem region of tRNA may be evolving by slightly deleterious mutant substitutions and compensatory substitutions [26]. Therefore, there is an overlap between the candidate mutants of Muller's ratchet and the nearly neutral mutations.

Another important aspect of genotype-phenotype relationship is punctuated equilibrium. The evolution of RNA secondary structure is characterized by a number of neutral networks that are interwoven, and a punctuation is a transition between networks that may occur when sequences drift and come to a space close to a neighboring network [9, 15]. Such a pattern results from the model in which

TABLE 2 The dispersion index under various sets of parameter values (from Ohta [31]). In each simulation, mean and variance of the number of substitutions per terms were calculated, and R was obtained The figures are the average of 100 replications with standard deviations. In all simulations, mutation rate = 0.0001 per site.

			N_e			
N	K	$\bar{s}$	50	100	200	400
24	0	0.25	1.710± 4.787	2.533± 7.500	2.572± 6.111	1.214±0.636
24	0	0.25	1.089± 0.434	1.180± 0.563	1.107± 0.604	1.371±0.679
24	4	0.25	1.140± 0.507	1.206± 0.511	1.403± 0.749	1.331±0.572
24	4	0.1	1.29 ± 0.485	1.076± 0.523	1.142± 0.477	1.310±0.607
48	0	0.1	4.493±16.114	5.203±20.635	8.523±23.775	0.978±0.479
48	2	0.1	0.875± 0.365	0.947± 0.400	1.003± 0.446	1.072±0.440
48	4	0.1	0.934± 0.431	1.021± 0.52	1.089± 0.562	1.129±0.501

TABLE 3 Average number of substitutions and dispersion index under changing population size. Calculation of the mean and R of the number of substitutions is the same as in table 2.

		Cyclic		Stochastic	
N_e	$\bar{s}$	no. subst.	R	no. subst.	R
50	1/4	5.521	1.983±1.242	5.555	2.314±1.579
|	1/6	6.846	2.420±2.306	6.935	2.727±1.836
200	1/8	7.468	4.518±6.063	7.637	3.063±2.518
	1/10	8.282	4.494±5.347	8.035	3.023±2.994
100	1/4	4.592	1.493±0.908	4.623	1.971±2.081
|	1/6	5.418	1.927±2.332	5.874	1.952±1.435
200	1/8	6.093	1.835±0.792	6.823	1.825±1.288
	1/10	7.195	2.794±5.428	7.331	2.655±3.114

the secondary structure is the only target of selection. However, on the one hand classical evolutionary study has focused on the subsequent modification of the molecule after establishing a network; i.e., the properties of hemoglobin molecule of various organisms have been investigated in great detail and have the same higher-order structure [13]. Therefore, the neutral network of RNA secondary structure would correspond to the nearly neutral class. On the other hand, a mutation that results in transition between different secondary structures has a large effect and may correspond to a lethal mutation if the gene product is essential for the organism.

The NK model also shows a similar pattern of evolution as shown by Newman and Engelhardt [25]. They used a generalized NK model, in which the number of fitness levels are assigned, and found the pattern of punctuated equilibrium. The reciprocal of the number of fitness levels, $1/F$, is the power of resolution of natural selection and corresponds to the magnitute of random genetic drift, $1/2N_e$, in our study.

Turning to the phenotypic evolution in paleontological time scale, the famous punctuated equilibrium [14] must come from the interaction among proteins and regulatory elements. If such an interaction is analyzable by a similar network model as used for RNA secondary structure, one would expect punctuated equilibrium in phenotypic evolution, and there would be nearly neutral drift at the level of on-off of gene regulation. For precise formulation of such an interaction, one needs more information about networks on gene regulation.

ACKNOWLEDGMENT

I thank Ms. Yuriko Ishii for her secretarial assistance.

REFERENCES

[1] Andersson, S. G. E., and C. G. Kurland. "Reductive Evolution of Resident Genomes." *Trends Microbiol.* **6** (1998): 263–268.

[2] Andersson, S. G. E., A. Zomorodipour, J. O. Andersson, T. Sicheritz-Ponten, U. C. M. Alsmark, R. M. Podowski, A. K. Näslund, A.-S. Eriksson, H. H. Winkler, and C. G. Kurland. "The Genome Sequence of *Rickettia prowazekü* and the Origin of Mitochondria." *Nature* **396** (1998): 133–140.

[3] Araki, H., and H. Tachida. "Bottleneck Effect on Evolutionary Rate in the Nearly Neutral Mutation Model." *Genetics* **147** (1997): 907–914.

[4] Bulmer, M. "The Selection-Mutation-Drift Theory of Synonymous Codon Usage." *Genetics* **129** (1991): 897–907.

[5] Chao, L. "Fitness of RNA Viruses Decreased by Muller's Ratchet." *Nature* **348** (1990): 454–455.

[6] Chao, L. "Evolution of Sex and the Molecular Clock in RNA Viruses." *Gene* **205** (1997): 301–308.

[7] Chao, L., and D. E. Carr. "The Molecular Clock and the Relationship between Population Size and Generation Time." *Evolution* **47** (1993): 688–690.

[8] Felsenstein, J. "The Evolutionary Advantage of Recombination." *Genetics* **78** (1974): 337–356.

[9] Fontana, W., and P. Schuster. "Continuity in Evolution: On the Nature of Transitions." *Science* **280** (1998): 1451–1455.

[10] Gavrilets, S. "Evolution and Speciation on Holey Adaptive Landscapes." *Trends Ecol. & Evol.* **12** (1997): 307–312.

[11] Gillespie, J. H. "Molecular Evolution over the Mutational Landscape." *Evolution* **38** (1984): 1116–1129.

[12] Gillespie, J. H. *The Causes of Molecular Evolution.* New York, Oxford: Oxford University Press, 1991.

[13] Goodman, M., J. Czelusniak, B. F. Koop, D. A. Tagle, and J. L. Slightom. In *Globins: A Case Study in Molecular Phylogeny*, vol. 52, 875–890. Proc. Cold Spring Harbor Symp. on Quant. Biol. New York: Cold Spring Harbor Laboratory, 1987.

[14] Gould, S. J., and N. Eldredge. "Punctuated Equilibria: The Tempo and Mode of Evolution Reconsidered." *Paleobiology* **3** (1977): 115–151.

[15] Huynen, M. A., P. F. Stadler, and W. Fontana. "Smoothness within Ruggedness: The Role of Neutrality in Adaptation." *Proc. Natl. Acad. Sci.* **93** (1996): 397–401.

[16] Kauffman, S. A. *The Origins of Order*. New York: Oxford Univeristy Press, 1993.

[17] Kimura, M. "Evolutionary Rate at the Molecular Level." *Nature* **217** (1968): 624–626.

[18] Kimura, M. "A Model of Effectively Neutral Mutations in which Selective Constraint is Incorporated." *Proc. Natl. Acad. Sci. USA* **76** (1979): 3440–3444.

[19] Lambert, J. D., and N. A. Moran. "Deleterious Mutations Destabilize Ribosomal RNA in Endosymbiotic Bacteria." *Proc. Natl. Acad. Sci.* **95** (1998): 4458–4462.

[20] Li, W.-H. "Maintenance of Genetic Variability under the Pressure of Neutral and Deleterious Mutations in a Finite Population." *Genetics* **92** (1979): 647–667.

[21] Lynch, M. "Mutation Accumulation in Nuclear, Organelle, and Prokaryotic Transfer RNA Genes." *Mol. Biol. Evol.* **14** (1997): 914–925.

[22] Moran, N. A. "Accelerated Evolution and Muller's Rachet in Endosymbiotic Bacteria." *Proc. Natl. Acad. Sci.* **93** (1996): 2873–2878.

[23] Muller, H. J. "The Relation of Recombination to Mutational Advance." *Mutat. Res.* **1** (1964): 2–9.

[24] Muse, S. V., and B. S. Gaut. "A Likelihood Approach for Comparing Synonymous and Nonsynonymous Nucleotide Substitution Rates, with Application to Chloroplast Genome." *Mol. Biol. Evol.* **11** (1994): 715–724.

[25] Newman, M. E. J., and R. Engelhardt. "Effects of Selective Neutrality on the Evolution of Molecular Species." *Proc. Roy. Soc. Lond. B* **265** (1998): 1333–1338.

[26] Ohta, T. "Slightly Deleterious Mutant Substitutions in Evolution." *Nature* **246** (1973): 96–98.

[27] Ohta, T. "Extension to the Neutral Mutation Random Drift Hypothesis." In *Molecular Evolution and Polymorphism*, edited by M. Kimura, 148–167. Mishima: National Institute of Genetics, 1977.

[28] Ohta, T. "The Nearly Neutral Theory of Molecular Evolution." *Ann. Rev. Ecol. & Syst.* **23** (1992): 263–286.

[29] Ohta, T. "Synonymous and Nonsynonymous Substitutions in Mammalian Genes and the Nearly Neutral Theory." *J. Mol. Evol.* **40** (1995): 56–63.

[30] Ohta, T. "Role of Random Drift in the Evolution of Interactive Systems." *J. Mol. Evol.* **44** (1997): S9–S14.

[31] Ohta, T. "The Meaning of Near-Neutrality at Coding and Non-coding Regions." *Gene* **205** (1997): 261–271.

[32] Ohta, T., and H. Tachida. "Theoretical Study of Near Neutrality. I. Heterozygosity and Rate of Mutant Substitution." *Genetics* **126** (1990): 219–229.

[33] Peek, A. S., R. C. Vrijenhoek, and B. S. Gaut. "Accelerated Evolutionary Rate in Sulfur-Oxidizing Endosymbiotic Bacteria Associated with the Mode of Symbiont Transmission." *Mol. Biol. Evol.* **15** (1998): 1514–1523.

[34] Tachida, H. "A Study on a Nearly Neutral Mutation Model in Finite Populations." *Genetics* **128** (1991): 183–192.

[35] Tajima, F. "Statistical Method for Testing the Neutral Mutation Hypothesis by DNA Polymorphism." *Genetics* **123** (1989): 585–595.

[36] Takano, T. S. "Rate Variation of DNA Sequence Evolution in *Drosophila* Lineages." *Genetics* **149** (1998): 959–970.

[37] Wright, S. "Evolution in Mendelian Populations." *Genetics* **16** (1931): 97–159.

[38] Wright, S. "The Shifting Balance Theory and Macroevolution." *Ann. Rev. Genet.* **16** (1982): 1–19.

[39] Zeng, L.-W., J. M. Comeron, B. Chen, and M. Kreitman. "The Molecular Clock Revisited: The Rate of Synonymous vs. Replacement Change in *Drosophila*." *Genetica* **102/103** (1998): 369–382.

Spectral Landscape Theory

Peter F. Stadler

The notion of an *adaptive landscape* has proved to be a valuable concept in theoretical investigations of evolutionary change, combinatorial optimization, and the physics of disordered systems. Landscape theory has emerged as an attempt to devise suitable mathematical structures for describing the "static" properties of landscapes as well as their influence on the dynamics of adaptation. Here we focus on the connections of landscape theory and algebraic combinatorics that form the basis of spectral approach to understanding landscape structure.

1 INTRODUCTION

Evolutionary change is believed to be caused by the spontaneously generated genetic variation and its subsequent fixation by drift or selection. Consequently, the main focus of evolutionary theory has been to understand the genetic structure and dynamics of populations; see, e.g., Nagylaki [104]. In recent years, however, alternative approaches have gained increasing prominence in evolutionary theory. This development has been stimulated to some extent by the application of evolutionary models to designing *evolutionary algorithms*, such as genetic algorithms (GA), evolution strategies, and genetic programming, as well as by the theory of complex adaptive systems [40, 71, 81].

The generic structure of an evolutionary model is

$$x' = S(x, \mathbf{w}) \circ T(x, \mathbf{t}) , \tag{1}$$

where x is, for example, the vector of haplotype frequencies, and $S(x, \mathbf{w})$ is a term describing the selection forces acting on x. The parameters $\mathbf{w}$ form the so-called fitness function, since they can be regarded as a mapping from the set of types into the real numbers. The second term, $T(x, \mathbf{t})$, describes the transmission processes by determining the probability of transforming one type into another one by mutation or recombination [3]. Hence, evolution models can be seen as dynamical systems of genotype frequencies that live on an algebraic structure [99] which is determined by genetic processes such as mutation and recombination.

Metaphorically, the dynamics of evolutionary adaptation can be seen as a walk on a landscape, where uphill moves are preferred. The realization that the topological features of fitness landscapes crucially influence the time course of natural and simulated evolution led to what is now called *landscape theory*. It has several roots. In evolutionary theory it can be traced back to Wright's ideas about adaptive landscapes (see Provine [119, pp. 304–317]) and became important in theories of molecular evolution and the origin of life [33, 34, 44, 50, 77, 82, 115, 132, 134] and in evolutionary computer science [78, 79]. Similar developments exist in physics [46], where free-energy landscapes of disordered systems, such as spin glasses, are considered [101], and in search theory [114]. The main challenge to landscape theory is to control which features of the fitness landscape control the evolvability of the systems on the landscape.

From the mathematical point of view, a landscape consists of three ingredients: (i) a set V of "configurations," which we shall assume to be finite but very large, (ii) a cost or fitness function $f : V \to \mathbb{R}$ that evaluates the configurations, and (iii) some sort of additional geometrical, topological, or algebraic structure $\mathcal{X}$ on V that allows us to define notions of closeness, similarity, or dissimilarity among the configurations. The structure $\mathcal{X}$, which turns the set V into the *configuration space* $(V, \mathcal{X})$, is determined by the particular application, e.g., a heuristic search procedure for a combinatorial optimization problem or by the mechanisms of mutation and recombination in biological evolution.

In this contribution we shall be concerned mostly with two aspects of landscapes *ruggedness* and *neutrality*. Clearly, both ruggedness and neutrality depend on details of the structure of the configuration space $(V, \mathcal{X})$. The structure of a landscape thus may vary substantially when the structure $\mathcal{X}$ of V is changed, even if the cost function f remains the same. Furthermore, we wish to compare different cost functions on the same configuration space in order to determine which one is more rugged or neutral than another.

A very promising approach in landscape theory is the decomposition of the fitness function $f : V \to \mathbb{R}$ in terms of a basis (of the vector space $\mathbb{R}^V$) that is induced in some natural way by $\mathcal{X}$. In other words, we search for a suitable spectral theory of the combinatorial space $(V, \mathcal{X})$, which we then use to "Fourier transform" f with respect to a suitable set eigenfunctions of $(V, \mathcal{X})$. The resulting "Fourier coefficients," so one hopes, will reveal the important features of the landscape much more readily than f itself.

2 CONFIGURATION SPACES

2.1 GRAPHS, HYPERGRAPHS, P-STRUCTURES, AND FINITE TOPOLOGIES

We argued in the introduction that the "additional structure" $\mathcal{X}$ makes the boring "bag of numbers" $f : V \to \mathbb{R}$ a "landscape"—and an interesting mathematical object. The structure of the set V is oftentimes related to, or derived from, the internal structure of the objects $x \in V$. In this section, we shall explore a few possibilities of imposing structure onto the set V of configurations.

2.1.1 Graphs. The simplest case is based on the notion of a *move set.* For each $x \in V$ we define as set $\mathcal{N}(x)$ of *neighbors* of x. The elements of $\mathcal{N}(x)$ are those configurations that can be reached in a single step starting from x. It will be convenient to assume $x \notin \mathcal{N}(x)$ for all x and to define $\overline{\mathcal{N}}(x) = \mathcal{N}(x) \cup \{x\}$. This definition allows us to regard the set

$$\mathcal{E} = \{(x, y) \,|\, x \in V,\, y \in \mathcal{N}(x)\} \tag{2}$$

as the edge set of a directed graph with vertex set V. Equivalently, $\mathcal{E}$ is a *neighborhood relation* on V, that is, a relation satisfying $(x, x) \notin \mathcal{E}$ for all $x \in V$.

In many cases, one is interested in symmetric neighborhood relations, i.e., in move sets in which each step is "reversible" and $(x, y) \in \mathcal{E}$ implies $(y, x) \in \mathcal{E}$. We may then regard V as an undirected graph with edges $\{x, y\} \in E$ if and only if $(x, y) \in \mathcal{E}$. The undirected graph case is by far the most studied one.

The tours of a traveling salesman problem, for example, can be encoded as the list of cities in the order in which they are visited. In other words, a particular tour is a permutation π of the cities $\{1, \ldots, n\}$. It seems natural to make use of the fact that these permutations form the symmetric group S_n: choose a move set $\Omega \subset \mathsf{S}_n$ and define that y is a neighbor if x if y is obtained from x by multiplication with an element of $t \in \Omega$, $y = xt$. Of course, we require that Ω does not contain the group identity. Thus, $(x, y) \in \mathcal{E}$ if and only if $x^{-1}y \in \Omega$. The resulting graph is a so-called *Cayley digraph* of the S_n. In most cases one assumes that $t \in \Omega$ implies $t^{-1} \in \Omega$, in which case the neighborhood relation is symmetric and the Cayley graph has undirected edges.

In molecular biology, for instance, we may consider sequences as configurations and mutation as the move set. We have to distinguish two types of mutation.

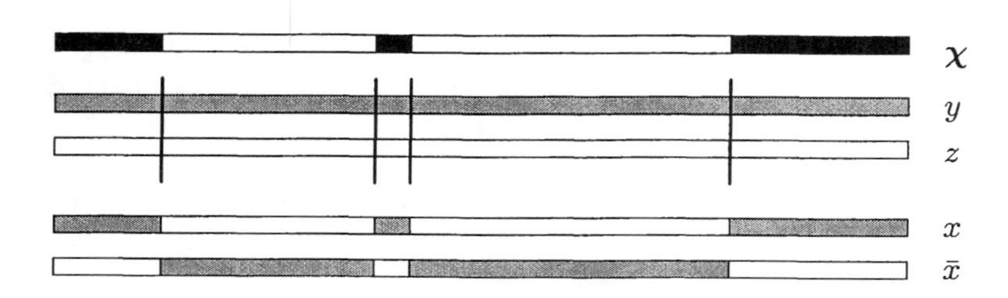

FIGURE 1 String Recombination. The children x and $\bar{x}$ are obtained by a multi-point crossover between the parents y and z. The bar on the top marks the positions that are transmitted together from a parent to an offspring in black and white, respectively.

On the one hand, point mutations change a letter in a sequence without affecting its length. Insertions and deletions, on the other hand, change the length of a sequence. Editing operations such as insertions and deletions are used, for example, in sequence alignment algorithms and can be generalized to trees [68, 110, 154]. Tree editing also provides a suitable analogue for mutation in genetic programming; see O'Reilly [108, 109]. Tree editing procedures are furthermore used in phylogenetic reconstruction [36]. Editing operations related to sorting are used to model genomic rearrangements [83].

2.1.2 Recombination, P-Structures, and Hypergraphs. Recombination, or crossover, is another way of imposing a sense of closeness on V. For strings, the meaning of crossover is easily defined.

A *crossover operator* is a map $\chi : V \times V \to V \times V$ with the following property: Suppose $\chi(y,z) = (u,v)$. Then for each k either $y_k = u_k \wedge z_k = v_k$ or $z_k = u_k \wedge y_k = v_k$. By abuse of notation, we write $x \in \chi(y,z)$, if $x = u$ or $x = v$; i.e., if x is an *offspring* of (y,z). As an immediate consequence, we see that $\chi(x,x) = (x,x)$. We follow here the spirit of Jones [79] when we regard a crossover operator as producing pairs of sequences rather than a single sequence from a pair of "ancestors."

We write $\boldsymbol{\chi} = \{k|x_k = u_k \wedge y_k = v_k\}$ and $\bar{\boldsymbol{\chi}} = \{k|y_k = u_k \wedge x_k = v_k\}$ for the two subsets of loci (sequence positions) that are *separated* by the crossover operator χ. There is, of course, a one-to-one correspondence between a crossover operator $\chi : V \times V \to V \times V$ and the associated set $\boldsymbol{\chi} \subseteq \{1, \ldots, n\}$: the set lists exactly those loci that are inherited from the first parent x by the first child u; see figure 1. Analogously, $\bar{\boldsymbol{\chi}}$ is the list of loci that the first child u inherits from the second parent y. Of course, $\bar{\boldsymbol{\chi}} = \{1, \ldots, n\} \setminus \boldsymbol{\chi}$.

Note that ancestors and offsprings have the same length. More general types of recombination, often called *unequal crossover*, do not adhere to this restriction: recombination events may occur between independent positions in the two sequences whence the chain lengths of the recombinants vary [137]. Crossover operators may also be defined for trees and permutations, with applications in genetic programming [91] and GAs for the traveling salesman problem [166], respectively.

A *recombination operator* (in the sense of much of the GA literature) is a family $\mathcal{F}$ of crossover operators that act on $V \times V$ with probabilities $\pi(\chi)$. In the following we restrict ourselves to equal-length crossover on strings. The two most important recombination operators are uniform recombination (derived $[\infty]$), consisting of all 2^n possible crossover operators, and one-point recombination (derived [1]), which contains all crossover operators χ for which the characteristic set is of the form $\chi = \{1, \ldots, k\}$.

Let V be a finite set with power set $\mathcal{P}(V)$. A *P-structure* [150] is a pair $(V, \mathcal{R})$ where $\mathcal{R} : V \times V \to \mathcal{P}(V)$. We say that the P-structure is *symmetric* if $\mathcal{R}(x, y) = \mathcal{R}(y, x)$ for all $x, y \in V$. In a weighted P-structure we attach a positive weight $\mathbf{H}_{x,(y,z)}$ to each triple (x, y, z) for which $x \in \mathcal{R}(y, z)$, and we set $\mathbf{H}_{x,(y,z)} = 0$ if $x \notin \mathcal{R}(y, z)$. We call $\mathbf{H}$ the incidence matrix of the P-structure.

In particular, there is a weighted P-structure associated in a natural way with each crossover operator χ:

$$\begin{aligned} \mathcal{R}^{\chi}(y, z) &= \{x \in V | x \in \chi(y, z)\}, \\ \mathbf{H}^{\chi}_{x,(y,z)} &= \begin{cases} 2 & \text{if} \quad x = y = z, \\ 1 & \text{if} \quad x \in \chi(y, z) \quad \text{and } y \neq z, \\ 0 & \quad\text{otherwise}. \end{cases} \end{aligned} \tag{3}$$

We observe that $\mathbf{H}^{\chi}_{x,(y,z)} > 0$ if and only if x is an offspring of (y, z). The doubled weight in the "diagonal," $\mathbf{H}^{\chi}_{x,(x,x)} = 2$, is mostly a technical convenience: It implies immediately $\sum_x \mathbf{H}_{x,(y,z)} = 2$, since any crossover operator produces exactly two offsprings from a pair of parents. If $y = z$, we simply count the offspring $y = z$ twice. The weighted P-structure associated with a recombination operator $\mathcal{F}$ is then

$$\begin{aligned} \mathbf{H} &= \sum_{\chi \in \mathcal{F}} \pi(\chi) \mathbf{H}^{\chi}, \\ \mathcal{R}(y, z) &= \bigcup_{\chi \in \mathcal{F}} \mathcal{R}^{\chi}(y, z) = \{x \in V | \exists \chi \in \mathcal{F} : x \in \chi(y, z)\}. \end{aligned} \tag{4}$$

The interpretation of this definition is straightforward: $\mathbf{H}_{x,(y,z)}$ is the chance that x is an offspring of the parents y and z under $\mathcal{F}$-recombination [160].

The *recombination hypergraph* imag$\mathcal{R}$ has vertex set V and hyperedges $\mathcal{R}(y, z)$, $y, z \in V$. A spectral theory of hypergraphs is described in Runge [129]. Gitchoff and Wagner [51] introduced a set of axioms to describe the action of recombination in terms of P-structures. In Stadler [150, Lemma C2] we showed that any recombination operator forms a recombination structure if and only if the identity map on $V \times V$ is a member of the family $\mathcal{F}$ of crossover operators.

2.1.3 Finite Topological Spaces. The shape space of RNA secondary structure has been treated as a finite metric space, with a distance measure that is based on "structure editing" [43, 76]. It has become apparent, however, that distance measures of this type are not useful for explaining the features of evolutionary trajectories [41, 42]. In these contributions, a notion of "continuity" is introduced, and the evolutionary transitions are classified as continuous or discontinuous based on how easily one shape can be accessed from a previous one. Continuity is a topological property. Taking this idea seriously, one may regard shape space as (finite) topological space, assuming that the "natural" topology is obtained by declaring the sets $\overline{\mathcal{N}}(x)$ of structures that are accessible from x as open sets. This approach has been developed elsewhere [21]. We just mention here that finite topological spaces have a unique nonredundant basis consisting of the sets

$$\mathcal{B}(x) = \bigcap_{y:x\in\overline{\mathcal{N}}(y)} \overline{\mathcal{N}(y)}\,, \tag{5}$$

which may be translated into the directed graph Υ with vertex set V and edges $\mathcal{B}(x) \setminus \{x\}$, $x \in V$. Topological properties, such as separation properties, can be then expressed as graph-theoretical properties of Υ, and we are back to the graph case.

2.2 MATRIX REPRESENTATIONS

2.2.1 Markov Chains. Not surprisingly, we shall encounter a close relationship between spectral graph theory [13, 23, 22, 24] and landscapes on graphs in the course of this survey. A graph is faithfully represented by its adjacency matrix $\mathbf{A}$, which has the entries

$$\mathbf{A}_{xy} = \begin{cases} 1 & \text{if} \quad (x,y) \in \mathcal{E}\,, \\ 0 & \text{if} \quad (x,y) \notin \mathcal{E}\,. \end{cases} \tag{6}$$

Of course, $\mathbf{A}$ is symmetric if and only if the graph is undirected.

The most straightforward way to search on a possibly weighted (di)graph is a random walk; that is with a Markov process with state space V. The most natural transition matrix is

$$\text{Prob}(y \to x) = \mathbf{S}_{xy} = \mathbf{A}_{xy} \Big/ \sum_{z\in V} \mathbf{A}_{xz}\,. \tag{7}$$

Such a random walk is usually called *simple* since each edge leaving y is chosen with the same probability. The denominator in equation (7) is the out-degree of vertex x. The matrix $\mathbf{S}$ is the transition matrix of the random walk. Note that this is the transpose of the convention in most of the literature on Markov chains; see, e.g., Bhattacharya [12] and Lovasz [98]. The most important feature of random walks is the existence of a stationary distribution $\wp$ such that $\wp = \mathbf{S}\wp$ to which all initial distributions converge under fairly general conditions.

A Markov process is called reversible if its stationary distribution $\wp$ satisfies the *balance equation* $\mathbf{S}_{xy}\wp(y) = \wp(x)\mathbf{S}_{xy}$. In particular, a simple random walk on an undirected graph is reversible. Let $\mathbf{P}$ be the diagonal matrix with diagonal $\wp$. If $\mathbf{S}$ is the transition matrix of a reversible chain, then $\mathbf{T} = \mathbf{P}^{-1/2}\mathbf{S}\mathbf{P}^{1/2}$ is a bistochastic symmetric matrix. The "regularized" transition matrix $\mathbf{T}$ still essentially describes the graph Γ since $\mathbf{T}_{xy} > 0$ if and only if (x, y) is an edge of Γ. Since $\mathbf{T}$ is a symmetric nonnegative matrix it serves as the starting point for the spectral theory of Markov processes [12].

Let us briefly consider the case of Hamming graphs in its most general setting. The configuration space consists of "genomes" with n loci (or positions) $k = 1, \ldots, n$. The are α_k alleles (or letters) at each position, which we denote by $x_k \in \mathcal{A}_k = \{0, 1, \ldots, \alpha_k - 1\}$. With $x_k \in \mathcal{A}_k$ we associate the root of unity

$$\hat{x}_k = \exp(2\pi\imath\, x_k/\alpha_k)\,. \tag{8}$$

Furthermore, for $I \in \prod_k \mathcal{A}_k$, set $\tilde{I} = \{k | I_k \neq 0\}$. For each "index" I we define the *generalized Walsh function*

$$\varepsilon_I : \prod_k \mathcal{A}_k \to \mathbb{C} : \varepsilon_I(x) = \prod_k \hat{x}_k^{I_k} = \prod_{k \in \tilde{I}} \hat{x}_k^{I_k}\,. \tag{9}$$

We remark that $\{\varepsilon_I | I \in V\}$ is the standard Fourier basis of the Abelian group $\prod Z_{\alpha_k}$; see section 2.3.6 below. These functions are eigenvectors of the adjacency matrix of the Hamming graphs $\prod \mathcal{Q}_{\alpha_k}$, i.e., the graphs obtained by considering point mutations; see e.g., Stadler [144]. Note that the formal association of the index sets I with the vertices in V is a mere bookkeeping device. If the number of alleles is the same for all loci, $\alpha_k = \alpha$, then the eigenvalue of $\mathbf{S}$ associated with ε_I is $\lambda_I = 1 - \frac{\alpha}{\alpha-1}|\tilde{I}|/n$.

Useful Markov processes on V can be defined, however, without any reference to a graph structure. The string recombination structures introduced in section 2.1.2 may serve as an example. A *crossover walk* [72, 73] on V is the Markov process based on the following rule: The "father" y is mated with a randomly chosen "mother" z. The offsprings are the "son" x and the "daughter" $\bar{x}$. The "son" x becomes the "father" of the next mating. We regard the sequence of "fathers" as a random walk on V. It is straightforward to derive the transition matrix of this Markov process:

$$\mathbf{S}_{xy} = \frac{1}{2} \sum_{z \in V} \mathbf{H}_{x,(y,z)} \wp(z)\,. \tag{10}$$

The factor $1/2$ stems from the fact that the offspring x is the "son" and not the "daughter" of the parents y and z with probability $1/2$. By $\wp(z)$ we denote the probability that z is the "mother" of the mating; i.e., $\wp(z)$ is the frequency of genotype z in the *population* $\wp$ under random mating. The uniform population case $\wp(z) = 1/|V|$, which is discussed in Stadler [150] and Wagner [160], is generalized to the *Wright manifold* $\mathcal{W} = \{\wp | \wp(z) = \prod_k p_k(z_k)\}$, where $p_k(a)$ denotes the frequency of allele a at locus k, in Stadler et al. [151]. It is not hard to verify that linkage equilibrium is maintained under recombination, i.e., that any $\wp \in \mathcal{W}$ is a stationary distribution of $\mathbf{S}$ as defined in equation (10). This fact was first proven for two alleles and arbitrary number of loci by Robbins [126] and for multiple alleles by Bennett [10].

The "population-weighted" Walsh functions

$$\psi_I^{\wp}(x) = \prod_{k \in \tilde{I}} \frac{1}{p_k(x_k)} \hat{x}_k^{I_k} \tag{11}$$

are defined for all $\wp \in \mathcal{W}$. For a uniform population they coincide with the generalized Walsh functions introduced in equation (9). In Stadler et al. [151] we show that the population-weighted generalized Walsh function $\psi_I^{\wp}$ is a left eigenvector of $\mathbf{S}^{\chi,\wp}$ with eigenvalue

$$\lambda_I^{\chi} = \begin{cases} 1 & \text{if} \quad \tilde{I} = \emptyset\,, \\ 1/2 & \text{if} \quad \emptyset \neq \tilde{I} \subseteq \chi \text{ or } \emptyset \neq \tilde{I} \subseteq \bar{\chi}\,, \\ 0 & \qquad \text{otherwise}\,. \end{cases} \tag{12}$$

This observation not only sets the stage for a spectral analysis of recombination landscapes, it also shows that recombination and mutation on strings are compatible operations that can therefore be compared directly in a meaningful way. The close relation between Hamming graphs and recombination spaces was noted with different methods by various groups, e.g., Culberson [20], Gitchoff [51], and Lipins [95]; and appendix A.

2.2.2 Schrödinger Operators and Graph Laplacians. Let Γ be a simple graph (without loops and multiple edges), and let a be a weight function on the edges of Γ, conveniently defined as $a : V \times V \to \mathbb{R}_0^+$ such that $a(x,y) = a(y,x) > 0$ if $\{x,y\} \in E$ and $a(x,y) = 0$ otherwise. We say that Γ is unweighted if $a(x,y) \in \{0,1\}$, i.e., if and only if $a(x,y) = \mathbf{A}_{xy}$. Furthermore, let $v : V \to \mathbb{R}$ be an arbitrary *potential*. The linear operator $\mathbf{H}$ defined by the action

$$\mathbf{H}f(x) = \sum_{y \sim x} a(x,y)\,[f(x) - f(y)] + v(x)f(x) \tag{13}$$

is a discrete *Schrödinger operator* associated with Γ [26]. This definition includes the transition matrices of random walks on graph discussed in the previous section. The quantity

$$\deg(x) = - \sum_{y:\{x,y\} \in E} \mathbf{H}_{xy}$$

is the (generalized) *degree* of a vertex $x \in V$. The degree matrix $\mathbf{D}$ is the diagonal matrix of the vertex degrees.

We call $-\mathbf{\Delta} = \mathbf{D} - \mathbf{A}$ the *Laplacian* of the edge-weighted graph Γ [102, 103]. A slightly different definition is explored in Chung [18]. Hence, any Schrödinger operator is of the form $\mathbf{H} = -\mathbf{\Delta} + \mathbf{diag}(v(x))$. The Laplacian is therefore a Schrödinger operator without potential.

The analogy between discrete and continuous Schrödinger operators is a close one because the discrete Laplacian $-\mathbf{\Delta}$ resembles the Laplacian differential operator Δ in many ways. To see this, one introduces an arbitrary *orientation* on Γ by choosing one of the two vertices u or v of the edge $h = \{v, w\}$ as the "positive end" and the other one as the "negative end." The matrix

$$\nabla^{+}_{xh} = \begin{cases} +\sqrt{a(x,y)} & \text{if } x \text{ is the positive end of } h = \{x,y\}\,, \\ -\sqrt{a(x,y)} & \text{if } x \text{ is the negative end of } h = \{x,y\}\,, \\ 0 & \text{otherwise}\,, \end{cases} \tag{14}$$

is called the (weighted) *incidence matrix of* Γ. The choice of the symbol ∇ is intentional. In fact, let $f : V \to \mathbb{R}$ be an arbitrary function. Then $(\nabla f)(h) = \sqrt{a(v,w)}[f(v) - f(w)]$, where h is the edge $\{v, w\}$ and v is the positive end of the edge h. This is as close to a first derivative as one can get on a graph. Note that $1/\sqrt{a(v,w)}$ takes the role of the distance between the vertices v and w.

The discrete Laplacian $-\mathbf{\Delta}$ is symmetric, nonnegative definite, and singular. The eigenvector $(1, \ldots, 1)$ belongs to the eigenvalue $\tilde{\Lambda}_0 = 0$. Λ_0 has multiplicity 1 if and only if Γ is connected. A few simple computations verify that $\mathbf{\Delta} = -\nabla^{+}\nabla$ and hence corresponds to "second derivatives" on Γ. Let $\langle\, . \,,\, . \,\rangle$ denote the standard scalar product on $\mathbb{R}^{|V|}$, and let $f, g\, V \to \mathbb{R}$ be arbitrary landscapes. Then *Green's formula* holds in the following form:

$$\langle \nabla f, \nabla g \rangle = -\langle f, \mathbf{\Delta} g \rangle = -\langle g, \mathbf{\Delta} f \rangle\,. \tag{15}$$

Graph Laplacians appear in very diverse fields of pure and applied mathematics. Their earliest use goes back to Kirchhoff's theory of electrical networks [86]; see, e.g., Biggs [13, ch. 5].

2.2.3 Courant's Nodal Domain Theorem. A well-known feature of Schrödinger operators on Riemannian manifolds is that the nodal domains—that is, the connected components of $M \setminus \psi^{-1}(0)$—of their eigenfunctions are severely constrained. In order to formulate *Courant's theorem* for graphs, we define for any function $f : V \to \mathbb{R}$ on Γ: $\text{supp}_{+}(f) = \{x \in V | f(x) > 0\}$, $\text{supp}_{-}(f) = \{x \in V | f(x) > 0\}$, $\text{zero}(f) = \{x \in V | f(x) > 0\}$, $\text{supp}^{0}_{+}(f) = \text{supp}_{+}(f) \cup \text{zero}(f)$, and $\text{supp}^{0}_{-}(f) = \text{supp}_{-}(f) \cup \text{zero}(f)$. A (strong) *nodal domain* of f is a maximal connected component of either $\text{supp}_{+}(f)$ or $\text{supp}_{-}(f)$. A *weak nodal domain* is a maximal connected component of $\text{supp}_{+}(f) \cup \text{zero}(f)$ or $\text{supp}_{-}(f) \cup \text{zero}(f)$, respectively.

Let $\lambda_1 \leq \lambda_2 \leq \cdots \leq \lambda_{|V|}$ be the eigenvalues of a Schrödinger operator on Γ with corresponding eigenvectors φ_i. Define $M(i) = \max\{k | \lambda_k = \lambda_i\}$ and

$m(i) = \min\{k\colon \lambda_k = \lambda_i\}$. Hence, $m(i) \leq i \leq M(i)$, $M(i) = m(i) + \text{mult}(\lambda_i) - 1$, and $m(i) = M(i) = i$ if and only if λ_i is a simple eigenvalue of $\mathbf{H}$.

The main result on discrete Schrödinger operators is the following version of *Courant's Nodal Domain Theorem*, which motivates why the eigenfunctions of a Laplacian form particularly interesting basis sets for our purposes:

Let ψ_i be an eigenvector of $\mathbf{H}$ with eigenvalue λ_i. Then:

i. There are at most $M(i)$ (strong) nodal domains of ψ.
ii. There are at most $m(i)$ weak nodal domains of ψ.
iii. If ψ_i has $m(i) + k$, $k > 0$, (strong) nodal domains, then no two of them meet at a nonvertex point of the geometric representation of the graph Γ, and every vertex meets at least $k + 1$ (strong) nodal domains.

The proofs of these results were obtained independently by different authors [25, 26, 47, 157], beginning with Fiedler [38] who showed that the number of components of $\text{supp}_+^0(\psi_i)$ is at most $M(i)$. Some closely related results on the component structure of $\text{supp}_+(\psi_i) \cup \text{supp}_-(\psi_i)$ can be found in Powers [117].

2.3 SYMMETRIES, PARTITIONS, AND MATRIX ALGEBRAS

In many cases of practical interest there is a substantial amount of symmetry in the ways the set of configurations is constructed. Below we shall briefly explore a few approaches, which haven been used to exploit these regularities in search of a workable spectral theory.

2.3.1 Relations and Automorphisms. A relation μ on V is simply a subset $\mu \subseteq V \times V$. The adjacency relation of a graph Γ may serve as an example. An *automorphism* of μ is a permutation $\mathsf{g} \in \mathsf{S}_{|V|}$ such that $(x, y) \in \mu$ if and only if $(\mathsf{g}(x), \mathsf{g}(y)) \in \mu$. The automorphisms of μ form the (permutation) group $\mathsf{Aut}[\mu]$, the *automorphism group* of μ. The automorphism group of a set $\mathcal{R}$ of relations on V is

$$\mathsf{Aut}[\mathcal{R}] = \bigcap_{\mu \in \mathcal{R}} \mathsf{Aut}[\mu]\,. \tag{16}$$

To each relation μ on V there is an associated characteristic $|V| \times |V|$ matrix $\mathbf{R}^{(\mu)}$ with entries $\mathbf{R}^{(\mu)}_{xy} = 1$ if $(x, y) \in \mu$ and $\mathbf{R}^{(\mu)}_{xy} = 0$ if $(x, y) \notin \mu$.

Let G be an arbitrary permutation group acting on V. By $\mathsf{2orb}(\mathsf{G}, V)$ we denote the set of orbits of G acting on $V \times V$. Of course, the $\mathsf{2orb}(\mathsf{G}, V)$ corresponds to a partition of $V \times V$, and each element of $\mathsf{2orb}$ may be regarded as a relation on V. These relations encapsulate the information about the symmetries that are most relevant for us.

A *matrix representation* of a finite group G is a map ρ from G into the group of $d \times d$ invertible matrices with complex coefficients such that $\rho(\mathsf{gh}) = \rho(\mathsf{g})\rho(\mathsf{h})$

for all g, h in G. The *permutation representation* $\mathbf{G}$ of (G, V) consists of the $|V| \times |V|$ permutation matrices $\mathbf{G}(\mathsf{g})$ whose nonzero entries are $\mathbf{G}_{xy}(\mathsf{g}) = 1$, if and only if $x = \mathsf{g}(y)$.

A permutation group G on V is intimately connected with its *centralizer algebra*

$$\mathfrak{V} = \mathfrak{V}_{\mathbb{C}}(\mathsf{G}, V) = \left\{ \mathbf{M} \in \mathbb{C}^{|V|\times|V|} \,\middle|\, \forall \mathsf{g} \in \mathsf{G} : \mathbf{MG}(\mathsf{g}) = \mathbf{G}(\mathsf{g})\mathbf{M} \right\}. \tag{17}$$

The set $\mathfrak{V}$ is closed w.r.t. addition and multiplication of matrices and w.r.t. multiplication with scalars from the underlying field $\mathbb{C}$. Its dimension (as vector space) equals the rank of its permutation group, $\dim(\mathfrak{V}) = \mathrm{rank}(\mathsf{G}, V)$. The characteristic matrices $\mathbf{R}^{(\mu)}$ of the orbits $\mu \in \mathsf{2orb}(\mathsf{G}, V)$ form the *standard basis* of the vector space $\mathfrak{V}$. From $\mathbf{R}^{(\mu)} \circ \mathbf{R}^{(\nu)} = \delta_{\mu,\nu}\mathbf{R}^{\mu}$ we see that $\mathfrak{V}$ is also closed under componentwise (Schur or Hadamard) multiplication. Finally, $\mathfrak{V}$ is closed under transposition since the transpose $\mu^{+} = \{(x, y)|(y, x) \in \mu\}$ of an orbit is again an orbit.

2.3.2 Coherent Algebras. A set of complex matrices that is closed under (i) scalar multiplication with complex numbers, (ii) componentwise addition, (iii) ordinary matrix multiplication, (iv) componentwise multiplication, and (v) transposition is called a *coherent algebra* or *cellular algebra*. Equivalently, a matrix algebra $\mathfrak{W} \subseteq \mathbb{C}^{|V|\times|V|}$ is coherent if and only if it satisfies the following axioms:

i. As a linear space over $\mathbb{C}$, $\mathfrak{W}$ has a basis of $\{\mathbf{R}^{(1)}, \ldots, \mathbf{R}^{(r)}\}$ of 0–1 matrices.
ii. $\sum_{j=1}^{n} \mathbf{R}^{(j)} = \mathbf{J}$, the all-one matrix.
iii. For every $i \in \{1, \ldots, r\}$ there is an i' such that $\mathbf{R}^{(i)\mathsf{T}} = \mathbf{R}^{(i')}$.
iv. $\mathbf{I} \in \mathfrak{W}$.

Sometimes coherent algebras without unity are considered; i.e., axiom (iv) is disregarded. The centralizer algebras of permutation groups form the most prominent class of coherent algebras (with identity).

Axiom (ii) above implies that the relations associated with the basis matrices $\mathbf{R}^{(j)}$ form a partition of $V \times V$. Such partitions are known as *coherent configurations* [65, 66, 67]. Table 2.3.2 gives an overview of various properties of partitions of $V \times V$ that are of interest in the context of landscapes. For details see, e.g., Stadler [144, 145].

For each collection $\mathcal{M} = \{\mathbf{M}_1, \ldots, \mathbf{M}_k\}$ of $|V| \times |V|$ matrices, there is a smallest coherent algebra $\langle\!\langle \mathcal{M} \rangle\!\rangle$, which is the defined as the intersection of all coherent algebras that contain $\{\mathbf{M}_1, \ldots, \mathbf{M}_k\}$. Since the centralizer algebra is coherent, we have

$$\langle\!\langle \mathcal{M} \rangle\!\rangle \subseteq \mathfrak{V}_{\mathbb{C}}(\mathsf{Aut}[\mathcal{M}], V). \tag{18}$$

Equality holds if and only if there is a permutation group that has $\langle\!\langle \mathcal{M} \rangle\!\rangle$ as its centralizer algebra [87]. The coherent algebra $\langle\!\langle \mathcal{M} \rangle\!\rangle$ can therefore be regarded

TABLE 1 Regularity properties of partitions. $\mathcal{I} = \{(x,x)|x \in V\}$ is called the diagonal of $V \times V$. The symbol • indicates properties that are used for definition, while ∘ marks additional properties that are implied by the definition.

i. $\mu \cap \mathcal{I} \neq \emptyset \implies \mu \subseteq \mathcal{I}$ for any $\mu \in \mathcal{R}$.
ii. $\mathcal{I} \in \mathcal{R}$.
iii. $\{x|\exists y \in V : (x,y) \in \mu\} = \{y|\exists x \in V : (x,y) \in \mu\} = V$ for all $\mu \in \mathcal{R}$.
iv. $\mu \in \mathcal{R} \implies \mu^{+} \in \mathcal{R}$.
v. $\mu = \mu^{\mathsf{T}}$ for all $\mu \in \mathcal{R}$.
vi. $\big|\{y \in V|(x,y) \in \mu\}\big|$ and $\big|\{y \in V|(y,x) \in \mu\}\big|$ depend only on $\mu \in \mathcal{R}$ but not on $x \in V$.
vii. The numbers $p^{\kappa}_{\mu,\nu} = \big|\{z \in V \,\big|\, (x,z) \in \mu \wedge (z,y) \in \nu\}\big|$ are the same for all pairs $(x,y) \in \kappa$.
viii. The matrices $\mathbf{R}^{(\mu)}$ and $\mathbf{R}^{(\nu)}$ commute for all $\mu, \nu \in \mathcal{R}$.

Property	i	ii	iii	iv	v	vi	vii	viii
homogeneous	∘	•						
transitive	•	∘	•					
precoherent configuration	∘	•		•				
symmetric					•			
class degree regular			∘			•		
homogeneous class deg. regular	•	∘	∘			•		
coherent	•			•			•	
homogeneous coherent configuration	∘	•	∘	•		∘	•	
class degree regular cc	•	∘	∘	•		•	•	
association scheme	•	∘	∘	•		∘	•	•
symmetric association scheme	•	∘	∘	∘	•	∘	•	∘

as a "combinatorial approximation" of the centralizer algebra [35, 89]. This is of particular importance in the graph case: given the adjacency matrix $\mathbf{A}$ of Γ, there is a polynomial time algorithm that determines the coherent algebra $\mathfrak{W}(\Gamma) = \langle\!\langle \mathbf{A} \rangle\!\rangle$; see Babel [6, 7] and Weisfeiler [164].

Let $\mathcal{R} = \{\mathbf{R}^{(1)}, \ldots, \mathbf{R}^{(r)}\}$ be the standard basis of a coherent algebra $\mathfrak{W}$. We have $\mathbf{R}^{(\mu)}\mathbf{R}^{(\nu)} = p^{\kappa}_{\mu,\nu}\mathbf{R}^{(\kappa)}$ where *intersection numbers*

$$p^{\kappa}_{\mu,\nu} = \big|\{z \in V \,\big|\, (x,z) \in \mu \wedge (z,y) \in \nu\}\big| \in \mathbb{N}_0 \tag{19}$$

are the same for all pairs $(x,y) \in \kappa$. The $r \times r$ matrices $\hat{\mathbf{R}}^{\kappa}$ with entries $\hat{\mathbf{R}}^{(\nu)}_{\mu,\kappa} = p^{\kappa}_{\mu,\nu}$ generate a matrix algebra $\hat{\mathfrak{W}}$ that is isomorphic to $\mathfrak{W}$ [65]. This observation makes coherent algebras appealing objects for our purposes because $\hat{\mathfrak{W}}$ is small enough in many cases to allow for explicit computations.

TABLE 2 Permutation Groups and Their Centralizer Algebras

G		$\mathfrak{W}_{\mathbb{C}}(\mathsf{G}, V)$
transitive	$\Longleftrightarrow$	homogeneous
multiplicity free	$\Longleftrightarrow$	commutative
generously transitive	$\Longleftrightarrow$	symmetric

The action of a permutation group (G, V) is *transitive* if, for all $x, y \in V$, there is $\mathsf{g} \in \mathsf{G}$ such that $y = \mathsf{g}(x)$. If $\mathsf{g} \in \mathsf{G}$ can be chosen such that $y = \mathsf{g}(x)$ and $x = \mathsf{g}(y)$, then (G, V) is *generously transitive*. Generously transitive permutations groups have symmetric (and therefore commutative) centralizer algebras; see Higman [65] and Wielandt [167, Thm. 29.3]. The group case is summarized in table 2.3.2.

2.3.3 Association Schemes. If the coherent algebra $\mathfrak{W}$ with standard basis $\mathcal{R} = \{\mathbf{R}^{(1)}, \ldots, \mathbf{R}^{(r)}\}$ is commutative, we obtain a symmetric association scheme $\mathfrak{W}^{\sigma}$ by taking as basis elements $\mathbf{R}^{(\mu)}$ if $\mu = \mu^{\mathsf{T}}$ and $\mathbf{R}^{(\mu)} + \mathbf{R}^{(\mu^{\mathsf{T}})}$ if $\mu \neq \mu^{\mathsf{T}}$, respectively. In particular, if a graph has a generously transitive or at least multiplicity-free group of automorphisms, then $\mathfrak{W}(\Gamma) = \langle\!\langle \mathbf{A} \rangle\!\rangle$ is a (symmetric) association scheme.

The situation becomes particularly simple in this case. Since $\mathfrak{W}$ is a commutative algebra of symmetric matrices, a so-called *Bose-Mesner algebra* [15, 27], there is a common basis $\mathbf{\Phi} = \{\varphi_i : V \to \mathbb{C}, i = 1, \ldots, |V|\}$ of eigenvectors of all matrices $\mathbf{M} \in \mathfrak{W}$.

The following observation is also of interest in this context [145, Lemma 9]: The coherent algebra $\mathfrak{W}[\Gamma]$ of a graph is always a refinement of the *distance partition* of Γ which has the classes $\delta_d = \big\{(x, y) \in V \big| d(x, y) = d\big\}$, i.e., $\mathbf{A}^{(d)} \in \langle\!\langle \mathbf{A} \rangle\!\rangle$, where $\mathbf{A}^{(d)}$ is the characteristic matrix of δ_d. The class of *distance regular graphs*, which contains important examples such as the Hamming graphs $\mathcal{Q}_{\alpha}^{n}$ and the Johnson graphs, is characterized by the fact that the distance partition forms a (symmetric) association scheme. These graphs have received considerable attention; see e.g., Brouwer [17].

2.3.4 Adjacency Algebra and Hoffman Algebras. The adjacency algebra of a graph Γ is the matrix algebra generated by the adjacency matrix, $\mathfrak{A}[\Gamma] = \langle \mathbf{A} \rangle$. Clearly, $\mathfrak{A}[\Gamma] \subseteq \mathfrak{W}[\Gamma]$. Higman [66] showed that $\mathfrak{A}[\Gamma] = \mathfrak{W}[\Gamma]$, if and only if $\mathfrak{W}[\Gamma]$ is commutative; i.e., it is an association scheme. It is interesting to note in this context that a homogeneous coherent algebra with rank $r \leq 5$ is always commutative [66]. So-called "orbit-polynomial graphs" characterized by $\mathfrak{A}[\Gamma] = \mathfrak{V}[\mathsf{Aut}[\Gamma], V]$ are considered in Beezer [9].

A *Hoffman algebra* is matrix algebra $\mathfrak{H} \subseteq \mathbb{C}^{|V|\times|V|}$ such that (i) there is a basis consisting of nonnegative integer matrices and (ii) $\mathbf{J} \in \mathfrak{H}$ [88]. This notion is of interest as a generalization of coherent algebras, since $\mathfrak{A}[\Gamma]$ is a Hoffman algebra; i.e., $\mathbf{J} \in \mathfrak{A}[\Gamma]$ if and only if Γ is a connected regular graph [69].

2.3.5 Equitable Partitions. Consider the set $\mathcal{R}$ of relations associated with the coherent algebra $\mathfrak{W}[\Gamma]$ of a graph Γ. Fix $x_0 \in V$ and define

$$\mu[x_0] = \{y \in V | (x, x_0) \in \mu\} . \tag{20}$$

Clearly, $\Pi(x_0) = \{\mu[x_0] | \mu \in \mathcal{R}\}$ is a partition of V. If $\mathfrak{W}$ is homogeneous, then $\mu[x_0]$ is nonempty for all $\mu \in \mathcal{R}$ (see table 2.3.2); i.e., $\mathcal{R}$ and $\Pi[x_0]$ have the same number of classes for all $x_0 \in V$. Higman [66] showed that

$$\hat{\mathbf{R}}^{(\nu)}_{\mu\kappa} = p^{\kappa}_{\mu,\nu} = \sum_{x\in\nu[x_0]} \hat{\mathbf{R}}^{(\nu)}_{xz} \qquad \text{for each } z \in \kappa[x_0] . \tag{21}$$

Noting that the adjacency matrix $\hat{\mathbf{A}}$ of Γ is a sum of $\mathbf{R}^{(\mu)}$-matrices, it is shown in Stadler [145] that

$$\sum_{x\in\mu[x_0]} \mathbf{A}_{xy} = \tilde{\mathbf{A}}_{\mu[x_0],\nu[x_0]} = \sum_{\kappa\subseteq\mathcal{E}} p^{\kappa}_{\mu,\nu} = \hat{\mathbf{A}}_{\mu,\nu} \tag{22}$$

holds for any $y \in \nu[x_0]$. Partitions of V that satisfy the first equality in equation (22) are called *equitable.* Equitable partitions have been introduced by Schwenk [135]; more recently they have been used by Powers and coworkers as "colorations"; see, e.g., Powers [118, 116]. In Cvetković [23, ch. 4] they appear as "divisors" of graphs.

The most important property of an equitable partition Π is that all eigenvalues of the *collapsed adjacency matrix* $\tilde{\mathbf{A}}$ are also eigenvalues of $\mathbf{A}$. If Π contains a class that consists of single vertex, then the minimal polynomials of $\tilde{\mathbf{A}}$ and $\mathbf{A}$ are the same [14, Thm.8.6]; i.e., the relevant spectral information is already contained in $\tilde{\mathbf{A}}$. More information about equitable partitions can be found in Godsil [52, 53, 54].

2.3.6 Fourier Transform on Finite Groups and Cayley Graphs. Let $\mathbf{G}$ be a finite group and let $f : \mathsf{G} \to \mathbb{C}$. Let ρ be matrix representation of G. Then

$$\hat{f}(\rho) = \frac{1}{\sqrt{|\mathsf{G}|}} \sum_{\mathsf{g}\in\mathsf{G}} f(\mathsf{g})\rho(\mathsf{g}) \tag{23}$$

is called the *Fourier transform*[1] *of f at ρ.* The Fourier transform on a complete set $\mathcal{R}$ of irreducible representations is inverted by

$$f(\mathsf{g}) = \frac{1}{\sqrt{|\mathsf{G}|}} \sum_{\rho\in\mathcal{R}} \dim\rho \, \mathrm{Tr}\left[\hat{f}(\rho)\rho(\mathsf{g}^{-1})\right] . \tag{24}$$

[1] In most of the literature the normalization factor $|\mathsf{G}|^{-1/2}$ is omitted.

Fast Fourier Transform algorithms are known for a variety of finite groups. For a recent overview see, e.g., Maslen [100] and Rockmore [127].

It is not surprising that the spectral properties of Cayley graphs are intimately related to the Fourier transform on the underlying group. The crucial observation is the following. Let δ_Ω be the characteristic function of the set of generators Ω. Then $\sqrt{|\mathsf{G}|}\hat{\delta}_\Omega(\boldsymbol{\rho}_{\mathrm{reg}})$, the Fourier transform of δ_Ω at the regular representation of G, equals the adjacency matrix of $\Gamma(\mathsf{G},\Omega)$ up to a reordering of the group elements. Its spectrum is, therefore, the union of the spectra of $\sqrt{|\mathsf{G}|}\hat{\delta}_\Omega(\boldsymbol{\rho}_i)$, where $\boldsymbol{\rho}_i$ are the irreducible representations of G. If Ω is a union of conjugacy classes of G, the situation simplifies further [31, 128].

The irreducible representations are all one-dimensional if G is commutative. Since G can be written as a direct product of cyclic groups, $\mathsf{G} = \prod_{k=1}^{m} \mathsf{C}_{n_k}$, the characters are

$$\chi_{\mathsf{g}}(\mathsf{x}) = \exp\left(2\pi i \sum_{k=1}^{m} \frac{x_k g_k}{n_k}\right), \tag{25}$$

where we use the additive representation of C_{n_k} as $\{0, 1, \ldots, n_k - 1\}$ with addition modulo n_k. It is not hard to verify that the characters χ_{g} are eigenvectors of the adjacency matrix of each Cayley graph of G. The corresponding eigenvalue are $\sum_{\mathsf{x}\in\Omega} \chi_{\mathsf{g}}(\mathsf{x})$; see, e.g., Lovasz [97]. The Fourier transform on C_2^n is also known as the *Walsh-Hadamard* transform. Note that the Boolean hypercube can be regarded as a Cayley graph on this group. An FFT algorithm for this case is due to Yates [170]. Further material about the Cayley graphs on commutative groups can be found in Alspach [1].

3 LANDSCAPES

3.1 FOURIER DECOMPOSITION AND ELEMENTARY LANDSCAPES

Having derived a set of basis functions $\{\varphi_k | V \to \mathbb{C}\}$ from the structure of a configuration space $(V, \mathcal{X})$ by means of one of the approaches outlined in the previous section, it is natural to expand the fitness function f in terms of this basis:

$$f(x) = \sum_k a_k \varphi_k(x). \tag{26}$$

We shall use the following convention:

i. The index 0 is reserved for the "ground state." If the basis is derived from a Laplacian, for instance, then φ_0 is constant, the associated eigenvalue is zero, and

$$a_0 = \sum_x \varphi_0(x) f(x) = |V|^{-1} \sum_{x\in V} f(x). \tag{27}$$

TABLE 3 Elementary Landscapes.

Problem	Graph	D	λ	State
p-spin glass	$\mathcal{Q}_2^n$	n	$2p$	p
NAES$^{(1)}$	$\mathcal{Q}_2^n$	n	4	2
Weight Partitioning	$\mathcal{Q}_2^n$	n	4	2
Graph α-Coloring	$\mathcal{Q}_2^\alpha$	$(\alpha-1)/n$	2α	2
XY-spin glass	$\mathcal{Q}_\alpha^n$	$(\alpha-1)/n$		2
for $\alpha > 2$:	$\mathcal{C}_\alpha^n$	2		2
TSP symmetric	$\Gamma(\mathcal{S}_n, \mathcal{T})$	$n(n-1)/2$	$2(n-1)$	2
	$\Gamma(\mathcal{S}_n, \mathcal{J})$	$n(n-1)/2$	n	2
	$\Gamma(\mathcal{A}_n, \mathcal{C}_3)$	$n(n-1)(n-2)/6$	$(n-1)(n-2)$	?
antisymmetric	$\Gamma(\mathcal{S}_n, \mathcal{T})$	$n(n-1)/2$	$2n$	3
	$\Gamma(\mathcal{S}_n, \mathcal{J})$	$n(n-1)/2$	$n(n+1)/2$	$\mathcal{O}(n)$
Graph Matching	$\Gamma(\mathcal{S}_n, \mathcal{T})$	$n(n-1)/2$	$2(n-1)$	2
Graph Bipartitioning	$J(n, n/2)$	$n^2/4$	$2(n-1)$	2

$^{(1)}$Not-All-Equal-Satisfiability

Similarly, the index 0 will refer to the stationary distribution in the case of a Markov chain on V.

ii. The distinct eigenvalues of $-\mathbf{\Delta}$ will be denoted by Λ_p; in the Markov chain case, we write λ_p. It will be convenient to define the index sets $J_p = \{k| - \mathbf{\Delta}\varphi_k = \Lambda_p \varphi_k\}$ that collect all eigenfunctions belonging to the same (Laplacian) eigenvalue.

iii. We write $\tilde{f}(x) = f(x) - a_0$. If φ_0 is constant, this is the "nonflat" part of fitness function.

Grover and others [19, 59, 145] observed that $\tilde{f}$ is in many cases an eigenfunction of the graph Laplacian $-\mathbf{\Delta}$; see table 3 for a list of examples. We say that f is *elementary* w.r.t. $-\mathbf{\Delta}$ if $\tilde{f}$ is an eigenfunction of $-\mathbf{\Delta}$ with an eigenvalue $\lambda_p < 1$. In Stadler et al. [151] this notion is extended to calling f elementary w.r.t. a random walk transition operator if and only if $\mathbf{S}\tilde{f} = \lambda_p \tilde{f}$ with an eigenvalue $\lambda_p < 1$.

If f is elementary, then $\tilde{f}$ satisfies the conditions of Courant's nodal domain theorem; see section 2.2.3. Elementary landscapes thus can be expected to have few nodal domains if they belong to a small Laplacian eigenvalue (or to an eigenvalue of Markov transition matrix close to 1), while landscapes that are far away from the ground state will in general have many nodal domains. Such landscapes will appear "rugged." Grover [59] showed that

$$f(\hat{x}_{\min}) \le a_0 \le f(\hat{x}_{\max}), \tag{28}$$

where $\hat{x}_{\min}$ and $x_{\max}$ are arbitrary local minima and maxima, respectively. This *maximum principle* shows that elementary landscapes are well behaved: There are no local optima with worse than average fitness. We shall return to local optima as a measure of ruggedness in section 4.4.

In section 2.2.1, we have seen that p-spin (or Walsh-) functions are the eigenfunctions not only of mutation operators but also of recombination operators. Indeed, there is an intriguing relationship between elementary landscapes for string recombination and schemata sensu Holland [2, 11, 70, 71]; see also appendix A. Each recombination-elementary landscape corresponds to a partitioning of the set of strings. Each equivalence class in this partitioning is a schema in the sense of Holland [71] and all the schema which make up this partitioning have the same positions fixed. An elementary landscape in this context is a landscape which assumes that only the fixed positions in the schema actually influence fitness. This was first noted by Weinberger in his seminal paper on Fourier and Taylor series of fitness landscapes [162]. In Stadler [150] and Wagner [160], it is shown rigorously that this is a legitimate way of decomposing the configuration space of string recombination.

3.2 CORRELATION MEASURES

3.2.1 Random Walk Autocorrelation Functions. The ruggedness of a landscape is most easily quantified by measuring the correlation of fitness values in "neighboring" positions. Weinberger [161, 162] suggested the following procedure. Given a Markov process on V, we sample the fitness values $f(x^{(t)})$, interpret them as a time series, and compute the autocorrelation function of this time series. Let $\mathbf{T}$ be the transition matrix of a such reversible Markov process with stationary distribution φ_0. We define the scalar product

$$\langle f, g\rangle_{\varphi_0} = \sum_{x\in V} f(x)\varphi_0(x)g^*(x)\,, \tag{29}$$

where a^* denotes the complex conjugate of a. The (expected) autocorrelation function along a $\mathbf{T}$-random walk on V is then

$$r(t) = \left(\sum_{x\in V} |\tilde{f}^2(x)|\varphi_0(x)\right)^{-1} \sum_{y\in V} \tilde{f}(x)(\mathbf{T}^t)_{xy}\tilde{f}^*(y)\varphi_0(y) = \frac{\langle \tilde{f}, \mathbf{T}^t\tilde{f}\rangle_{\varphi_0}}{\langle \tilde{f}, \tilde{f}\rangle_{\varphi_0}}\,. \tag{30}$$

Expanding f w.r.t. eigenvectors of $\mathbf{T}$, it can be shown [145] that

$$r(t) = \sum_{p\neq 0} B_p\lambda_p^t \qquad \text{with} \qquad B_p = \frac{\sum_{i\in J_p} |a_i|^2}{\sum_{i\neq 0} |a_i|^2}\,. \tag{31}$$

Thus, a landscape f is elementary w.r.t. a transition operator $\mathbf{T}$, if and only if the "random walk" autocorrelation function is exponential: $r(t) = \lambda_p^t$. In this case, the *order* p indicates to which eigenvalue (not counting multiplicities) $\tilde{f}$ belongs. On a Boolean hypercube $\mathcal{Q}_\alpha^2$ we have eigenfunctions of the form $\sum_I a_I \prod_{k\in I} x_k$, where $p = |I|$ is constant. These are exactly Derrida's [28] p-spin models. The order of the elementary landscape thus equals the "interaction order" of the underlying spin glass model.

3.2.2 Amplitude Spectra. Equation (31) decomposes nonelementary landscapes in a natural way into a superposition of elementary ones. The amplitudes B_p measure the relative variance contributions of the different eigenspaces (or "modes"). Instead of the random walk correlation function $r(t)$, we can therefore use the *amplitude spectrum* B_p, $p \geq 1$, as a measure for the ruggedness of a landscape. In many cases, it is much easier to interpret than the correlation function; see, e.g., Stadler [145], García-Pelayo [48], Hordijk [74], and Rockmore et al. [128]. This technique was applied successfully to realistic landscapes, such as those arising from RNA folding; see figure 2 for an example. The RNA secondary structure folding model is described in detail in Schuster's contribution to this book [131].

3.2.3 Distance Correlation Functions. Most early work on RNA landscapes, e.g., Fontana [44] and Tacker [153], uses a different type of correlation measure based on the Hamming distance. In Stadler [145, 148] a more general version starting with a collection of relations on V is introduced. Let μ be a relation on V. Then we set

$$\varrho(\mu) = \frac{|V|^2}{|\mu|} \frac{\sum_{(x,y)\in\mu}(f(x)-\bar{f})(f(y)-\bar{f})}{\sum_{x,y\in V}(f(x)-\bar{f})(f(y)-\bar{f})}, \tag{32}$$

where $\bar{f} = |V|^{-1}\sum_x f(x)$. Thus, $\varrho(\mu)$ is the variation of points of vertices within a relation μ compared to the variance of f over all configurations $x \in V$. On Hamming graphs, for instance, it is natural to consider the distance classes; i.e., $(x,y) \in \mu_d$, if and only if $d_H(x,y) = d$, a predefined value. Such distance-dependent correlation functions have been considered also for some combinatorial optimization problems [5, 4, 138, 140]. Given a partition of $V \times V$, we may of course regard ϱ as a function of the classes of this partition. Furthermore, if this partition is sufficiently "nice," then the correlation function ϱ itself also has useful algebraic properties. The main result of Stadler [143], for instance, is the following theorem: Let f be a landscape on a regular graph Γ that has a homogeneous coherent algebra $\mathfrak{W}[\Gamma]$. Then $r(s)$ is exponential, if and only if ϱ is a left eigenvector of the collapsed adjacency matrix $\hat{\mathbf{A}}$.

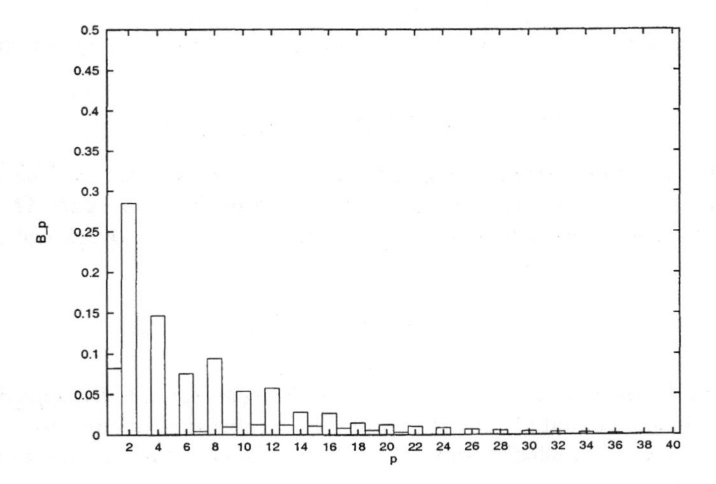

FIGURE 2 The estimated amplitude spectrum for a GC landscape with $n = 100$ under mutation [74]. The configuration space is the Hamming graph $\mathcal{Q}_2^{100}$ of sequences taken from the two-letter alphabet {G,C}. The most striking feature of the amplitude spectrum of RNA landscapes is a strong difference between even and odd modes. This can easily be explained in terms of the physics underlying RNA folding: The major contribution of the folding energy comes from stacking of base pairs. Hence, the major changes in free energy caused by a point mutation will arise from these contributions. Since stacking energies are influenced by even number of nucleotides depending on the location of the affected base pair within a stack. A recent comparison of amplitude spectra for different landscapes based on folding short RNA chains indicates that the amplitude spectra of the free-energy landscapes are typical [146].

4 RANDOM LANDSCAPES

4.1 BASIC DEFINITIONS

In many cases, for instance in applications to spin glasses, the definition of the landscape contains a number of random parameters. We, therefore, define *random landscapes* as elements of an appropriate probability space, following the presentation in Reidys [122].

Let V be a finite set and let W be a predicate of landscapes $f : V \to \mathbb{R}$. A *random W-landscape* over V is the probability space

$$\boldsymbol{\Omega} = (\{f : V \to \mathbb{R} \mid f \text{ has property } W\}, \mathcal{A}, \mu) \ , \tag{33}$$

where $\mathcal{A}$ is a σ-field and $\mu : \mathcal{A} \to [0,1]$ a measure. Let $\xi : \boldsymbol{\Omega} \to \mathbb{R}$ be an Ω-random variable; we denote expectation value and variance of ξ by $\mathbb{E}[\xi]$ and $\mathbb{V}[\xi]$, respectively. In particular, we shall write $\mathbb{E}[f(x)]$ for the expected value of $f(x)$, i.e., of $\mathbb{E}[f]$ evaluated at $x \in V$.

The *covariance matrix* $\mathbf{C}$ of the random landscape $\boldsymbol{\Omega}$ is given componentwise by

$$\mathbf{C}_{xy} = \mathbb{E}[f(x)f(y)] - \mathbb{E}[f(x)]\mathbb{E}[f(y)]\,. \tag{34}$$

Clearly, $\mathbf{C}$ is a symmetric nonnegative definite $|V| \times |V|$ matrix. Taking the set of *all* maps, $\{f : V \to \mathbb{R}\}$, as basis space of the probability space $\boldsymbol{\Omega}$, a basis is formed by a set of orthonormal eigenvectors $\{\psi_k\}$ of the covariance matrix $\mathbf{C}$. An expansion of the form

$$f(x) \doteq \sum_k b_k \psi_k(x) \tag{35}$$

is know as *Karhunen-Loève series* or *principal component decomposition*. The symbol $\doteq$ denotes equality almost surely. The importance of equation (35) comes from the following classical result [75]: The coefficients $\{b_k\}$ in equation (35) are uncorrelated random variables satisfying

$$\mathrm{Cov}[b_k, b_j] = \theta_k \delta_{kj}, \qquad 1 \le k, l \le |V|\,, \tag{36}$$

where $\theta_k = \mathbb{V}[b_k]$ is the eigenvalue of $\mathbf{C}$ belonging to the eigenvector ψ_k.

A random landscape is elementary if $\mathbb{E}[r(t)] = \lambda_p^s$ or, equivalently, if $\mathbb{E}[B_p] > 0$ only for a single mode $p > 0$. While Ising spin glasses, TSPs with random coefficients, and other random parameter variants of combinatorial optimization problems are elementary, Kauffman's NK models are not [148]. As a consequence, there are many landscapes that cannot be constructed as a superposition of NK-models, for instance, the Sherrington-Kirkpatrick spin glass [136]; see Stadler [144], Happel [62], and Heckendorn [63] for a detailed discussion of the amplitude spectrum $\mathbb{E}[B_p]$ of NK models.

A random landscape is *pseudo-isotropic* [148] if there are constants a_0, v, and w such that for all $x \in V$ holds (i) $\mathbb{E}[f(x)] = a_0$, (ii) $\mathbb{V}[f(x)] = v^2$, and (iii) $|V|^{-1} \sum_{y \in V} \mathbf{C}_{xy} = w$. Pseudo-isotropy is a fairly weak regularity property that is satisfied by many random landscape models of practical importance; see table 4.

4.2 ADDITIVE RANDOM LANDSCAPES

Many important random landscapes can be written as a sum of components with random coefficients. More precisely, let M be finite index set; let c_j, $j \in M$ be independent, real-valued random variables over appropriate probability spaces $\Omega_j = (\mathbb{R}, \mathcal{A}_j, \mu_j)$; and let $\Theta = \{\vartheta_j : V \to \mathbb{R} \,|\, j \in M\}$ be a family of real-valued functions on V. An *additive random landscape* (ARL) is the probability space $(\boldsymbol{\Omega}_V, \otimes_j \mathcal{A}_j, \otimes_j, \mu_j)$ with

$$\boldsymbol{\Omega}_V = \{f : V \longrightarrow \mathbb{R} \mid f(x) = \sum_{j=1}^{M} c_j \vartheta_j(x)\}\,. \tag{37}$$

TABLE 4 Examples of Additive Random Landscapes. The component landscapes ϑ_I and the index set M, equation (37), are listed together with information whether the models are uniform (U), strictly uniform (S), or pseudo-isotropic (P). As in table 2.3.2, properties that are implied by stronger ones are shown as ∘.

Model	Component Landscapes and Index Set		U	S	P
Ising spin glass	$\vartheta_I(x) = \prod_{k\in I} x_k$	$I \subseteq \{1,\ldots,n\}$	•		•
SK model	as above with $\|I\| = 2$		•		•
NK Landscapes	see [122]		∘	•	∘
Graph Bipartitioning	$\vartheta_{ij}([A,B]) = \begin{cases} 1 & \text{if } \{i,j\} \not\subseteq A, B \\ 0 & \text{otherwise} \end{cases}$	$i < j$	∘	•	∘
Asymmetric TSP	$\vartheta_{kl}(\tau) = \sum_i \delta_{k,\tau(i)}\delta_{l,\tau(i-1)}$	$k \neq l$	∘	•	∘

In other words, the random landscape is constructed as a linear combination of nonrandom landscapes ϑ_j with independent random coefficients c_j.

In particular, any Gaussian random landscape is additive. Using the Karhunen-Loève decomposition, equation (35), any random landscape can be written as linear combination with uncorrelated random coefficients; uncorrelated Gaussian random variables are independent.

The most important additive random landscapes exhibit further regularities. An ARL is *uniform* if and only if (i) the random variables c_i, $i \in M$, are i.i.d. and (ii) there exist constants $a, b \in \mathbb{R}$ such that $\sum_{x\in V} \vartheta_i(x) = |V|a$ and $\sum_{x\in V} \vartheta_i^2(x) = |V|b$. A uniform random landscape is *strictly uniform* if there exist constants $d, e \in \mathbb{R}$ such that $\sum_j \vartheta_i(x) = d$ and $\sum_j \vartheta_i^2(x) = e$. In Reidys [122] we show that a uniform random landscape is pseudo-isotropic if and only if at least one of the following two conditions is satisfied: (i) $\mathcal{F}$ is strictly uniform, or (ii) $a = 0$, $\mathbb{E}[c_i] = 0$, and there is a constant $e \in \mathbb{R}$ such $\sum_i \vartheta_i^2(x) = e$ for all $x \in V$.

4.3 ISOTROPY AS A MAXIMUM ENTROPY CONDITION

Uniformity and pseudo-isotropy are still rather weak properties. In Stadler [144, 148] the notion of an *isotropic* random landscape was introduced as a "statistically symmetric model": a random landscape with a covariance matrix that shares the symmetries of the underlying configuration space. More precisely, a random landscape is *isotropic w.r.t. a partition* $\mathcal{R}$ of $V \times V$ if there are constants a_0 and s and a function $c : \mathcal{R} \to \mathbb{R}$ such that (i) $\mathbb{E}[f](x) = a_0$ and $\mathbb{V}[f](x) = s^2$ for all $x \in V$, and (ii) $\mathbf{C}_{xy} = c(\mu)$ for all $(x, y) \in \mu$; i.e., the covariance matrix $\mathbf{C}$ is constant on the classes $\mu \in \mathcal{R}$.

The notion of isotropy for random landscapes is the analog of *stationarity* for stochastic processes. Following the conventions of Karlin and Taylor [80], our notion of isotropy would be called "covariance isotropic," "weakly isotropic," or "wide sense isotropic." For a Gaussian random landscape the notions of (weak) isotropy and strict isotropy coincide, of course.

Not surprisingly, a useful theory does not arise by considering arbitrary partitions $\mathcal{R}$; see table 2.3.2. Isotropy is a stronger concept than pseudo-isotropy only if $\mathcal{R}$ is sufficiently regular. Transitivity, for instance, ensures that the classes of $\mathcal{R}$ are large enough to be interesting. In Stadler [148, Thm. 4] the following result is proved: Let $\mathcal{F}$ be isotropic w.r.t. a homogeneous class degree regular partition $\mathcal{R}$ of $V \times V$. Then $\mathcal{F}$ is pseudo-isotropic. Furthermore, suppose $\mathbb{E}[f(x)] = a_0$ for all $x \in V$. Then the random landscape is isotropic w.r.t. a homogeneous coherent configuration if and only if $\mathbf{C} \in \langle\langle\mathcal{R}\rangle\rangle$ [148].

If $\mathbf{A}$ is the adjacency matrix of an undirected graph (or, more generally, the asymmetric transition matrix of a Markov process on V), then we say that a random landscape is **-isotropic w.r.t.* $\mathbf{A}$ if $\mathbb{E}[f(x)] = a_0$ and $\mathbf{C} \in \langle\mathbf{A}\rangle$. For association schemes, such as those arising from distance regular graphs including the hypercube, isotropy and *-isotropy are equivalent. In Stadler [148] we show that a random landscape is *-isotropic if and only if the Fourier coefficients (w.r.t. an orthonormal basis of eigenvectors of $\mathbf{A}$) satisfy: (i) $\mathbb{E}[a_k] = 0$ for $k \neq 0$, (ii) $\mathrm{Cov}[a_k, a_j] = \delta_{kj}\mathbb{V}[a_k]$, and (iii) $\mathbb{V}[a_k] = \mathbb{V}[a_j]$ if $k, j \in J_p$. These conditions mean that the Fourier coefficients are uncorrelated and that they have the same mean and variance whenever they belong to the same mode (eigenspace of $\mathbf{A}$). Hence, Fourier and Karhunen-Loève series coincide for *-isotropic landscapes.

For a random landscape with measure μ, we define the *entropy*

$$S = -\int \mu(f) \ln \mu(f) df \,. \tag{38}$$

In appendix B we review some well-known properties of the entropy functional. In particular, S can be decomposed into a "homogeneous" part S_{σ^2} that only depends on $\sigma^2 = \mathrm{Tr}\ \mathbf{C}$, the total variance of landscape, and terms $S_{\mathbf{C}}$ that depends only on the variations among the eigenvalues of $\mathbf{C}$, and a third term that measures the effect of deviations from the normal distribution given a fixed covariance matrix $\mathbf{C}$. Given $\mathbf{C}$, a random landscape maximizes entropy if and only if its Gaussian.

In the following we assume a Gaussian random landscape. Now suppose the values $\beta_p = \sum_{k \in I_p} \mathbb{V}[a_k]$ are prescribed, too. It follows from the discussion in appendix B that the entropy is maximized if and only if the covariance matrix restricted to J_p is a multiple of the identity; i.e., if and only if $\mathbb{V}[a_k]$ is constant on J_p. Given its amplitude spectrum, a random landscape therefore maximizes entropy if and only if it is Gaussian and *-isotropic. This is of practical interest since the class of *-isotropic models (on their natural configuration spaces) includes, among others, Derrida's p-spin Hamiltonians, the graph-bipartitioning problem, and the TSP.

Most variants of Kauffman's NK model—the XY-Hamiltonians, short-range Ising models, or the graph-matching problem—are not isotropic. This has important implications for the structure of these landscapes, as we shall see below.

4.4 LOCAL OPTIMA

Palmer [113] used the existence of a large number of local optima to define ruggedness. We say that $x \in V$ is a *local minimum* of the landscape f if $f(x) \leq f(y)$ for all neighbors y of x. The use of $\leq$ instead of $<$ is conventional [84, 130]; it does not make a significant difference for spin glass models. Local maxima are defined analogously. The number $\mathcal{N}$ of local optima of a landscape, however, is much harder to determine than its autocorrelation function $r(s)$ or its correlation length

$$\ell = \sum_{k=0}^{\infty} r(s) = \sum_{p \neq 0} \frac{B_p}{1 - \lambda_p} . \tag{39}$$

As it appears that $\mathcal{N}$ and ℓ are two sides of the same coin, we search for a connection between the two quantities. In a random landscape setting it is customary to determine $\mathbb{E}[\ln \mathcal{N}]$. The only known case in which $\mathbb{E}[\ln \mathcal{N}] \neq \ln \mathbb{E}[\mathcal{N}]$ is the linear spin chain [29].

For the case of short-range spin glasses, in which only a small number z of coupling constants J_{ij} are nonzero for any given spin i, a slightly larger number of local optima has been found [16, 155] than for the long-range Sherrington-Kirkpatrick model [136]. Since all Ising models have the same correlation length $\ell = n/4$ [163, 142] but somewhat different values of $\mathcal{N}$, we cannot hope for a general, exact formula relating $\mathbb{E}[\ln \mathcal{N}]$ and $\mathbb{E}[\ell]$.

From the maximum entropy interpretation of isotropy, however, we know that the expected density of metastable states in an isotropic Gaussian random landscape is determined completely by the expected correlation function $\mathbb{E}[r(s)]$ because such a model simply does not contain any further information. In the case of an elementary isotropic random landscape, the correlation length ℓ already determines $r(s)$ and, hence, there must be a direct relationship between $\mathbb{E}[\ell]$ and the expected number of metastable states $\mathbb{E}[\ln \mathcal{N}]$. Its functional form will, of course, depend on the geometric properties of Γ.

Stadler and Schnabl [139] conjectured that $\mathbb{E}[\ln \mathcal{N}]$ can be estimated as follows: For a typical elementary landscape we expect that the correlation length ℓ gives a good description of its structure because the landscape does not have any other distinctive features. By construction ℓ determines the size of the mountains and valleys. As there are many directions available at each configuration, we expect that there are only very few metastable states besides the summit of each of these ℓ-sized mountains—almost all of the configurations will be saddle points with at least a few superior neighbors. We measure ℓ along a random walk but the radius $R(\ell)$ of a mountain is more conveniently described in terms of the distance between vertices on Γ. Here $R(\ell)$ is the average distance that is reached by

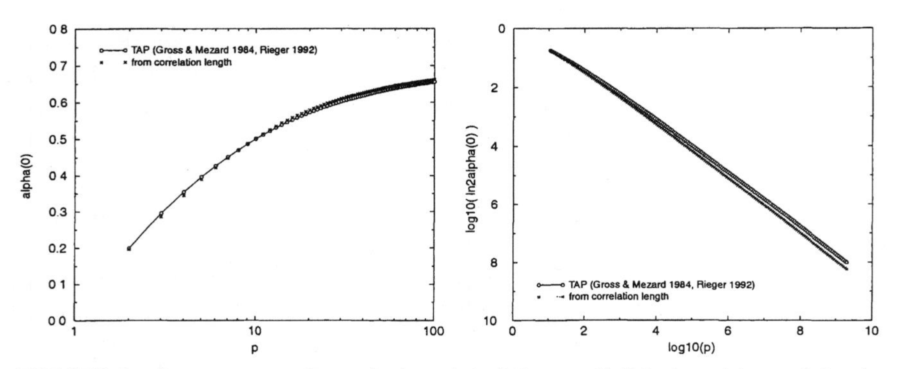

FIGURE 3 Comparison of eqs. (40) and (41) for small (l.h.s) and large (r.h.s.) values of p.

the random walk in ℓ steps. With the notation $B(R)$ for the number of vertices contained in a ball of radius R in Γ, we expect approximately $|V|/B(R(\ell))$ local optima.

As an example we consider a comparison of the correlation length conjecture with an exact computation based on the TAP equations [156] for Derrida's p-spin Hamiltonian [28]. The TAP approach yields [58, 125]:

$$\begin{aligned} \lim_{n\to\infty} \frac{1}{n} \ln \mathbb{E}[\mathcal{N}] = \alpha(0) = \ln 2 - \frac{\delta(p)^2}{2(p-1)} + \ln \Phi(\delta(p)) \qquad \text{with} \\ \Phi(x) = (1 + \operatorname{erf}(\delta(p)/\sqrt{2}))/2\,, \\ \delta(p) = \frac{p-1}{\sqrt{2\pi}} \frac{\exp(-\delta(p)^2/2)}{\Phi(\delta(p))}\,. \end{aligned} \tag{40}$$

The last equation must be solved numerically for $\delta(p)$.

An explicit evaluation of the correlation length conjecture for the p-spin Hamiltonian [149] yields

$$\begin{aligned} \alpha(0) = \ln 2 + (1-\zeta)\ln(1-\zeta) + \zeta \ln \zeta\,, \qquad \text{where} \\ \zeta = \frac{1}{2}\left(1 - \mathrm{e}^{-1/p}\right). \end{aligned} \tag{41}$$

Equation (41) compares very well with data numerical simulations for $p = 2, 3, \ldots, 6$ [149]. Figure 3 shows that equation (41) is in excellent agreement with the TAP result in equation (40). It is interesting to note that even for very large p there is a good qualitative agreement between the value of $\alpha^* = \ln 2 - \alpha(0)$ obtained by the two methods. We find $\alpha^*_{\text{TAP}} \sim p^{-1} \ln p$ and $\alpha^*_{\text{clc}} \sim (1/2p) \ln p$.

On the one hand, the correlation length conjecture works very well for isotropic random landscapes on various other configuration space besides Boolean hypercubes [92, 139]. On the other hand, the correlation length conjecture yields sometimes very poor estimates if the landscapes deviate significantly from isotropy [37, 48].

4.5 NEUTRALITY

We say two configurations $x, y \in V$ are *neutral* if $f(x) = f(y)$. We colloquially refer to a landscape as "neutral" if a substantial fraction of adjacent pairs of configurations are neutral. This should not be confused with the *flat* landscape, in which f is constant. Extensive computer simulations, based on RNA secondary structures [60, 61, 133], have revealed that neutrality plays an important role in understanding the dynamics of RNA evolution [77, 123, 124].

Kimura proposed a theory of biological evolution that focuses exclusively on the aspects of neutrality [85] by assuming a flat fitness landscape. Very recently, landscapes with a large degree of neutrality have also been described in computational models such as cellular automata [73], for the mapping of sequences in combinatorial random structures [121], and in the context of sequential dynamical systems [8].

Combinatorial aspects of neutral landscapes are discussed in more detail in Kopp and Reidys [90]. Here we restrict ourselves to indicating how a rigorous investigation of neutrality can be linked to the techniques described above. In the case of additive random landscapes, a promising starting point is provided by the random variables $X_{\{y,x\}}(f)$, which take the value 1 if $f(x) = f(y)$ and 0 otherwise.

The number of *neutral neighbors* of a configuration $x \in V$ is then

$$\nu_x(f) = \sum_{y \in \mathcal{N}(x)} X_{\{y,x\}}(f) . \tag{42}$$

The following parameters have turned out to be particularly important for understanding the neutrality in an additive random landscape:

$$\begin{aligned} c_x(y) &= \left|\{j \in M \mid \vartheta_j(x) \neq \vartheta_j(y)\}\right| , \\ w_x(y', y'') &= \left|\{j \in M \mid \vartheta_j(x) \neq \vartheta_j(y') \ \wedge \vartheta_j(x) \neq \vartheta_j(y'')\}\right| , \end{aligned} \tag{43}$$

where $x \in V$ is an arbitrary vertex and $y, y', y'' \in \mathcal{N}(x)$. Note that $c_x(y)$ and $w_x(y', y'')$ only depend on the properties of the component landscapes ϑ_k but not on the distribution of the coefficients c_k.

In Reidys [122] explicit expressions for the mean and variance of neutrality are derived for the simplest possible case, namely for a distribution of the coefficients c_j satisfying

$$\mu\{c_j = \xi\} = \begin{cases} \mu_0 > 0 & \text{if } \xi = 0 , \\ 0 & \text{otherwise} . \end{cases} \tag{44}$$

For any additive random landscape with coefficients c_i satisfying equation (44), we obtain

$$\begin{aligned}\mathbb{E}[\nu_x] &= \sum_{y\in\mathcal{N}(x)} \mu_0^{c_x(y)}\,, \\ \mathbb{V}[\nu_x] &= \sum_{y',y''\in\mathcal{N}(x)} \mu_0^{c_x(y')+c_x(y'')}\left[\mu_0^{-w_x(y',y'')}-1\right].\end{aligned} \tag{45}$$

More explicitly, the expected number of neutral neighbors of a p-spin landscape is therefore $\mathbb{E}[\nu] = n\,\mu_0^{\binom{n-1}{p-1}}$. Depending on μ_0, the expected fraction of vanishing interaction coefficients, the fraction of neutral mutations $\mathbb{E}[\nu]/n$ may take any value between 0 and 1. This fact is independent of the order p of the spin glass. Thus, ruggedness (as measured by p) and neutrality (as measured by μ_0) are independent properties of (random) landscapes.

In many spin glass models the spins are arranged on a finite-dimensional lattice. Hence, each spin has only a finite number of other spins to which it couples in such a *short-range spin glass.* All but $\mathcal{O}(n)$ coefficients therefore vanish, and we have $\mu_0 \sim 1 - z/n^{p-1}$, where $z > 0$ is a parameter determined by the connectivity of the lattice. The fraction of neutral spin flips is constant in such systems, $\mathbb{E}[\nu]/n \sim e^{-z}$; see Reidys [122] for more details. We remark further that the fraction of neutral mutations is the crucial input parameter for random graph models of neutral landscapes [49, 120, 123]; see also Kopp and Reidys [90].

The study of neutrality in more general classes of random landscapes requires the determination of the distributions of the random variables

$$\delta(x,x') := \sum_{i:\,\vartheta_i(x)\neq\vartheta_i(x')} c_i\left[\vartheta_i(x)-\vartheta_i(x')\right]. \tag{46}$$

Since the c_i are by definition independent in an additive landscape, we have to compute the convolutions

$$g_{x,x'}(\delta) = \mathop{\star}_{i:\vartheta_i(x)\neq\vartheta_i(x')} \rho_i\big(c_i/[\vartheta_i(x)-\vartheta_i(x')]\big)\,, \tag{47}$$

where $\star_j$ denotes convolution of all functions indexed by j, $g_{x,x'}(\delta)$, which is the density of the values of $\delta(x,x')$, and $\rho_i(.)$ is the density function of c_i, which in this case has to be evaluated with the argument $c_i/[\vartheta_i(x) \neq \vartheta_i(x')]$. Then we have

$$\mathrm{Prob}[X_{x'}(x)=1] = \lim_{\epsilon\to 0}\int_{-\epsilon}^{\epsilon} g_{x,x'}(\delta)d\delta\,. \tag{48}$$

If $\{i|\,\vartheta_i(x)\neq\vartheta_i(x')\} = \varnothing$, x and x' are neutral for any distribution of the c_i. This is observed for instance in the *graph matching problem* [141].

We conclude from equation (48) that a continuous density $g_{x,x'}$, which necessarily arises if the individual densities ρ_i are continuous, does not lead to neutrality. Neutrality hence depends on a "discrete" contribution to the probability densities of the coefficients c_k, $k \in M$. Indeed, only these discrete components influence neutrality. In practice, evaluation of the convolution (47) therefore boils down to a combinatorial exercise, as for instance in the case of the integer-valued NK model proposed by Newman and Engelhardt [105].

Finally, we remark that $g(.)$ would also be an appropriate starting point for a theory of *nearly neutral landscapes* [111], in which the condition such as $|f(x) - f(x')| < \varepsilon$ for some finite $\varepsilon > 0$ could replace the condition $f(x) = f(x')$.

5 DISCUSSION

The exposition above has been focussed almost entirely on the "static" properties of a landscape. A mathematical language has been introduced that allows us to view a cost or fitness function as it is seen by a search operator. This formalism lends a precise meaning to notions such as ruggedness, neutrality, or isotropy. Intuitively, the dynamics of (evolutionary) adaptation and the performance of optimization heuristics should be determined by exactly these properties.

The spectral approach described here has been useful in distinguishing different types of landscapes. RNA folding landscapes, for instance, are very different from spin glasses in both ruggedness and neutrality. A more detailed analysis of the distribution of the Fourier coefficients that belong to a particular mode might help to understand and quantify the structure of anisotropies.

Dynamics on landscapes, unfortunately, is much less understood at present. Apart from a few global results such as the "No Free Lunch Theorem" [169] and detailed studies on simple landscapes, such as those described by van Nimwegen and Crutchfield [106], very few exact results are known. Various dynamical phenomena have been described for special classes of landscapes. There is an error threshold limiting the mutation rate in biological evolution [34], which is well understood at least on landscapes with a few peaks. A tunneling effect was described between two separated peaks [32]. A diffusionlike process has also been observed on landscapes with a high degree of neutrality [77, 124], similar to the situation in a flat landscape [30]. A theory that could treat all these aspects within a common formalism, however, is still missing.

ACKNOWLEDGMENTS

Stimulating discussion with Jim Crutchfield, Jan Cupal, Walter Fontana, Silvio Franz, Ricardo García-Pelayo, Ivo Hofacker, Wim Hordijk, Stuart Kauffman, Richard Palmer, Christian Reidys, Peter Schuster, Gottfried Tinhofer, Henri Waelbroeck, Günter Wagner, and many others made this work possible. Section 4.4 originated during a workshop at the ICTP in Trieste in 1997.

APPENDIX

A. SCHEMATA AND DECEPTIVENESS

A.1 Introduction. Walsh functions and "schemata" have been used extensively in the analysis of GA behavior [39, 55, 56, 57, 94, 112, 158, 159]. A *schema* is simply a hyperplane in sequence space. It is defined by the set H of "fixed" bits and their values h_i, $i \in H$. In symbols

$$\mathcal{H} = H[h] = \{x \in V | \forall i \in H : x_i = h_i\}. \tag{49}$$

For a discussion of the Schema Theorem and the Building Block Hypothesis, we refer to the literature [2, 11, 45, 70, 71, 152]. Instead, we briefly consider a few properties of landscapes that are naturally defined in terms of schemata. For simplicity we restrict ourselves to landscapes on the set of binary strings of length n. Notions such as "local optimum" in the following subsections consequently refer to the graph structure of the Boolean hypergraph. Schemata and Walsh functions are linked by means of

$$f(\mathcal{H}) = \sum_{I \subseteq H} a_I \varepsilon(h) \quad \text{and} \quad \text{var}(\mathcal{H}) = \sum_{I \subseteq H} \left\{ \sum_{K \not\subseteq H} a_K a_{K \triangle I} \right\} \varepsilon_I(h), \tag{50}$$

where $I \triangle J$ denotes the symmetric difference of the sets I and J. Note that these quantities are superpositions of Walsh functions with index set $I \subseteq H$ evaluated at the fixed bits of the schema. It is interesting that $f(\mathcal{H})$ depends only on Fourier coefficients a_I with $I \subseteq H$, while $\text{var}(\mathcal{H})$ depends only on coefficients with $I \not\subseteq H$.

Let Q be any property of a landscape f on V. We say that f is robustly Q, if there is an $\epsilon > 0$ such that any landscape g satisfying $|f(x) - g(x)| < \epsilon$ for all $x \in V$ also has property Q. The condition $|f(x) - g(x)| < \epsilon$ may be replaced by $|a_k - \tilde{a}_k| < \epsilon$ for the Fourier coefficients of f and g, respectively. The results in following sections have not been published before; nevertheless, we omit their (rather simple) proofs in this survey.

A.2 Funnels. A landscape is called a *funnel* if there is a string h^* such that $H \subseteq H'$ implies $f(H[h^*]) \leq f(H'[h^*])$ for all $H \subseteq \{1, \ldots, n\}$. If f is a funnel with peak h^*, then (i) $f(h^*) \geq \bar{f}$ and (ii) h^* is a local maximum on $\mathcal{Q}_2^n$. It is not hard to construct simple examples of funnels with just $n = 3$ bits showing that the peak of a funnel need not be globally optimal and that there may be multiple local optima. With some more work it is also possible to show that a landscape is a robust funnel if and only if $H \subset H'$ implies $f(H[h^*]) < f(H'[h^*])$ for all H. The peak h^* of a robust funnel is unique.

A.3 GA-Easy Functions. Let us call a function *GA-easy* if there is a global optimum x^* such that $f(H[x^*]) \geq f(H[x])$ for all $x \in V$ and all $H \subseteq [n]$. In the GA literature a more common definition of easy is what we call *robustly GA-easy*, namely a function f with a global optimum x^* satisfying $f(H[x^*]) > f(H[x])$ for all $x \in V$ and all $H \subseteq [n]$ for which $H[x^*] \neq H[x]$. By setting $H = \{1, \ldots, n\}$ we see that the global optimum of a robustly GA-easy function is unique. Robustly GA-easy is called "fully easy" in Liepins et al. [93]. We prefer to say that f is *fully GA-easy* if for each global optimum x^*, for each $H \subseteq [n]$, and for each $x \in V$ holds $f(H[x^*]) \geq f(H[x])$. Naturally, a fully GA-easy function is GA-easy, but the converse is not true. Note also that "robustly fully GA-easy" is the same as robustly GA-easy. A short computation shows that a (robustly) GA-easy function with global optimum x^* is a (robust) funnel with peak x^*.

Linear functions f are, of course, GA-easy. However, linear functions do not form a generic class of landscapes in the sense that linearity is not a robust property. Wilson [168] introduced a slightly larger class of landscapes: Given a string $x \in V$ let $\hat{x}$ be a string satisfying $\hat{x}_k = x_k$ if $f(x) > f(x^{(k)})$ and $\hat{x}_k = \bar{x}_k$ if $f(x) < f(x^{(k)})$. That is, $\hat{x}$ is obtained from x by keeping the best bit among all one-error mutants in each position. A function f on V is *bit setting optimizable* (b.s.o.) if $\hat{x}$ is a global optimum for each $x \in V$. Of course, linear functions are b.s.o., and the set of robustly b.s.o. functions is nonempty: all sufficiently small perturbations of linear functions are b.s.o. It can be shown that a (robustly) b.s.o. function is (robustly) GA-easy. It is shown in Wilson [168] that the converse is not true for all $n \geq 3$.

A.4 Deceptive Functions. The literature on deceptive functions uses a variety of slightly different notions of deceptiveness (and sometimes does not even precisely define the notion at all). In a deceptive landscape, an optimal schema of some size is "contradicted" by one of its sub-schemata. Intuitively, this is just the converse of GA-easy. Following Whitley [165], we use the following formal definition: A landscape f is *deceptive* if there are vertices $x, y \in V$ and index sets $H \subset K \subset \{1, \ldots, n\}$ with the following properties:

i. $K[x] \neq K[y]$,
ii. $f(H[x]) > f(H[z])$ for all $z \in V$ with $H[x] \neq H[z]$, and
iii. $f(K[y]) > f(K[z])$ for all $z \in V$ with $K[y] \neq K[z]$.

As expected, it can be shown that a GA-easy function f is not deceptive. However, the converse is not true since a "symmetric function," i.e., a function fulfilling $f(x) = f(\bar{x})$ where $\bar{x}$, is the complement of x, is never deceptive according to Whitley's definition. An example of a symmetric function that is not GA-easy can be constructed, for example, on Q_2^4. We will say that f is *GA-hard* if it is not GA-easy.

We say that f is *weakly deceptive* if there are vertices $x, y \in V$ and index sets $H \subset K$ such that (i) $f(H[x]) \geq f(H[z])$ for all $z \in V$ and $f(H[x]) > f(H[y])$, and (ii) $f(K[y]) \geq f(K[z])$ for all $z \in V$ and $f(K[y]) > f(K[x])$. A deceptive landscape is of course weakly deceptive. Since symmetric functions can be weakly deceptive, the converse is not true in general.

Let Ω denote the set of global optima. For each $x^* \in \Omega$ and each index set H, we define the set of vertices that belong to an H-schema that is superior to $H[x^*]$:

$$\Psi(x^*, H) = \{y | f(H[y]) > f(H[x^*]) \text{ and} \forall z : f(H[y]) \geq f(H[z])\} . \tag{51}$$

Clearly $x^* \notin \Psi(x^*, H)$. The following propositions are easily verified:

i. f is GA-hard if and only if for each $x^* \in \Omega$ there is an index set H such that $\Psi(x^*, H)$ is nonempty.
ii. f is not fully GA-easy if and only if there is a $x^* \in \Omega$ and an index set H such that $\Psi(x^*, H)$ is nonempty.
iii. f is weakly deceptive if and only if there is a $x^* \in \Omega$ and an index set H such that $\Psi(x^*, H) \setminus \Omega$ is nonempty.

Both weak deceptiveness and GA-hardness imply that f is not fully GA-easy. If f has a unique global optimum, however, then "weakly deceptive," "GA-hard," and not "fully GA-easy" are equivalent properties.

B. MAXIMUM ENTROPY CONDITIONS

It is well known that the Gaussian distributions maximize entropy. The proof for the one-dimensional case can be found, for example, in Hida [64, prop. 1.15]. For the convenience of the reader, a short proof of the general case is included here as it is not readily accessible in the literature.

The starting point is the following inequality that holds for arbitrary probability spaces:

$$\int p(x) \ln p(x) dx - \int p(x) \ln q(x) dx \geq 0 . \tag{52}$$

Equality holds in equation (52) if and only if $p = q$ almost everywhere. Let $\mathbf{C}$ be the covariance matrix of p. We assume that $\mathbf{C}$ is invertible. Without loosing generality we furthermore assume $\mathbb{E}[x] = 0$. Substituting the Gaussian distribution

$$q(x) = \frac{1}{(2\pi)^{|V|/2}\sqrt{\det \mathbf{C}}} \exp\left(-\frac{1}{2} x \mathbf{C}^{-1} x\right) , \tag{53}$$

equation (52) translates into a general inequality for the the entropy of p:

$$S \leq \frac{|V|}{2} \ln(2\pi) + \frac{1}{2} \ln \det \mathbf{C} + \frac{1}{2} \int_{\mathbb{R}^n} (x \mathbf{C}^{-1} x) p(x) dx .$$

The integral is a simple constant independent of p as the following computation shows

$$\int_{\mathbb{R}^n} \sum_{k,l} (\mathbf{C}^{-1})_{kl} x_k x_l p(x) dx = \sum_{k,l} (\mathbf{C}^{-1})_{kl} \int_{\mathbb{R}^n} x_k x_l p(x) dx =$$

$$\sum_{k,l} (\mathbf{C}^{-1})_{kl} \mathbf{C}_{kl} = \sum_{k} (\mathbf{C}^{-1}\mathbf{C})_{kk} = \mathrm{Tr}\mathbf{I} = |V|.$$

Let $\{\Lambda_k,\ k = 1, \ldots, |V|\}$ be the eigenvalues of $\mathbf{C}$. Since $\mathbf{C}$ is invertible by assumption, we have $\Lambda_k > 0$ for all k. Using $\sigma^2 = \mathrm{Tr}\mathbf{C} = \sum_k \Lambda_k$ we obtain

$$S \leq S_{\mathbf{C}} = \frac{1}{2}|V| \ln \frac{2\pi e}{|V|} + \frac{1}{2} \sum_k \ln \frac{\Lambda_k |V|}{\sigma^2}. \tag{54}$$

It is easy to verify that $S_{\mathbf{C}}$ is indeed the entropy of a Gaussian distribution with covariance matrix $\mathbf{C}$.

The two terms in equation (54) allow for a direct interpretation. The Gaussian entropy $S_{\mathbf{C}}$ attains its maximum subject to a given variance σ^2 if and only if $\Lambda_k = \sigma^2/|V|$, in which case the second term vanishes. We may then split the entropy of a random landscape into three contributions

$$S = S_{\sigma^2} + \Delta S_{\mathbf{C}} + \Delta S_{ng}, \tag{55}$$

where $\Delta S_{ng} = S - S_{\mathbf{C}}$ is the entropy loss due to deviations from a Gaussian distribution, S_{σ^2} is the maximal entropy with given variance σ^2s, and $\Delta S_{\mathbf{C}}$, the second term in equation (54), measures the entropy loss due to variations in the spectrum of $\mathbf{C}$. In particular, whenever there are correlations between different vertices, then $\mathbf{C}$ is nondiagonal and, hence, $\Delta S_{\mathbf{C}} < 0$. More precisely, $\Delta S_{\mathbf{C}} = 0$ if and only if the corresponding Gaussian random landscape is i.i.d.

REFERENCES

[1] Alspach, Brian. "Isomorphism and Cayley Graphs on Abelian Groups." In *Graph Symmetry: Algebraic Methods and Applications*, edited by G. Hahn and G. Sabidussi, 1–22. NATO ASI Series C, vol. 497. Dordrecht: Kluwer, 1997.

[2] Altenberg, Lee. "The Schema Theorem and the Price's Theorem." In *Foundations of Genetic Algorithms 3*, edited by L. D. Whitley and M. D. Vose, 23–49. San Francisco, CA: Morgan Kauffman, 1995.

[3] Altenberg, Lee, and Marc W. Feldman. "Selection, Generalized Transmission, and the Evolution of Modifier Genes. I. The Reduction Principle." *Genetics* **117** (1987): 559–572.

[4] Angel, E., and V. Zissimopoulos. "Autocorrelation Coefficient for the Graph Bipartitioning Problem." *Theor. Comp. Sci.* **191** (1998): 229–243.

[5] Angel, E., and V. Zissimopoulos. "On the Quality of Local Search for the Quadratic Assignment Problem." *Discr. Appl. Math.* **82** (1998): 15–25.

[6] Babel, L., S. Baumann, M. Lüdecke, and G. Tinhofer. "`STABCOL`: Graph Isomorphism Testing Based on the Weisfeiler-Leman Algorithm." TU München, Garching, Germany, 1997.

[7] Babel, L., I. V. Chuvaeva, M. Klin, and D. V Pasechnik. "Algebraic Combinatorics in Mathematical Chemistry. Methods and Algorithms. II. Program Implementation of the Weisfeiler-Leman Algorithm." TU München, Garching, Germany, 1995.

[8] Barrett, C. L., H. Mortveit, and C. M. Reidys. "Elements of a Theory of Simulation II: Sequential Dynamical Systems." *Appl. Math. & Comp.* **107** (2000): 121–136.

[9] Beezer, R. A. "Trivalent Orbit-Polynomial Graphs." *Lin. Alg. Appl.* **73** (1986): 133–146.

[10] Bennett, J. H. "On the Theory of Random Mating." *Ann. Eugen.* **18** (1954): 311–317.

[11] Bethke, A. D. *Genetic Algorithms and Function Optimizers.* Ann Arbor: University of Michigan, 1991.

[12] Bhattacharya, R. N., and E. C. Waymire. *Stochastic Processes with Applications.* New York: Wiley, 1990.

[13] Biggs, Norman. *Algebraic Graph Theory,* 2d ed. Cambridge UK: Cambridge University Press. 1994.

[14] Bollobás, Bela. *Graph Theory—An Introductory Course.* New York: Springer-Verlag, 1979.

[15] Bose, R. C., and D. M. Mesner. "On Linear Associative Algebras Corresponding to Association Schemes of Partially Balanced Designs." *Ann. Math. Statist.* **30** (1959): 21–38.

[16] Bray, A. J., and M. A. Moore. "Metastable States in Spin Glasses with Short-Ranged Interactions." *J. Phys. C* **14** (1981): 1313–1327.

[17] Brouwer, A. E., A. M. Cohen, and A. Neumaier. *Distance-Regular Graphs.* Berlin, New York: Springer Verlag, 1989.

[18] Chung, Fan R. K. "Spectral Graph Theory." *Am. Math. Soc.* **92** (1997).

[19] Codenotti, Bruno, and Luciano Margara. "Local Properties of Some NP-Complete Problems." Technical Report 92–021, International Computer Science Institute, Berkeley, CA, 1992.

[20] Culberson, Joseph C. "Mutation-Crossover Isomorphism and the Construction of Discriminating Functions." *Evol. Comp.* **2** (1995): 279–311.

[21] Cupal, Jan, and Peter F. Stadler. "RNA Shape Space Topology." *Alife* **6** (2000): 3–23.

[22] Cvetković, D. M., Michael Doob, Ivan Gutman, and Aleksandar Torgašev. "Recent Results in the Theory of Graph Spectra." *Annals of Discrete Mathematics*, vol. 36. Amsterdam, New York, Oxford, Tokyo: North Holland, 1988.

[23] Cvetković, D. M., Michael Doob, and H. Sachs. *Spectra of Graphs—Theory and Applications.* New York: Academic Press, 1980.
[24] Cvetković, D. M., Peter Rowlinson, and Slobodan Simić. *Eigenspaces of Graphs.* Cambridge, UK: Cambrigdge University Press, 1997.
[25] Davies, E. B., G. M. L. Gladwell, J. Leydold, and P. F. Stadler. "Discrete Nodal Domain Theorems." *Lin. Alg. Appl.* (2000): to appear.
[26] de Verdière, Yves Colin. "Multiplicités des Valeurs Propres Laplaciens Discrete at Laplaciens Continus." *Rendiconti di Matematica* **13** (1993): 433–460.
[27] Delsarte, P. "An Algebraic Approach to Association Schemes of Coding Theory." Phillips Research Reports Supplements, vol. 10. Phillips, 1973.
[28] Derrida, B. "Random Energy Model: Limit of a Family of Disordered Models." *Phys. Rev. Lett.* **45** (1980): 79–82.
[29] Derrida, B., and E. Gardner. "Metastable States of a Spin Glass Chain at Zero Temperature." *J. Physique* **47** (1986): 959–965.
[30] Derrida, B., and L. Peliti. "Evolution in a Flat Fitness Landscape." *Bull. Math. Biol.* **3** (1991): 355–382.
[31] Diaconis, Persi. *Group Representations in Probability and Statistics* Hayward, CA: Institute of Mathematical Statistics, 1989.
[32] Ebeling, W., A. Engel, B. Esser, and R. Feistel. "Diffusion and Reaction in Random Media and Models of Evolution Processes." *J. Stat. Phys.* **37** (1984): 369–384.
[33] Eigen, Manfred, and Peter Schuster. *The Hypercycle.* New York, Berlin: Springer-Verlag, 1979.
[34] Eigen, M., J. McCaskill, and P. Schuster. "The Molecular Quasispecies." *Adv. Chem. Phys.* **75** (1989): 149–263.
[35] Faradžev, I. A., M. H. Klin, and M. E. Muzychuk. "Cellular Rings and Groups of Automorphisms of Graphs." In *Investigations in Algebraic Theory of Combinatorial Objects*, edited by I. A. Faradžev, A. A. Ivanov, M. H. Klin, and A. J. Woldar. Mathematics and Its Applications, Soviet Series, vol. 84. Dordrecht: Kluwer, 1994.
[36] Felsenstein, J. "Evolutionary Trees from DNA Sequences: A Maximum Likelihood Approach." *J. Mol. Evol.* **17** (1981): 368–376.
[37] Ferreira, F. F., J. F. Fontanari, and P. F. Stadler. "Landscape Statistics of the Low Autocorrelated inary String Problem." *J. Phys. A: Math. Gen.* **33** (2000): 8635–8647.
[38] Fiedler, Miroslav. "A Property of Eigenvectors of Nonnegative Symmetric Matrices and Its Application to Graph Theory." *Czechoslovak Math. J.* **25** (1975): 619–633.
[39] Field, Paul. "Non-binary Transformations of Genetic Algorithms." *Complex Systems* **9** (1995): 11–28.
[40] Fogel, D. B. *Evolutionary Computation.* New York: IEEE Press, 1995.
[41] Fontana, Walter, and P. Schuster. "Continuity in Evolution: On the Nature of Transitions." *Science* **280** (1998): 1451–1455.

[42] Fontana, Walter, and P. Schuster. "Shaping Space: The Possible and the Attainable in RNA Genotype-Phenotype Mapping." *J. Theor. Biol.* **194** (1988): 491–515.

[43] Fontana, Walter, D. A. M. Konings, P. F. Stadler, and P. Schuster. "Statistics of RNA Secondary Structures." *Biopolymers* **33** (1993): 1389–1404.

[44] Fontana, Walter, Peter F. Stadler, Erich G. Bornberg-Bauer, Thomas Griesmacher, Ivo L. Hofacker, Manfred Tacker, Pedro Tarazona, Edward D. Weinberger, and Peter Schuster. "RNA Folding and Combinatory Landscapes." *Phys. Rev. E* **47** (1993): 2083–2099.

[45] Forrest, Stephanie, and Melanie Mitchell. "Relative Building Block Fitness and the Building Block Hypothesis." In *Foundations of Genetic Algorithms 2*, edited by L. Darrell Whitley, 109–126. San Mateo, CA: Morgan Kaufmann, 1993.

[46] Frauenfelder, H., A. R. Bishop, A. Garcia, A. Perelson, P. Schuster, D. Sherrington, and P. J. Swart, eds. *Landscape Paradigms in Physics and Biology: Concepts, Structures, and Dynamics.* Amsterdam: Elsevier, 1997. Special Issue of *Phyica D* **107(2–4)**.

[47] Friedman, Joel. "Some Geometric Aspects of Graphs and Their Eigenfunctions." *Duke Math. J.* **69** (1993): 487–525.

[48] García-Pelayo, Ricardo, and Peter F. Stadler. "Correlation Length, Isotropy, and Meta-stable States." *Physica D* **107** (1997): 240–254.

[49] Gavrilets, S., and J. Gravner. "Percolation on the Fitness Hypercube and the Evolution of Reproductive Isolation." *J. Theor. Biol.* **184** (1997): 51–64.

[50] Gillespie, J. H. *The Causes of Molecular Evolution.* New York, Oxford: Oxford University Press, 1991.

[51] Gitchoff, Paul, and Günter P. Wagner. "Recombination Induced Hypergraphs: A New Approach to Mutation-Recombination Isomorphism." *Complexity* **2** (1996): 47–43.

[52] Godsil, Chris D. *Algebraic Combinatorics.* New York: Chapman & Hall, 1993.

[53] Godsil, Chris D. "Equitable Partitions." In *Combinatorics, Paul Erdős in Eighty*, edited by D. Miklós, V. T. Sós, and T. Szőnyi, 173–192. Budapest: János Bolyai Mathematical Society, 1993.

[54] Godsil, Chris D. "Tools from Linear Algebra." In *Handbook of Combinatorics*, edited by R. Graham, M. Grötschel, and L. Lovász, 1105–1748. Amsterdam: North-Holland, 1995.

[55] Goldberg, David E. "Genetic Algorithms and Walsh Functions. Part I: A Gentle Introduction." *Complex Systems* **3** (1989): 129–152.

[56] Goldberg, David E. "Genetic Algorithms and Walsh Functions. Part II: Deceptiveness and Its Analysis." *Complex Systems* **3** (1989): 153–176.

[57] Goldberg, David E., and Mike Rudnik. "Genetic Algorithms and the Variance of Fitness." *Complex Systems* **5** (1991): 265–278.

[58] Gross, D. J., and M. Mézard. "The Simplest Spin Glass." *Nucl. Phys. B* **240** (1984): 431–452.

[59] Grover, L. K. "Local Search and the Local Structure of NP-Complete Problems." *Oper. Res. Lett.* **12** (1992): 235–243.

[60] Grüner, Walter, Robert Giegerich, Dirk Strothmann, Christian M. Reidys, Jacqueline Weber, Ivo L. Hofacker, Peter F. Stadler, and Peter Schuster. "Analysis of RNA Sequence Structure Maps by Exhaustive Enumeration. I. Neutral Networks." *Monath. Chem.* **127** (1996): 355–374.

[61] Grüner, Walter, Robert Giegerich, Dirk Strothmann, Christian M. Reidys, Jacqueline Weber, Ivo L. Hofacker, Peter F. Stadler, and Peter Schuster. "Analysis of RNA Sequence Structure Maps by Exhaustive Enumeration. II. Structures of Neutral Networks and Shape Space Covering." *Monath. Chem.* **127** (1996): 375–389.

[62] Happel, Robert, and Peter F. Stadler. "Canonical Approximation of Fitness Landscapes." *Complexity* **2** (1996): 53–58.

[63] Heckendorn, R. B., and D. Whitley. "A Walsh Analysis of NK-Landscapes." In *International Conference on Genetic Algorithms*, edited by T. Baeck. San Francisco, CA: Morgan Kaufmann, 1997.

[64] Hida, T. *Brownian Motion.* New York: Springer-Verlag, 1980.

[65] Higman, D. G. "Intersection Matrices for Finite Permuation Groups." *J. Algebra* **6** (1967): 22–42.

[66] Higman, D. G. "Coherent Configurations. Part I: Ordinary Representation Theory." *Geometriae Dedicata* **4** (1975): 1–32.

[67] Higman, D. G. "Coherent Configurations. Part II: Weights." *Geometriae Dedicata* **5** (1976): 413–424.

[68] Hofacker, Ivo L., Walter Fontana, Peter F. Stadler, Sebastian Bonhoeffer, Manfred Tacker, and Peter Schuster. "Fast Folding and Comparison of RNA Secondary Structures." *Monatsh. Chemie* **125(2)** (1994): 167–188.

[69] Hoffman, A. J. "On the Polynomial of a Graph." *Am. Math. Monthly* **70** (1963): 30–36.

[70] Holland, J. H. "Genetic Algorithms and Classifier Systems: Foundations and Future Directions." In *Proceedings of the 2nd International Conference on Genetic Algorithms*, edited by J. J. Grefenstette, 82–89. Hilldale, NJ: Lawrence Erlbaum Assoc., 1987.

[71] Holland, J. H. *Adaptation in Natural and Artificial Systems.* Cambridge, MA: MIT Press, 1993.

[72] Hordijk, Wim. "A Measure of Landscapes." *Evol. Comp.* **4(4)** (1996): 335–360.

[73] Hordijk, Wim. "Correlation Analysis of the Synchronizing-CA Landscape." *Physica D* **107** (1997): 255–264.

[74] Hordijk, Wim, and Peter F. Stadler. "Amplitude Spectra of Fitness Landscapes." *Adv. Complex Syst.* **1** (1998): 39–66.

[75] Hotelling, H. "Analysis of a Complex of Statistical Variables into Principal Components." *J. Educ. Psych.* **24** (1933): 417–441 and 498-520.

[76] Huynen, Martin A., and P. Hogeweg. "Pattern Generation in Molecular Evolution. Exploitation of the Variation in RNA Landscapes." *J. Mol. Evol.* **39** (1994): 71–79.

[77] Huynen, Martijn A., Peter F. Stadler, and Walter Fontana. "Smoothness within Ruggedness: The Role of Neutrality in Adaptation." *Proc. Natl. Acad. Sci. USA* **93** (1996): 397–401.

[78] Jones, Terry. "Evolutionary Algorithms, Fitness Landscapes, and Search." Ph.D. Thesis, University of New Mexico, Albuquerque, NM, 1995.

[79] Jones, Terry. "One Operator, One Landscape." Working paper 95-02-025, Santa Fe Institute, Santa Fe, New Mexico, 1995.

[80] Karlin, S., and H. M. Taylor. *A First Course in Stochastic Processes.* New York: Academic Press, 1975.

[81] Kauffman, Stuart A. *The Origin of Order.* New York, Oxford: Oxford University Press, 1993.

[82] Kauffman, Stuart A., and S. Levin. "Towards a General Theory of Adaptive Walks on Rugged Landscapes." *J. Theor. Biol.* **128** (1987): 11–45.

[83] Kececioglu, J., and D. Sankoff. "Exact and Approximate Algorithms for Sorting by Reversals, with Applications to Genome Rearrangement." *Algorithmica* **13** (1995): 180–210.

[84] Kern, W. "On the Depth of Combinatorial Optimization Problems." *Discrete Appl. Math.* **43** (1993): 115–129.

[85] Kimura, Motoo. *The Neutral Theory of Molecular Evolution.* Cambridge, UK: Cambridge University Press, 1983.

[86] Kirchhoff, G. "Über die Auflösung der Gleichungen, auf welche man bei der Untersuchung der lineare Verteilung galvanischer Ströme geführt wird." *Ann. Phys. Chem.* **72** (1847): 487–508.

[87] Klin, Mikhail H., R. Pöschel, and K. Rosenbaum. *Angewandte Algebra.* Braunschweig: Vieweg, 1988.

[88] Klin, Mikhail H., Akihiro Munemasa, Mikhail E. Muzychuk, and Paul-Hermann Zieschang. "Directed Strongly Regular Graphs via Coherent (Cellular) Algebras." Unpublished manuscript, Ben Gurion University, 1997.

[89] Klin, Mikhail H., C. Rücker, G. Rücker, and G. Tinhofer. "Algebraic Combinatorics in Mathematical Chemistry. Methods and Algorithms. I. Permutation Groups and Coherent (Cellular) Algrebras." TU München, Garching, Germany, 1997.

[90] Kopp, S., C. M. Reidys. "Neutral Networks: A Combinatorial Perspective." *Adv. Complex Syst.* **2** (1999): 283–301.

[91] Koza, John R. *Genetic Programming: On the Programming of Computers by Means of Natural Selection.* Cambridge, MA: MIT Press, 1992.

[92] Krakhofer, Bärbel, and Peter F. Stadler. "Local Minima in the Graph Bipartitioning Problem." *Europhys. Lett.* **34** (1996): 85–90.

[93] Liepins, Gunar E., and Michael D. Vose. "Deceptiveness and Genetic Algorithm Dynamics." In *Foundations of Genetic Algorithms*, edited by Gregory J. E. Rawlins, 36–50. San Mateo, CA: Morgan Kaufmann, 1991.

[94] Liepins, Gunar E., and Michael D. Vose. "Polynomials, Basis Sets, and Deceptiveness in Genetic Algorithms." *Complex Systems* **5** (1991): 45–61.

[95] Liepins, Gunar E., and Michael D. Vose. "Characterizing Crossover in Genetic Algorithms." *Ann. Math. Art. Intel.* **5** (1992): 27–34.

[96] Lin, S., and B. W. Kernighan. "An Effective Heuristic Algorithm for the Traveling Salesman Problem." *Oper. Res.* **21** (1965): 498–516.

[97] Lovász, L. "Spectra of Graphs with Transitive Groups." *Periodica Math. Hung.* **6** (1975): 191–195.

[98] Lovász, L. "Random Walks on Graphs: A Survey." In *Combinatorics, Paul Erdős in Eighty*, edited by D. Miklós, V. T. Sós, and T. Szőnyi, vol. 2, 1–46. Budapest: János Bolyai Mathematical Society, 1993.

[99] Lyubich, Yuri I. *Mathematical Structures in Population Genetics.* Berlin: Springer-Verlag, 1992.

[100] Maslen, D., and D. Rockmore. "Generalized FFTs—A Survey of Some Recent Results." In *Groups and Computation II*, edited by L. Finkelstein and W. Kantor, vol. 28, 183–238. Providence, RI: American Mathmatical Society, 1996.

[101] Mézard, M., G. Parisi, and M. A. Virasoro. *Spin Glass Theory and Beyond.* Singapore: World Scientific, 1987.

[102] Mohar, Bojan. "The Laplacian Spectrum of Graphs." In *Graph Theory, Combinatorics, and Applications*, edited by Y. Alavi, G. Chartrand, O. R. Ollermann, and A. J. Schwenk, 871–898. New York: John Wiley & Sons, 1991.

[103] Mohar, Bojan. "Some Applications of Laplace Eigenvalues of Graphs." In *Graph Symmetry: Algebraic Methods and Applications*, edited by G. Hahn and G. Sabidussi, vol. 497, 227–275. NATO ASI Series C. Dordrecht: Kluwer, 1997.

[104] Nagylaki, T. *Introduction to Theoretical Population Genetics.* Berlin: Springer-Verlag, 1992.

[105] Newman, Mark E. J., and R. Engelhardt. "Effects of Selective Neutrality on the Evolution of Molecular Species." *Proc. Roy. Soc. Lond. B* **265** (1998): 1333–1338.

[106] van Nimwegen, Erik, and James P. Crutchfield. "Optimizing Epochal Evolutionary Search: Population-Size Independent Theory." *Machine Learning* **45** (2001): 77–114.

[107] de Oliveira, V. M., J. F. Fontanari, and P. F. Stadler. "Metastable States in High Order Short-Range Spin Glasses." *J. Phys. A: Math. Gen.* **32** (1999): 8793–8802.

[108] O'Reilly, Una-May. "Using a Distance Metric on Genetic Programs to Understand Genetic Operators." Genetic Programming 1997, I.E.E.E. Systems, Man and Cybernetics, Orlando FL.

[109] O'Reilly, Una-May, and Franz Oppacher. "Program Search with a Hierarchical Variable Length Representation: Genetic Programming, Simulated Annealing, and Hill Climbing." In *Parallel Problem Solving from Nature III*, edited by Y. Davidor, H. P. Schwefel, and R. Manner, vol. 866, 39–46. Lecture Notes in Computer Science. Berlin: Springer, 1994.

[110] Ohmori, K., and E. Tanaka. "A Unified View on Tree Metrics." In *Syntactic and Structural Pattern Recognition*, edited by G. Ferrate, 85–100. Berlin, Heidelberg: Springer-Verlag, 1988.

[111] Ohta, T. "The Current Significance of Neutral and Near Neutral Theories." *BioEssays* **18** (1996): 673–677.

[112] Page, Scott E., and David E. Richardson. "Walsh Functions, Scheme Variance, and Deception." *Complex Systems* **6** (1992): 125–136.

[113] Palmer, R. "Optimization on Rugged Landscapes." In *Molecular Evolution on Rugged Landscapes: Proteins, RNA, and the Immune System*, edited by A. S. Perelson and S. A. Kauffman, 3–25. Santa Fe Institute Studies in the Sciences of Complexity, Proc. Vol. IX. Redwood City, CA: Addison Wesley, 1991.

[114] Pearl, J. *Heuristics: Intelligent Search Strategies for Computer Problem Solving*. Reading MA: Addison-Wesley, 1984.

[115] Perelson, Alan S., and Stuart A. Kauffman, eds. *Molecular Evolution on Rugged Landscapes: Proteins, RNA, and the Immune System*. Santa Fe Institute Studies in the Sciences of Complexity, Proc. Vol. IX. Reading, MA: Addison-Wesley, 1991.

[116] Powers, D. L. "Eigenvectors of Distance-Regular Graphs." *SIAM J. Matrix Anal. Appl.* **9** (1988): 399–407.

[117] Powers, D. L. "Graph Partitioning by Eigenvectors." *Lin. Algebra Appl.* **101** (1988): 121–133.

[118] Powers, D. L., and M. M. Sulaiman. "The Walk Partition and Colorations of a Graph." *Lin. Algebra Appl.* **48** (1982): 145–159.

[119] Provine, W. B. *Sewall Wright and Evolutionary Biology*. Chicago, London: University of Chicago Press, 1986.

[120] Reidys, Christian M. "Random Induced Subgraphs of Generalized n-Cubes." *Adv. Appl. Math.* **19** (1997): 360–377.

[121] Reidys, Christian M. "Random-Structures." *Ann. Comb.* **4** (2000): 375–382.

[122] Reidys, Christian M., and Peter F. Stadler. "Neutrality in Fitness Landscapes." *Appl. Math. & Comput.* **117** (2001): 321–350.

[123] Reidys, Christian M., Peter F. Stadler, and Peter Schuster. "Generic Properties of Combinatory Maps: Neural Networks of RNA Secondary Structures." *Bull. Math. Biol.* **59** (1997): 339–397.

[124] Reidys, Christian M., C. V. Forst, and P. Schuster. "Replication and Mutation on Neutral Networks of RNA Secondary Structures." *Bull. Math. Biol.* **63** (2001): 57–94.

[125] Rieger, H. "The Number of Solutions of the Thouless-Anderson-Palmer Equations for p-Spin Interaction Spin Glasses." *Phys. Rev. B* **46** (1992): 14655–14661.

[126] Robbins, R. B. "Some Applications of Mathematics to Breeding Problems. III." *Genetics* **3** (1918): 375–389.

[127] Rockmore, D. "Some Applications of Generalized FFTs." In *Groups and Computation II*, edited by L. Finkelstein and W. Kantor, vol. 28, 329–370. Providence, RI: American Mathmatical Society, 1995.

[128] Rockmore, D., P. Kostelec, W. Hordijk, and P. F. Stadler. "Fast Fourier Transform for Fitness Landscapes." *Appl. Comput. Harmonic Anal.* (2001): to appear.

[129] Runge, Fritz, and Horst Sachs. "Berechnung der Anzahl der Gerüste von Graphen und Hypergraphen mittels deren Spektren." *Math. Balkanica (Belgrade)* **4** (1974): 529–536.

[130] Ryan, Jennifer. "The Depth and Width of Local Minima in Discrete Solution Spaces." *Discrete Appl. Math.* **56** (1995): 75–82.

[131] Schuster, Peter. "Molecular Insights Into Evolution of Phenotypes." This book.

[132] Schuster, Peter, and Peter F. Stadler. "Landscapes: Complex Optimization Problems and Biopolymer Structures." *Comp. & Chem.* **18** (1994): 295–314.

[133] Schuster, Peter, Walter Fontana, Peter F. Stadler, and Ivo L. Hofacker. "From Sequences to Shapes and Back: A Case Study in RNA Secondary Structures." *Proc. Roy. Soc. Lond. B* **255** (1994): 279–284.

[134] Schuster, Peter, Peter F. Stadler, and Alexander Renner. "RNA Structures and Folding: From Conventional to New Issues in Structure Predictions." *Curr. Opin. Struc. Biol.* **7** (1997): 229–235.

[135] Schwenk, A. J. "Computing the Characteristic Polynomial of a Graph." In *Graphs and Combinatorics*, edited by R. A. Bari and F. Harary, vol. 406, 153–162. Lecture Notes in Mathematics. Berlin: Springer-Verlag, 1974.

[136] Sherrington, David, and Scott Kirkpatrick. "Solvable Model of a Spin-Glass." *Phys. Rev. Lett.* **35(26)** (1975): 1792–1795.

[137] Shpak, Max, and Günter P. Wagner. "Asymmetry of Configuration Spaces Induced by Unequal Crossover: Implications for the Mathematical Theory of Evolutionary Innovation." Preprint, Department of Biology, Yale University, New Haven CT, 1999.

[138] Sorkin, G. B. "Combinatorial Optimization, Simulated Annealing, and Fractals." IBM Research Report RC13674 (No.61253).

[139] Stadler, Peter F., and Wolfgang Schnabl. "The Landscape of the Traveling Salesman Problem." *Phys. Lett. A* **161** (1992): 337–344.

[140] Stadler, Peter F., and Robert Happel. "Correlation Structure of the Landscape of the Graph-Bipartitioning-Problem." *J. Phys. A: Math. Gen.* **25** (1992): 3103–3110.

[141] Stadler, Peter F. "Correlation in Landscapes of Combinatorial Optimization Problems." *Europhys. Lett.* **20** (1992): 479–482.

[142] Stadler, Peter F. "Linear Operators on Correlated Landscapes." *J. Physique I (France)* **4** (1994): 681–696.

[143] Stadler, Peter F. "Random Walks and Orthogonal Functions Associated with Highly Symmetric Graphs." *Discrete Math.* **145** (1995): 229–238.

[144] Stadler, Peter F. "Towards a Theory of Landscapes." In *Complex Systems and Binary Networks*, edited by R. Lopéz-Peña, R. Capovilla, R. García-Pelayo, H. Waelbroeck, and F. Zertuche, 77–163. Berlin, New York: Springer-Verlag, 1995.

[145] Stadler, Peter F. "Landscapes and Their Correlation Functions." *J. Math. Chem.* **20** (1996): 1–45.

[146] Stadler, Peter F. "Fitness Landscapes Arising from the Sequence-Structure Maps of Biopolymers." *J. Mol. Struc. (THEOCHEM)* **463** (1999): 7–19.

[147] Stadler, Peter F., and Gottfried Tinhofer. "Equitable Partitions, Coherent Algebras and Random Walks: Applications to the Correlation Structure of Landscapes." *MATCH* **40** (1999): 215–261.

[148] Stadler, Peter F., and Robert Happel. "Random Field Models for Fitness Landscapes." *J. Math. Biol.* **38** (1999): 435–478.

[149] Stadler, Peter F., and Barbel Krakhofer. "Local Minima of p-Spin Models." *Rev. Mex. Fis.* **42** (1996): 355–363.

[150] Stadler, Peter F., and Günter P. Wagner. "The Algebraic Theory of Recombination Spaces." *Evol. Comp.* **5** (1998): 241–275.

[151] Stadler, Peter F., Rudi Seitz, and Günter P. Wagner. "Evolvability of Complex Characters: Population Dependent Fourier Decomposition of Fitness Landscapes over Recombination Spaces." *Bull. Math. Biol.* **62** (2000): 399–428.

[152] Stephens, C. R., and H. Waelbroeck. "Effective Degrees of Freedom in Genetic Algorithms." *Phys. Rev. E* **57** (1998): 3251–3264.

[153] Tacker, Manfred, Peter F. Stadler, Erich G. Bornberg-Bauer, Ivo L. Hofacker, and Peter Schuster. "Algorithm Independent Properties of RNA Structure Prediction." *Eur. Biophy. J.* **25** (1996): 115–130.

[154] Tai, K. "The Tree-to-Tree Correction Problem." *J. ACM* **26** (1979): 422–433.

[155] Tanaka, F., and S. F. Edwards. "Analytic Theory of Ground State Properties of a Spin Glass: I. Ising Spin Glass." *J. Phys. F* **10** (1980): 2769–2778.

[156] Thouless, D. J., P. W. Anderson, and R. G. Palmer. "Solution of 'Solvable Model of a Spin Glass.'" *Phil. Mag.* **35** (1977): 593–601.

[157] van der Holst, Hein. "Topological and Spectral Graph Characterizations." Universiteit van Amsterdam, Amsterdam, 1996.

[158] Vose, Michael D., and Alden H. Wright. "The Simple Genetic Algorithm and the Walsh Transform. Part I: Theory." *Evol. Comp.* **6** (1998): 253–274.

[159] Vose, Michael D., and Alden H. Wright. "The Simple Genetic Algorithm and the Walsh Transform. Part II: The Inverse." *Evol. Comp.* **6** (1988): 275–289.

[160] Wagner, Günter P., and Peter F. Stadler. "Complex Adaptations and the Structure of Recombination Spaces." In *Algebraic Engineering*, edited by Chrystopher Nehaniv and Misami Ito, 96–115. Singapore: World Scientific, 1999.

[161] Weinberger, Edward D. "Correlated and Uncorrelated Fitness Landscapes and How to Tell the Difference." *Biol. Cyber.* **63** (1990): 325–336.

[162] Weinberger, Edward D. "Fourier and Taylor Series on Fitness Landscapes." *Biol. Cyber.* **65** (1991): 321–330.

[163] Weinberger, Edward D., and Peter F. Stadler. "Why *Some* Fitness Landscapes are Fractal." *J. Theor. Biol.* **163** (1993): 255–275.

[164] Weisfeiler, B. Y., and A. A. Leman. "Reduction of a Graph to a Canonical Form and an Algebra Arising during This Reduction." *Naucho—Techn. Inf.; Ser. 2* **9** (1968): 12–16.

[165] Whitley, L. Darrel "Fundamental Principles of Deception in Genetic Search." In *Foundations of Genetic Algorithms*, edited by G. Rawlins, 221–241. San Mateo, CA: Morgan Kaufmann, 1991.

[166] Whitley, L. Darrel, and J. Dzubera. "Advance Correlation Analysis of Operators for the Traveling Salesman Problems." In *Parallel Problem Solving from Nature—PPSN III*, edited by Y. Davidor, H. P. Schwefel, and R. Manner, 68–77. Berlin: Springer-Verlag, 1994.

[167] Wielandt, H. *Finite Permutation Groups.* New York: Academic Press, 1964.

[168] Wilson, Stewart W. "GA-Easy Does not Imply Steepest-Ascent Optimizable." In *Proceedings of the Fourth International Conference on Genetic Algorithms*, edited by Richard K. Belew and Lashon B. Booker, 85–89. San Mateo, CA: Morgan Kaufmann, 1991.

[169] Wolpert, David H., and William G. Macready. "No Free Lunch Theorems for Search." Working paper 95-02-010, Santa Fe Institute, Santa Fe, NM, 1995.

[170] Yates, F. "The Design and Analysis of Factorial Experiments." Imperial Bureau of Soil Science, Technical Communication No. 35, Harpenden, 1937.

Quasispecies Evolution on Dynamic Fitness Landscapes

Nigel Snoad
Martin Nilsson

In this chapter we investigate the behavior of a quasispecies model of evolution that includes a changing environment, i.e., a dynamic fitness landscape. We derive new expressions for the error threshold and show that there exist lower and upper thresholds, both of which represent limits to the copying fidelity of simple replicators. The lower bound can be expressed as a correction term to the error threshold present on a static landscape. The upper threshold is a new limit that only exists on dynamic fitness landscapes. We calculate and observe some of the effects of finite-population size on these thresholds. We also show that, for long genomes on rapidly moving fitness landscapes, there exists a lower bound on the selection pressure, which enables effective maintenance of genomes with superior fitness independent of mutation rates. That is, there are distinct limits to the evolvable evolutionary parameters in dynamic environments.

1 INTRODUCTION

When we observe the richness and diversity of the dynamics of living organisms, it often seems amazing that a theory based on assumptions as simple as evolution's is able to provide an explanatory framework for what we see. It is the huge range of interactions extant and possible between organisms struggling to survive that provides the diverse selective forces required to generate this complexity. Each organism experiences a rich and diverse environment that provides the selective background against which its own evolution, both directed and random, will occur. In many cases the evolutionary environment is relatively static; i.e., the rate of environmental change is significantly slower than the time scale with which populations evolve, and so organisms become adapted to their current environment. Changes to selective forces will affect the evolving population either catastrophically or gradually, depending on the continuity of the experienced environmental transition.

In many evolutionary systems, however, the picture of adaptation to a relatively static environment is not appropriate. All co-evolving organisms—pathogens, and hosts engaged in evolutionary arms races, symbionts, mutualists, and organisms engaged in an ecosytem—have a selective environment that continually changes at a rate that forces them (or more properly their lineages) to evolve in response to the rate and nature of this change.

The field of population genetics has, over many years, developed a range of tools and standard models describing changes in the frequencies of alleles (fitness determining and not) in response to selective forces. Work on fluctuating selection has, since the 1970s, resulted in an understanding of the way in which rates of mutational change should evolve in response to the rate of environmental change [11, 12, 13, 14]. In this chapter, we utilize a different model, Eigen's quasispecies concept [6], to provide descriptions of some phenomena not easily addressed by this existing work. We are able to outline several limits to the evolvability (e.g., mutation rate and genome length) of simple haploid organisms that arise only in dynamic environments. This approach opens the door to a number of potential new directions for theoretical studies of evolving populations in fluctuating environments.

2 QUASISPECIES MODEL

Ever since Eigen's work on replicating molecules in 1971 [6], the quasispecies model has proven to be a very fruitful way of studying fundamentals of evolutionary dynamics. A quasispecies[1] is an equilibrium distribution of closely related gene sequences, localized around one or a few sequences with high fitness. The combination of simplicity and mathematical preciseness makes it possible to isolate the effects of different fundamental parameters in the model.

[1] The term derives from the concept of chemical species rather than biological ones.

Much of this preciseness results from the existence of a direct correspondence between the physics of spin-glass models in statistical mechanics and the quasispecies [1, 15, 16, 26].

This tractability and the intuition so developed (though only applicable to haploid evolution) has made it possible for the model to capture some general phenomena in nature, notably the critical relation between mutation rate and information transmission [6, 8]. The dynamics of these evolutionary models has been the subject of much fruitful research—see, for instance, [1, 2, 3, 6, 7, 8, 18, 21, 23, 24, 25].

One of the main results of this work was the statement of a lower bound on the copying accuracy required for evolution to effectively select for the genome with highest fitness—in effect a rigorous multilocus restatement of Wright's [33] mutation-selection balance. This lower bound is usually referred to as the error threshold (or catastrophe) and has important implications for biological systems. Some viruses, for example, seem to have evolved mutation rates that are close to the error threshold [17]. Indeed, it is argued that the error threshold provides an optimal compromise between adaptedness and adaptability in changing evolutionary environments; though this rests on a group-selection type of argument.

The studies of quasispecies have focused on static fitness landscapes. Many organisms in nature, however, live in quickly changing environments [27]. This is especially important for viruses and other microbial pathogens that must survive in a host with a quickly adapting immune system for which there only exist tight and temporary niches with high fitness (for the pathogen). In such environments, the standard quasispecies model is obviously inapplicable—dynamic environments require the development of intuitions and models that are substantially different from those that result from static equilibria. In the following work and that presented in Nilsson and Snoad [19], we take the simplest quasispecies model and explicitly build in time dependence on the replication rate (or fitness). Other models including time dependence in the quasispecies model have been studied [22, 28, 29, 30, 31, 32].

3 THE MODEL

For our investigations we use a quasispecies model where the fitness landscape is time dependent. Our population of replicating genomes are each represented by a sequence of n bases s_k, $(s_1, s_2, \cdots, s_n)$, where each base is a member of an index set of size λ. Obviously $\lambda = 4$ for RNA and DNA. For the sake of simplicity from here on we will assume binary bases $\{0, 1\}$ though the arguments presented can include larger base sets set without too much difficulty. We also require that all sequences have equal length; i.e., they are all evolving on the same landscape graph (which is unchanging). Every genome is then a binary string $(011001 \cdots)$, which can be represented by an integer k $(0 \leq k < 2^n)$.

To describe how mutations affect a population we define W_k^l as the probability that replication of genome l results in the production of genome k. For perfect copying accuracy, W_k^l is the identity matrix. Mutations, however, give rise to off-diagonal elements in W_k^l. Since the genome length is fixed to n, we only consider mutations that conserve the genome length, i.e., point mutations with a rate $p = 1 - q$ (where q is the copying accuracy per base). We assume that this rate is constant in time and independent of position in the genome; though both of these restrictions may perhaps be relaxed in future work. We can then write an explicit expression for W_k^l in terms of the copying fidelity:

$$W_k^l = p^{h_{kl}} q^{n-h_{kl}} = q^n \left(\frac{1-q}{q}\right)^{h_{kl}}, \tag{1}$$

where h_{kl} is the Hamming distance between genomes k and l, and n is the genome length. The Hamming distance h_{kl} is defined as the number of base positions at which genomes k and l differ.

The equations describing the population dynamics may be written in relatively simple form. Let x_k denote the relative concentration of genome k and A_k its fitness (replication rate). We then obtain the rate equations:

$$\dot{x}_k = \sum_l W_k^l A_l x_l - f x_k \,, \tag{2}$$

where $f = \sum_l A_l x_l$ is the total fitness of the population and the dot denotes a time derivative. The second term ensures the total normalization of the population (as $\sum_l x_l = 1$) so that x_k describes relative concentrations.

3.1 MOVING LANDSCAPES

To create a dynamic landscape, we consider a single-peaked fitness landscape [9] whose optimum moves, i.e., one with a single high-fitness genotype (the master-sequence) that changes over time. Formally, we can write $A_{k(t)} = \sigma > 1$ and $A_l = 1 \ \forall, \ l \neq k(t)$ where the (changing) genome $k(t)$ describes how the peak moves through sequence space. If $k(t)$ is constant in time the rate equation [eq. (2)] corresponds to the classical (static) theory of quasispecies studied by Eigen and others. When the peak in the fitness landscape moves, we assume that it is to one of its closest neighbors (chosen randomly).

The fact that the mutation matrix W describes position-independent point mutations imposes a symmetry on the rate equations, dividing the relative concentrations into error classes Γ_I described by their Hamming distance i from the master sequence (Γ_0). This reduces the effective dimension of the sequence space from 2^n to $n+1$, thereby making the problem significantly more tractable. The use of asymmetric evolution operators (such as most forms of recombination) or fitness landscapes (i.e., with epstasis) is obviously significantly more problematic but is beyond both the scope of current (static) quasispecies theory

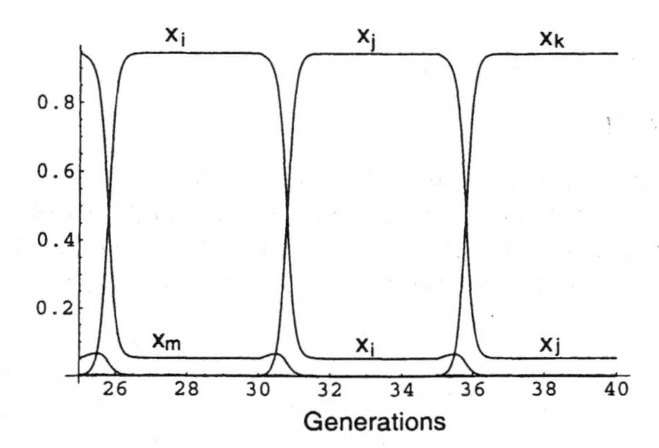

FIGURE 1 This plot shows the population dynamics for gene sequences of length 50 when the fitness peak moves every fifth generation ($\tau = 5$) and the height of the fitness peak σ is 10. The result is shown from $t = 25$ to remove initial transients.

and our model. The movement of the peak breaks this useful symmetry, as one sequence in the set of one mutant neighbors (Γ_1) will be singled out as the new master sequence. This would affect the results we present below if the mean time between shifts in the fitness landscape were small. In this case there would be a substantial concentration of the old master sequence present when the peak moved again. We make a kind of adiabatic elimination by assuming that the dynamics are slow enough to allow us to neglect this asymmetry.

Moving the fitness peak then corresponds to relabeling the fitness landscape or, more simply and equivalently, the concentration vector, i.e., $\mathbf{x}' = R\mathbf{x}$ where

$$R = \begin{pmatrix} 0 & \frac{1}{n} & 0 & \cdots \\ 1 & 0 & \frac{2}{n} & \cdots \\ 0 & \frac{n-1}{n} & 0 & \cdots \\ \vdots & \vdots & \vdots & \ddots \end{pmatrix} . \tag{3}$$

Numerical solutions to the (static peak) rate equations [eq. (2)] enable us to describe the population dynamics of the quasispecies distribution from time 0 to τ (where τ is a parameter determining the number of generations between shift of the fitness peak). At this time we apply the R transformation to the concentration vector. The resulting concentration distribution is used as the initial condition for a repeat of this regular stasis-shift cycle. Figure 1 shows the relative concentrations of the three most common species (at any given time) for a number of peak shifts (after the initial transient) where $\tau = 5$, $\sigma = 10$, $q = 0.999$, and string-length $n = 50$.

4 ERROR THRESHOLDS

4.1 HOLDING STILL

A simple approximation to the model presented above enables us to derive analytical expressions for the error thresholds on a dynamic fitness landscape. Assuming that back mutations (from γ_1 into the master sequence) are negligible, we write the rate equation for the master sequence during the static part of a cycle

$$\dot{x}_{\text{mas}} = Q\sigma x_{\text{mas}} - f x_{\text{mas}}\,, \tag{4}$$

where $Q = q^n$ is the copying fidelity of the whole genome and $f = (\sigma x_{\text{mas}} + 1 - x_{\text{mas}})$ is, as before, the total fitness of the population. The stationary-state occupancy of the master sequence is then

$$x_{\text{mas}}(t) \rightarrow \frac{Q\sigma - 1}{\sigma - 1} \text{ when } t \rightarrow \infty\,. \tag{5}$$

Thus, the error threshold of a population on a static fitness landscape occurs when

$$Q^{\text{stat}} = \frac{1}{\sigma}\,. \tag{6}$$

(See, e.g., Eigen [6] and Eigen and Schuster [8].) At this copying fidelity the superior fitness, and hence the growth rate, of the master sequence just compensates for the loss of γ_0 individuals due to mutations that occur during replication. It should be noted that one of the most important features of equation (6) is the scaling with the *genomic* rather than the *point* mutation rate. As DNA replication and repair are local processes, this is somewhat surprising—but it has been observed (and puzzled over) in the biological literature for some time [4, 5, 17]. We have applied a version of this dynamic-landscape model to the question of the optimization of mutation rates elsewhere [20] with some success, notably reproducing this scaling, and we demonstrate reasonable values for the optimal mutation rates given potentially biologically realistic parameters.

4.2 GOING DYNAMIC

Our approach to the dynamic landscape is different from that for the static: what determines the loss of the fitness optimum is whether its occupancy (which is reduced when a shift occurs) will have time to regrow between the shifts of the fitness peak. In this dynamical case the fitness peak moves into error-class one every τ generations and so, to find an analytical approximation for the error threshold, we must include the first error class as well as the master sequence in the dynamics [eq. (4)].

The difficulties that this introduces may be reduced by assuming that the occupancy of Γ_0 is small near the error threshold, and hence we can neglect the nonlinearity in equation (4); i.e., the growth is exponential and not in any way saturated. We can thus write an approximation of the rate equations for the master sequence and a representative member of error-class one (we can assume this to be the individual to which the peak will shift):

$$\dot{x}_{\text{mas}} = (Q\sigma - 1)x_{\text{mas}}\,, \tag{7}$$

$$\dot{x}_{1j} = \tilde{Q}\sigma x_{\text{mas}} + (Q-1)x_{1j}\,, \tag{8}$$

where mutations into x_{1j} are neglected and $\tilde{Q} = (1-q)q^{n-1}$ describes mutation from x_{mas} into x_{1j}. We now assume $x_{1j}(0) = 0$, which is a good approximation since the new x_{1j} is (almost always) in Γ_2 before the shift. The solutions to equations (7) and (8) using this boundary condition are

$$\begin{aligned} x_{\text{mas}}(t) &= x_{\text{mas}}(0)e^{(q^n\sigma-1)t}\,, \\ x_{1j}(t) &= x_{\text{mas}}(0)\left(\frac{(e^{(q^n\sigma-1)t} - e^{(q^n-1)t}(1-q)\sigma}{(\sigma-1)q}\right). \end{aligned} \tag{9}$$

The initial concentration of master sequences at the beginning of a shift cycle is $x_{\text{mas}}(0) = x_{1j}(\tau)$, since at this time ($\tau$) the change in the landscape moves the fitness peak to one of the sequences in error-class one.

We may now establish the condition on the mutation rate that allows for the continued occupancy of the master sequence as it moves, i.e., the error threshold. We note that this will occur when $x_{\text{mas}}(0) > x_1(\tau)$. At this point the concentration of the master sequence after the shift is lower than immediately after the previous shift and, thus, over a number of cycles the concentration of the master sequence will diminish until it is no higher than any other sequence. While not precise in the sense of a statistical mechanical definition of a phase transition [10, 15, 26], it suffices as a phenomenological description of a definite change in the evolutionary dynamics in the deterministic case.

We may thus derive an expression for the growth rate κ of the master sequence over a full shift cycle:

$$\kappa \equiv \frac{(e^{(q^n\sigma-1)\tau} - e^{(q^n-1)\tau)}(1-q)\sigma}{(\sigma-1)q} > 1\,. \tag{10}$$

With the roots of $\kappa = 1$ giving the critical values of q, i.e., the error threshold.

Figure 2 shows the region where $\kappa \geq 1$ can be expected to hold. The figure also shows the existence of two error thresholds, $q_{\text{lo}}^{\text{dyn}}$ and $q_{\text{hi}}^{\text{dyn}}$ corresponding to the real roots of $\kappa = 1$. The lower threshold is a perturbed version of the static error threshold, resulting from the movement of the fitness landscape. The upper threshold is, however, a new phenomenon that appears only on dynamic fitness landscapes. It possesses an intuitive interpretation: The peak moves out from under the not-sufficiently-diverse quasispecies and, hence, the population cannot track shifts in the fitness landscape.

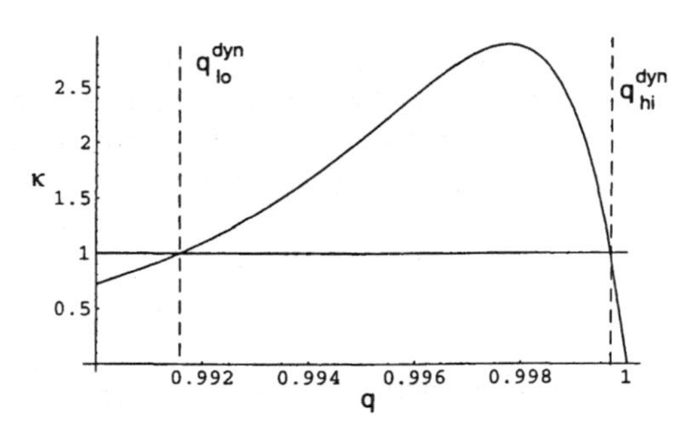

FIGURE 2 The left-hand side of equation (10) is plotted as a function of the copying fidelity q. The genome length $n = 50$, $\tau = 2$, and $\sigma = 5$. The lower threshold is located at $q_{\text{lo}}^{\text{dyn}} = 0.988$ and the upper threshold at $q_{\text{hi}}^{\text{dyn}} = 0.9997$.

The biological implications of this are obvious. For example, if the niche afforded a pathogen by an immune system (assuming that one always exists) changes too quickly, then the pathogen cannot maintain a suitable growth rate and eventually dies out. This shifting could be due to the adaptive immune system identifying the pathogen or, perhaps, due to repeated multiple drug therapy. Admittedly the links between such complex scenarios and the simple model we have presented are tenuous. However, there is little doubt that by including the environmental dynamics explicitly, we have gained a good deal of insight. The exact interpretive or predictive value of this is, of course, open to question—but a question well worth exploring.

4.3 THE LOWER THRESHOLD

While it is not possible to find exact analytical solutions for the roots of equation (10), we may derive approximations to the error thresholds by assuming different dominant terms for the two thresholds. To find the lower threshold $q_{\text{lo}}^{\text{dyn}}$ we assume q^n to dominate the behavior. Solving for q^n gives

$$q^n \approx \frac{\tau - \ln\left(\frac{\sigma}{\sigma-1} \cdot \frac{1-q}{q}\right)}{\sigma\tau}. \tag{11}$$

We can use equation (11) to find a first-order correction in τ to the static threshold by putting $q_{\text{crit}}^{\text{dyn}} = \sigma^{-1/n}$ on the right-hand side

$$Q_{\text{lo}}^{\text{dyn}} \approx \frac{1}{\sigma} - \frac{\ln(\sigma^{1/n} - 1)}{\tau\sigma} \tag{12}$$

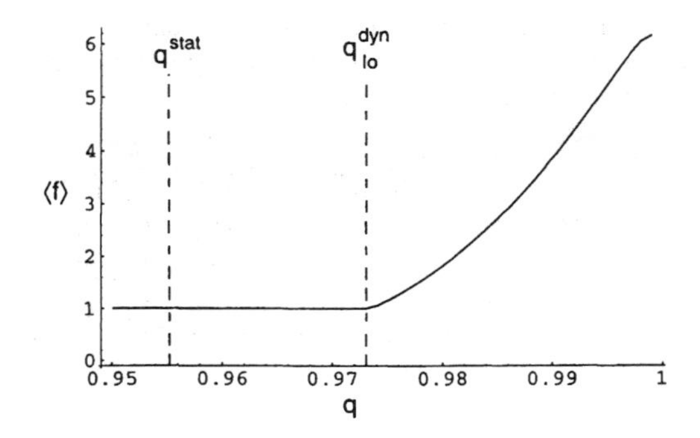

FIGURE 3 The mean fitness is plotted as a function of the copying fidelity per base q. The fitness peak moves every second generation ($\tau = 2$), the stringlength n is 0, and the growth superiority of the master sequence σ is 10. The error threshold occurs at the predicted value $q_{\mathrm{lo}}^{\mathrm{dyn}} = 0.973$. The static error threshold is located at $q^{\mathrm{stat}} = 0.955$.

where we also made the approximation $\sigma/(\sigma - 1) \approx 1$. This is an expression for the lower error threshold on a dynamic fitness landscape. Not that $Q_{\mathrm{lo}}^{\mathrm{dyn}} \to Q_{\mathrm{crit}}^{\mathrm{stat}}$ when $\tau \to \infty$; i.e., we recover the stationary landscape limit.

Figure 3 shows the mean fitness of a population as a function of the copying fidelity. When q is below $q_{\mathrm{lo}}^{\mathrm{dyn}}$, the concentration of master sequence is approximately zero and, therefore, the mean fitness will be 1. The figure is based on numerical solutions of the full rate equations [eq. (2)]. Note that the analytic approximation to $q_{\mathrm{lo}}^{\mathrm{dyn}}$ given by equation (12) is quite accurate. Further comparisons to numerical solutions to the full dynamics are shown in table 1.

Unlike the static threshold, the critical copying fidelity $Q_{\mathrm{lo}}^{\mathrm{dn}}$ *does* depend on the genome length—with the perturbation from the static error threshold increasing with genome length. This is not surprising since the fitness peak shifts into a specific member of Γ_1, which consists of n different gene sequences. For reasonable values of $\tau \gg 1$ and $\sigma \gg 1$, however, the static and dynamic error threshold are of the same order of magnitude and $Q_{\mathrm{lo}}^{\mathrm{dyn}}$ is relatively independent of the genome length.

4.4 THE UPPER THRESHOLD

An analytical approximation to the new upper threshold can be found by assuming q to be very close to 1 and, therefore, the $(1-q)$-term dominates the behavior of equation (10). Again assuming $\sigma \gg 1$ and in this case putting $q^n = 1$, give

$$q_{\mathrm{hi}}^{\mathrm{dyn}} \approx 1 - e^{-(\sigma-1)\tau} \,. \tag{13}$$

As for the lower threshold, explicit numerical solutions of the full dynamics confirm that this threshold exists and is as predicted by equation (13). For most

TABLE 1 The table shows results of numerical solutions of the error threshold ($q_{\text{threshold}}$) compared to predicted values given by equation (12) and the threshold for the corresponding static fitness landscape.

τ	σ	n	$q_{\text{threshold}}$	$q_{\text{lo}}^{\text{dyn}}$	q^{stat}
2	10	25	0.940	0.941	0.912
2	10	50	0.973	0.973	0.955
2	5	50	0.988	0.988	0.968
5	10	50	0.963	0.964	0.955

values of σ and τ, $q_{\text{hi}}^{\text{dyn}}$ is very close to 1 (e.g., $(\sigma-1)\tau = 50$ gives 10^{-22} as a lower bound on the mutation rate per base pair). It is important to note that $q_{\text{hi}}^{\text{dyn}}$ is independent of the genome length. The genomic copying fidelity $Q_{\text{hi}}^{\text{dyn}} = (q_{\text{hi}}^{\text{dyn}})^n$ will then depend strongly on the genome length. This means that, as the genome length increases, the region of evolvability (i.e., between the two error thresholds) narrows.

4.5 FINITE-POPULATION SIZE EFFECTS

The relations we have derived above for the critical upper and lower values of the copying fidelity assume an infinite-population size; i.e., they are deterministic. Obviously, the introduction of finite-population sizes creates stochastic fluctuations around these (expected[2]) values. Most notably, sampling effects will obviously require a limit of $\kappa > 1 + \varepsilon$ for the continued occupancy of the peak over many shift cycles. Nowak and Schuster [21] demonstrated that the lower (stationary) threshold scales with the population size as $N^{-1/2}$. Computer simulations of finite populations evolving on a dynamic landscape performed by the authors but not reproduced here have demonstrated both this scaling and the appropriate variation of $q_{\text{lo}}^{\text{dyn}}$ with n and σ.

An finite-population size limit exists for the lower threshold, notably that when the expected occupancy of the new peak position is less than one, the peak will almost certainly be lost. As a first approximation we might assume that in such situations which will be far below the deterministic copying fidelity limit, the master sequence approaches equilibrium between shifts (see also the simulation examples given in figure 5). In this case, we may use equation (5) as the equilibrium master-sequence occupancy and solve for the stationary state of equation (8), giving the condition

$$x_{1I}^{\text{stat}} = \frac{\tilde{Q}\sigma(Q\sigma - 1)}{Q(\sigma - 1)^2} \geq \frac{1}{N} \,. \tag{14}$$

[2] In a statistical sense.

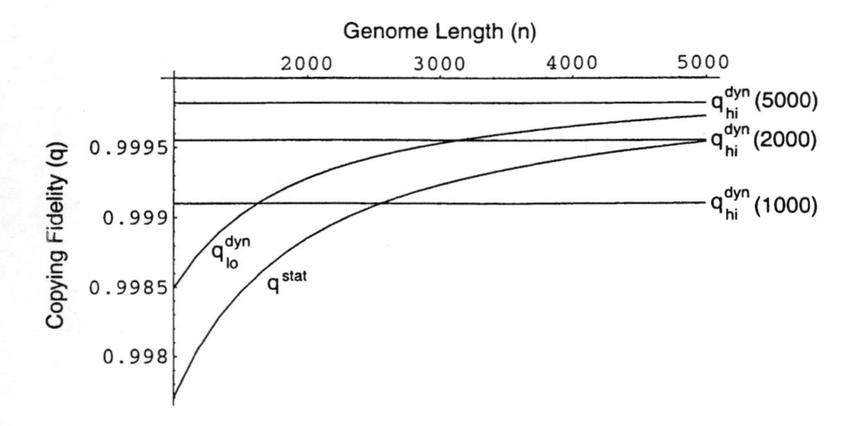

FIGURE 4 The dependence of various (per base) error thresholds on the genome length. The notation $q_{\rm hi}^{\rm dyn}(1000)$ represents the higher dynamic threshold with a population size of $N = 1000$, etc. Note the similar scaling of $q^{\rm stat}$ and $Q_{\rm lo}^{\rm dyn}$.

While this equation is not analytically soluble for $q_{\rm hi}^{\rm dyn}(N)$, we may approximate it in the same way as for equation (13)

$$q_{\rm hi}^{\rm dyn}(N) \approx 1 - \frac{\sigma - 1}{\sigma N} . \tag{15}$$

This expression for the finite-population size upper limit gives remarkably good correspondence with computer simulations.

Figure 4 shows how all of the thresholds we have derived scale with the genome length n.

4.6 SIMULATION

To help visualize this discussion of a multitude of thresholds, figure 5 presents the time evolution of the mean fitness of a computer simulation of genomes evolving on a dynamic landscape as we have described. The simulation involves the use of a generational rather than a continuous-time reproductive model, but nonetheless it demonstrates the same qualitative features as we have described. Indeed, similar analytic error thresholds may be derived.

The peak is held static for 25 generations and then shifts every 5 generations. During the static phase we observe convergence to the stationary state occupancy as predicted by equation 5. For mutation rates greater than the static error threshold [fig. 5(a)], no significant population is able to be established on the master sequence, moving or not. There exists (b) and (c) copying fidelities above $q^{\rm stat}$ and below $q_{\rm lo}^{\rm dyn}$, such that the master sequence is occupied while stationary

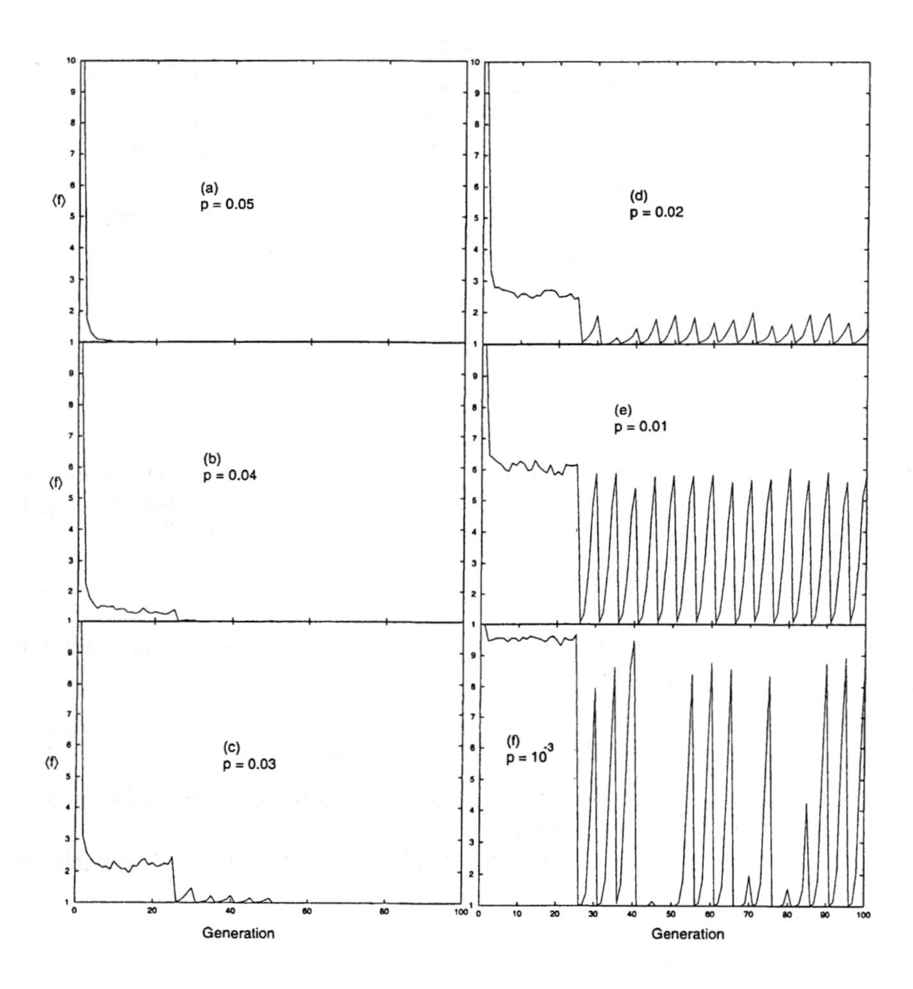

FIGURE 5 Mean population fitness $\langle f_t \rangle$ over time for a range of per-base mutation rates $p = 1 - q$. Population size $M = 1000$, $\nu = 50$, $\tau = 5$, and $\sigma = 10$. The first movement of the peak comes at generation 25. The dashed horizontal lines indicate the theoretical stationary state mean fitness.

TABLE 2 The minimum selection pressure required for an infinite population to track the peak is listed for different values of the genome length n and the number of generations between shifts of the fitness peak τ.

$\tau \backslash n$	50	500	5000	50000	10^9
1	7.8	10.4	13.0	15.5	25.9
10	1.7	2.0	2.2	2.4	3.5
50	1.1	1.2	1.2	1.3	1.5

but is lost when it begins moving. In (d) and (e) the peak is easily held; indeed, the occupancy of Γ_0 approaches the stationary equilibrium value between shifts. In (f), however, we see a destabilization due to the small number of individuals present when the peak shifts; i.e., $q_{\text{hi}}^{\text{dyn}}(N)$ is approached. The peak is often lost entirely when the shift occurs but is quickly found again by mutants. The peak is lost completely for any value of $p \leq 9 \cdot 10^{-4}$, which is the analytically predicted value.

5 LIMITS TO EVOLVABILITY

On a static fitness landscape it is always possible to find copying fidelities high enough for evolution to be effective. It turns out that this is no longer the case for dynamic fitness landscapes. There exist regions in parameter space (spanned by σ, τ, and n) where solutions to equation (10) cease to exist. This occurs when the upper and lower error thresholds coincide or, to put it differently, when the maximum (taken over q) of the left-hand side of equation (10) become less than 1. Such a convergence is seen in figure 4.

To find this convergence (in the infinite-population limit) point, we will search for a direct approximation of q that maximizes the left-hand side of equation (10). To this end, we assume the leading behavior is determined by the factor $e^{q^n\sigma-1)\tau}(1-q)$. Taking the derivative of this expression and setting it to zero gives the equation $q^{n-1}(1-q) = 1/(n\sigma\tau)$. Assuming q to be very close to 1 (and hence $q^{n-1} \approx 1$) gives

$$q_{\max} \approx 1 - \frac{1}{\sigma\tau n}. \tag{16}$$

This approximation for $q_{\max}$ can be substituted into equation (10). It is easy to find points in phase space where this inequality starts to hold by fixing two parameters (e.g., τ and n) and then numerically solving for the third (σ). Table 2 shows the minimal height of the fitness peak for different values of τ and n. The required selective pressure becomes large for fast-moving fitness landscape and large genome lengths. Obviously, finite-population size makes a large difference to these values.

One corollary of this result is that we may equivalently define a maximal genome length maintainable in a given environment. Thus we have limits on the amount of information for which it is possible to evolve encodings. Indeed, we could perhaps rephrase the discussion we have just given in terms of the amount of information about the environment (or fitness function) that is necessary for a population to encode by way of its diversity and genome length. An additional point to make is that $q_{\max}$ defines the maximal (deterministic) growth rate of the population and hence gives the optimal mutation rate. By this we mean a strain with this rate will, over the long term, outgrow all other strains. For large n this value of $Q_{\max} = q_{\max}^n$ becomes effectively independent of the genome length. This same rate may be arrived at independently by examining the relative growth rates of master sequences without assuming anything about exponential growth regimes or even quasispecies distribution (see Nilsson and Snoad [20]).

6 SUMMARY

In conclusion, we have shown existence of, and derived analytic expressions for, two error thresholds on a simple dynamic fitness landscape. These thresholds have the effect of placing limits on the mutation rates with which replicators can stably maintain a cohesive population in a dynamic environment.

The lower of these two thresholds is a perturbation of the well-known error catastrophe that exists in a static fitness landscape and accounts for the destabilizing effect of the changing environment. The existence of an upper bound on the copying fidelity is a new phenomenon, only existing in dynamic environments. The presence of this upper bound results in the existence of critical regions of the landscape parameters (σ, τ, and n) where the two thresholds coincide (or cross), and therefore no effective selection can occur. This places explicit upper bounds on the sustainable genome length (or equivalently limits on the other landscape parameters). Thus, dynamical landscapes impose strong constraints on evolvability, which arise from qualitatively different processes to those occurring in stable environments.

Given the simplicity of the model, it would be surprising to gain any direct biological insight from it. It explicitly applies to haploid replicators and refers exclusively to the effective or coding length of the genome; i.e., it excludes eukaryotes bearing introns. Nonetheless, we feel that, given the interest shown in the quasispecies model as a method of (at least conceptually) studying the evolutionary dynamics of viruses and microbial pathogens, the extension to include variable environments may provoke further work and even biological insight in those situations when it seems most valuable, notably co-evolutionary or rapidly changing environments.

We pointed out the existence of a surprising scaling of mutation rate with genome length amongst DNA-based microbes [5], which may potentially be explained by an explicitly dynamical quasispecies model. A toy example may be the (extremely complex) question as to why yeast *S. cerevisiae* has a genome double the size of *E. coli.* We could hypothesize that in part it is a result of a slower evolutionary environment or perhaps stronger selection.

ACKNOWLEDGMENTS

The authors would wish to thank Claes Andersson and Erik van Nimwegen for early discussions of this work, Michael Lachmann, Mats Nordahl and Marc Feldman for their later suggestions, and Jim Crutchfield for inviting us to prepare it for publication. This work was supported in part by SFI grants.

Nigel Snoad would like to acknowledge the support of DARPA/ONR grant N00014-95-1-1000 at the SFI and the centre for Computational Genetics and Biological Modeling at Stanford University while drafting this chapter.

REFERENCES

[1] Alves, D., and J. F. Fontanari. "Error Thresholds in Finite Populations." *Phys. Rev. E* **57** (1998): 7008–7013.

[2] Bergstrom, C. T., P. McElhany, and L. A. Real. "Transmission Bottlenecks as Determinants of Virulence in Rapidly Evolving Pathogens." *Proc. Natl. Acad. Sci. USA* **96** (1999): 5095–5100.

[3] Bonhoeffer, L. S., and P. F. Stadler. "Error Thresholds on Correlated Fitness Landscapes." *J. Theor. Biol.* **164** (1993): 359–372.

[4] Drake, J. W. "Rates of Spontaneous Mutation among RNA Viruses." *Proc. Natl. Acad. Sci. USA* **90(9)** (1993): 4171–4175.

[5] Drake, J. W., B. Charlesworth, D. Charlesworth, and J. F. Crow. "Rates of Spontaneous Mutation." *Genetics* **148(4)** (1998): 1667–1686.

[6] Eigen, M. "Self-Organization of Matter and How the Evolution of Biological Macromolecules." *Naturwissenschaften* **58** (1971): 465–523.

[7] Eigen, M., J. McCaskill, and P. Schuster. "The Molecular Quasispecies." *Adv. Chem. Phys.* **75** (1989): 149–263.

[8] Eigen, M., and P. Schuster. "The Hypercycle. A Principle of Natural Selection. Part A: Emergence of the Hypercycle." *Naturwissenschaften* **64** (1977): 541–565.

[9] Eigen, M., and P. Schuster. *The Hypercycle—A Principle of Natural Self-Organization.* Berlin: Springer, 1979.

[10] Franz, S., and L. Peliti. "Error Threshold in Simple Landscapes." *J. Phys. A: Math. & Gen.* **30** (1997): 4481–4487.

[11] Gillespie, J. H. "Mutation Modification in a Random Environment." *Evolution* **30** (1981): 468–476.

[12] Ishii, K., H. Matsuda, Y. Iwasa, and A. Saskai. "Evolutionarily Stable Mutation Rate in a Periodically Changing Environment." *Genetics* **121** (1989): 163–174.
[13] Kimura, M. "On the Evolutionary Adjustment of Spontaneous Mutation Rates." *Genet. Res.* **9** (1967): 23–24.
[14] Leigh, E. G. "The Evolution of Mutation Rates." *Genetics Suppl.* **73** (1973): 1–18.
[15] Leuthaäusser, I. "Statistical Mechanics of Eigen's Evolution Model." *J. Stat. Phys.* **48(1/2)** (1987): 343–360.
[16] Leuthaäusser, I. "An Exact Correspondence between Eigen's Evolution Model and a Two-Dimensional Ising System." *J. Chem. Phys.* **84(3)** (1986): 1884–1885.
[17] Maynard-Smith, J., and E. Szathmáry. *The Major Transitions in Evolution.* New York: Oxford University Press, 1995.
[18] Bonhoeffer, S., M. C. Boerlijst, and M. A. Nowak. "Viral Quasi-Species and Recombination." *Proc. Roy. Soc. Lond. B* **263** (1996): 1577–1584.
[19] Nilsson, M., and N. Snoad. "Error Thresholds on Dynamic Fitness Landscapes." *Phys. Rev. Lett.* **84** (2000): 191–195.
[20] Nilsson, M., and N. Snoad. "Optimal Mutation Rates for a Simple Organism in a Dynamic Environment." Working Paper 99-04-030, Santa Fe Institute, Santa Fe, NM, 1999.
[21] Nowak, M., and P. Schuster. "Error Thresholds of Replication in Finite Populations Mutation Frequencies and the Onset of Muller's Ratchet." *J. Theor. Biol.* **137** (1989): 375–395.
[22] Ronnewinkel, C., C. O. Wilke, and T. Martinez. "Genetic Algorithms in Time-Dependent Environments." In *Proceedings of the 2nd EvoNet Summerschool*, Natural Computing SEries. Berlin: Springer, 2000.
[23] Schuster, P. "Dynamics of Molecular Evolution." *Physica D* **16** (1986): 100–119.
[24] Schuster, P., and K. Sigmund. "Dynamics of Evolutionary Optimization." *Ber. Bunsenges. Phys. Chem.* **89** (1985): 668–682.
[25] Swetina, J., and P. Schuster. "Stationary Mutant Distribution and Evolutionary Optimization." *Bull. Math. Biol.* **50** (1988): 635–660.
[26] Tarazona, P. "Error Thresholds for Molecular Quasispecies as Phase Transitions: From Simple Landscapes to Spin-Glass Models." *Phys. Rev. A* **45(8)** (1992): 6038–6050.
[27] Van Valen, L. "A New Evolutionary Law." *Evol. Theory* **1** (1973): 1–30.
[28] Wilke, C. O., and T. Martinez. "Adaptive on Time-Dependent Fitness Landscapes." Los Alamos National Laboratory e-print archive: physics/99903024, Los Alamos, NM, 1999.
[29] Wilke, C. O., C. Ronnewinkel, and T. Martinez. "Molecular Evolution in Time Dependent Environments." In *Proceedings of ECAL '99*, edited by H. Lund and R. Kortmann, 471. Lecture Notes in Computer Science. Hiedelberg: Springer-Verlag, 1999.

[30] Wilke, C. O. "Evolutionary Dynamics in Time-Dependent Environments." Ruhr-Universität ochum, 1999.

[31] Wilke, C. O., C. Ronnewinkel, and T. Martinez. "Dynamnic Fitness Landscapes in the Quasispecies Model." Los Alamos National Laboratory e-print archive: physics/9912012, Los Alamos, NM, 1999.

[32] Wilke, C. O. "Evolution in Time Dependent Fitness Landscapes." Los Alamos National Laboratory e-print archive: physics/9811021, Los Alamos, NM, 1998.

[33] Wright, S. "The Roles of Mutations, Inbreeding, Crossbreeding and Selection in Evolution." *Proceedings of the Sixth International Cognress on Genetics* **1** (1932): 356–366.

Recombination and Bistability in Finite Populations

Lionel Barnett

In this chapter we analyze the phenomenon of "bistability" in finite-population evolutionary dynamics, especially with regard to recombination. Bistability, where the steady-state population distribution depends on the initial state of the population, has recently been observed in an (infinite-population) quasispecies model of viral recombination [1]. We analyze a comparable finite-population model using a birth and death process due to Moran [8]. Bistability (or its stochastic analogue) is revealed in the bimodality of the stationary probability distribution of the birth and death process and long mean transition times between the modes. These effects are demonstrated to be exaggerated by recombination.

1 INTRODUCTION

In Boerlijst et al. [1] a mathematical model for an asexual (haploid) quasi-species [4] evolving with recombination is introduced to study recombination in retrovirus populations. The model is analyzed on several simple fitness landscapes. A striking feature of the model is the appearance of "bistability" or "hysteresis" in the steady-state population distribution for particular combinations of mutation rate and recombination rate; i.e., the steady-state distribution of genotypes depends on the distribution of genotypes in the initial population. This phenomenon has also been observed in various diploid models, both with and without recombination. In Boerlijst et al. [1] bistability in their (infinite-population, hence deterministic) model is explained in terms of bifurcation of the differential equations describing the time evolution of the quasispecies. A recent empirical study [10] suggests strongly that many qualitative features of the infinite-population model are preserved in the corresponding finite-population dynamics. In this chapter we investigate bistability in finite-population stochastic population dynamics. The model we use is based on a birth and death process originally devised by Moran [8] and previously deployed in a situation analogous to ours (but without recombination) by Nowak and Schuster [9].

2 THE MORAN MODEL

In Moran [8] a model for the evolution of a fixed-size finite population of genotypes was introduced, based on the idea of fitness as the expected (reproductive) lifetime of a genotype. Here we extend the model to arbitrary fitness landscapes and include recombination.

Let $\mathbf{Q}^\nu$ represent the ν-dimensional binary hypercube; i.e., an element of $\mathbf{Q}^\nu$ is a binary sequence of length ν, which we identify with a haploid genotype. We specify a fitness landscape on $\mathbf{Q}^\nu$ by assigning to each $g \in \mathbf{Q}^\nu$ a (real-valued) *fitness* $f(g) > 0$. Consider a population comprising N such genotypes. We may identify such a population with an integer vector $\mathbf{n} = (n_g)$, $g \in \mathbf{Q}^\nu$, where n_g represents the number of copies of genotype g in the population, $n_g \geq 0 \; \forall g$ and $\sum_{g\in\mathbf{Q}^\nu} n_g = N$. We now define a *birth-death event* on the population $\mathbf{n}$ to be a transformation of $\mathbf{n}$ into a new population $\mathbf{n}'$ as follows: a copy of some genotype g_1 "dies," and a copy of another (possibly the same) genotype g_2 is "born." In terms of the population vectors, we have $n'_{g_1} = n_{g_1} - 1$ and $n'_{g_2} = n_{g_2} + 1$; the population size thus remains constant.

Suppose now that we have a stochastic process $\{\mathbf{n}(t) \mid t \geq 0\}$ of populations (t represents a continuous time parameter) that evolves according to the following scheme: in any time interval $[t, t+h]$ the probability that a copy of genotype g dies is given by Moran [8]:

$$\mathbf{P}(\text{a copy of } g \text{ dies in the interval } [\mathrm{t}, \mathrm{t+h}]) = \frac{\theta}{f(g)} \frac{n_g(t)}{N} h + \mathbf{o}(h)\,, \tag{1}$$

where θ is a fixed timescale parameter. It is straightforward to verify that the "lifetime" of a genotype g is exponentially distributed with expectation $(N/\theta) \times$

$f(g)$. A death triggers an immediate birth, thus defining a birth-death event. Candidates for a birth are selected as follows: with probability $1-\rho$, the birth is *asexual* and with probability ρ, *sexual*, where $0 \le \rho \le 1$ is the recombination rate. For asexual reproduction, a parent is selected uniformly at random and with replacement from the population. The offspring is taken to be a copy of the parent mutated with per-allele probability μ where $0 \le \mu \le 0.5$ is the *(per-allele) mutation rate.* In the sexual case, two parents are independently selected uniformly at random and with replacement from the population. The parents are mated by *uniform recombination*; i.e., independently for each locus on the genotype, one of the parents is chosen at random, and its allele (0 or 1) becomes the allele of the offspring at that locus. After recombination the offspring is mutated with mutation rate μ, as in the asexual case. It may also be verified that the expected number of offspring of a given genotype during its lifetime is proportional to its fitness. To see this, note that for a given genotype the times between its successive selections as a parent are (identically and independently) exponentially distributed. Thus, the number of offspring of a genotype from its birth up to a given time t constitutes a Poisson process [11]. From this and the exponential distribution of lifetimes the result follows by a straightforward calculation. In this sense, selection in the Moran model is *fitness proportional.*

It is clear that $\{\mathbf{n}(t) \mid t \ge 0\}$ thus defined is a Markovian birth and death process with state space the (vast!) set of all possible populations of size N. Because of the huge size and awkward structure of the state space, it is difficult to say anything useful about such a process. In the next section we specialize to a specific simple fitness landscape and, with the help of some judicious approximations, reduce the state space to a tractable form.

3 THE SINGLE-PEAK FITNESS LANDSCAPE

We specify a single-peak fitness landscape as follows: all genotypes have fitness 1 except for a single genotype, the "peak" or *optimal genotype*,[1] which has fitness σ where $\sigma > 1$ is the *selection coefficient.* Without loss of generality we take the optimal genotype to be the sequence of ν zeroes. For $\alpha = 0, 1, 2, \ldots, \nu$ let us define [4] the *error class* $E_\alpha \subset \mathbf{Q}^\nu$ to be the set of all genotypes Hamming distance α from the optimum, i.e., with exactly α bits set. The E_α with $\alpha > 0$ are said to constitute the *error tail.*

Given a Moran birth and death process as described above on such a landscape, we will be interested in the number $X(t)$ of copies of the optimal genotype present in the population at time t. It would be convenient if the process $\{X(t) \mid t \ge 0\}$ were also a Markov process. Unfortunately, it is clear that the Markov property does not hold. This is because the probability that a nonoptimal genotype (i.e., a genotype in the error tail) mutates to the optimal genotype

[1] Generally known as the *master sequence* in the quasispecies literature.

depends on the distribution of genotypes over the error classes; without knowing this distribution, we cannot know the probability that a birth will be optimal. In Nowak and Schuster [9] this issue is addressed by making a "maximum entropy" approximation. Specifically, it is assumed that a genotype in the error tail is as likely to be any one (nonoptimal) genotype as another; i.e., that the distribution of genotypes in the error tail is always uniformly random. This implies that at any time, $\alpha = 0, 1, 2, \ldots, \nu$:

$$\mathbf{P}(g \in E_\alpha | g \in \text{ error tail}) = \kappa \binom{\nu}{\alpha} \tag{2}$$

and

$$\mathbf{P}(\text{bit } i \text{ of } g \text{ is set } | g \in E_\alpha) = \frac{\alpha}{\nu}, \tag{3}$$

where, following Nowak and Schuster[9], we have set $\kappa \equiv 1/(2^\nu - 1)$. Note that equation (3) implies the absence of *linkage disequilibrium* between loci.

Under the assumptions (2) and (3), $\{X(t) \mid t \geq 0\}$ is indeed a Markovian continuous-time birth and death process with state space the set of integers from 0 to N and retaining barriers at 0 and N. If the mutation rate μ is non-zero, then the process is also *irreducible* [11] and, thus, has a unique stationary distribution [6]. Such processes are quite well understood and tractable to analysis; the question remains as to how well our approximation agrees with the original Moran birth and death process. It is, in fact, well known that equation (2) does not hold in general [1, 9, 10]. In particular, at low mutation rates the distribution of genotypes over the error tail is skewed toward the optimum. This is more or less the defining characteristic of a quasispecies! Furthermore, equation (3) will not in general hold due to *neutral drift* of the population [2, 3, 7] *within* the individual error classes. These issues will be addressed in a future paper. It is sufficient at this stage to note that preliminary research suggests that the behavior of the model using the maximum entropy approximation agrees surprisingly well with the full model over a wide range of parameter values, and that, in particular, it appears to preserve at least qualitatively the features addressed in this chapter.[2]

4 ANALYSIS OF THE BIRTH AND DEATH PROCESS

We are now in a position to calculate the infinitesimal generators [6] of the simplified birth and death process $\{X(t) \mid t \geq 0\}$. To this end it suffices to know the probabilities:

$$m_1 \equiv \mathbf{P}\,(\text{optimum mutates to optimum})\,,$$
$$m_2 \equiv \mathbf{P}\,(\text{nonoptimum mutates to optimum})\,,$$

[2]It is also worth pointing out that near the *error threshold* [4] the approximation (2) becomes more accurate. This is reflected in the accuracy of the error threshold approximation calculated in Nowak and Schuster [9].

$$\begin{aligned} r_{11} &\equiv \mathbf{P}\,(\text{optimum recombined with optimum is optimum})\,, \\ r_{12} &\equiv \mathbf{P}\,(\text{optimum recombined with nonoptimum is optimum})\,, \\ r_{22} &\equiv \mathbf{P}\,(\text{nonoptimum recombined with nonoptimum is optimum})\,. \end{aligned} \tag{4}$$

Using equations (2) and (3), and the definition of uniform recombination, a straightforward, if tedious, computation yields:

$$\begin{aligned} m_1 &= Q\,, \\ m_2 &= \kappa(1-Q)\,, \\ r_{11} &= 1\,, \\ r_{12} &= \eta\,, \\ r_{22} &= \kappa(1-2\eta)\,, \end{aligned} \tag{5}$$

where we have set $Q \equiv (1-\mu)^\nu$, and $\eta \equiv ((3/2)^\nu - 1)/(2^\nu - 1)$. The infinitesimal generators λ_i and μ_i of the birth and death process are defined by Karlin and Taylor [6]:

$$\mathbf{P}(X(t+h) = i+1 | X(t) = i) = \lambda_i + \mathbf{o}(h)\,, \qquad i = 0, 1, \ldots, N-1\,; \tag{6}$$

$$\mathbf{P}(X(t+h) = i-1 | X(t) = i) = \mu_i + \mathbf{o}(h)\,, \qquad i = 1, 2, \ldots, N\,. \tag{7}$$

By convention we define $\lambda_N \equiv \mu_0 \equiv 0$. For compactness of notation let us also define, for $a,\ b = 1,\ 2$: $\bar{m}_a \equiv 1 - m_a$, $\bar{r}_{ab} \equiv 1 - r_{ab}$, $u_{ab} \equiv m_1 r_{ab} + m_2 \bar{r}_{ab}$, and $\bar{u}_{ab} \equiv 1 - u_{ab}$. Then, using equation (1) and the definition of the Moran process, we calculate:

$$\begin{aligned} \lambda_i = \theta \frac{N-i}{N} \Bigg\{ &(1-\rho) \left[m_1 \frac{i}{N} + m_2 \frac{N-i}{N} \right] \\ &+ \rho \left[u_{11} \left(\frac{i}{N} \right)^2 + 2u_{12} \frac{i}{N} \frac{N-i}{N} + u_{22} \left(\frac{N-i}{N} \right)^2 \right] \Bigg\}\,, \end{aligned} \tag{8}$$

$$\begin{aligned} \mu_i = \frac{\theta}{\sigma} \frac{i}{N} \Bigg\{ &(1-\rho) \left[\bar{m}_1 \frac{i}{N} + \bar{m}_2 \frac{N-i}{N} \right] \\ &+ \rho \left[\bar{u}_{11} \left(\frac{i}{N} \right)^2 + 2\bar{u}_{12} \frac{i}{N} \frac{N-i}{N} + \bar{u}_{22} \left(\frac{N-i}{N} \right)^2 \right] \Bigg\}\,. \end{aligned} \tag{9}$$

These equations correspond to equations (17) and (18) in Nowak and Schuster [9].[3]

We can also now calculate the (unique) stationary probability distribution p_i, $i = 0, 1, 2, \ldots, N$ of the process [6, 11] as follows. Set:

$$\begin{aligned} \pi_0 &= 1\,, \\ \pi_i &= \frac{\lambda_{i-1}}{\mu_i} \pi_{i-1}\,, \qquad i = 1, 2, \ldots, N\,. \end{aligned} \tag{10}$$

[3]Nowak and Schuster [9] use a slightly different version of the Moran birth and death process, perhaps to match the quasispecies formalism more closely. The resulting models are qualitatively similar.

Then we have, for $i = 0, 1, \ldots, N$:

$$p_i = \frac{\pi_i}{\sum_{j=0}^{N} \pi_j}. \tag{11}$$

We will be interested in the *mean first passage time* (*mfpt*) [6] of the process from state i to state j. This may be calculated as follows: let U_i denote the *mfpt* from state i to state $i+1$ ($i = 0, 1, \ldots, N-1$) and V_i the *mfpt* from state i to state $i-1$ ($i = 1, 2, \ldots, N$). We then have the recurrence relations:

$$\begin{aligned} U_0 &= \frac{1}{\lambda_0}, \\ U_i &= \frac{1}{\lambda_i}(1 + \mu_i U_{i-1}), \qquad i = 1, 2, \ldots, N, \end{aligned} \tag{12}$$

and

$$\begin{aligned} V_N &= \frac{1}{\mu_N}, \\ V_i &= \frac{1}{\mu_i}(1 + \lambda_i V_{i+1}), \qquad i = 0, 1, 2, \ldots, N-1. \end{aligned} \tag{13}$$

The *mfpt* from state i to state j for $i < j$ is then given by $U_i + U_{i+1} + \ldots + U_{j-1}$ and for $i > j$ by $V_i + V_{i-1} + \ldots + V_{j+1}$. Note that these *mfpt* cannot be expressed solely in terms of the stationary probabilities; knowledge of the actual infinitesimal generators is required.

Finally, to simulate the birth and death process, we make use of the following [6]: the process maintains the state i for a period of time distributed exponentially with parameter $\lambda_i + \mu_i$. It then makes a transition to state $i+1$ (if $i < N$) with probability $(\lambda_i)/(\lambda_i + \mu_i)$, or to state $i-1$ (if $i > 0$) with probability $(\mu_i)/(\lambda_i + \mu_i)$.

5 BEHAVIOR OF THE MODEL

In the results that follow, we have used a short sequence length ($\nu = 10$) and population size ($N = 100$) to make the pertinent features of the model clear. All results extend to higher sequence lengths and larger populations. Figure 1 below plots the stationary distribution of the birth and death process for a few values of the mutation rate, all other parameters remaining fixed.

We see that at low mutation rates the optimum genotype frequency is generally high; the process spends most of its time with a high proportion of the population "on the spike." It appears unimodal, but there is actually another mode at 0, not visible at this scale. At a slightly higher mutation rate, the bimodality becomes more pronounced, and the position of the rightmost mode shifts to a lower optimum genotype frequency. At a critical mutation rate above

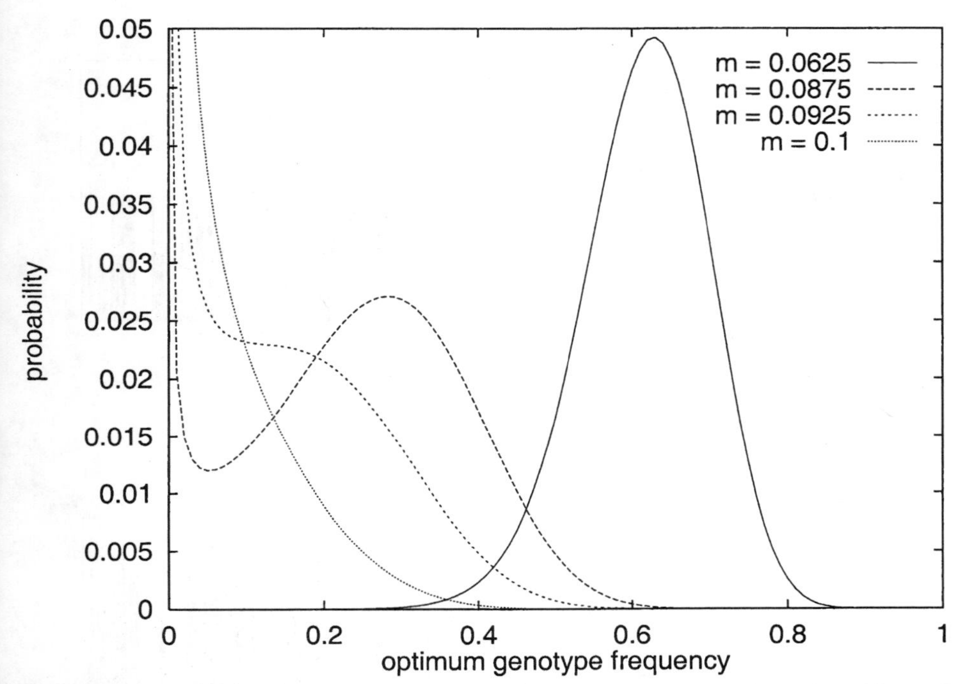

FIGURE 1 Stationary distribution of the birth and death process for a few values of mutation rate μ. Other parameters are: $\nu = 10$, $N = 100$, $\sigma = 4$, and $\rho = 0.3$.

this, an inflexion point appears, and the distribution becomes unimodal. Following Nowak and Schuster [9], we identify this critical mutation rate with the error threshold.[4] Beyond the error threshold the distribution is unimodal, and the process spends most of the time with optimum genotype frequency close to zero. Figures 2, 3, and 4 illustrate the effects of increasing recombination rate on the dynamics of the process in the suberror threshold regime.

In each case the left-hand figure shows the stationary probability distribution while the right-hand figure plots the results of a simulation of the process with the same parameter values. In all figures the (arbitrary) timescale $\theta = 100$, $\sigma = 2$, $\nu = 10$, and $N = 100$. A subtlety in comparing the dynamics for different values of ρ is that changing the recombination rate alters the shape of the stationary distribution. Indeed, in Boerlijst et al. [1] and Ochoa and Harvey [10], it is demonstrated that increasing the recombination rate lowers the error threshold.

[4]Note that this is not the only possible definition of the error threshold for finite populations. See e.g., Forst et al. [5].

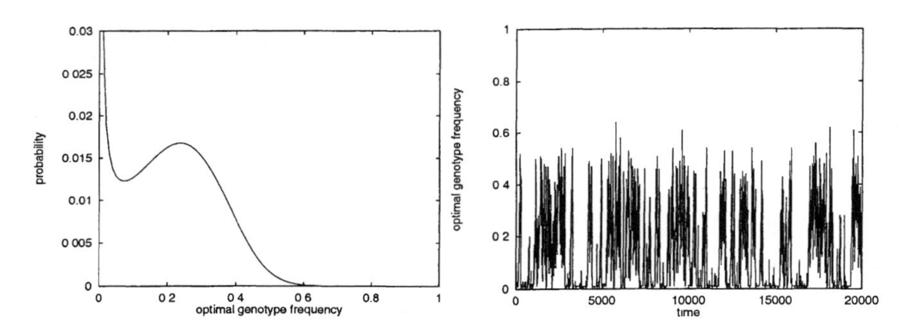

FIGURE 2 Parameters are $\mu = 0.0527$, $\rho = 0$.

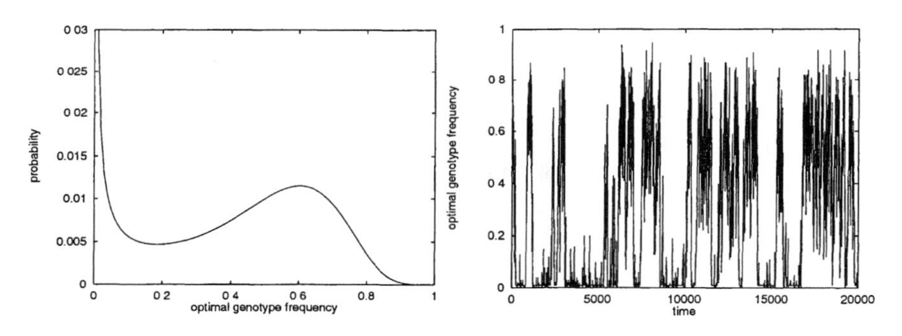

FIGURE 3 Parameters are $\mu = 0.01612$, $\rho = 0.45$.

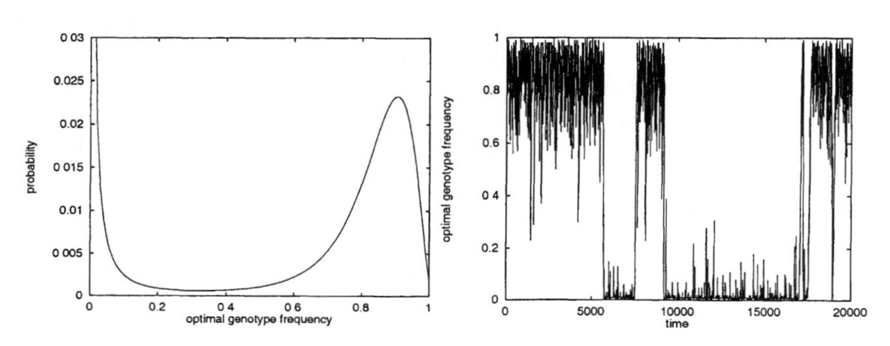

FIGURE 4 Parameters are $\mu = 0.005174$, $\rho = 0.6$.

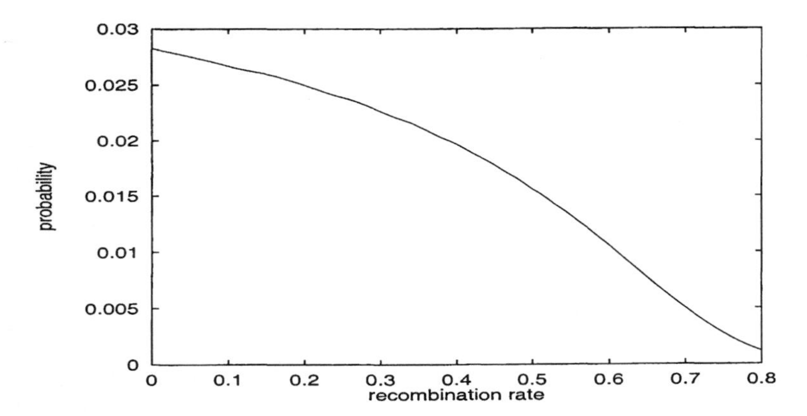

FIGURE 5 Between-mode minimum probability.

Thus, to establish a baseline for comparison, we followed the following procedure: for each value of ρ the mutation rate μ was adjusted so that the *median* of the stationary distribution coincides with the optimum genotype frequency at which the stationary distribution takes on its minimum value between the modes. Thus, the process spends equal amounts of time in states above and below the optimum genotype frequency dividing the modes.

Figures 5 and 6 plot the between-mode minimum probability and right-hand mode optimum genotype frequency respectively against recombination rate. Figure 7 plots the *mfpt* of the process from the right-hand mode down to the zero state. Again, in all figures the mutation rate is adjusted so that the stationary median coincides with the between-modes minimum. In all plots $\theta = 1$, $\sigma = 4$, $\nu = 10$, and $N = 100$.[5]

The effects of increasing the recombination rate now become clear:

1. The time spent by the process in states *between* the modes decreases.
2. The optimum genotype frequency of the right-hand mode increases; the modes are "pulled apart."
3. The expected waiting time for transitions between the modes "blows up" rapidly.

Consider now, for example, the right-hand plot in figure 4 and suppose we were to watch the process over a period of time very small compared with the mean between-mode transition time. If we happened to observe the state at a given time to be near one mode, it is unlikely that we should ever see it make

[5]The "wobbliness" of these plots is due to the fact that there is, for a given recombination rate, a (small) range of mutation rates for which the median equals the between-modes minimum. There was thus some leeway in the precise choice of mutation rate.

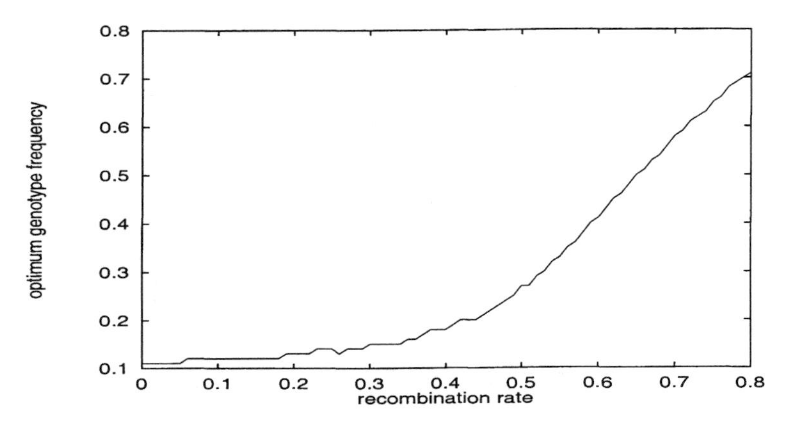

FIGURE 6 Right-hand mode optimum genotype frequency.

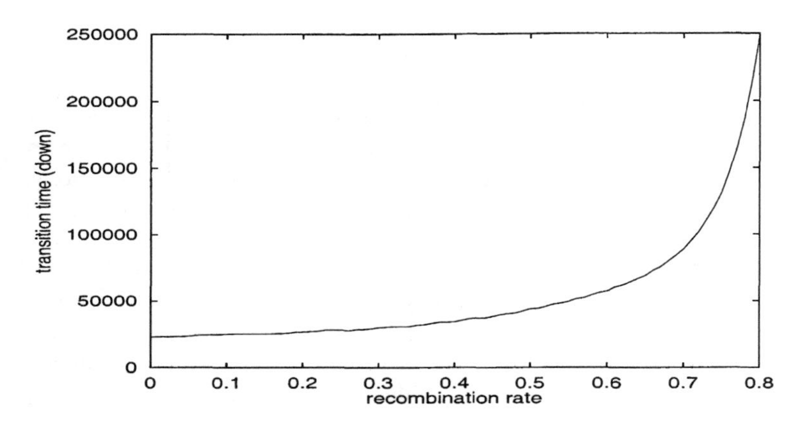

FIGURE 7 Mean first passage time from right-hand mode down to zero state.

a transition to the other mode; as far as we could tell the process would be settled in a unimodal steady state. If, however, as is equally likely (recall that by construction the process spends more or less equal amounts of time near each mode), we happened to observe the process near the *other* mode, we would consider the process to be settled in a *different* steady state. In short, on a time scale small compared to the mean between-modes transition time, the process appears bistable. We term this phenomenon *stochastic bistability*.

We see from points 1 and 2 above that at low recombination rates it is more difficult to "separate" the modes. It is thus more difficult to discern a bistable situation (cf. the right-hand plots in figs. 2 and 3). From point 3 we see that for any given observational time scale we are less and less likely to see a transition between modes as the recombination rate is increased.

6 CONCLUSIONS

We have demonstrated that stochastic bistability arises in a finite population as the result of two factors: bimodality of the steady-state distribution and between-mode transition times that are long compared to the observer's time scale. We have seen that increasing the recombination rate accentuates both of these factors (in the sense of points 1–3 of the previous section). It is of interest to note that, strictly speaking, this form of bistability is present in our model even with *no* recombination present, albeit not readily discernible even at very short time scales due to the poor separation of the modes. We note that for any (Markovian) stochastic evolutionary process that is irreducible (and this would seem to include most finite-population models in population biology) there is a unique stationary distribution. For such processes, therefore, it seems likely that bi- (or multi-) stability must always arise in a similar fashion to our model. Of course, there are many stochastic evolutionary scenarios that cannot be modeled by an irreducible Markov process or, indeed, by a Markov process. Nonetheless, the phenomenon would appear to be very general.

It would be of great interest to connect the stochastic bistability observed in our model with the bistability observed in infinite-population deterministic models. We speculate that there is a limiting procedure whereby the dynamical equations describing the time evolution of our birth and death model (the forward or backward Chapman-Kolmogorov equations [6]) converge to quasispecies-like differential equations, which bifurcate as in the simplified model presented in Boerlijst et al. [1].

We also note that, in principle at least, our model allows us to calculate approximations for the error threshold in a finite population where recombination is present, along the lines of Nowak and Schuster [9]. In that paper the optima of the stationary distribution are approximated by treating the frequency of optimal genotypes i/N as a continuous variable x. The positions of the optima are then revealed as the solutions of a quadratic equation for x, the discriminant of which (a quadratic in the quantity Q) yields the mutation rate at which the optima coalesce and the distribution becomes unimodal, i.e., the error threshold. In our case recombination introduces a cubic term to the equations, making (analytical) solution more difficult. We hope to carry out a mathematical analysis in a future paper.

ACKNOWLEDGMENTS

The author wishes to thank Inman Harvey, Gabriela Ochoa, and the Sussex University Neutral Networks group for helpful discussions.

REFERENCES

[1] Boerlijst, M. C., S. Bonhoeffer, and M. A. Nowak. "Viral Quasi-species and Recombination." *Proc. Roy. Soc. Lond. B.* **263** (1996): 1577–1584.

[2] Crow, J. F., and M. Kimura. *An Introduction to Population Genetics Theory.* New York: Harper and Row, 1970.

[3] Derrida, B., and L. Peliti. "Evolution in a Flat Fitness Landscape." *Bull. Math. Biol.* **53(3)** (1991): 355–382.

[4] Eigen, M., J. McCaskill, and P. Schuster. "The Molecular Quasispecies." *Adv. Chem. Phys.* **75** (1989): 149–263.

[5] Forst, C. V., C. Reidys, and J. Weber. "Evolutionary Dynamics and Optimization: Neutral Networks as Model-Landscapes for RNA Secondary-Structure Folding-Landscapes." In *Proc. ECAL '95*, edited by F. Moran, A. Moreno, J. J. Merelo, and P. Chacon, vol. 929. Lecture Notes in Artificial Intelligence, Advances in Artificial Life. Berlin: Springer-Verlag, 1995.

[6] Karlin, S., and H. M. Taylor. *A First Course in Stochastic Processes*, 2d ed. New York: Academic Press, 1975.

[7] Kimura, M. *The Neutral Theory of Molecular Evolution.* Cambridge: Cambridge University Press, 1983.

[8] Moran, P. A. P. "The Effect of Selection in a Haploid Genetic Population." *Proc. Cambridge Phil. Soc.* **54** (1958): 463–467.

[9] Nowak, M., and P. Schuster. "Error Thresholds of Replication in Finite Populations: Mutation Frequencies and the Onset of Müller's Ratchet." *J. Theor. Biol.* **137** (1989): 375–395.

[10] Ochoa, G., and I. Harvey. "Recombination and Error Thresholds in Finite Populations." In *Proceedings of the Foundations of Genetic Algorithms (FOGA) 5*, edited by W. Banzhaf and C. Reeves. San Mateo: Morgan Kauffman, 1999. ⟨ftp://ftp.cogs.susx.ac.uk/pub/users/inmanh/fogat.ps.gz⟩ (Sept. 1998).

[11] Stirzaker, D. *Elementary Probability.* Cambridge: Cambridge University Press, 1994.

Evolution of Cooperation

On the Dynamic Persistence of Cooperation: How Lower Individual Fitness Induces Higher Survivability

Guy Sella
Michael Lachmann

This chapter is reprinted, with minor changes, by permission of *J. Theor. Biol.* **206** (2000): 465–485.

We study a model in which cooperation and defection coexist in a dynamical steady state. In our model, subpopulations of cooperators and defectors inhabit sites on a lattice. The interactions among the individuals at a site, in the form of a Prisoners' Dilemma (PD) game, determine their fitnesses. The chosen PD payoff allows cooperators, but not defectors, to maintain a homogeneous population. Individuals mutate between types and migrate to neighboring sites with low probabilities. We consider both density-dependent and density-independent versions of the model. The persistence of cooperation in this model can be explained in terms of the life cycle of a population at a site. This life cycle starts when one cooperator establishes a population. Then defectors invade and eventually take over, resulting finally in the death of the population. During this life cycle, single cooperators migrate to empty neighboring sites to found new cooperator populations. The system can reach a steady state where cooperation prevails if the global "birth" rate of a population is equal to its global "death" rate. The dynamic persistence of cooperation ranges over a large section of the model's parameter space. We compare these dynamics to those from other models for the persistence of altruism and to predator-prey models.

1 INTRODUCTION

Explaining the evolution and persistence of cooperation is a central problem in evolutionary biology and the social sciences [1]. A cooperating individual has higher fitness within a group of cooperators than it has in isolation. Nevertheless, cooperative behavior often entails a fitness cost. Consequently, a defecting individual—one that enjoys the cooperation of others but abstains from cooperative behavior—will have an immediate selective advantage over cooperators. This advantage renders a population of cooperators susceptible to invasion and takeover by defecting individuals. Therefore, the persistence and evolution of cooperation seems to face an intrinsic instability.

The interaction between cooperators and defectors is often formalized in terms of the game known as the Prisoners' Dilemma (PD) [24]. The PD payoff matrix appears in table 1. This is a symmetric game between two players, where each player has two possible strategies: defect or cooperate. The game is set up such that, for any strategy of the opponent, a defector has a greater payoff than a cooperator but, if both players cooperate, they have a greater payoff than if both defect. In the framework of evolutionary game theory [24], this game can be used to study the evolution and persistence of cooperation. In the following we assume that individuals in a population of cooperators and defectors interact randomly in pairs, and that the fitness of an individual is determined by the payoffs it receives in its interactions. The population dynamics under these assumptions have two principal characteristics:

1. A population consisting only of cooperators grows faster than a population consisting only of defectors. This follows from the relation $\alpha > \beta$ in the payoff matrix.
2. In any population with both types, a defector has a higher fitness than a cooperator. This follows from the relations $\delta > \alpha$ and $\beta > \gamma$.

How can cooperation evolve and persist if defection is always the locally favored strategy? We mention three main categories of answers to this question, though the distinction between them is not always sharp. The first category is the individual-centered approach, where cooperation persists because it eventually confers a fitness advantage at the level of the individual. Models incorporating reciprocity [1], partnership [4], or the handicap principle [23] fall into this class.

TABLE 1 The Prisoners' Dilemma payoff matrix: Each box describes the payoff for a possible two player interaction. The left entry refers to the player employing the strategy above, while the second refers to the player employing the strategy listed on the side. The payoffs are set such that: $\delta > \alpha > \beta > \gamma$. Under this condition, the best strategy is to defect independent of the other player's strategy. If both players defect, however, they both receive lower payoff than if they both cooperate.

	c	**d**
c	(α, α)	(δ, γ)
d	(γ, δ)	(β, β)

The second category is kin selection [13]. This includes models of kin recognition [1], models where kin interaction results from individual behavior in a spatial context, and more generally models of statistical kinship [6]. The third category consists of structured population dynamic models. It includes among others, the hay stack model [18, 25], models of the founder effect [3], and models of the neighbor effect [7]. Nowak and May [21] introduced a family of models that combine aspects of all three categories mentioned above. This family has been studied intensively during the last decade. In these models [10, 19, 20, 22] individuals occupy lattice sites and play the iterated or noniterated PD game with their neighbors on the lattice. The model we present here falls into the third category. The dynamics that maintain cooperation in our model, which we describe below, distinguish it from earlier models in this category.

In a review of group selection [17], Maynard Smith mentions a predator-prey model [15] and claims that it is analogous to a model for the persistence of altruism (or cooperation in our terminology). In that model, isolated patches may be in one of three states: E—empty, containing neither prey nor predator, H—containing prey only, and M—containing both prey and predator. An empty patch may be colonized by prey that migrate from a different patch, thus changing its state from E to H. A patch in state H may be colonized by migrating predators, changing its state from H to M. In a patch in state M, the predators eventually exhaust the prey and die, thus changing the state of the system from M to E. Having studied such systems with computer simulations, Maynard Smith concludes, "such models can rather easily give persistent coexistence of predator and prey; that is, persistence does not require a particularly careful choice of parameters." In such a state of persistence, each patch goes through a series of transitions $E \to H \to M \to E \to \ldots$, indefinitely. In this chapter we show that the underlying dynamics of interaction between cooperators and defectors can lead to the persistence of cooperation in a dynamic mode similar to that described by Maynard Smith.

In the model presented in section 2, subpopulations of cooperators and defectors cohabit sites on a lattice. The random interaction among the individuals at a site determine their fitness, based on PD payoffs. The standard PD condition that $2\alpha > \gamma + \delta$, meaning that a group of cooperators has a higher average fitness than any other group, does not affect the dynamics, and therefore we do not impose it. Individuals mutate between types and migrate to a neighboring site with low probabilities. The fitness function at a site may also depend on the population density at the site. It is important to stress that in this model the population size at a site is finite and varies over time. The population at a given site may die out, thereby leaving the site empty until it is settled again by a migrating individual.

We seek the conditions for the persistence of cooperation when subpopulations consisting solely of defectors are doomed to extinction, while subpopulations consisting solely of cooperators are capable of persisting. In terms of the PD payoffs, this qualitative asymmetry occurs when $\alpha > 1 > \beta$. We show that under this assumption, the population dynamics at a site takes the form of a life cycle. The life cycle begins when a cooperator migrates into an empty site and founds a cooperator population. This population is later invaded by a defector, which is either a mutant or a migrant. Defectors then take over, after which the population dies, and the life cycle ends. In section 3, we study the life cycles that emerge in both the density-dependent and density-independent models. The life cycles are analyzed in terms of two variables that are later used to characterize the conditions for the dynamic persistence of cooperation in the system: R—the ratio of the total number of cooperators to the total number of defectors over the duration of a life cycle; and M—the total number of individuals that migrate out of a site over the duration of a life cycle.

Over a large range of the parameter values, cooperation persists in a steady state where, on average, one cooperator migrates to an empty neighboring site during a single life cycle, thus initiating a new cycle before the original cycle ends. In such a steady state, the global "birth" rate of cooperating populations balances their "death" rate due to defector takeover. In section 4, we use computer simulations to study the regions of dynamical persistence in terms of the life-cycle variables R and M. We also construct a simplified model for the density-dependent model, which is similar to the predator-prey model described above. Using this simplified model, we construct a mean-field approximation in which we analytically derive the conditions for dynamical persistence and a higher-order approximation incorporating spatial correlations. These approximations explain the shape of the boundary that separates the regions (in the R and M space) in which cooperation persists from the regions in which it does not. For the density-independent model we show that increasing the payoff α for the interaction between cooperators can increase the number of cooperators in the system, the number of sites they occupy, and their number relative to defectors. Moreover, increasing this payoff beyond a critical value results in the extinction of cooperation. These results are explained both intuitively and on the basis of analytical derivations.

2 THE MODEL

We introduce an evolutionary model of finite subpopulations inhabiting sites on an infinite two-dimensional rectangular lattice. Each individual in a subpopulation is a cooperator or a defector, where these behaviors are genetically determined (variables referring to these will be marked with subscript c for cooperators and d for defectors). The dynamics proceed from selection with absolute fitnesses, mutation, and diffusion.

The fitness of an individual stems from its interactions with other individuals at the same site on the lattice, according to a PD payoff matrix, as in table 1. We assume a PD-type interaction, with the asymmetry described in the introduction. The payoffs must therefore satisfy the following conditions:

$$\delta_{(dc)} > \alpha_{(cc)} > 1 > \beta_{(dd)} > \gamma_{(cd)} \geq 0 \,. \tag{1}$$

(Henceforth we omit the subscripts on these parameters). Consider a site with n_c cooperators and n_d defectors interacting at random. The absolute fitness functions of cooperators and defectors at time t, which measure the average growth in a time step Δ, are[1]:

$$f_c(t) = g(n(t))\left(\alpha\frac{n_c(t)}{n(t)} + \gamma\frac{n_d(t)}{n(t)}\right), \quad \text{and} \tag{2}$$

$$f_d(t) = g(n(t))\left(\delta\frac{n_c(t)}{n(t)} + \beta\frac{n_d(t)}{n(t)}\right), \tag{3}$$

where $n = n_c + n_d$ is the total population at the site, and $g(n)$ is a function reflecting the density dependence. We will assume $g(n) \leq 1$, but that $\alpha g(n_c) > 1$ for a population size smaller than the carrying capacity $n^* > 0$ (the density-independent model corresponds to $g(n) = 1$).

These fitness functions are characterized by two principal features:

1. A homogeneous cooperator population grows faster than a homogeneous defector population. Moreover, a homogeneous cooperator population is capable of maintaining itself while a homogeneous defector population is not. These properties follow from the relations $f_d(0, n_d) = \beta g(n_d) < 1$ for any $n_d \neq 0$, and $f_c(n_c, 0) = \alpha g(n_c) > 1$ for n_c below the carrying capacity n^*.
2. A defector has a higher fitness than a cooperator in any population structure, since $\delta > \alpha$ and $\beta > \gamma$ imply $f_d(n_c, n_d) > f_c(n_c, n_d)$ for all n_c and n_d.

The population growth at a site is described by a stochastic process, resulting in the absolute fitness functions of equation (3). The following analytic

[1]Note that the fitness functions include the interaction of an individual with itself. Not including an individual in the calculation of its fitness requires a separate definition of the fitness of an individual in isolation, but does not change qualitatively any of the results in this chapter.

results depend only on the average growth rates and not on the specific stochastic process that realizes them. In our simulation we assume that an individual with fitness $f = n + x$, where n is a nonnegative integer and $0 < x < 1$, is represented in the next time step by $n + 1$ individuals with probability x, and by n individuals with probability $1 - x$.

Mutation and migration are incorporated in each time step as follows:

1. The subpopulations at all sites grow stochastically according to the fitness functions.
2. Each individual may mutate to become the other type with probability $\mu \ll 1$.
3. Each individual may migrate to one of its neighboring sites with probability $D \ll 1$. We assume the von Neumann neighborhood of four neighbors on a rectangular lattice.

The expected values for $n_c(t + \Delta)$ and $n_d(t + \Delta)$ at a site given $n_c(t)$ and $n_d(t)$ are then:

$$\begin{aligned} E\Big(n_c(t+\Delta) \mid n_c(t), n_d(t)\Big) &= (1-D)\left[(1-\mu) f_c(t) n_c(t) + \mu f_d(t) n_d(t)\right] \\ &+ D\left[\frac{1}{4}\sum_{n.n.} f_c^{n.n.}(t) n_c^{n.n.}(t) - f_c(t) n_c(t)\right] + O\left[(D+\mu)^2\right], \qquad (4) \\ E\left(n_d(t+\Delta) \mid n_c(t), n_d(t)\right) &= (1-D)\left[(1-\mu) f_d(t) n_d(t) + \mu f_c(t) n_c(t)\right] \\ &+ D\left[\frac{1}{4}\sum_{n.n.} f_d n.n.(t) n_d^{n.n.}(t) - f_d(t) n_d(t)\right] + O\left[(D+\mu)^2\right], \end{aligned}$$

where superscript $n.n.$ denotes values at nearest neighboring sites. We take $\Delta \ll 1$ where 1 denotes the duration of an average generation. This implies that the fitness coefficients α, β, γ, δ, and the corresponding fitness functions are close to 1. Furthermore, it means the stochastic process approaches a process continuous in time. When $\Delta \ll 1$, $\mu \ll 1$, and $D \ll 1$, however, our scheme is equivalent to any other reasonable scheme that incorporates selection, mutation, and diffusion, and to other reasonable schemes for the stochastic selection.

3 LOCAL BEHAVIOR—THE LIFE CYCLE

We begin the analysis by considering the local behavior at one site. This behavior can be described in terms of a typical *life cycle*. A life cycle for the density-independent model (d.i.) is described in figure 1. It begins when one cooperator migrates to an empty site. The population of cooperators then begins growing. The growth rate depends on $f_c(n_c, 0) = \alpha g(n_c)$. At some time, which we denote by t_f, the first defector appears from mutation or migration, and defectors begin to take over. At some stage when defectors dominate the population, the fitness

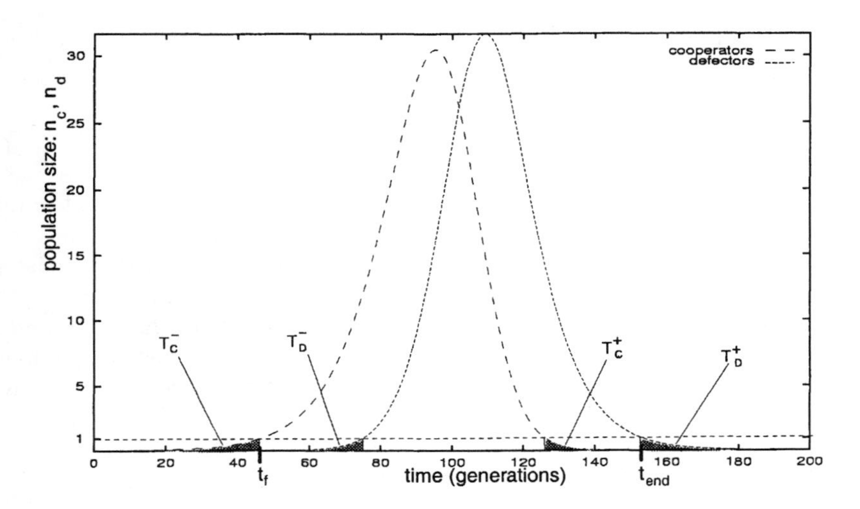

FIGURE 1 A schematic representation of a life cycle: population size vs. time. Changing mutation rate scales the whole cycle, leaving the relations between the the number of cooperators and defectors the same.

of cooperators drops below 1, and their number starts decreasing. Some time afterwards, as the frequency of cooperators decreases, the fitness of defectors approaches β and in the process becomes less than 1. The life cycle ends at time t_e when the last defector dies, some time after the last cooperator disappeared.

Every life cycle ends with the death of the whole population at a site after a finite time. Therefore, for cooperation to persist, new life cycles have to be founded at a rate that balances their termination. The conditions for the existence of such a steady state are studied in section 4. These conditions will be stated in terms of parameters characterizing a life cycle. A natural choice of parameters relevant for the global dynamics is the number of cooperators and defectors that migrate out of a site during a life cycle; we denote them M_c and M_d. In order to derive these parameters from a description of the life cycle, we define:

$$S_c \equiv \int_0^{t_e} n_c(t)dt\,, \tag{5}$$

$$S_d \equiv \int_0^{t_e} n_d(t)dt\,. \tag{6}$$

M_c and M_d are then given by:

$$M_c = DS_c\,, \quad \text{and} \tag{7}$$

$$M_d = DS_d\,. \tag{8}$$

Two equivalent parameters that turn out to be useful are:

$$M \equiv M_c + M_d = D(S_c + S_d)\,, \quad \text{and} \tag{9}$$

$$R \equiv \frac{M_c}{M_d} = \frac{DS_c}{DS_d} = \frac{S_c}{S_d}\,, \tag{10}$$

where M corresponds to the total number of units migrating from a site during a life cycle and R corresponds to the ratio of cooperators to defectors in the life cycle. The dependence of the global dynamics and in particular of the persistence of cooperation on R and M, will be studied in section 4. In this section we study the population dynamics at a site, focusing on the factors determining R and M. We find that in the d.i. model, R depends on α, β, γ, and δ, whereas S_d depends mainly on μ and D; while in the density-dependent (d.d.) models, R depends on μ and D, and S_d is essentially constant.

3.1 THE DENSITY-INDEPENDENT MODEL

In the d.i. models the dynamics (eq. (5)) are homogeneous to the first order in n_c and n_d. Thus scaling n_c and n_d by a factor Λ at some time will just scale S_c and S_d by the same factor, leaving R unchanged. It is not hard to show, that taking two life cycles that vary only in the time in which the first defector invades, i.e., taking $\tau_f > t_f$, is equivalent to scaling S_c and S_d by the factor[2]:

$$\Lambda = \alpha^{\left[1+\frac{\log \alpha}{\log \frac{\alpha}{\delta}}\right](\tau_f - t_f)}\,. \tag{11}$$

The only significant effect (i.e., not $O[\mu, D]$) that mutation and migration has on the life cycle is in determining t_f. Therefore, we conclude that $R = R(\alpha, \beta, \gamma, \delta) + O[\mu, D]$. Result 1 provides an explicit expression for $R(\alpha, \beta, \gamma, \delta)$:

Result 1. *In the density-independent model the ratio R of the average number of migrating cooperators to the average number of migrating defectors in a life cycle is:*

$$R(\alpha, \beta, \gamma, \delta, \mu, D) = \frac{E(M_c)}{E(M_d)} = \frac{DE(S_c)}{DE(S_d)} = \frac{1-\beta}{\alpha-1}\frac{\alpha-\gamma}{\delta-\beta} + O[\mu, D]\,. \tag{12}$$

We prove this in appendix 6.

A careful look at equation (12) reveals that reducing α (but maintaining the condition $\alpha > 1$) while leaving every other parameter fixed can yield a larger ratio of cooperators to defectors, since

$$\frac{\partial R}{\partial \alpha}|_{\beta,\gamma,\delta} < 0 \quad \text{for} \quad \delta > \alpha > 1 > \beta > \gamma\,. \tag{13}$$

[2] In this argument we are ignoring the tails T_c and T_d shown in figure 1. These tails represent the parts of an extrapolated life cycle in which n_c and n_d drop below one. Changing t_f moves parts of these tails into S_c and S_d. However, this effect can be shown to change R only by $O[\mu, D]$.

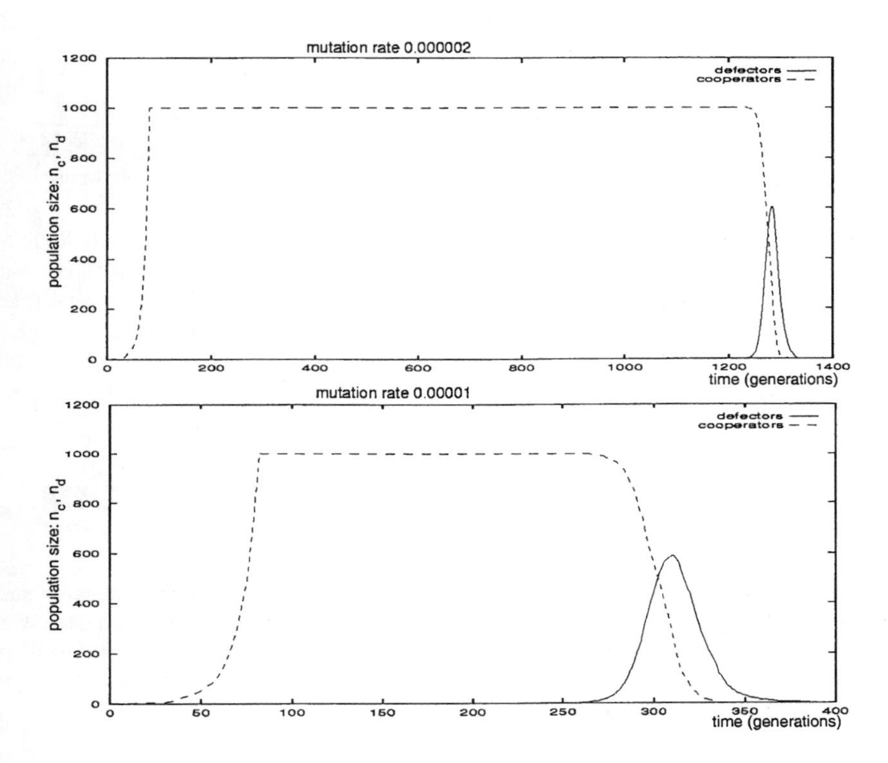

FIGURE 2 The density-dependent life cycle at a site for the model in the example. The graph shows population size vs. time for two different mutation rates $\mu = 0.00001$ and $\mu = 0.000002$.

It seems reasonable, and it will be shown later, that the larger R is, the more likely it is that cooperation persists in the system. This hints at the possibility that, in certain parameter regions of the d.i. model, decreasing α while leaving all the other parameters fixed will transform the global behavior from a state where cooperation cannot persist to a state where it can. Such behavior is seen in simulation results in figures 3, 4, 5, and 9, in section 4. Thus, in this model lower cooperator fitness may induce higher survivability!

This effect would not have been anticipated from an individual-centered perspective. However, when the life cycle is considered, there is a simple explanation for this effect: The integral number of defectors during the life cycle S_D strongly depends on the number of cooperators at the time the first defector invades $n_c(t_f)$. Therefore, cooperators can increase their fraction by maximizing their integral S_c, thus keeping $n_c(t_f)$ fixed. This explains why R increases when α

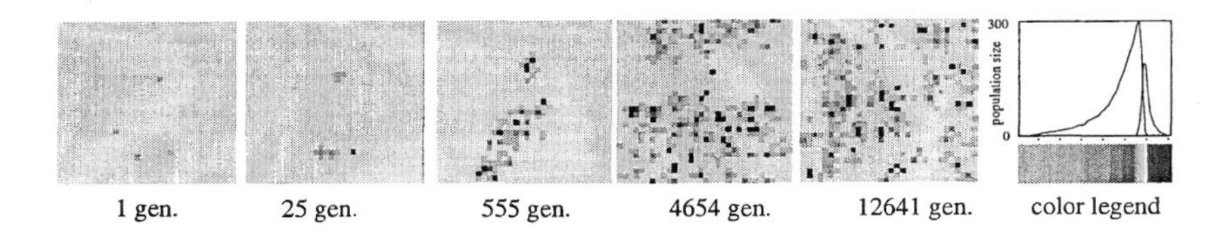

FIGURE 3 In this simulation $\alpha = 1.06$, all the other parameters are identical to those in the simulation in figure 5. In this case, the system reaches the trivial—all empty steady state. This is a case where taking higher cooperator fitness results in the extinction of cooperation. This effect is discussed further in section 4.3.

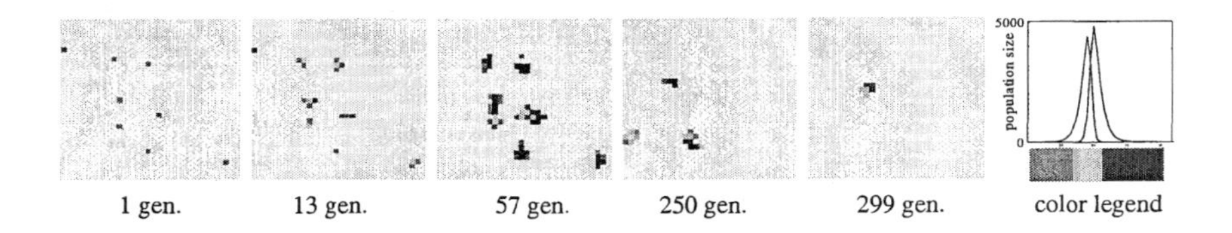

FIGURE 4 The population at a specific site as a function of time, for the same simulation presented in figure 5. The life cycles vary in size due to the stochasticity in the t_f. This stochasticity is caused by variations in the environment and by mutation. However, the shape of the different life cycles is similar, in correspondence with the scaling properties discussed in section 3.1.

is smaller. Roughly speaking, the moral is that when surrounded by defectors, keeping a low profile might be a good idea.

3.2 THE DENSITY-DEPENDENT MODEL

We consider an example of a density-dependent (d.d.) model such that:

$$g(n) = \begin{cases} 1 & n < \frac{n_{\max}}{\delta} \\ \frac{n_{\max}}{\delta n} & n \geq \frac{n_{\max}}{\delta} \end{cases} .$$

In this model the population size is bounded by $n_{\max}$.

The life cycle for this model is described in figure 2. Unlike the d.i. life cycle, in this case, the number of cooperators stabilizes after a finite time $t(\alpha)$ on $c(\alpha)n_{\max}$, where $c(\alpha) \equiv \alpha/\delta$. The shape of this life cycle implies that as long as $t_f > t(\alpha)$, changing parameters so that t_f increases will increase S_c, leaving S_d constant. This means that one could make R bigger than any $R^* > 0$ by picking a large enough t_f. Taking a large t_f simply means taking a small enough μ and D. Hence, for this example we conclude that R can assume any large value and S_d remains essentially constant if μ and D are taken to be small enough.

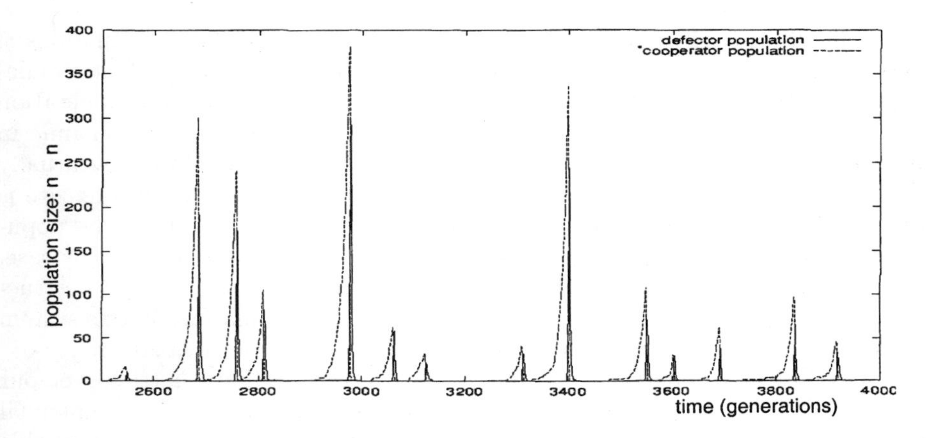

FIGURE 5 Simulation of the density-independent model on a 32×32 lattice with periodic boundary conditions. The parameters for this simulation were: $\alpha = 1.01$, $\beta = 0.95$, $\gamma = 0.8$, $\delta = 1.2$, $\mu = 0.0001$, and $D = 0.0003$. The state of the sites of the lattice are given according to the key on the right, which corresponds to the different stages in the life cycle for these parameter values. The simulation begins with single cooperators inhabiting a few sites. The system reaches a nontrivial steady state, where cooperation persists dynamically.

This behavior is generic; it characterizes a class of d.d. models we define in appendix 7 as models of type 1. In appendix 7 we apply the same reasoning used in the example, to prove the following result:

Result 2. *Given a density-dependent model of type 1 and $R^* \geq 0$, taking μ and D such that*

$$D + \mu \leq \frac{c(\alpha)}{c(\alpha)n_{max}t(\alpha) + \widetilde{S_d}R^*}$$

ensures that R is bounded from below by R^*, where $\widetilde{S_d}$ is $E(S_d)$ for the density-independent model with the same parameters α, β, γ, δ, and initial conditions $\widetilde{n_c}(0) = n_{\max}$, $\widetilde{n_d}(0) = 1$.

This means that for any such model and parameters α, β, γ, and δ, any desired ratio R may be attained by taking small enough μ and D.

4 GLOBAL BEHAVIOR

In the systems we studied there are two types of steady states for the global behavior: the trivial steady state where all sites are empty, and nontrivial steady states in which cooperation persists globally. The behavior of two simulations of systems with d.i. dynamics is presented in figures 5 and 3. In the first simulation we start with a few sites inhabited by one cooperator each, and cooperation spreads to establish a nontrivial steady state. This steady state is dynamic in nature; cooperation persists even though each cooperator population eventually dies out. This can be seen in figure 4, where the population size at one site is described as a function of time. In the system described in figure 3, the population also starts with a few sites inhabited by one cooperator but, in this case, populations do not seed new ones at a rate that balances their rate of destruction by defectors from within and without. All the subpopulations in this system eventually die out, leaving it in a steady state where all the sites are empty.

A well-defined nontrivial steady state requires an infinite lattice. Yet our simulations occur on a finite lattice. We argue that, when it exists, the nontrivial steady state is the only steady state and, thus, it will attain in any reasonable choice of initial conditions. We do not prove this claim for our system. We show, however, that the fixed-point analog to the nontrivial steady state in the mean-field approximation to the d.d. dynamics is the only stable fixed point in the system when it exists. Due to the stochasticity, any finite realization of the model on a finite lattice will always end up in the state where all the sites are empty. Nevertheless, the nontrivial steady state has a pronounced signature in the finite realizations of the model. In the parameter ranges corresponding to the nontrivial steady state, the duration in which cooperation persists in the finite system grows very fast with the size of the lattice. In appendix 9 we describe the criteria we use to determine when the finite simulations reach a state corresponding to the steady state. In the parameter ranges where this criteria holds, the system never reaches the empty state, in thousands of simulations lasting hundreds of thousands of generations each. Consequently, we can study the regions in which cooperation persists using simulations on finite systems and assume that large systems have dynamical behavior that is independent of initial conditions.

In these models, the global dynamics derive from an interplay between the local dynamics at a site and the interaction of the subpopulations in this site with its environment. The local dynamics at a site were described in the previous section. They are affected by the environment through the inflow of cooperators and defectors. When a site is empty, this inflow determines when it will become inhabited by a cooperator; and when a site is inhabited by cooperators with no defectors, this inflow will affect how long it will take it to be invaded by a defector—t_f. The environment, however, is generated by the local dynamics at sites. Essentially, the more intricate are the population dynamics at a site, the more complex is the analysis of the global dynamics. In the d.d. models, the rough temporal structure of the life cycle is rather simple, and it is possible to approximate its global behavior by dividing the life cycle into three main stages: *empty*, *cooperation*, and *defection*, where in each the population could be considered as being in one state. In each of these states, the internal dynamics at

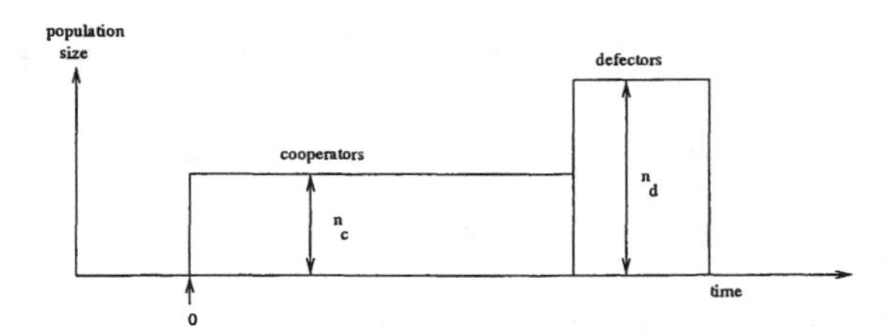

FIGURE 6 A life cycle in the simplified model: population as a function time. Notice the similarity to the density-dependent life cycle (fig. 2).

each site and its interactions with its environment can be described as a Markov process switching between states. Note that this simplified model is very similar to Maynard Smith's [15] predator-prey model described in the introduction. For a simplified model of this nature, we can obtain an analytical approximation of the conditions for the existence of a nontrivial steady state. We briefly outline a simplified model and its analysis in sections 4.1 and 4.2 and in appendix 8. The life cycle for the d.i. model described in figures 1 and 4 consists of many different states, since each population structure—n_c, n_d—affects the evolution at the site and the site's effect on its environment differently. Each of these states is characterized by a different outflow of cooperators and defectors. This makes the global analysis of these models much more complicated. For this reason, we restrict the study of their global behavior to simulations. The results from the analysis and simulations of the simplified d.d. model and from the simulations of the d.i. model are presented in section 4.3.

4.1 A SIMPLIFIED DENSITY-DEPENDENT MODEL

The life cycle at a site for the simplified d.d. model is described in figure 6. In this model, for which the life cycle is a simplification of the d.d. life cycle shown in figure 2, when a cooperator enters an empty site, it immediately establishes a population of n_c cooperators. After some time, the population is invaded by a defector that is either a mutant from within or a migrant from without. Once a defector invades, it instantaneously takes over and establishes a constant population of n_d defectors. This population has a probability P_d per unit time to die and leave the site empty. Diffusion and mutation are stochastic as in the nonsimplified models.

The simplified model can be seen as an interacting particle system, where a site (i, j) (which corresponds to the particle) can be in one of three states: *empty* ($S_{ij} = e$), *cooperation* ($S_{ij} = c$), and *defection* ($S_{ij} = d$). The dynamics of this system can be described as a Markov process, written here in terms of the transition probabilities for a site (i,j) during a time step Δ:

$$A_{e\to c}^{ij}(t) \equiv P\left(S_{ij}(t+\Delta) = c | S_{ij}(t) = e\right) = \frac{D}{4} n_c I_c^{ij}(t)\,, \tag{14}$$

$$A_{c\to d}^{ij}(t) \equiv P\left(S_{ij}(t+\Delta) = dt | S_{ij}(t) = c\right) = \frac{D}{4} n_d I_d^{ij}(t) + \mu n_c\,, \text{ and} \tag{15}$$

$$A_{d\to e}^{ij}(t) \equiv P(S_{ij}(t+\Delta) = e | S_{ij}(t) = d) = P_d\,, \tag{16}$$

and

$$\begin{aligned} A_{e\to e}^{ij}(t) &= 1 - A_{e\to c}^{ij}(t)\,, \\ A_{c\to c}^{ij}(t) &= 1 - A_{c\to d}^{ij}(t)\,, \quad \text{and} \\ A_{d\to d}^{ij}(t) &= 1 - A_{d\to e}^{ij}(t)\,. \end{aligned} \tag{17}$$

Here, the number of (i,j)'s nearest neighbors in state c is denoted I_c^{ij}, and the number of nearest neighbors in state d was denoted I_d^{ij}. In writing these dynamics it was assumed that the time step $\Delta \ll 1$, so that effects that are second order in μ and D can be ignored. Equation (14) describes how a site changes its state from e to c, by way of a nearest-neighbor interaction corresponding to diffusion. Equation (15) describes how a site changes its state from c to d, either by nearest-neighbor interaction corresponding to diffusion, or spontaneously in a way that corresponds to mutation. Finally, equation (16) describes how a site changes its state from d to e, spontaneously, in a way that corresponds to the death of the defector population.

Equations (14) and (17) have four parameters: Dn_c, Dn_d, μn_c, and P_d. As D, μ, and P_d are all homogeneous to the first order in the time scale Δ, so are the right-hand sides in equations (14) and (17). This means one of the four parameters, such as P_d, could be taken to determine the time scale. The other three could be taken to be independent of the time scale, for example Dn_d/P_d, D/μ, and n_c/n_d (which are independent and homogeneous with degree 0 in Δ).

This model captures the qualitative features of the local behavior of the explicit d.d. models of section 2. The establishment of cooperation, the defector takeover, and the populations' extinction, which derive from the population dynamics at a site in the explicit d.d. models, are assumed in the simplified model. The interactions between a site and its environment, however, are of same form in both simplified and general d.d. models. On the one hand, the environment affects the times when the empty site becomes inhabited and when the defector takeover occurs. On the other hand, a site affects its environment by diffusing out cooperators and defectors.

4.2 A MEAN-FIELD APPROXIMATION TO THE SIMPLIFIED DENSITY-DEPENDENT MODEL

We would like to find the region in the model's parameter space in which a nontrivial steady state exists. One way to do this is to solve the model analytically.

A solution is a stationary probability distribution on the space of all possible lattice configurations $P(\{S_{ij}\}_{i,j\in Z})$ as a function of the model's parameters. Using a mean-field approximation, one can find the best solution within a restricted class of distributions. Roughly speaking, as the class of distributions becomes larger the approximations become better. In this chapter we will not evaluate the accuracy of the approximations, other than by comparing their predictions with simulations. A systematic evaluation of these approximations, as well as a more accurate analysis using Renormalization Groups, has been done for other particle systems [2, 12].

The first-order mean-field approximation is restricted to probability distributions of the form:

$$P\Big(\{S_{i,j}(t)\}_{i,j\in Z}\Big) = \prod_{i,j\in Z} P\Big(S_{i,j}(t)\Big). \tag{18}$$

This means that the probability of finding the system in a certain configuration can be decomposed into a product of the probabilities of finding each site in its state. One further assumes that the probabilities of finding a site in state c, d, or e are uniform across the lattice. Under these assumptions, the system's description reduces to the probabilities of finding any site in each one of the possible states.[3] Denoting these probabilities, which are independent of the site, by p_e, p_c, and p_d, the system's dynamics reduces to:

$$\begin{aligned} p_e(t+\Delta) &= p_e(t)\left(1 - A_{e\to c}(t)\right) + p_d(t)A_{d\to e}(t)\,, \\ p_c(t+\Delta) &= p_c(t)\left(1 - A_{c\to d}(t)\right) + p_e(t)A_{e\to c}(t)\,, \quad \text{and} \\ p_d(t+\Delta) &= p_d(t)\left(1 - A_{d\to e}(t)\right) + p_c(t)A_{c\to d}(t)\,, \end{aligned} \tag{19}$$

where $A_{e\to c}$, $A_{c\to d}$, and $A_{d\to e}$ denote the transition probabilities, which can be derived from equations (14):

$$\begin{aligned} A_{e\to c}(t) &= \frac{D}{4} n_c 4p_c(t)\,, \\ A_{c\to d}(t) &= \frac{D}{4} n_d 4p_d(t) + \mu n_c\,, \quad \text{and} \\ A_{d\to e}(t) &= P_d\,. \end{aligned} \tag{20}$$

Here we set $I_d = 4p_d(t)$ and $I_c = 4p_c(t)$.

The fixed point and stability analysis for this system is straightforward. A nontrivial fixed point (one where $p_e \neq 1$) exists if:

$$\frac{D}{\mu} > 1\,. \tag{21}$$

[3]Note that both this and the "second order" mean-field approximations can be treated as models for the persistence of cooperation in their own right.

When this condition holds, the system has two meaningful fixed points, one trivial ($p_e = 1$), the other not. In this case, only the nontrivial fixed point is stable, and thus the persistence of cooperation is obtained for any initial condition in which $p_c(0) \neq 0$. For this nontrivial fixed point, expressions for R, M, or any other dynamic parameter of the system, as functions of Dn_d/P_d, D/μ, n_c/n_d, and P_d, can be derived.

Deriving condition 21 from general considerations will help in understanding the scope of the first-order approximation. For a nontrivial steady state to be maintained, every life cycle has to establish on average exactly one new life cycle. This requirement takes the form:

$$DS_c\rho_e^c = \frac{M}{1+\frac{1}{R}}\rho_e^c = 1\,, \tag{22}$$

where ρ_e^c denotes the density of empty sites near a site in state c. This density equals the probability that a cooperator leaving a site will establish a new life cycle. Condition 21 could be derived from equation (22) by putting trivial bounds on ρ_e^c and S_c:

$$\rho_e^c \leq 1\,, \tag{23}$$

$$S_c = n_c t_f \leq n_c \frac{1}{\mu}\,. \tag{24}$$

The bound on ρ_e^c is realized only when all the neighboring sites are empty. The bound on t_f is also realized when all the neighboring sites are empty, i.e., when the first defector is always a mutant. These two bounds imply that condition 21 is equivalent to the requirement that at least one cooperator diffuses out in a life cycle at a site surrounded by empty neighbors. As the number of cooperators in a life cycle at an isolated site depends only on μ, the number of cooperators diffusing from it depends only on μ and D.

Condition 21 indicates that the first-order mean-field approximation cannot incorporate the harmful effects of migration, an important feature of the model. In the first-order mean-field approximation, the density ρ_e^c can approach 1, enabling cooperators to survive as long as, on average, one cooperator migrates during a life cycle. This means that in this approximation defector migration does not really affect whether cooperation prevails or not because the density of inhabited sites can always be so low that no defector ever invades it. In the spatial model the density ρ_e^c can never reach 1 because near a population of cooperators there is always a finite probability of having the population from which the founding cooperator migrated. The neighboring population, in this case, will be in either state c or d during some part of the life cycle of its daughter subpopulation. This discrepancy between the spatial model and the first-order mean-field approximation is demonstrated in figures 5 and 3. The second picture in figure 3 (25 generations) indicates that a life cycle at a site surrounded by empty sites produces more than 1 diffusing cooperator. Yet cooperation does not prevail due to the effects of extensive defector migration into sites inhabited by cooperators. An approximation incorporating such effects would have to describe the correlations between the states of nearest neighbors. Such an approximation is outlined in appendix 8. Results presented in the next section will hint at the possibility that as the phase transition between persistence and nonpersistence of cooperation is approached the correlation length in the system goes to infinity. This would imply that near the parameters at which the transition happens, the reliability of such mean-field approximations is questionable.

4.3 RESULTS I: REGIONS CHARACTERIZED BY THE DYNAMICAL PERSISTENCE OF COOPERATION

The d.i. model has six parameters: α, β, γ, δ, μ, and D, while the simplified d.d. model has four: D/μ, Dn_d/P_d, n_c/n_d, and P_d. A point at which a nontrivial steady state is maintained is characterized by a stationary probability distribution on all possible lattice configurations. We would like to present these 6/4-dimensional phase spaces in a comprehensible way, which permits comparison with the global behavior of models that derive from different local parameters. In doing so, we will necessarily lose some information, information that can be further explored using different representations. To the extent that the fine temporal and spatial structure of a steady state in these models can be ignored, the basic variables characterizing the global dynamics would be M—the average number of migrants during a life cycle, and R—the cooperator-to-defector ratio among these migrants.

Phase spaces in the R–M coordinates, which were derived from analytical approximations to the simplified d.d. model from simulations of the simplified d.d. model and from simulations of the d.i. model, are presented in figures 7 and 8. The solid lines in figures 7 and 8 correspond to the boundaries in R–M space below which a nontrivial steady state does not exist according to first/second-order approximations. We will refer to such a boundary as a *phase boundary*. The thick lines in figures 7 and 8 represent the phase boundaries derived from simulations. They were derived as described in appendix 9. In the simplified d.d. model, $DS_d = Dn_d/P_d$ is one of the basic parameters of the model, while DS_c derives from the dynamics (see sections 3.2 and 4.1). As $M = DS_d + DS_c$ and $R = DS_c/DS_d$, one component of R and M is a parameter; whereas, the other is an outcome of the dynamics depending on the other parameters. In the d.i. model, the situation is similar (see section 3.1); R is a function of α, β, γ, and δ, and thus can be considered to be a parameter; whereas, M derives from the dynamics, which depends on the other parameters.

The shape of the phase boundaries from the analysis and simulations can be roughly understood from the heuristic derivation in the last section, equation (22), which predicts the phase boundary takes the form:

$$M = C\left(1 + \frac{1}{R}\right), \tag{25}$$

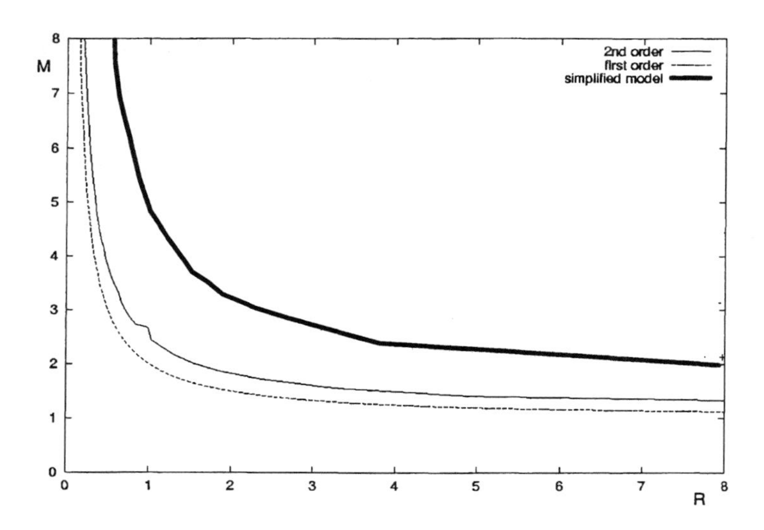

FIGURE 7 The R–M phase space for the simplified d.d. model: The graph presents regions of cooperation persistence according to the first- and second-order mean-field approximations (above the phase boundaries pictured) and according to simulations. For details on how the phase boundaries were derived from simulations, see appendix 9.

where C is some constant. The differences in the shape and position of the phase boundaries reflect the effects of the fine spatiotemporal dynamic structure. As we discussed at the end of the last section, one can state roughly that the effect of spatial correlations, i.e., spatio-temporal structure, is to increase the damage defectors inflict—thus imposing stronger restrictions on the region in the R–M space where a nontrivial steady state can be maintained. This causes the phase boundaries resulting from the simulations to be above those resulting from the second-order approximation, as well as for the second-order phase boundary to be above the first order. A systematic study of the factors effecting the phase boundaries requires the study of higher-order correlations.

4.4 RESULTS II: HOW LOWER INDIVIDUAL FITNESS INDUCES HIGHER SURVIVABILITY

The drawback of using R–M phase spaces to study a specific model is a loss of information about the relation between the system's behavior and its basic parameters. In the d.i. model the persistence of cooperation depends on the parameter α, which controls the persistence of cooperation on the fitness of cooperators. In section 3.1 we explain why a reduction in α leads to a larger cooperator-to-defector ratio R and derive the functional dependence of R on α.

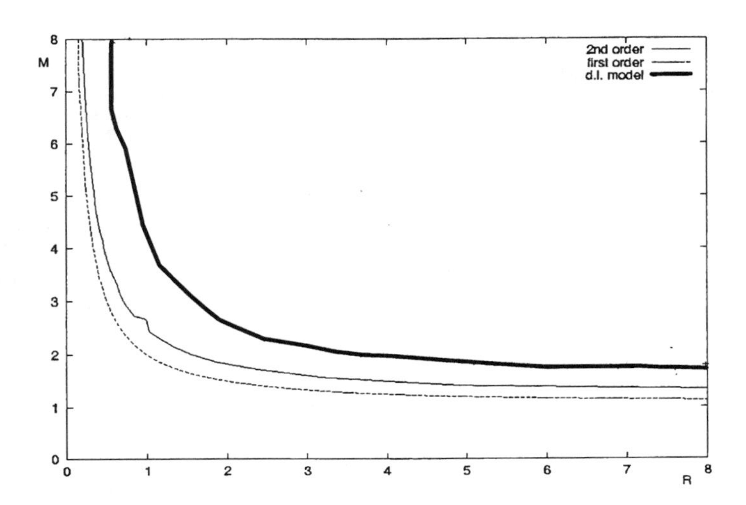

FIGURE 8 The R–M phase space for the d.i. model: The graph presents regions of cooperation persistence according to simulations of the d.i. model. The phase boundaries according to the mean-field approximations to the simplified d.d. model are also presented as reference. For details on how the phase boundaries were derived from simulations, see appendix 9.

As α increases and R decreases, we expect that the disturbance from defectors will grow to a point where cooperation cannot be maintained. This effect is illustrated in figures 5 and 3. Across two systems, we took all the parameters other than α to be equal. The system with the smaller α reached a nontrivial steady state in which cooperation persisted; whereas, the system with the larger α reached the trivial steady state without cooperation.

Figures 9 (a)–(c) illustrate the behavior of several dynamic variables as a function of α, while all the other parameters are fixed. When α increases, the number of migrating defectors also increases, while the number of migrating cooperators remains approximately constant (fig. 9(a)). Thus both R (fig. 9(b)) and the density of occupied sites (fig. 9(c)) decrease. Hence, a decrease in the individual fitness of cooperators leads to increases in the total number of cooperators in the system, the density of sites they occupy, and their numbers relative to defectors. Note that the measured R (fig. 9(b)) is very close to the analytically derived value. This supports the scaling argument described in section 3.1 and demonstrated in figure 4.

Around $\alpha = 1.0175$, the rate of destruction by migrating defectors reaches a level that precludes the maintenance of a nontrivial steady state, like the example described in figure 3. Increasing the individual fitness of cooperators therefore leads to a condition in which cooperation can no longer persist. Note that near

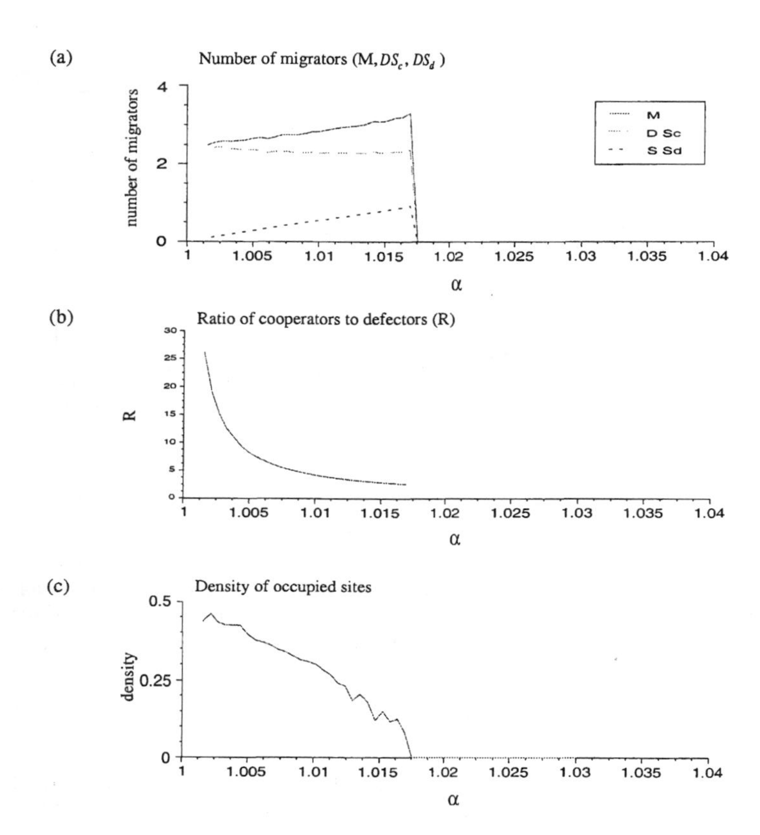

FIGURE 9 Dynamic behavior in the d.i. model as a function of α: Figures (a)–(c) describe different dynamic variables of the system as they result from simulations of the d.i. model with $\beta = 0.95$, $\gamma = 0.8$, $\delta = 1.2$, $\mu = 0.0001$, and $D = 0.0003$ where α varies between 1.001–1.035. (a) The average number of migrating cooperators (DS_c), defectors (DS_d), and their sum (M) in a life cycle are given as a function of α. (b) The average ratio of cooperators to defectors migrating during a life cycle (R) is given as a function of α. (c) The density of occupied sites is given as a function of α.

the phase boundary, the density of occupied sites drops (fig. 9(c)). This suggests that the correlation distance in the system grows at this vicinity. As mentioned in the previous section, this sheds doubts on the reliability of the mean-field approximations near the phase boundary.

5 DISCUSSION

We have demonstrated that cooperation may persist in a dynamic mode where populations of cooperators and defectors constantly appear and disappear. We explain the persistence of cooperation by considering the life cycle of a population at a site. The life cycle starts when one cooperator establishes a population, this population grows, defectors invade and take over, and ultimately the population goes extinct. During this life cycle, new populations of cooperators are founded by single cooperators that migrate to empty neighboring sites. The system reaches a steady state in which cooperation persists, if the global "birth" rate of populations is equal to their "death" rate, or equivalently, if on average every population gives rise to one other population during its life cycle. This steady state arises from a repeated turnover of populations. Cooperation persists although every single population of cooperators eventually dies out. In section 4, we demonstrate that these dynamics enable the persistence of cooperation in a large section of the model's parameter space. Furthermore, we demonstrate and explain that lowering the local fitness of cooperators in the d.i. model can enable the persistence of cooperation. Within the region of persistence, lowering the local fitness of cooperators can increase the number of cooperators, the density of sites inhabited by them, and their numbers relative to the defectors.

The dynamic mode we have described may appear in a variety of biological systems. In addition to Maynard Smith's predator-prey model reviewed in the introduction, we consider one model for the persistence of altruism and one model for the persistence of "prudent" predation in a predator-prey system. Epstein [5] observes oscillatory behavior in a spatial PD model, which is analogous to the dynamic mode that we describe. In his model, individuals that are either cooperators or defectors occupy sites on a two-dimensional lattice. Each individual plays the PD game with his neighbors, where the payoffs it accumulates determine its probability to produce an identical offspring or to die and leave its site empty. In one version, he sets the payoffs for cooperator-cooperator interactions to be positive and the payoffs for defector-defector interaction to be negative. This results in oscillation of the total number of cooperators and defectors over time, where the peaks in defectors appear to follow closely the peaks in cooperators. Epstein also notes that in this dynamical regime, decreasing the payoffs for cooperator-cooperator interactions may improve the cooperators-to-defectors ratio. This is similar to the mode that we describe, the localized subpopulation in our model being analogous to the spatially extended neighborhoods in Epstein's model.

Gilpin [11] studies a model for the persistence of altruistic behavior in the context of predator-prey systems. Through computer simulations of structured populations, in which subpopulations of predators and prey inhabit isolated patches, he studies the persistence of predator "prudence," restraint to not overexploit the food supply. In his model, "selfish" predators in a patch drive the prey to extinction, which in turn drives the predator population in the patch to extinction. Gilpin allows for migration between patches and genetic drift within them. Although he does not find parameter values where the "selfish" and "prudent" predators can coexists, we believe that introducing mutations that cause the "selfish" predators to reappear, or increasing the number and perhaps introducing spatial organization of patches, would produce the mode of dynamic coexistence we have described.

Maynard Smith [17] considers the implications of migration in a patchy environment on the persistence of altruism. He suggests a criterion based on the average number of new defector populations founded by migration from a patch with defectors before that patch goes extinct, which he denotes M.[4] In the following summary of his reasoning, note that in our system, M should be defined in terms of cooperator populations rather than defector populations.[5] He claims that if $M > 1$, then defectors would takeover the population, while if $M < 1$, then cooperators would prevail, and defectors would go extinct. If we define M for cooperators rather than defectors, then $M = 1$ corresponds to our steady state. From Maynard Smith's formulation, one may assume that the case $M = 1$ is a mathematical artifact requiring an exact parameter, and therefore it should not be considered seriously. This is not the true for our system, however, and it need not be true for other systems. In our system, the value of this variable M, which is defined for cooperators rather than defectors, is produced by the dynamics of the system. Within the parameter range where cooperation persists, changes in the underlying parameters affect other variables of the system—such as the density of sites that are empty or inhabited by cooperators or the average number of cooperators in the system (see fig. 9)—but leave $M = 1$. One mechanism underlying this stability holds when the parameters of the system are changed such that the extinction of subpopulations becomes faster and the number of migrating cooperators becomes smaller. Under these conditions, the density of empty sites may increase, thus increasing the chance of a migrating cooperator to colonize an empty site (see fig. 9). A similar process stabilizes the steady state with fixed parameters: when the density of sites occupied by cooperators drops below the steady-state level, a migrating cooperator has an increased probability of finding an empty site. A more precise structural and dynamical stability analysis would consider spatiotemporal patterns such as the correlation between newly inhabited sites with neighboring sites inhabited by defectors. We note, however, that the mode of persistence that we describe seems both structurally and dynamically stable, and that this may result from self-regulating processes of the type discussed above. Maynard Smith reaches a similar conclusion in the

[4]Note that we have used M for a different meaning.

[5]Maynard Smith considers systems where defectors can either persist in coexistence with cooperators or go extinct. In our system these options apply to cooperators rather than for defectors. Defectors and cooperators are not equivalent in our system because we have incorporated mutation and defectors may depend on repeated appearance via mutation in order to persist. We note, however, that even in the absence of mutation there are parameter regions in our system where cooperators and defectors coexist in a steady state. In these cases M can be defined for either cooperators or defectors

analysis of the predator-prey model described in the introduction to this chapter. The extent to which the dynamic mode we have described occurs in biological systems depends on the parameter values in these systems. Yet the fact that this mode is both structurally and dynamically stable and holds for a large range of parameters makes it likely to occur in natural systems.

We describe how cooperation persists but not how it originates in the first place, or continues to evolve once it is established. Although a proper treatment of these questions requires extensions to our model, we offer a few comments here. When considering the origin and evolution of cooperative behavior, one should remember that cooperative and defective behaviors are often relative terms. In an homogeneous population, the appearance of an individual behaving more altruistically than its peers may elicit a dynamic where the preexisting type is redefined as a defector. If cooperation is costly, the preexisting type benefits from interacting with the new type without having to pay the cost and will have the characteristics of defectors upon the appearance of cooperators.

We suggest two scenarios for the origination of the mode we have described. Consider a system consisting of reproductively isolated subpopulations, where new subpopulations are founded by individuals that leave existing subpopulations. Assume further that individuals with cooperative behavior appear in the system, after which defectors, which may have been the preexisting type, appear. The type of dynamic we describe requires one additional condition: subpopulations of defectors must not be self-maintaining. This condition seems unreasonable at first, if defectors were the preexisting type and therefore existed independently of cooperators. We suggest two ways to account for this seeming paradox. One is to consider a system with an inhomogeneous environment containing harsh areas where subpopulations of the preexisting type are not able to survive. Since cooperator subpopulations are more efficient, they can inhabit some of these niches. Once they do, the preexisting type may invade these areas, by taking advantage of the cooperators, and thus the conditions for the dynamic persistence of cooperation may arise upon invasion by cooperators. Another possibility is a system in which reproductively isolated subpopulations share common resources. Once cooperating subpopulations appear, the conditions for solitary subpopulation change for the worse, again due to the higher efficiency of subpopulations consisting of cooperators. Consequently, subpopulations of defectors are not self-maintaining, and the stage is set for a steady state of the type we describe.

Now consider the course of evolution after a population reaches a steady state with cooperation. Again, we assume that subpopulations share common resources. At some point "improved" cooperators appear, and consequently both the old cooperators and defectors assume the role of defectors. The long-term evolution [8, 9] of the system then becomes relevant. Specifically, one should consider the conditions for invasion by the new cooperators. "Improved" cooperators may invade by establishing the conditions, through the shared resources, under which subpopulations of the preceding cooperators cannot maintain themselves. If this is the case, then after the new cooperators takeover, the population reaches a new steady state in which the carrying capacity of the environment has increased, and the fitnesses within subpopulations are renormalized.

Taking a higher cooperator fitness while leaving everything else the same can lead, however, to a breakdown in the persistence of cooperation, as we describe in section 4. If an improved cooperator is characterized by a large α when it appears in a given patch, then a large number of defectors is generated during a life cycle, which annihilates the population in its vicinity (see fig. 3). This leads to extinction of the "improved" cooperators and to the continuing persistence of the preexisting cooperators. Therefore, invasibility conditions for "improved" cooperators in these systems can be subtle and deserve closer analysis. We conjecture that such factors dictate the rate at which cooperation evolves and may prevent it from improving significantly in a single transition. If this is true, it will be reflected in both the invasion criterion and in the renormalization of fitness after a takeover event.

ACKNOWLEDGMENT

This work originated while we were staying at the home of Arian and Amnon Tamir in Paris in the summer of 1995. We thank them for their kind hospitality. We have benefited greatly from comments of, and discussions with, Ilan Eshel. We are also grateful for editorial corrections suggested by Lauren Ancel, Susan Ptak, and Carl Bergstrom. Last but not least, we thank Marcus W. Feldman and the entire Feldman lab for their help. This work was supported by NIH grant GM 28016 to M. W. Feldman and by the Santa Fe Institute.

6 PROOF OF RESULT 1

Result 1. *In the density-independent model the ratio R of the average number of migrating cooperators to the average number of migrating defectors in a life cycle is:*

$$R(\alpha,\beta,\gamma,\delta,\mu,D)=\frac{E(M_c)}{E(M_d)}=\frac{DE(S_c)}{DE(S_d)}=\frac{1-\beta}{\alpha-1}\frac{\alpha-\gamma}{\delta-\beta}+O[\mu,D]\,. \tag{26}$$

Proof. The life cycle begins when the first cooperator enters an empty site. Denoting this time as $t=0$, we have:

$$n_c(0)=1\,, \tag{27}$$

$$n_d(0)=0\,. \tag{28}$$

The population at a site then begins evolving according to equation (5). These equations can be written as follows, separating zero and first-order terms in μ and D:

$$E\left(n_c(t+1) \mid n_c(t), n_d(t)\right) = f_c(n_c(t), n_d(t))n_c(t) + \left[\mu\Big(f_d\left(n_c(t), n_d(t)\right) n_d(t) - f_c\left(n_c(t), n_d(t)\right) n_c(t)\Big) + D\Big(\frac{1}{4}\sum_{n.n.} n_c^{n.n.}(t) - f_c(n_c(t), n_d(t))n_c(t)\Big)\right], \quad \text{and} \tag{29}$$

$$E\left(n_d(t+1) \mid n_c(t), n_d(t)\right) = f_d(n_c(t), n_d(t))n_d(t) + \left[\mu\Big(f_c(n_c(t), n_d(t))n_c(t) - f_d(n_c(t), n_d(t))n_d(t)\Big) + D\Big(\frac{1}{4}\sum_{n.n.} n_d^{n.n.}(t) - f_d(n_c(t), n_d(t))n_d(t)\Big)\right]. \tag{30}$$

The first-order terms in μ and D affect the dynamics in two ways. The first is by slightly changing the population sizes due to migration between sites and mutation between types. This changes R to the first order in μ and D. The second is by affecting t_f, the time the first defector appears in the life cycle. The first defector appearance has a dramatic effect on the life cycle, as it marks the beginning of defector takeover. Consequently, we will ignore the effects of mutation and migration at all times other than when the defector population size is 0. This will be done by incorporating a "source" to the zero-order defector dynamics, which is "on" as long as $n_d = 0$ and "off" otherwise. This source term adds one defector at time $t+1$ with the same probability with which it would appear as a result of cooperator mutation and defector migration.
The zero-order (in μ and D) dynamics with the source term are then described by:

$$E\left(n_c(t+1) \mid n_c(t), n_d(t)\right) = f_c(n_c(t), n_d(t))n_c(t)\,, = \left(\alpha\frac{n_c(t)}{n(t)} + \gamma\frac{n_d(t)}{n(t)}\right) n_c(t)\,, \quad \text{and} \tag{31}$$

$$E\left(n_d(t+1) \mid n_c(t), n_d(t)\right) = f_d(n_c(t), n_d(t))n_d(t) + I(n_d(t))P(n_d(t+1) \neq 0 \mid n_c(t), n_d(t)) = \left(\delta\frac{n_c(t)}{n(t)} + \frac{n_d(t)}{n(t)}\right) n_d(t) + I(n_d(t))P(n_d(t+1) \neq 0 \mid n_c(t), n_d(t))\,, \tag{32}$$

where:

$$I(n_d(t)) = \begin{cases} 1\,, & n_d(t) = 0\,; \\ 0\,, & \text{otherwise}\,. \end{cases} \tag{33}$$

Denoting $a \equiv (\alpha - \gamma)$ and $b \equiv (\delta - \beta)$ these equations could be written as:

$$E(n_c(t+1) \mid n_c(t), n_d(t)) = \alpha n_c(t) - a\frac{n_c(t)n_d(t)}{n(t)}, \quad \text{and} \tag{34}$$

$$\begin{aligned} E(n_d(t+1) \mid n_c(t), n_d(t)) = \beta n_d(t) + b\frac{n_c(t)n_d(t)}{n(t)} \\ +I(n_d(t))P(n_d(t+1) \neq 0 \mid n_c(t), n_d(t)) \,. \end{aligned} \tag{35}$$

Taking a linear combination of these equations (34 and 35) to eliminate the nonlinear term gives:

$$\begin{aligned} bE\left(n_c(t+1) \mid n_c(t), n_d(t)\right) &+aE(n_d(t+1) \mid n_c(t), n_d(t)) \\ &= b\alpha n_c(t) + a\beta n_d(t) + aI(n_d(t))P(n_d(t+1) \\ &\neq 0 \mid n_c(t), n_d(t)) \,. \end{aligned} \tag{36}$$

In order to turn the conditional averages and free variables to averages, we multiply equation (37) by $P(n_c(t), n_d(t))$ and sum over all possible values of $n_c(t)$ and $n_d(t)$, to get:

$$\begin{aligned} bE(n_c(t+1)) &+ aE(n_d(t+1)) \\ &= b\alpha E(n_c(t)) + a\beta E(n_d(t)) + aP(n_d(t+1) \\ &\neq 0; n_d(t) = 0) \,, \end{aligned} \tag{37}$$

where we used:

$$\begin{aligned} \sum_{n_c(t), n_d(t)} I(n_d(t))P(n_d(t+1) \neq 0 \mid n_c(t), n_d(t))P(n_c(t), n_d(t)) \\ = \sum_{n_c(t)} P(n_d(t+1) \neq 0 \mid n_c(t), n_d(t) = 0) \\ = P(n_d(t+1) \neq 0, n_d(t) = 0) \,. \end{aligned} \tag{38}$$

Summing these equations from $t = 0$ to $t = \infty$, we get:

$$b(E(S_c) - 1) + aE(S_d) = b\alpha E(S_c) + a\beta E(S_d) + a \,, \tag{39}$$

because:

$$\sum_{t=0}^{\infty} P(n_d(t+1) \neq 0; n_d(t) = 0) = 1 \,. \tag{40}$$

This sum is simply the probability that the first defector will appear sometime (ignoring cases in which defectors disappear and then appear again). Reorganizing equation (39) and resubstituting a and b, we get:

$$R = \frac{E(S_c)}{E(S_d)} = \frac{1-\beta}{\alpha - 1}\frac{\alpha - \gamma}{\delta - \beta} - \frac{1}{E(S_d)}\frac{\alpha + \delta - \beta - \gamma}{(\alpha - 1)(\delta - \beta)} \,. \tag{41}$$

It remains to be shown that $1/E(S_d) \approx O[\mu, D]$. We will not give a complete formal proof; instead, we provide the essence of the argument. It begins by showing that $1/E(S_c) \approx O[\mu, D]$. Assuming the first defector appears as a result of a mutation, it would appear when $\int_0^{t_f} n_c(t)dt \approx 1/\mu$, therefore $S_c \geq 1/\mu$. If, however, it appears as a result of diffusion, then $t_f \approx 1/D\langle n_d \rangle$ where $\langle n_d \rangle$ is the average number of defectors in a site's neighborhood. In this case, $\int_0^{t_f} n_c(t)dt \approx \alpha^{t_f} \approx \alpha^{1/D\langle n_d \rangle}$; therefore $S_d \geq O[1/D]$. If the probability of the first defector appearing from the two processes is comparable, both estimates are valid. Hence, we conclude that $E(S_c) \approx O[1/\mu, 1/D]$, and since S_d scales with S_c we get: $1/E(S_d) \approx O[\mu, D]$. Which concludes the proof that:

$$R(\alpha, \beta, \gamma, \delta, \mu, D) = \frac{DE(S_c)}{DE(S_d)} = \frac{1-\beta}{\alpha-1}\frac{\alpha-\gamma}{\delta-\beta} + O[\mu, D]\,. \tag{42}$$

■

7 PROOF OF RESULT 2

In section 3.2 we considered the life cycle in a specific d.d. model. We found that for any α, β, γ, and δ, R could be made larger than any R^* provided that μ and D are taken to be small enough. An expression relating R^* to μ and D was given. In this appendix, that expression will be derived for a class of d.d. referred to as d.d. models of type 1.

Definition. A model with absolute fitness functions of the form:

$$f_c(n_c, n_d) = g(n)\left(\alpha\frac{n_c}{n} + \gamma\frac{n_d}{n}\right) \quad \text{and}$$
$$f_d(n_c, n_d) = g(n)\left(\delta\frac{n_c}{n} + \beta\frac{n_d}{n}\right)$$

will be called a *density-dependent model of type 1*, if

1. $g(n) \leq 1$ for all n.
2. The average population size at each time step is bounded by some maximum population size $n_{\max}$. This means that for all $n_c(t)$ and $n_d(t)$:
$$E(n(t+1)\,|n_c(t),\, n_d(t)) \;\leq\; n_{\max}\,. \tag{43}$$
3. For $\alpha > 1$ there exists a time $t(\alpha)$ and a constant $0 < c(\alpha) \leq 1$ such that $E(n_c(t)) \geq c(\alpha)n_{\max}$ for $t > t(\alpha)$ when there are no defectors, taking $n_c(0) = 1$.

The first requirement means that the population size in each time step is bounded from above by the population resulting from the density-independent dynamics. The second condition is a requirement for limiting the population size. It is surely satisfied by requiring $g(n) \leq \min\{n_{\max}/\delta n\,,\ 1\}$ for all n, since then we have:

$$E(n(t+1)\,|n_c(t),\ n_d(t)) \leq g(n)\delta n(t) \leq n_{\max}.$$

Both conditions are trivially met by the function $g(n)$ described at the beginning of section 3.2. The third requirement captures the way the cooperator population assumes a certain size after time $t(\alpha)$ and remains in this size until t_f. In the example, this condition is satisfied by choosing $c(\alpha) = \alpha/\delta$ and $t(\alpha) = \lceil \log_\alpha n_{\max}/\delta \rceil + 1$.

Under these conditions we prove:

Result 2. *Given a density-dependent model of type 1 and $R^* \geq 0$, taking μ and D such that:*

$$D + \mu \leq \frac{c(\alpha)}{c(\alpha)n_{\max}t(\alpha) + \widetilde{S_d}R^*},$$

ensures that R is bounded from below by R^*, where $\widetilde{S_d}$ is $E(S_d)$ for the density-independent model with the same parameters α, β, γ, and δ, and the initial conditions $\widetilde{n_c}(0) = n_{\max}$, $\widetilde{n_d}(0) = 1$.

Proof. The proof follows the same line of reasoning applied to the example in section 3.2. By increasing t_f, one can increase S_c as much as required, leaving S_d bound. However, increasing t_f simply means taking smaller μ and D.

In order to find the condition that t_f must satisfy, we begin by finding a lower bound for R. We do so by finding a lower bound for $E(S_c)$ and an upper bound for $E(S_d)$:

$$\begin{aligned} E(S_c) &= E\left(\sum_{t=0}^{\infty} n_c(t)\right) \geq E\left(\sum_{t=t(\alpha)}^{t_f} n_c(t)\right) \\ &= \sum_{t=t(\alpha)}^{t_f} E(n_c(t)) \geq (t_f - t(\alpha))c(\alpha)n_{\max} \quad \text{and} \end{aligned} \tag{44}$$

$$\begin{aligned} E(S_d) &= E\left(\sum_{t_f}^{\infty} n_d(t)\right) \leq E\left(\sum_{t_f}^{\infty} \widetilde{n_d}(t) \mid \widetilde{n_c}(0) = n_{\max}, \widetilde{n_d}(0) = 1\right) \\ &\equiv t\widetilde{S_d}\,, \end{aligned} \tag{45}$$

where $\widetilde{n_d}(t)$ is the defector population at time t for the density-independent model with the same α, β, γ, and δ, and initial conditions: $\widetilde{n_c}(0) = n_{\max}$ and $\widetilde{n_d}(0) = 1$. In bounding $E(S_d)$, we have used condition 1 (from the definition of type 1), which implies the d.i. population bounds the d.d. population with the same parameters. From equation (45) we get:

$$R \geq \frac{c(\alpha)n_{\max}}{\widetilde{S_d}}(t_f - t(\alpha)). \tag{46}$$

Next we find a lower bound for t_f depending on μ and D. The first defector in the life cycle can appear either as a mutant or as a migrant from a neighboring site. The probability for a defector to appear as a mutant in one time step is bounded by $\mu n_{\max}$, while the probability a defector would migrate from a neighboring site could be bounded by $4((1/4)Dn_{\max}) = Dn_{\max}$. From these bounds, we get a lower bound on t_f in terms of μ and D:

$$t_f \geq \frac{1}{(D+\mu)n_{\max}}. \tag{47}$$

From equations (46) and (47), given some R^* we can ensure $R > R^*$ by taking:

$$R^* \leq \frac{c(\alpha)n_{\max}}{\widetilde{S_d}}\left(\frac{1}{(D+\mu)n_{\max}} - t(\alpha)\right) \leq \frac{c(\alpha)n_{\max}}{\widetilde{S_d}}(t_f - t(\alpha)) < R\,, \tag{48}$$

which means choosing μ and D such that:

$$D + \mu \leq \frac{c(\alpha)}{c(\alpha)n_{\max}t(\alpha) + \widetilde{S_d}R^*}. \tag{49}$$

■

8 THE SECOND-ORDER APPROXIMATION

The first-order mean-field approximation was presented in section 4.1. It was derived from the assumption that the probability of finding a site (i,j) in state S_{ij} is independent of the state of its neighbors and uniform across the lattice. This assumption may be expressed as stating that the probability for a lattice configuration $S_{ij}(t)$ satisfies the relation:

$$P(\{S_{ij}(t)\}_{i,j\in Z}) = \prod_{i,j} P(S_{i,j}(t))\,, \tag{50}$$

where the $P(S_{ij}(t))$ derive from a uniform single-site state distribution on the lattice. In section 4.2 we also argued that the first-order approximation falls short of providing conditions for the persistence of cooperation in the system, since it cannot capture the effects of nearest-neighbors (n.n.) correlations in state.

The second-order mean-field approximation, also referred to as the Bethe-Peierls approximation [14], considers the state of tuple consisting of a site and its n.n. (a von Neumann environment). Such a tuple will be denoted by $E_{i,j} \equiv (S_{i,j}, S_{i+1,j}, S_{i-1,j}, S_{i,j-1}, S_{i,j+1})$. It is assumed that the probability distribution of the tuples is independent of the environment and uniform across the lattice so that:

$$P(\{S_{i,j}(t)\}_{i,j\in Z}) = \prod_{i,j\in Z} P(E_{i,j}(t)) . \tag{51}$$

This approximation facilitates studying n.n. correlations but ignores higher-order correlations. The class of probability distributions considered in this approximation contains the probability distributions considered in the first-order approximation as special cases, where $P(E_{i,j})$ depends only on $S_{i,j}$. Higher-order mean-field approximations consider larger tuples, thus extending the class of probability distributions further. If the system does not have long-range correlations, then as the tuples in the approximation grow it becomes more accurate and approaches the true solution of the system.

The peripheral sites within the tuple do not affect each other, and they all affect the center site in the same way. Thus a state of a tuple can adequately described by the state of the center site and by how many of the peripheral sites are in each state. Therefore a tuple E could be in one of $3\binom{6}{2} = 45$ states. The dynamics for the probability of each state can be derived by finding the transition probabilities between states, as was done in equation (14). A similar derivation has been used for the first-order approximation in section 4.2. As this derivation is not very informative but is, nevertheless, incredibly tedious, we do not present it here. Results from the second-order approximation are presented in section 4.3.

9 DERIVING PHASE BOUNDARIES FROM SIMULATIONS

In this appendix, we describe how the simulated phase boundaries, from figures 7 and 8, were derived. As noted in section 4.1, the simplified d.d. model has three parameters that can be taken to be n_c/n_d, D/μ and $S_d = Dn_d/P_d$. We ran simulations on a 32×32 lattice, with $n_c/n_d = 2$ and D/μ varying from 1.1 to 5.0 (50 values) and Dn_d/P_d varying from 0.1 to 6.8 (50 values). The points on this parameter grid, in which a nontrivial steady state was established, appear in figure 10.

For each of the simulations that established a nontrivial steady state p_e, p_c, and p_d were measured. The measurement was averaged over a long time compared to the dimension of the lattice and the typical time of a life cycle, to control for the stochasticity of the simulation and the finite dimensions of the lattice. Even though we tried to control for the accuracy of the measurements in individual simulations, it is not homogeneous across the parameter space. Generally, it decreases when the parameters are closer to the phase boundaries, as the time it takes to obtain an accurate measurement diverges at the phase boundary.

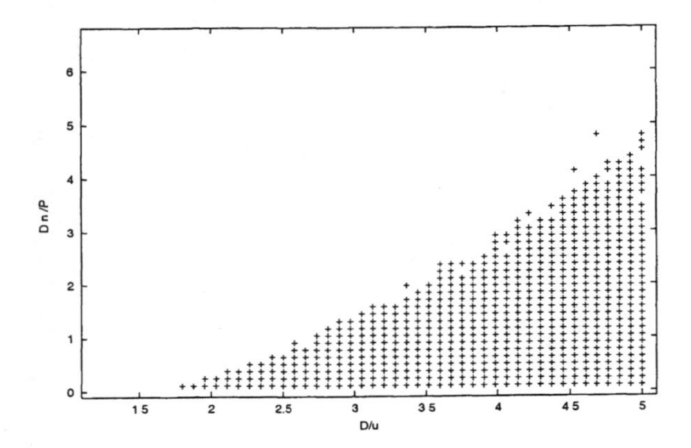

FIGURE 10 Parameter values for which the simplified d.d. model reached a nontrivial steady state. The parameters taken were D/μ varying from 1.1 to 5.0 (50 values) and Dn_d/P_d varying from 0.1 to 6.8 (50 values), where for all simulations $n_c/n_d = 2$.

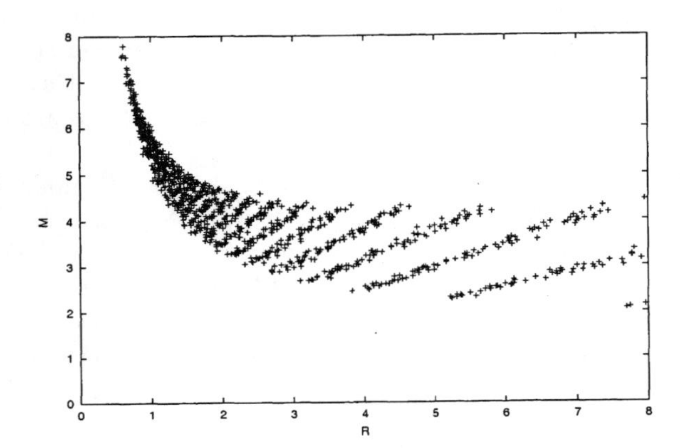

FIGURE 11 The R–M phase space for the simplified d.d. model: The graph presents regions of cooperation persistence according to to simulations. Simulations on a 32×32 lattice were run with the following parameters: $n_c/n_d = 2$, $P_d = 0.01$, $D/\mu = 1.1 - 5.0$ (50 values), and $S_d = Dn_d/P_d = 0.1 - 6.8$ (50 values). In each simulation in which a nontrivial steady state was established, the densities: p_e, p_c, and p_d were measured, and from them R and M were computed.

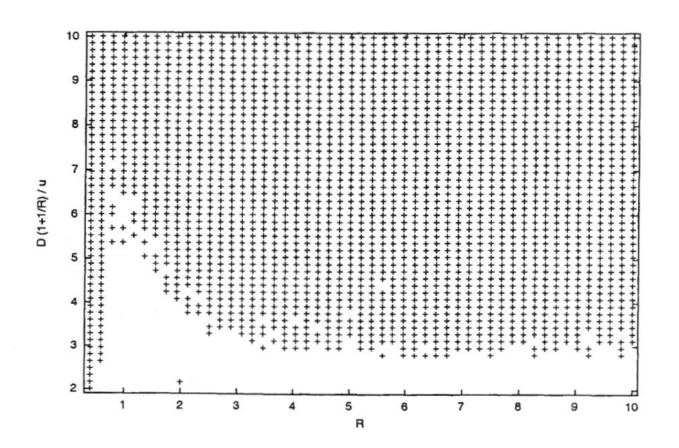

FIGURE 12 Parameter values for which the simplified d.d. model reached a nontrivial steady state. The parameters taken were D/μ varying from 1.1 to 5.0 (50 values) and Dn_d/P_d varying from 0.1 to 6.8 (50 values), where for all simulations $n_c/n_d = 2$.

From the measured p_e, p_c, and p_d, we derived $R = n_c/n_d p_c/p_d$ and $M = Dn_d/P_d(1+R)$. We then plotted the phase space in figure 11, where every point in the R–M space corresponds to a simulation that attains these values. The fact that no points are found below a certain contour means that none of the simulations attained a steady state where such R, M values were measured. Thus, within the accuracy of the simulations, a steady state with these R, M values cannot be maintained. Based on this premise, we draw the phase boundaries in figure 7.

The way the phase boundaries were drawn for the d.i. model is essentially similar. The d.i. model has six parameters: α, β, γ, δ, μ, and D. We ran simulations on a 32 × 32 lattice, with $\beta = 0.9$, $\gamma = 0$, $\delta = 1.6$, and $D = 0.01$, and varying $R = 1 - \beta/\alpha - 1\alpha - \gamma/\delta - \beta$ from 0.4 to 10 (50 values) and $\frac{D}{\mu}(1+1/R)$ from 2 to 10 (50 values), where R and $D/\mu(1+1/R)$ replace for parameters α and μ. The points on this parameter grid, in which a nontrivial steady state was established, are presented in figure 12.

The fact that groups of points in both phase spaces appear to be on a straight line has a simple explanation. In the d.i. model, R is a parameter and was chosen over a grid of values; whereas, M is a variable deriving from the dynamics and can therefore appear anywhere on the fixed R line. The explanation for the d.d. model is similar: $DS_d = M/1 + 1/R$ is a parameter, which explains the straight lines, and M and R are mutually dependent dynamical variables.

Noting that the points in figures 11 and 13 appear to be bounded from above, we performed simulations for the d.i. model with other parameter values, to inquire whether this is in fact the case. As in figure 13, $\beta = 0.9$, $\gamma = 0$, $\delta = 1.6$, $D = 0.01$, and R varies from 0.4 to 10 (50 values). In one run we

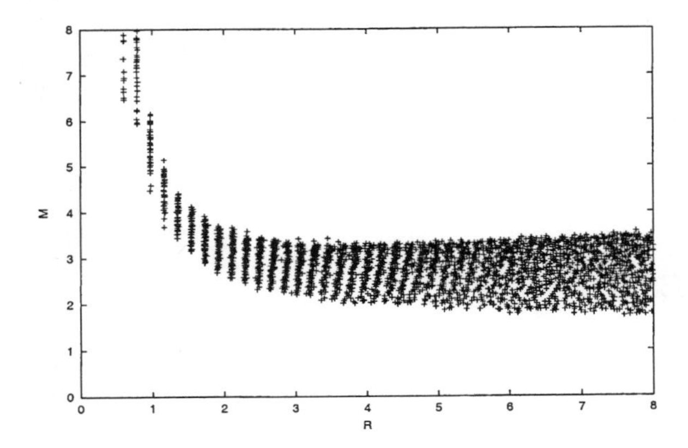

FIGURE 13 The R–M phase space for the d.i. model: The graph presents regions of cooperation persistence according to simulations of the d.i. model. The simulations were done on a 32×32 lattice, using the following parameters: $\beta = 0.9$, $\gamma = 0$, $\delta = 1.6$, $D = 0.01$, $R = \alpha - \gamma/1 - \alpha 1 - \beta/\delta - \beta = 0.4 - 10$ (50 values), and $D/\mu(1 + 1/R) = 2 - 10$ (50 values). Each simulation, in which a nontrivial steady state was established, appears in the phase space according to the R and M measured in the steady state.

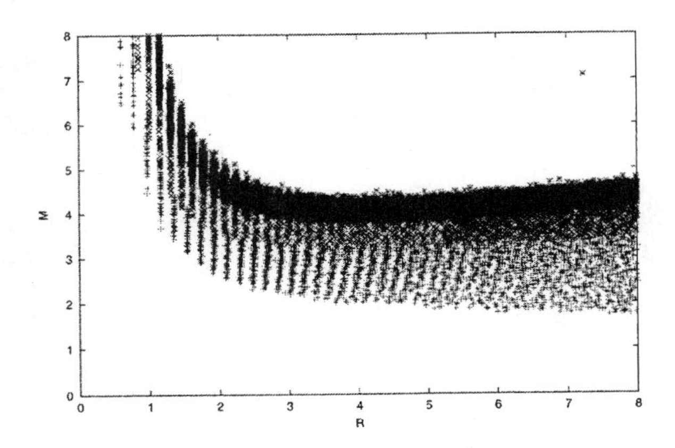

FIGURE 14 Experiment to study whether an upper phase boundary exists. The two additional simulation series (for parameters see text) appear above the previous one. Establishing that, to the extent we have checked, there does not seem to be an upper phase boundary.

took $D/\mu(1+1/R)$ to vary from 10 to 18 (50 values), and in a second run it varied between 18 to 26 (50 values). These values were chosen in a such way that they should appear above the points in figure 11. The points for the two new simulation series appear in figure 14 along with the points from the previous run, and indeed, all points established a nontrivial steady state. There appears to be no sign of an upper phase boundary.

REFERENCES

[1] Axelrod, Robert. *The Evolution of Cooperation.* New York: Basic Books, 1984.

[2] Baxter, R. J. *Exactly Solved Models in Statistical Mechanics.* London; New York: Academic Press, 1982.

[3] Cohen, D., and I. Eshel. "Founder Effect and Evolution of Altruistic Traits." *Theor. Pop. Biol.* **10** (1976): 276–302.

[4] Cooper, B., and C. Wallace. "Evolution, Partnership and Cooperation." *J. Theor. Biol.* **195** (1998): 315–328.

[5] Epstein, J. M. "Zones of Cooperation in Demographic Prisoner's Dilemma." *Complexity* **4(2)** (1998): 36–48.

[6] Eshel, I., and L. L. Cavalli-Sforza. "Assortment of Encounters and Evolution of Cooperation." *Proc. Natl. Acad. Sci.* **79** 1331–1335.

[7] Eshel, I. "On the Neighbor Effect and the Evolution of Altruistic Traits." *Theor. Pop. Biol.* **3** (1971): 258–277.

[8] Eshel, I., U. Motro, and E. Sansone. "Continuous Stability and Evolutionary Convergence." *J. Theor. Biol.* **185** (1997): 333–343.

[9] Eshel, I., M. W. Feldman, and A. Bergman. "Long-Term Evolution, Short-Term Evolution, and Population Genetic Theory." *J. Theor. Biol.* **191** (1998): 391–396.

[10] Ferriere, R., and R. E. Michod. "Invading Wave of Cooperation in a Spatial Iterated Prisoner's Dilemma." *Proc. Roy. Soc. Lond. B* **259(1354)** (1995): 77–83.

[11] Gilpin, M. E. *Group Selection in Preditor-Pray Communities.* Princeton, NJ: Princeton University Press, 1975.

[12] Goldenfeld, D. *Lectures on Phase Transitions and the Renormalization Group.* Reading, MA: Adison-Wesley, 1992.

[13] Hamilton, W. D. "The Genetical Evolution of Social Behavior, I and II." *J. Theor. Biol.* **7** (1964): 1–52.

[14] Huang, K. *Statistical Mechanics.* New York: John Wiley & Sons, 1987.

[15] Maynard Smith, J. *Models in Ecology.* UK: Cambridge University Press, 1974.

[16] Maynard Smith, J. *Evolutionary Genetics*, 2d ed. UK: Oxford University Press, 1998.

[17] Maynard Smith, J. "Group Selection." *Quart. Rev. Biol.* **51** (1976): 277–283.

[18] Maynard Smith, J. "Group Selection and Kin Selection." *Nature* **201** (1964): 1145–1147.

[19] Nakamaru, M., H. Matsuda. and Y. Iwasa. "The Evolution of Cooperation in a Lattice-Structured Population." *J. Theor. Biol.* **184** (1997): 65–81.

[20] Nakamaru, M., H. Nogami, and Y. Iwasa. "Score-Dependent Fertility Model for the Evolution of Cooperation in a Lattice." *J. Theor. Biol.* **194** (1998): 101–124.

[21] Nowak, M. A., and R. M. May. "Evolutionary Games and Spatial Chaos." *Nature* **359** (1992): 826–829.

[22] Oliphant, M. "Evolving Cooperation in the Non-iterative Prisoner's Dilemma: The Importance of Spatial Organization." In *Artificial Life IV*, edited by R. A. Brooks and P. Maes. Cambridge, MA: MIT University Press, 1999.

[23] Roberts, G. "Competitive Altruism: From Reciprocity to the Handicap Principle." *Proc. Roy. Soc. Lond. B* **265** (1998): 427–431.

[24] Weibull, W. J. *Evolutionary Game Theory.* Cambridge, MA: MIT University Press, 1995.

[25] Wilson, S. D. "Altruism in Mendelian Populations Derived from Sibling Groups: The Hey Stack Model Revisited." *Evolution* **41(5)** (1987): 1059–1070.

Coevolution of Strategies in n-Person Prisoners' Dilemma

Kristian Lindgren
Johan Johansson

The evolution of strategies in the iterated n-person Prisoners' Dilemma game is studied in various types of models. By varying the payoff parameters and other characteristics of the models, we investigate some circumstances under which cooperative behavior evolves, both in a mean-field situation (where all interact with all) and in a spatially extended system on a lattice using a cellular automaton dynamics (with only local interactions). In one class of mean-field models, cooperative behavior may dominate in a dynamics that avoids less cooperative stable fix points. We also present and briefly discuss models that use finite automata as representation for the strategies.

1 INTRODUCTION

The evolution of cooperative behavior has been studied extensively through the use of the two-person Prisoners' Dilemma (PD) game as a model for interaction between individuals. In such models reciprocity and "kin selection" are examples of mechanisms that allow for cooperation to be established. In multiperson games, the problem of avoiding exploitation, or free riders, is more difficult, and cooperation may be harder to achieve.

It is well known that, in the iterated PD game without noise, there are simple strategies that are capable of establishing cooperative behavior [5, 6]. Also when noise, in the form of misunderstanding or mistake, disturbs the game, there are strategies that may correct for the mistakes, or may deal with the misunderstandings, in order to re-establish cooperation [8, 14, 15, 26, 27]. This seems to be the case for a variety of model settings and parameter choices.

It turns out that there may be different mechanisms that are active in deciding which cooperative, or in some cases defecting, strategy will dominate in a certain environment. This depends on, for example, if the model is of the mean-field type—i.e., that all individuals interact with all—or if there is a spatial distribution of individuals where the game is played locally.

In the spatially extended models, coevolutionary dynamics may result in cooperative behavior of very simple strategies that are maintained due to a form of (localized) kin selection [16, 22, 23]. In some cases complicated spatiotemporal patterns like spiral waves may be sustained due to group selection, in which different types of groups compete for space. These types of mechanisms seem to be less common in mean-field models, where most of the cooperative strategies depend on reciprocal altruism, such as a tit-for-tat strategy. Other mechanisms may include choosing the partner to play with [25].

Much more complex games are given by the multiperson versions of the Prisoners' Dilemma [24]. In these games reciprocity may be less advantageous to use, since one would not only be punishing the defector but also all others that cooperate. Therefore, the multiperson game provides us with an interesting problem to be used in coevolutionary dynamics modeling.

In social and natural systems the action of an individual often affects a number of other individuals, and there are numerous examples of situations where "free riders" or defectors take an advantage of others cooperating for a common good [26]. This type of problem has been recognized as "the tragedy of the commons," an expression coined by Hardin [12]. Many environmental problems are of this type, e.g., where the costs for exploiting the assimilation capacity of ecosystems are not internalized in the economy. The largest problem of this type may be the increase of the greenhouse effect due to emission of fossil carbon into the atmosphere. This situation is even more problematic since some of those "participating" in the game, or affected by the choice of action, are not yet born.

When several agents are exploiting a common resource, such as a fish population in a lake, there is an almost inevitable risk that overexploitation will result in a lower overall yield. In this case there is no "best" solution on the individual level (unless individual optimization also takes into account how the choice of action influences the future composition of strategies in the population), but what is good depends on the behavior of the others. In such situations, coevolutionary dynamics is one approach to investigate under what circumstances individual strategies lead to the overall best solution (highest total yield).

N-person games may serve as one basis for understanding aggregation to higher system levels in evolutionary systems (in biology or society), e.g., the transition from unicellular to multicellular organisms, the formation of groups in societal systems, and the division of labor in predator inspection [10]. Evolutionary transitions involving the aggregation of previously separate units often depend on cooperative behavior that is made permanent as a result of the aggregation. For a discussion on major evolutionary transitions, we refer to Maynard Smith and Szathmary [19].

A number of evolutionary models using the (iterated) n-person Prisoners' Dilemma game (n-PD) have been proposed; as well as models on social dilemmas based on agents trying to predict their future benefits [11]. A variation of an iterated n-person PD game on a lattice is given by a local game where an individual's action is used simultaneously in all local groups in which the individual participate, implying that the iterated game is global. A certain action in one lattice site may spread over the whole system. This type of n-PD game on a lattice was studied by Matsuo and Adachi [1, 2, 17], and large-scale cellular automata simulations using this game were performed by Albin [4]. In the spatially extended models of the present study, we use the approach by Matsushima and Ikegami [18], in which all groups are closed; i.e., each player chooses a unique action for each of the local groups.

Several ways of representing strategies have been suggested and put into evolutionary models. In this chapter we focus on a very simple strategy set that has been previously analyzed by, e.g., Boyd and Richerson [9] and Molander [21]. The results of our simulations add new insight to their analysis. For example, the behavior in the coevolutionary dynamics is not always well described by the fixed-point characteristics. More advanced types of strategies for the n-PD are finite-memory strategies [7, 13, 17]. A more general class of strategies is given by finite automata [2, 18], and we propose such a representation and apply it to the two types of coevolutionary models studied: the mean-field models with global interactions and the cellular automata models with local interactions.

2 THE n-PERSON PRISONERS' DILEMMA

In a single round of the n-person Prisoners' Dilemma game, n players simultaneously choose an action—cooperate or defect. Depending on the number i of others cooperating, one receives the score $V(C|i)$ when one cooperates and the higher score $V(D|i)$ when one defects. The scores V increase with an increasing number of cooperators, and also the total score given to all players increases if one player switches from defection to cooperation. To summarize:

$V(C|i)$: score for playing C when i <u>others</u> cooperate.

$V(D|i)$: score for playing D when i others cooperate.

$$V(D|i) > V(D|i-1), \text{ and}$$
$$V(C|i) > V(C|i-1), i = 1, \ldots, n-1. \tag{1}$$
$$V(D|i) > V(C|i). \tag{2}$$
$$(i+1)V(C|i) + (N-i-1)V(D|i+1) > iV(C|i-1) + (N-i)V(D|i). \tag{3}$$

In this chapter we shall assume that the scores V can be calculated as a linear combination of the scores against the other players in $(n-1)$ ordinary two-player PD games. Note that this is still an n-person game since the same action is performed simultaneously in all games.

In the two-person game, the scores are R (reward) for mutual cooperation, T (temptation score) for defection against a cooperator, S (sucker's payoff) for cooperation against a defector, and P (punishment) for mutual defection, with the inequalities $S < P < R < T$ and (usually) $R > (T+S)/2$. It is known that in evolutionary models for the two-person PD on a lattice, there are only two independent parameters [16]. This also extends to the n-person game on the lattice, and we shall use fixed values on R and S in this study. Therefore, we assume that $R = 1$ and $S = 0$, while $1 < T < 2$ and, $0 < P < 1$. Also for the mean-field models we use this parameter space, together with a third independent parameter as a growth constant in the population dynamics equations. Then we define the score functions V as follows.

$$V(C|n_C) = \frac{n_C}{n-1}, \tag{4}$$
$$V(D|n_C) = \frac{Tn_C}{n-1} + \frac{P(n-n_C-1)}{n-1}, \tag{5}$$

where we have divided by $n-1$ in order to make it easier to compare results from different group sizes.

In the single-round game, the rational choice of action is to defect, leading to all players in the group defecting and scoring only $P < 1$ instead of 1, which they get if they all cooperate. If there is a high probability that the group will play again, we have the iterated n-person Prisoners' Dilemma, and then cooperation may develop under some circumstances. The purpose of this chapter is to study some coevolutionary models and to discuss some of the conditions for the evolution of cooperation. For simplicity, since we want to avoid simulating the game, we assume the games to be infinitely iterated, and we use the average score per player and per round as a fitness variable for the selection in the population dynamics.

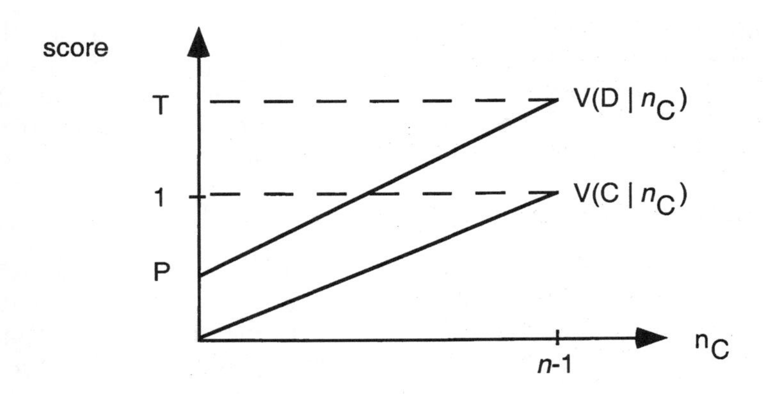

FIGURE 1 The scores V depend on the number of cooperating opponents n_C in the group playing the n-person PD game. This is a linear generalization of the two-person game.

3 STRATEGY TYPES USED IN THE n-PERSON PRISONERS' DILEMMA

Several different types of strategies have been tested in the n-person Prisoners' Dilemma. Here we will briefly describe the most commonly used classes of strategies. The strategies can be divided into different classes, depending on how they are constructed. The characteristics that divide the strategies into different classes are:

- Whether the strategies are deterministic or probabilistic, that is whether a strategy always plays C (or D) for a given history or whether it does it with a certain probability. (In this study we have only included deterministic strategies.)
- Whether the strategies have memory, i.e., whether an individual using a strategy remembers the decisions that itself and other individuals in the same group made in previous rounds of the game.
- Whether a strategy has internal states or not. Strategies with internal states can be described by finite automata.

3.1 STRATEGIES WITHOUT MEMORY

If the strategies are only AllC (always cooperate) or U (unconditional defection), i.e., they have no memory, the n-person game is equivalent to the two-person game for the linear payoff functions assumed in the previous section. Then it is well known that, regardless of parameter choice, any mean-field evolutionary model evolves to a population completely dominated by unconditional defectors U. In contrast, if the game is played locally on a lattice using a cellular automaton updating rule, coexistence in various forms of spatiotemporal patterns is possible between the C and D strategies for some parts of parameter space [16, 22, 23]. This restricted set will not be dealt with further in this chapter.

3.2 THE SIMPLE SET OF S_k STRATEGIES

A natural extension of the strategy set is to allow "triggering" strategies that switch to defection if a sufficient number in the group defect. These strategies can be viewed as having a one-step memory, remembering the number of cooperators in the previous round. Suppose that the game is played in groups of n players. Then the strategy set $\{S_k\}_{k=0,\ldots,n}$ can be defined by

S_k : Cooperate (C) if at least k others do, otherwise defect (D).

For example:

$$\begin{aligned} S_0 &= \text{ Always cooperates.} \\ &\vdots \\ S_{n-1} &= \text{ Cooperates if \underline{all} others do.} \\ S_n = U &= \text{ Unconditional defection.} \end{aligned}$$

These strategies can be interpreted in two ways. The first fits into a single round game, where it can be assumed that the players negotiate to find who shall defect and who shall cooperate. Then they are forced to do what they have promised. The second interpretation involves the iterated game where the players start with the assumption that the others will cooperate, but in the following rounds they adapt their actions according to their strategies; i.e., if too many of the others defect they also switch to defection (which S_n, or U, does anyway). This leads to the game settling down to a certain number of cooperators, and if the stationary score decides the fitness of the players, the two views are equivalent.

3.3 FINITE MEMORY

Hauert and Schuster [13] have looked at more general and probabilistic strategies with memory 1, which means that individuals remember the groups decisions from the previous round and act accordingly. Each strategy can be represented by a transition matrix in which each element represents the probability p_X for the player to cooperate (play C) given a certain history X, and $1 - p_X$ is the probability for the strategy to defect. Numerical simulations done by Hauert and Schuster for three and four players in each group show that cooperative solutions to the game exist, though they seem not to be evolutionarily stable. Defecting strategies can invade a cooperating population, but cooperation may be reestablished after several generations.

Another example of a memory 1 strategy class was proposed by Akimov and Soutchanski [3] who used a strategy that takes its own previous choice and the majority choice of the other players as input.

In an evolutionary model it may be desirable to allow the memory capacity to increase by mutations in order to allow for more advanced strategies to evolve. This approach has been used before, both for finite memory strategies [14, 16] and for the more powerful representation of finite automata [15, 20] described below.

3.4 FINITE AUTOMATA

We have studied strategies that have internal states and that can be described by finite automata (FA). The nodes represent the possible internal states (and corresponding actions, C or D) for an individual using the strategy, and the transition arcs show how the player switches between internal states. In figure 2, the structure of such an FA strategy for the three-person PD game is shown.

The start node is marked by a double circle. The next action depends on how many other players in the group played C last time, and the arcs are labeled by this number. For the three-player strategy shown in figure 2, the player starts with defection, but if no others cooperate the player switches to the second node (determined by the arc labeled 0 for no cooperators) to cooperate in the next round. Then the player will stay in the second node cooperating as long as all others do, and so on. (The strategy described by this automaton has no finite-memory representation.)

In the simulations we have used the following simple mutations: Alter node label, move the address of an arc, and change start node. In order to allow for evolution from simpler to more complex strategies, we have also used mutations that add new nodes. We construct such a growth mutation so that it is neutral; i.e., the arcs connecting the new node are introduced in such a way that the resulting strategy has exactly the same behavior as before. (This is done by

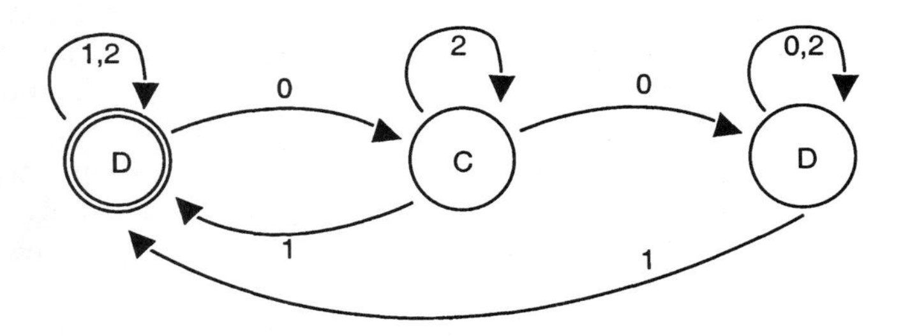

FIGURE 2 This is an example of a nontrivial finite automaton strategy, for the three-person PD game, which evolved in one of the simulations.

splitting a node that has two arcs leading to it. One of the arcs leads to the old node while the other one leads to the new node, which is a copy of the old one including both the label as well as all arcs leaving the node.)

4 EVOLUTIONARY MODELING

The iterated n-person Prisoners' Dilemma is used as a description of the interaction within groups of n individuals picked from a large population of N individuals. We use two different types of dynamics resulting in two different classes of coevolutionary models: (i) *mean-field* models in which all combinations of individuals are used in each generation, and (ii) *cellular automata* (CA) models in which local groups are playing the game on a square lattice. In both cases we illustrate the evolutionary behavior both for the simple S_k strategies and for the finite automata strategies.

4.1 MEAN-FIELD MODELS

In the mean-field model we assume a population of N individuals (here $N = 1000$), and in each generation all possible combinations of the present strategies are used in the formation of the groups of n players. The infinitely iterated n-PD is played in every group. Then the difference between the score for each strategy type and the average score in the population decides the change in population density for the different strategies. The change, from one generation to the next, in population density x_i for strategy i ($i = 1, \ldots, M$, where M is the number of different strategies present), can then be written

$$x_i' = x_i + dx_i \left(\sum_{j_1, j_2, \ldots, j_{n-1}} \left(\prod_{k=1}^{n-1} x_{j_k} \right) s(i, j_1, \ldots, j_{n-1}) - \sum_{j_1, j_2, \ldots, j_n} \left(\prod_{k=1}^{n} x_{j_k} \right) s(j_1, \ldots, j_n) \right), \tag{6}$$

where d is a growth constant. Here $s(i, j_1, \ldots, j_{n-1})$ is the score for a player of strategy i playing in a group with the strategies $j_1, \ldots, j_{n-1}$, and hence the second sum is the average score. These equations conserve total populations size; here $\sum x_i = 1$. The number M of different strategies may change due to mutations, which are randomly applied after each generation (step) of the equations (6). Even if the x_is are continuous variables, the model has a discrete component in that a certain population size N is assumed, implying that one individual is given by $x = 1/N$. If any x_i falls below this value, the corresponding strategy becomes extinct and is removed from the population. (The extinctions may change the total population size, and therefore the population is normalized to 1 after each generation.) The described dynamics means that generations are non-overlapping and that reproduction is asexual.

4.2 CELLULAR AUTOMATA MODELS

For the spatially extended system we have chosen the synchronous and local updating rule of a cellular automaton, with one player in each cell (or lattice site) on a square lattice with periodic boundary conditions. The world size used is 128×128. Groups are formed locally to play the five-person PD game, so that each group has a player in the center and the four nearest neighbors. This means that each individual participates in five different groups and games.

This leads to a CA with a next-nearest neighborhood interaction. The average score for each individual is calculated and compared to the nearest neighbors' scores. In each five-cell neighborhood the highest scored individual will put its offspring in the middle cell; see figure 3. A small random number is added to each individual's score in order to break ties. The offspring inherits the parent's strategy, possibly altered by mutations. (Other local group formations could also be chosen. For example, local three-player groups have been used by Matsushima and Ikegami [18].)

5 MATHEMATICAL ANALYSIS OF S_k STRATEGIES

In this section we review and discuss some results of the extensive study by Boyd and Richerson [9] on the S_k strategies. Even though they have some restrictions on their payoff parameters, parts of their results can be used in a qualitative way for discussing some of the behavior we have observed in the coevolutionary models.

Boyd and Richerson argue that the conditions that allow the evolution of reciprocal cooperation become extremely restrictive as group size increases. This statement is based on their investigation of strategies of the type S_k playing the iterated n-person Prisoners' Dilemma. They sample groups of n individuals from

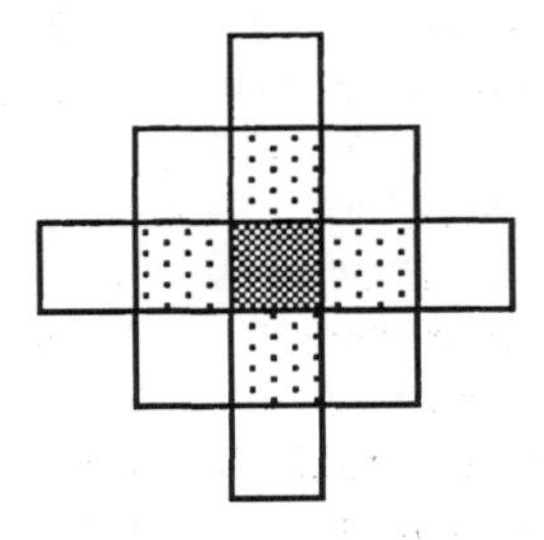

FIGURE 3 In the cellular automaton model each individual participates in five different five-person PD games, involving up to next-nearest neighbors. Then the player with the highest score in the five-cell neighborhood puts its offspring in the middle cell. This rule is applied in parallel over the whole lattice.

a large population and let them play the iterated game with a varying length determined by a parameter w, which is the probability for a sampled group to interact another time. This means that the expected number of iterations in a group is $1/(1-w)$. In the limit $w \to 1$, we have the infinitely iterated Prisoners' Dilemma as we have discussed above.

A population in which unconditional defectors are dominant can resist invasion by any reciprocating strategy. This is true for all values of w, except $w = 1$. This is straightforward because if S_n is dominant, a strategy $S_a(a < n)$ will most likely meet $n-1$ unconditional defectors and will get a lower score than S_n. Hence, individuals using strategy S_a cannot invade a population of unconditional defectors. For the infinitely iterated game, which corresponds to $w = 1$, an individual using a reciprocal strategy (except S_0) in a group of $n-1$ defectors will get the same payoff as the unconditional defectors and can therefore (initially) increase by genetic drift (and then, for example, S_{n-1} may increase by its own cooperative and reciprocating behavior).

A population in which strategy S_{n-1} dominates can resist invasion by unconditional defectors if, and only if, w is sufficiently large. S_{n-1} is the only strategy that has this property of resisting invasion from unconditionally defectors. With linear payoff, the domain of attraction of S_{n-1}becomes rapidly smaller as the group size increases. The domain of attraction of S_{n-1} is the set of initial strategy combinations that end up in an equilibrium with S_{n-1} dominant. There is a smallest initial frequency of S_{n-1} for which the population eventually will end up with only individuals using S_{n-1}. This frequency tends to one as the group size increases. If the initial frequency of S_{n-1} is lower than this value, the population will end up with only unconditional defectors.

In a population composed only of a strategy $S_a(0 < a < n-1)$ and the unconditional defector U, there exists one evolutionarily stable equilibrium if w is large enough. This stable equilibrium will eventually evolve if the initial frequency of S_a is larger than a value that depends on the parameters in the Prisoners' Dilemma game and on the group size. This initial frequency increases with the group size. If the initial frequency of S_a is less than this critical value, the population will eventually consist only of unconditionally defectors. A population at the stable equilibrium involving two strategies $S_a(0 < a < n-1)$ and U can resist invasion from other rare strategies $S_b(b \neq a)$.

The overall conclusion that can be drawn from Boyd and Richerson's paper is that it is more difficult for cooperation to evolve in larger groups. Although, they only show that this is true when evolutionary stable states involve only one or two types of strategies. Our numerical simulations show that there exist other evolutionarily stable states involving more than two strategies. Therefore, there may be parameter values for which this conclusion is not true. As we shall see in the simulation examples, there are also other dynamical effects that further complicate this picture.

Boyd and Richerson have also studied the case where the groups are not formed randomly. Rather reciprocal strategies are more likely to meet each other than by pure chance. In this case, reciprocating strategies are more likely to increase. This kind of social interaction could arise if individuals tend to interact with genetic relatives. Such a modification of the game will not be considered further in this article. Still, it is worth noticing that the CA models may involve such effects, since the local interactions may imply that there is a high probability that one's relatives are in the group.

6 SIMULATIONS AND RESULTS

Here we shall briefly show our preliminary results for the mean-field and the CA coevolutionary models, using both the S_k strategies and the finite automata representation. A more thorough investigation will be reported elsewhere.

6.1 THE SIMPLE SET OF THE S_k STRATEGIES

6.1.1 Mean-field Model. It is known from the results of Boyd and Richerson that in the S_k strategy set there are stable fixed points involving the U strategy (unconditional defection) and an $S_k (k < n - 1)$ strategy. In a coevolutionary model, however, the mutations may form a distribution of strategies that avoids the stable fix points in establishing a population of cooperating S_k strategies $(k < n)$. Depending on the distribution of these strategies, the mutant U strategy may increase and, in some cases, the dynamics is caught on a fixed point with mixed actions or the population may swing back to a fully cooperative (meta-stable) state again. An example of such a pattern is shown in figure 4 below, where oscillations between cooperative and defective dominance is seen for the first 120,000 generations, but after this the system settles to a stationary state involving all strategies except S_4. Two strategies (S_0 and S_2) are maintained due to mutations only, while there is a fixed point of the mutation-free dynamics involving the three strategies U, S_3, and S_1. This positively answers the question posed by Boyd and Richerson [9] of the existence of fixed points with more than two strategies in this dynamic, and it contradicts the claim by Molander [21] that such a fixed point does not exist.

In order to illustrate how the behavior in this simple system depends on the payoff parameters, we have run a series of simulations recording the average score for the four-person PD as functions of T and P. The statistics shown in figure 5, illustrate that there is a nontrivial pattern of regions where cooperation is established. From this picture we can sketch the phase diagram in figure 6.

In figures 7–9, we illustrate how these regions change (for a constant $T = 1.5$) when the number of players n increases, from 3 to 5. When $n = 2$ (not shown in the figures), we have the ordinary PD game, which has been extensively studied in various evolutionary models. In this case, cooperative behavior is established by the coexistence of S_1 and S_0, and the average score is close to 1 regardless of payoff parameters. (In this environment, S_1 behaves like tit-for-tat.) It should be

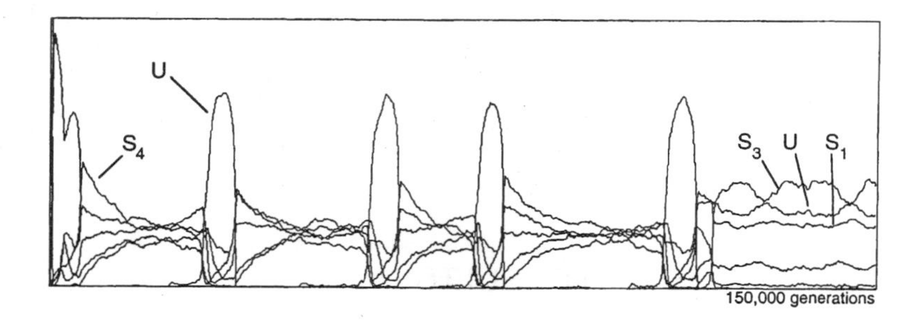

FIGURE 4 For the simple strategy set, there are payoff parameters leading to oscillatory behavior between cooperating strategies and the unconditional defectors. Here we have used the five-player game in the mean-field dynamics, with the parameter values $T = 1.5$, $P = 0.2$, $p_{\text{mut}} = 0.0001$, $d = 0.04$, and $N = 1000$.

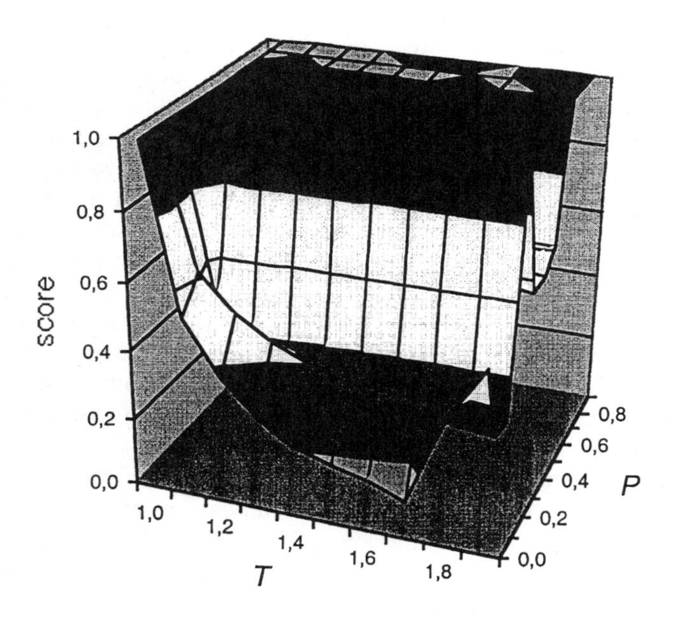

FIGURE 5 The average score for the four-person PD as function of T and P is calculated from, in each point, the average score in the population in 10 simulations after a 60,000 generations transient.

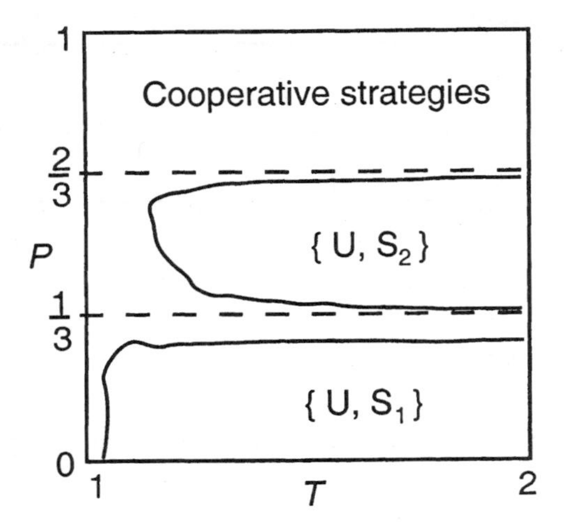

FIGURE 6 In this sketch of a phase diagram, based on figure 5 and simulation results, we illustrate how the parameter region (T, P) is divided into regions with cooperative behavior and regions characterized by fixed points involving U and one of the $S_k(0 < k < n-1)$. (We have not investigated the existence of other fixed points here, but, as shown in figure 4, they exist.)

noted that the cooperative state is not evolutionarily stable, since genetic drift may alter the ratio between S_1 and S_0 such that a temporary increase of U may occur.

When the number of players n increases, the corresponding cooperative region (characterized by the absence of stable fixed points between two strategies) is approximately $(n-2)/(n-1) < P < 1$. This means that a high score for mutual defection actually increases the possibilities for cooperation. It is also worth noting that for some values of P (e.g., $P = 0.2$), increasing population size may lead to an increased level of cooperation; see figures 7–9.

The change in cooperation level as P increases can be understood in the following way. First, for the regions where fixed points dominate the dynamics (less cooperative regions), we can use the analysis by Molander [21]. There exists a stable equilibrium distribution consisting of unconditional defectors (U) and individuals using strategy S_{k^*}, where k^* is the lowest number $k < n-1$ such that $V(C|k) > V(D|0)$. With n players in each group this condition can be written $k > (n-1)P$. For groups of four players $k^* = 1$ for $P < 1/3$ and $k^* = 2$ for $1/3 < P < 2/3$. (For $P > 2/3$, the lowest k equals $n-1$, and there is no stable coexistence.) These different S_{k^*} strategies correspond to different average scores of the less cooperative regions. Secondly, for the cooperative regions in between, the fixed point given by the analysis above is weakened, and the mutations may

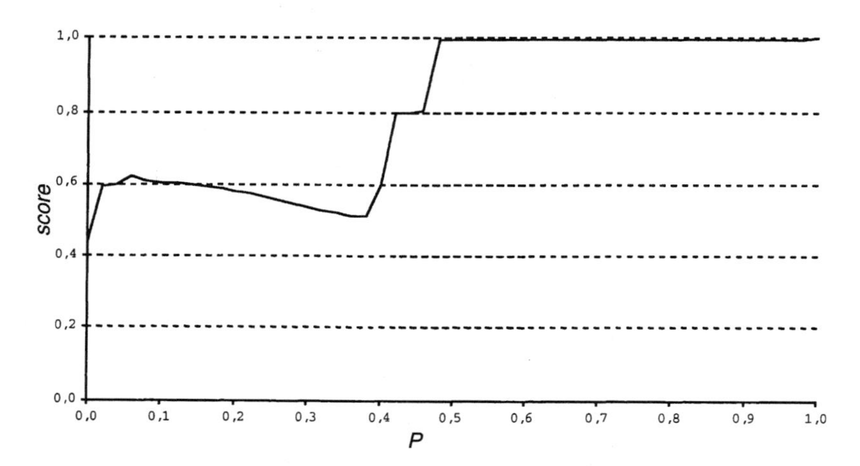

FIGURE 7 The average score as a function of P for the S_k strategies in the mean-field model using the three-person PD game. The average is calculated from 10 simulations using 100,000 generations following a transient of 50,000 generations. Other parameter values are $T = 1.5$, $p_{\text{mut}} = 0.0001$, and $d = 0.04$.

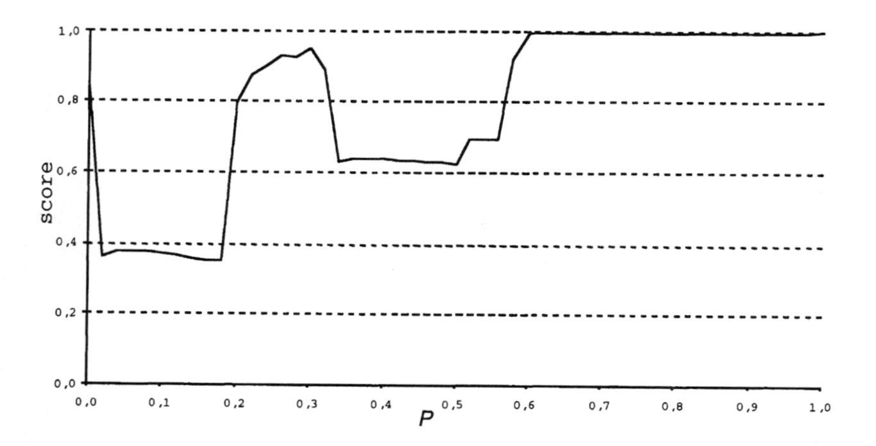

FIGURE 8 The average score as a function of P for the S_k strategies in the mean-field model using the four-person PD game. The average is calculated from 5 simulations using 200,000 generations following a transient of 100,000 generations. Other parameter values are $T = 1.5$, $p_{\text{mut}} = 0.0001$, and $d = 0.04$.

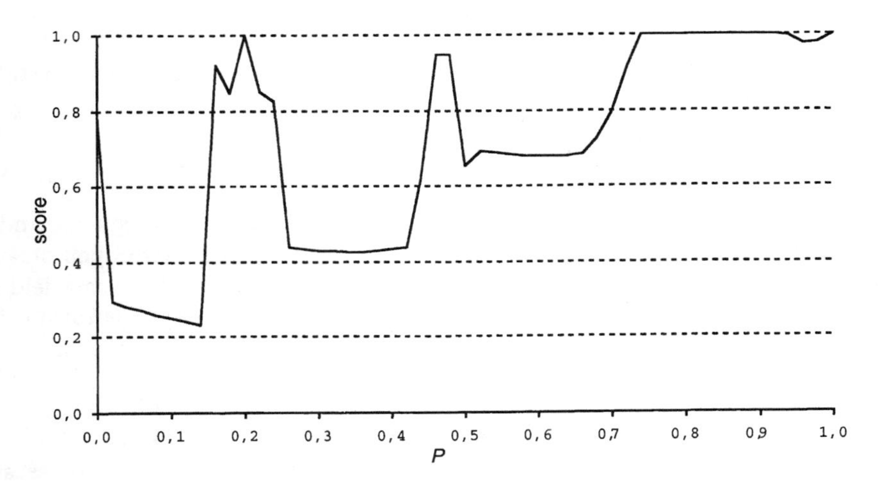

FIGURE 9 The average score as a function of P for the S_k strategies in the mean-field model using the five-person PD game. The average is calculated from 5 simulations using 200,000 generations following a transient of 100,000 generations. Other parameter values are $T = 1.5$, $p_{\text{mut}} = 0.0001$, and $d = 0.04$.

sustain a distribution of strategies that destabilizes the fixed point. This may result in either a distribution of cooperating strategies (which is not a fixed point), or a fixed point involving more than two strategies (as was observed in fig. 4).

6.1.2 CA Model with Simple Strategies. If the S_k-strategies are put on a lattice according to the coevolutionary CA dynamics described above, the locality of the interactions completely changes the behavior. It has been observed for the 2-person PD game that cooperative behavior is favored in the CA dynamics compared to the mean-field case [15, 16]. Here we find that in the interior of the parameter region ($1 < T < 2, 0 < P < 1$), the cooperative strategies dominate and the U strategy only occurs in minor bursts following a mutation. Depending on the payoff parameters, U may expand locally by exploiting S_k strategies (with $k < n - 1$), but when areas of S_{n-1} strategies are met, the U strategy will disappear, for any values of (P, T). This leads to the aggregated population dynamics pattern showing a quick shift from a population of defectors to a population dominated almost completely by cooperating strategies. In almost the entire parameter region, it is S_{n-1} that dominates. The spatial extension of the system does not allow for the type of coexistence between cooperative and defective strategies that the mean-field case exhibited in the form of stable fixed points. In principle, the spatial dynamics *may* give rise to spatiotemporal patterns that globaly stabilizes a mixture of cooperators and defectors, as we shall see below, but it does not seem to be the case for the very limited strategy set used here.

6.2 FINITE AUTOMATA STRATEGIES

Next we wish to illustrate the limitations of the previous strategy set and investigate the more complex evolutionary dynamics that a more open strategy set may generate. To do this, we show a few examples from the mean field and the CA models using finite automata strategies. These are only illustrations that are included for a qualitative discussion of the FA strategies in a coevolutionary context. A thorough investigation of these strategies is being done and will be presented elsewhere.

The simulation of figure 10, based on the three-player PD, uses the mean-field dynamics and illustrates how coevolution in a number of steps increases the level of cooperation in the system. The initial state consists of defectors only, but since there are 73 two-node FA that correspond to the unconditional defector, all of them are put in the group labeled U. At the end of the simulation, most of the strategies use three nodes, and the dominating strategy is cooperative in groups where both the opponents are.

For the simple strategy set the cellular automaton dynamics made cooperation extremely simple. This is also the case for finite automata, which is illustrated in figure 11. The transient from the initial state with unconditional defectors may involve several nontrivial strategies involved in various types of spatiotemporal patterns. At the end the population is mainly cooperating, dominated by a simple stategy that continues with cooperation as long as one or all of its neighbors defect.

For higher values of P, it seems, though, that cooperation is much harder to achieve than in the more limited class of the S_k strategies. Both the spatiotemporal behavior and the dependence of payoff parameters will be further discussed in our future work on these models.

7 SUMMARY AND POSSIBLE EXTENSIONS

In this chapter we have discussed some aspects of coevolutionary models of the multiperson Prisoners' Dilemma game. In particular, we have focused on a very simple strategy set that has previously been analyzed mathematically in the literature. It is found that the coevolutionary dynamics may avoid being trapped by stable fixed points which also involve defecting strategies, thereby establishing a more cooperative behavior. The structure of the "phase space" also implies that there are payoff parameters for which an increase in number of players in the PD group increases the chances for cooperative behavior to evolve. Another interesting result is that cooperation can be easier to achieve when the score P for mutual defection is large enough.

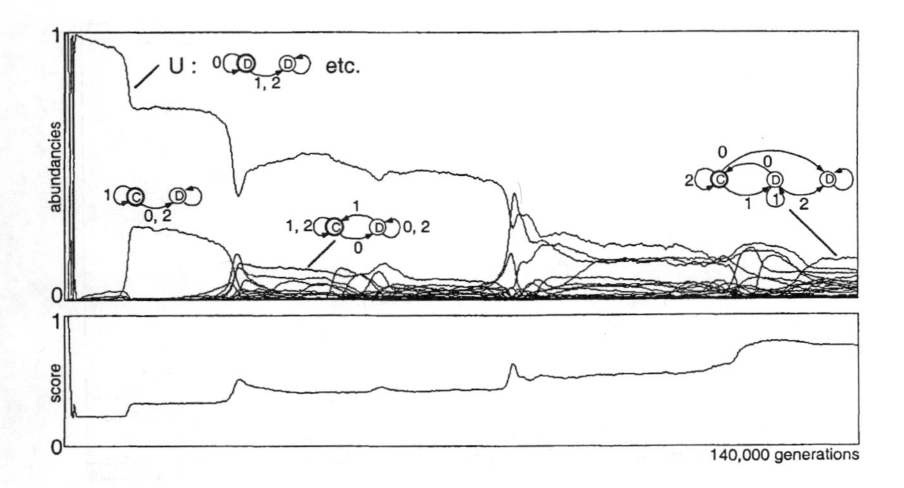

FIGURE 10 The coevolution of FA strategies in the mean-field $n = 3$ PD game. The top chart shows how the initial population of defectors is gradually replaced by more cooperative strategies, but in this simulation there are still some free riders left that manage to exploit the cooperative strategies. The bottom chart shows the increase in average score. Parameter values are $T = 1.5$, $P = 0.25$, $p_{\text{mut}} = 5 \times 10^{-5}$, $d = 0.05$, and the number of nodes is maximized to 3.

In the CA model, the introduction of a spatial dimension and the use of local interactions leads to the evolution of more cooperative strategies, as should be expected. For the simple strategy set S_k, we find that in almost all of the studied payoff region the strategy S_{n-1} dominates; i.e., this is the strategy that requires all others in group cooperate before it cooperates. This effectively keeps out mutant unconditional defectors.

The finite automata strategy representation is discussed in the coevolutionary model context in a qualitative way. The increased capacity gained by this representation may be used both by advanced free riders as well as by more cautious cooperators, and it is not clear to what extent cooperative behavior evolves in the different model types.

An immediate extension that would be interesting to try is the introduction of mistakes or misunderstanding. In the two-person PD game this has been used as one complication in order to increase the selection pressure and hence evolutionary activity.

The games studied here have all been "few-person" PD games. In principle the models can be extended to arbitrarily large games, but for practical (computational) reasons one is limited to a number of players in the order of 10. To be able to study more general problems of "the tragedy of the commons" type

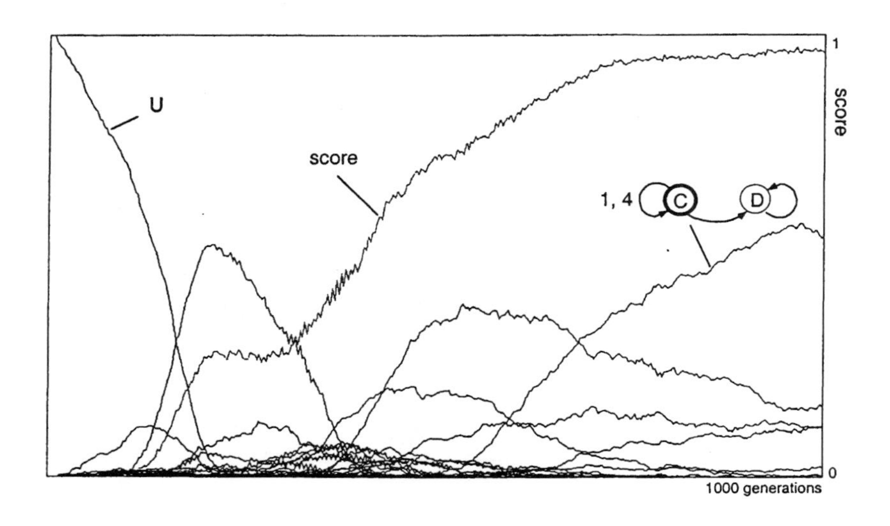

FIGURE 11 In this illustration of the cellular automaton model, the system evolves from an initial state of only defectors to an almost fully cooperating population within a few thousand generations. Here $T = 1.5$, $P = 0.25$, and $p_{\text{mut}} = 0.002$. The final strategy cooperates as long as one or all its neighbors have cooperated in the previous round (otherwise it switches to unconditional defection). There are several possible strategies that may bring the population to the cooperating state.

in coevolutionary contexts, one needs to find both a suitable extension of the game and an appropriate *simple* strategy set. Some studies along these lines are in progress, involving harvesting and prey population dynamics.

REFERENCES

[1] Adachi, N., and K. Matsuo. "Ecological Dynamics under Different Selection Rules in Distributed and Iterated Prisoners' Dilemma Games." In *Parallel Problem Solving From Nature*, 388–394. Lecture Notes in Computer Science, vol. 496. Berlin: Springer-Verlag, 1991.

[2] Adachi, N., and K. Matsuo. "Ecological Dynamics of Strategic Species in Game World." *FUJITSU Sci. Tech. J.* **28** (1992): 543–558.

[3] Akimov, V., and M. Soutchanski. "Automata Simulation of N-Person Social Dilemma Games." *J. Conflict Res.* **38** (1994): 138–148.

[4] Albin, P. "Approximations of Cooperative Equilibria in Multi-Person Prisoner's Dilemma Played by Cellular Automata." *Math. Soc. Sci.* **24** (1992): 293–319.

[5] Axelrod, R., and D. H. Hamilton. "The Evolution of Cooperation." *Science* **211** (1982): 1390–1396.
[6] Axelrod, R. *The Evolution of Cooperation.* New York: Basic Books, 1984.
[7] Axelrod, R. "The Evolution of Strategies in the Iterated Prisoner's Dilemma." In *Genetic Algorithms and Simulated Annealing*, edited by L. Davis, 32–41, Los Altos, CA: Morgan Kaufmann, 1987.
[8] Boerlijst, M. C., M. A. Nowak, and K. Sigmund. "The Logic of Contrition." *J. Theor. Biol.* **185** (1997): 281–293.
[9] Boyd, R., and P. Richerson. "The Evolution of Reciprocity in Sizable Groups," *J. Theor. Biol.* **132** (1988): 337–356.
[10] Dugatkin, L. A. "N-Person Games and the Evolution of Co-operation: A Model based on Predator Inspection in Fish." *J. Theor. Biol.* **142** (1990): 123–135.
[11] Glance, N. S., and B. A. Huberman. "The Dynamics of Social Dilemmas." *Sci. Am.* **270** (1994): 76–81.
[12] Hardin, G. "The Tragedy of the Commons." *Science* **162** (1968): 1243–1248.
[13] Hauert, Ch., and H. G. Schuster. "Effects of Increasing the Number of Players and Memory Size in the Iterated Prisoner's Dilemma, a Numerical Approach." *Proc. Roy. Soc. Lond. B* **264** (1997): 513–519.
[14] Lindgren, K. "Evolutionary Phenomena in Simple Dynamics." In *Artificial Life II*, edited by C. G. Langton, C. Taylor, J. D. Farmer, and S. Rasmussen, 295–311. Redwood City, CA: Addison-Wesley, 1992.
[15] Lindgren, K. "Evolutionary Dynamics in Game-Theoretic Models." In *The Economy as an Evolving Complex System II*, edited by B. Arthur, S. Durlauf, and D. Lane, 337–367. Reading, MA: Addison-Wesley, 1997.
[16] Lindgren, K., and M. G. Nordahl. "Evolutionary Dynamics of Spatial Games." *Physica D* **75** (1994): 292–309.
[17] Matsuo, K. "Ecological Characteristics of Strategic Groups in 'Dilemmatic World'." In *Proceedings of the IEEE International Conference on Systems and Cybernetics*, 1071–1075. 1985.
[18] Matsushima, M., and T. Ikegami. "Evolution of Strategies in the Three-Person Iterated Prisoner's Dilemma Game." *J. Theor. Biol.* **195** (1998): 53–67.
[19] Maynard Smith, J., and E. Szathmary. *The Major Transitions in Evolution.* New York: Oxford University Press, 1997.
[20] Miller, J. H. "The Coevolution of Automata in the Repeated Iterated Prisoner's Dilemma." Working paper 89-003, Santa Fe Institute, Santa Fe, NM, 1989.
[21] Molander, P. "The Prevalence of Free Riding." *J. Conflict Res.* **36** (1992): 756–771.
[22] Nowak, M. A., and R. M. May. "Evolutionary Games and Spatial Chaos." *Nature* **359** (1993): 826–829.
[23] Nowak, M. A., and R. M. May. "The Spatial Dilemmas of Evolution." *Intl. J. Bifur. & Chaos* **3** (1993): 35–78.

[24] Schelling, T. C. *Micromotives and Macrobehaviour.* New York: Norton, 1978.
[25] Stanley, E. A., D. Ashlock, and L. Tesfatsion. "Iterated Prisoner's Dilemma with Choice and Refusal of Partners." In *Artificial Life III*, edited by C. G. Langton, 131–175. Reading, MA: Addison-Wesley, 1993.
[26] Sugden, R. *The Economics of Rights, Co-operation and Welfare.* Oxford: Basil Blackwell, 1986.
[27] Wu, J., and R. Axelrod. "How to Cope with Noise in the Iterated Prisoner's Dilemma." *J. Conflict Res.* **39** (1995): 183–189.

Evolutionary Design of Collective Computation in Cellular Automata

James P. Crutchfield
Melanie Mitchell
Rajarshi Das

We investigate the ability of a genetic algorithm to design cellular automata that perform computations. The computational strategies of the resulting cellular automata can be understood using a framework in which "particles" embedded in space-time configurations carry information, and interactions between particles effect information processing. This structural analysis can also be used to explain the evolutionary process by which the strategies were designed by the genetic algorithm. More generally, our goals are to understand how machine-learning processes can design complex decentralized systems with sophisticated collective computational abilities and to develop rigorous frameworks for understanding how the resulting dynamical systems perform computation.

Evolutionary Dynamics, edited by
J. P. Crutchfield and P. Schuster. Oxford University Press.

1 INTRODUCTION

From the earliest days of computer science, researchers have been interested in making computers and computation more like information-processing systems in nature. In the 1940s and 1950s, von Neumann viewed the new field of "automata theory" as closely related to theoretical biology and asked questions such as "How are computers and brains alike?" [76] and "What is necessary for an automaton to reproduce itself?" [77]. Turing was deeply interested in the mechanical roots of humanlike intelligence [73], and Weiner looked for links among the functioning of computers, nervous systems, and societies [80]. More recently, work on biologically and sociologically inspired computation has received renewed interest; researchers are borrowing information-processing mechanisms found in natural systems such as brains [8, 29, 64], immune systems [26, 32], insect colonies [10, 22], economies [79, 81], and biological evolution [3, 30, 41]. The motivation behind such work is both to understand how systems in nature adaptively process information and to construct fast, robust, adaptive computational systems that can learn on their own and perform well in many environments.

Although there are some commonalities, natural systems differ considerably from traditional von Neumann-style architectures.[1] Biological systems such as brains, immune systems, and insect societies consist of myriad relatively homogeneous components that are extended in space and operate in parallel without central control and with only limited communication among components. Information processing in such systems arises from coordination among large-scale patterns that are distributed across components (e.g., distributed activations of neurons or activities of antibodies). Such decentralized systems, being highly nonlinear, often exhibit complicated, difficult-to-analyze, and unpredictable behavior. The result is that they are hard to control and "program." It seems clear that in order to design and understand decentralized systems and to develop them into useful technologies, engineers must extend traditional notions of computation to encompass these architectures. This has been done to some extent in research on parallel and distributed computing (e.g., Crichlow [11]) and with architectures such as systolic arrays [48]. However, as computing systems become more parallelized and decentralized and employ increasingly simple individual processors, it becomes harder and harder to design and program such systems.

Cellular automata (CAs) are a simple class of system that captures some of the features of systems in nature listed above: large numbers of homogeneous components (simple finite-state machines) extended in space, no central control, and limited communication among components. Given that there is no programming paradigm for implementing parallel computations in CAs, our research investigates how genetic algorithms (GAs) can evolve CAs to perform computations requiring coordination among many cells. In other words, the GA's job is

[1] It should be noted that although computer architectures with central control, random access memory, and serial processing have been termed "von Neumann style," von Neumann was also one of the inventors of "non von Neumann-style" architectures such as cellular automata.

to design ways in which the actions of simple components with local information and communication give rise to coordinated global information processing. In addition, we have adapted a framework—*computational mechanics*—that can be used to discover how information processing is embedded in dynamical systems [13] and thus to analyze how computation emerges in evolved CAs. Our ultimate motivations are twofold: (i) to understand collective computation and its evolution in natural systems and (ii) to explore ways of automatically engineering sophisticated collective computation in decentralized multiprocessor systems.

In previous work, we described some of the mechanisms by which genetic algorithms evolve CAs to perform computations as well as some of the impediments faced by the GA [57]. We also briefly sketched our adaptation of the computational mechanics approach to understanding computation in the evolved CAs [18, 20, 21]. In this chapter, we give a more fully developed account of our research to date on these topics, report on new results, and compare our work with other work on GAs, CAs, and distributed computing.

This chapter is organized as follows. In sections 2–4, we review cellular automata, define a computational task for CAs—"density classification"—that requires global coordination, and describe how we used a GA to evolve CA to perform this task. In sections 5–8, we describe the results of the GA evolution of CAs. We first describe the different types of CA computational strategies discovered by the GA for performing the density classification task. We then make the notion of computational "strategies" more rigorous by defining them in terms of embedded particles, particle interactions, and geometric "subroutines" consisting of these components. This high-level description enables us to explain how the space-time configurations generated by the evolved CAs give rise to collective computation and to predict quantitatively the CAs' computational performance. We then use embedded-particle descriptions to explain the evolutionary stages by which the successful CAs were produced by the GA. Finally, in section 9, we compare our research with related work.

2 CELLULAR AUTOMATA

A one-dimensional cellular automaton consists of a lattice of N identical finite-state machines (*cells*), each with an identical topology of local connections to other cells for input and output, along with boundary conditions. Let Σ denote the set of states in a cell's finite-state machine, and let $k = |\Sigma|$ denote the number of states per cell. Each cell is indexed by its site number $i = 0, 1, \ldots, N-1$. A cell's state at time t is denoted by s_i^t, where $s_i^t \in \Sigma$. The state s_i^t of cell i, together with the states of the cells to which it is connected, is called the *neighborhood* η_i^t of cell i. Each cell obeys the same transition rule $\phi(\eta_i^t)$, which gives the update state $s_i^{t+1} = \phi(\eta_i^t)$ for cell i as a function of η_i^t. We will drop the indices on s_i^t and η_i^t when we refer to them as general (local) variables.

Rule table ϕ:

Neighborhood η:	000	001	010	011	100	101	110	111
Output bit $\phi(\eta)$:	0	1	1	1	0	1	1	0

Lattice configuration:

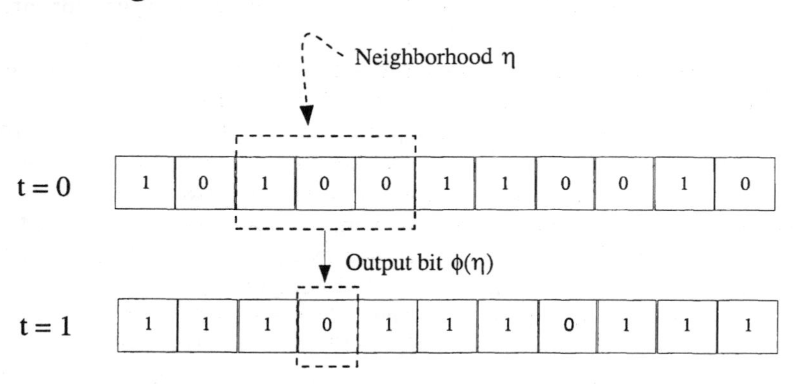

FIGURE 1 The components of one-dimensional, binary-state, $r = 1$ ("elementary") CA 110 shown iterated one time step on a configuration with $N = 11$ lattice sites and periodic boundary conditions (i.e., $s_N = s_0$).

We use $\mathbf{s}^t$ to denote the *configuration* of cell states:

$$\mathbf{s}^t = s_0^t s_1^t \ldots s_{N-1}^t .$$

A CA $\{\Sigma^N, \phi\}$ thus specifies a global map Φ of the configurations,

$$\Phi : \Sigma^N \to \Sigma^N ,$$

with

$$\mathbf{s}^{t+1} = \Phi(\mathbf{s}^t) .$$

In some cases in the discussion below, Φ will also be used to denote a map on subconfigurations of the lattice. Whether Φ applies to global configurations or subconfigurations should be clear from context.

In a *synchronous* CA, a global clock provides an update signal for all cells: at each t, all cells synchronously read the states of the cells in their neighborhood and then update their own states according to $s_i^t = \phi(\eta_i^t)$.

The neighborhood η is often taken to be spatially symmetric. For one-dimensional CAs, $\eta_i = s_{i-r}, \ldots, s_0, \ldots, s_{i+r}$, where r is the CA's *radius*. Thus, $\phi : \Sigma^{2r+1} \to \Sigma$. For small-radius, binary-state CAs, in which the number of possible neighborhoods is not too large, ϕ is often displayed as a look-up table, or *rule table*, which lists each possible η together with its resulting *output bit* s^{t+1}.

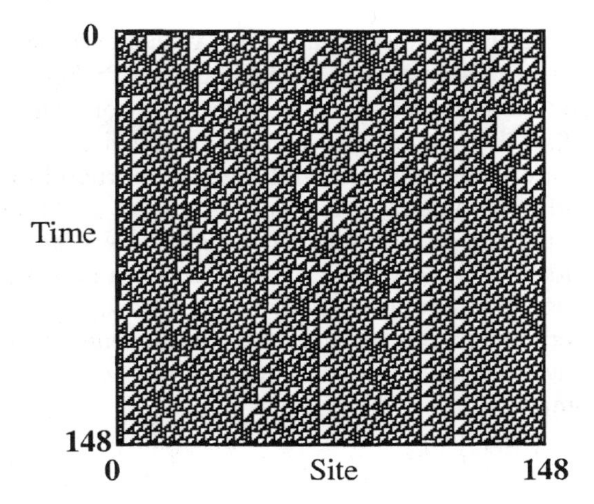

FIGURE 2 A space-time diagram illustrating the typical behavior of elementary CA (ECA) 110. The lattice of 149 sites, displayed horizontally at the top, starts with s_0, being an arbitrary initial configuration. Cells in state 1 are displayed as black, and cells in state 0 are displayed as white. Time increases down the page.

The architecture of a one-dimensional, $(k, r) = (2, 1)$ CA is illustrated in figure 1. Here, the neighborhood of each cell consists of itself and its two nearest neighbors, and the boundary conditions are periodic: $s_N = s_0$.

The 256 one-dimensional, $(k, r) = (2, 1)$ CAs are called *elementary CAs* (ECAs). Wolfram [82] introduced a numbering scheme for one-dimensional CAs. The output bits can be ordered lexicographically, as in figure 1, and are interpreted as the binary representation of an integer between 0 and 255 with the leftmost bit being the least significant digit and the rightmost the most significant digit. In this scheme, the elementary CA pictured here is number 110.

In this chapter, we will restrict our attention to synchronous, one-dimensional, $(k, r) = (2, 3)$ CAs with periodic boundary conditions. This choice of parameters will be explained below. For ease of presentation, we will sometimes refer to a CA by its transition rule ϕ (e.g., as in "the CA ϕ ... ").

The behavior of CAs is often illustrated using space-time diagrams in which the configurations $\mathbf{s}^t$ on the lattice are plotted as a function of time. Figure 2 shows a space-time diagram of the behavior of ECA 110 on a lattice of $N =$ 149 sites and periodic boundary conditions, starting from an arbitrary initial configuration (the lattice is displayed horizontally) and iterated over 149 time steps with time increasing down the figure. A variety of local structures are apparent to the eye in the space-time diagram. They develop over time and move in space and interact.

ECAs are among the simplest spatial dynamical systems: discrete in time, space, and local state. Despite this, as can be seen in figure 2, they generate quite complicated, even apparently aperiodic behavior. The architecture of a CA can be modified in many ways—increasing the number of spatial dimensions, the number k of states per cell, and the neighborhood radius r; modifying the boundary conditions; making the local CA rule ϕ probabilistic rather than deterministic; making the global update Φ asynchronous; and so on.

CAs are included in the general class of "iterative networks" or "automata networks." (See Fogelman-Soulie et al. [31] for a review.) They are distinguished from other architectures in this class by their homogeneous and local ($r \ll N$) connectivity among cells, homogeneous update rule across all cells, and (typically) relatively small k.

For quite some time, due to their appealingly simple architecture, CAs have been successfully employed as models of physical, chemical, biological, and social phenomena, such as fluid flow, galaxy formation, earthquakes, chemical pattern formation, biological morphogenesis, and vehicular traffic dynamics. They have been considered as mathematical objects about which formal properties can be proved. They have been used as parallel computing devices, both for the high-speed simulation of scientific models and for computational tasks such as image processing. In addition, CAs have been used as abstract models for studying "emergent" cooperative or collective behavior in complex systems. For discussions of work in all these areas, see e.g., [5, 27, 31, 38, 47, 60, 45, 56, 72, 84].

3 A COMPUTATIONAL TASK FOR CELLULAR AUTOMATA

It has been shown that some CAs are capable of universal computation [4, 51, 68]. The constructions either embed a universal Turing machine's tape states, read/write head location, and finite-state control in a CA's configurations and rule, or they design a CA rule, supporting propagating and interacting particles, which simulates a universal logic circuit. These constructions are intended to be in-principle demonstrations of the potential computational capability of CAs, rather than implementations of practical computing devices; they do not give much insight about the computational capabilities of CAs in practice. Also, in such constructions, it is typically very difficult to design initial configurations that perform a desired computation. Moreover, these constructions amount to using a massively parallel architecture to simulate a serial one.

Our interest in CA computation is quite different from this approach. In our work, CAs are considered to be massively parallel and spatially extended pattern-forming systems. Our goal is to use machine-learning procedures, such as GA stochastic search, to automatically design CAs that implement parallel computation by taking advantage of the patterns formed via collective behavior of the cells.

To this end, we chose a particular computation for a one-dimensional, binary-state CA—density classification—that requires collective behavior. The task is to determine whether ρ_0, the fraction of 1s in the initial configuration (IC) $\mathbf{s}_0$, is greater than or less than a critical value ρ_c. If $\rho_0 > \rho_c$, the entire lattice should relax to a fixed point of all 1s (i.e., $\Phi(1^N) = 1^N$) in a maximum of $T_{\max}$ time steps; otherwise, it should relax to a fixed point of all 0s (i.e., $\Phi(0^N) = 0^N$) within that time. The task is undefined for $\rho_0 = \rho_c$. In our experiments, we set $\rho_c = 1/2$ and $T_{\max} = 2N$. The *performance* $\mathcal{P}_N^I(\phi)$ of a CA ϕ on this task is calculated by randomly choosing I initial configurations on a lattice of N cells, iterating ϕ on each IC for a maximum of $T_{\max}$ time steps, and determining the fraction of the I ICs that were correctly classified by ϕ—a fixed point of all 1s for $\rho_0 > \rho_c$ and a fixed point of all 0s otherwise. No partial credit is given for final configurations that have not reached an all-1s or all-0s fixed point. As a shorthand, we will refer to this task as the "$\rho_c = 1/2$" task. Defining the task for other values of ρ_c is of course possible; e.g., Chau et al. showed that it is possible to perform the task for rational densities ρ_c using two one-dimensional elementary CAs in succession [7].

This task is trivial for a von Neumann-style architecture that holds the IC as an array in memory: it simply requires counting the number of 1s in $\mathbf{s}_0$. It is also trivial for a two-layer neural network presented with each s_i^0 on one of its N input units, all of which feed into a single output unit: it simply requires weights set so that the output unit fires when the activation reaches the desired threshold ρ_c. In contrast, it is nontrivial to design a CA of our type to perform this task: all cells must agree on a global characteristic of the input even though each cell communicates its state only to its neighbors.

The $\rho_c = 1/2$ task for CAs can be contrasted with the well-studied tasks known as "Byzantine agreement" and "consensus" in the distributed computing literature [23, 28]. These are tasks requiring a number of distributed processors to come to agreement on a particular value held initially by one of the processors. Many decentralized protocols have been developed for such tasks. They invariably assume that the individual processors have more sophisticated computational capabilities and memory than the individual cells in our binary-state CAs or that the communication topologies are more complicated than that of our CAs. Moreover, to our knowledge, none of these protocols addresses the problem of classifying a global property (such as initial density) of all the processors.

Given this background, we asked whether a GA could design CAs in which collective behavior allowed them to perform well above chance ($\mathcal{P}_N^I(\phi) > 0.5$) on this task for a range of N. To minimize local processor and local communication complexity, we wanted to use the smallest values of k and r for which such behavior could be obtained. Over all 256 ECAs ϕ, the maximum performance $\mathcal{P}_N^{10^4}(\phi)$ is approximately 0.5 for $N \in \{149, 599, 999\}$. For all CAs ϕ evolved in 300 runs of the GA on $(k, r) = (2, 2)$ CAs, the maximum $\mathcal{P}_N^{10^4}(\phi)$ was approximately 0.58 for $N = 149$ and approximately 0.5 for $N \in \{599, 999\}$. (The GA's details

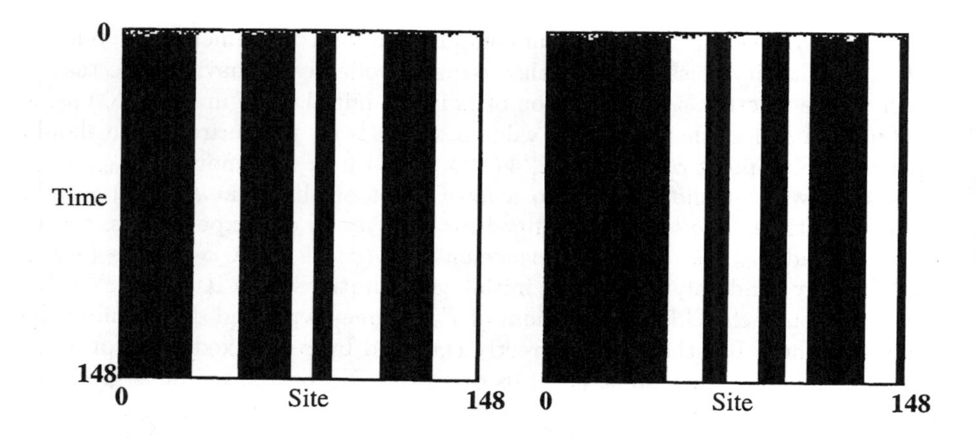

FIGURE 3 Space-time diagrams for ϕ_{maj}, the $r = 3$ local-majority-vote CA. In the left diagram, $\rho_0 < 1/2$; in the right diagram, $\rho_0 > 1/2$.

will be given in the next section.) Increasing the radius to $r = 3$, though, resulted in markedly higher performance and more sophisticated collective behavior. As a result, all of the experiments described in this chapter were performed on one-dimensional, $(k, r) = (2, 3)$ CAs with $N \in \{149, 599, 999\}$ and periodic boundary conditions. Note that for $r = 3$, the neighborhood size $|\eta| = 7$.

One naïve candidate solution to the $\rho_c = 1/2$ task, which we will contrast with the GA-evolved CAs, is the $r = 3$ "majority vote" CA. This CA, denoted ϕ_{maj}, maps the center cell in each seven-cell neighborhood to the majority state in that neighborhood. Figure 3 gives two space-time diagrams illustrating the behavior of ϕ_{maj} on two ICs, one with $\rho_0 < 1/2$ and the other with $\rho_0 > 1/2$. As can be seen, small high- and low-density regions are mapped to regions of all 1s and 0s, respectively. But when an all-1s region and an all-0s region border each other, there is no way to decide between them and both persist. Thus, ϕ_{maj} does not perform the $\rho_c = 1/2$ task. In particular, $\mathcal{P}_N^{10^4}(\phi_{\text{maj}})$ was measured to be zero for $N \in \{149, 599, 999\}$. At a minimum more sophisticated coordination in the form of information transfer and decision making is required. And, given the local nature of control and communication in CAs, the coordination among cells must emerge in the absence of any central processor or central memory directing the cells.

Other researchers, building on our work, have examined variations of the $\rho_c = 1/2$ task that can be performed by simple CAs or by combinations of CAs. Capcarrere et al. [6] noted that changing the output specification makes the task significantly easier. For example, ECA 184 classifies densities of initial conditions within $\lceil N/2 \rceil$ time steps by producing a final configuration of a checkerboard pattern $(01)^*$ interrupted by one or more blocks of at least two consecutive 0s for low-density ICs or at least two consecutive 1s for high-density ICs. Fukś [33] noted that by using the final configuration of ECA 184 as the initial configuration of ECA 232, the correct final configuration of either all-0s or all-1s is obtained. Note that Fukś' solution requires a central controller that counts time up to $\lceil N/2 \rceil$ steps in order to shift from a CA using rule 184 to one using rule 232.

Both solutions always yield a correct density classification; whereas, the single-CA $\rho_c = 1/2$ task is considerably more difficult. In fact, it has been proven that no single, finite-radius two-state CA can perform the $\rho_c = 1/2$ task perfectly for all N [19, 49].

Our interest is not focused on developing a new and better parallel method for performing this specific task. Clearly, one-dimensional, binary-state cellular automata are far from the best architectures to use if one is interested in performing density classification efficiently. As we have emphasized before, the task is trivial within other computational model classes. Instead, our interest is in investigating how GAs can design CAs that have interesting collective computational capabilities and how we can understand those capabilities. Due to our more general interest, we have been able to adapt this paradigm to other spatial computation tasks—tasks for which the above specific solutions do not apply and for which even approximate hand-designed CA solutions were not previously known [20].

4 EVOLVING CELLULAR AUTOMATA WITH GENETIC ALGORITHMS

Genetic algorithms are search methods inspired by biological evolution [41]. In a typical GA, candidate solutions to a given problem are encoded as bit strings. A population of such strings ("chromosomes") is chosen at random and evolves over several generations under selection, crossover, and mutation. At each generation, the fitness of each chromosome is calculated according to some externally imposed fitness function, and the highest-fitness chromosomes are selected preferentially to form a new population via reproduction. Pairs of such chromosomes produce offspring via crossover, where an offspring receives components of its chromosome from each parent. The offspring chromosomes are then subject at each bit position to a small probability of mutation (i.e., being flipped). After several generations, the population often contains high-fitness chromosomes—approximate solutions to the given problem. (For overviews of GAs, see Goldberg [36] and Mitchell [55].)

We used a GA to search for $(k, r) = (2, 3)$ CAs to perform the $\rho_c = 1/2$ task.[2] Each chromosome in the population represented a candidate CA: it consisted of the output bits of the rule table, listed in lexicographic order of neighborhood

[2]The basic framework was introduced in Packard [62] to study issues of phase transitions, computation, and adaptation. For a review of the original motivations and a critique of the results, see Mitchell et al. [58].

(cf. ϕ in figure 1). The chromosomes representing CAs were of length $2^{2r+1} = 128$ bits. The size of the space in which the GA searched was thus 2^{128}—far too large for exhaustive enumeration and performance evaluation.

Our version of the GA worked as follows. First, an initial population of M chromosomes was chosen at random. The *fitness* $F_N^I(\phi)$ of a CA ϕ in the population was computed by randomly choosing I ICs on a lattice of N cells, iterating the CA on each IC either until it arrived at a fixed point or for a maximum of $T_{\max}$ time steps. It was then determined whether the final configuration was correct—i.e., the all-0s fixed point for $\rho_0 < 1/2$ or the all-1s fixed point for $\rho_0 > 1/2$. $F_N^I(\phi)$ was the fraction of the I ICs on which ϕ produced the correct final behavior. No credit was given for partially correct final configurations.

In each generation, (1) a new set of I ICs was generated; (2) $F_N^I(\phi)$ was computed for each CA ϕ in the population; (3) CAs in the population were ranked in order of fitness (with ties broken at random); (4) a number E of the highest fitness CAs (the "elite") was copied to the next generation without modification; and (5) the remaining $M - E$ CAs for the next generation were formed by crossovers between randomly chosen pairs of the elite CAs. With probability p_c, each pair was crossed over at a single randomly chosen locus l, forming two offspring. The first child inherited bits 0 through l from the first parent and bits $l + 1$ through 127 from the second parent—vice versa for the second child. The parent CAs were chosen for crossover from the elite with replacement—that is, an elite CA was permitted to be chosen any number of times. The two offspring chromosomes from each crossover (or copies of the parents, if crossover did not take place) were mutated ($0 \to 1$ and $1 \to 0$) at each locus with probability p_m. This process was repeated for G generations during a single GA run. Note that since a different sample of ICs was chosen at each generation, the fitness function itself is a random variable.

We ran experiments with two different distributions for choosing the M chromosomes in the initial population and the set of I ICs at each generation: (i) an "unbiased" distribution in which each bit's value is chosen independently with equal probability for 0 and 1, and (ii) a density-uniform distribution in which strings were chosen with uniform probability over $\lambda \in [0, 1]$ or over $\rho_0 \in [0, 1]$, where λ is the fraction of 1s in ϕ's output bits and ρ_0 is the fraction of 1s in the IC. Using the density-uniform distribution for the initial CA population and for the ICs considerably improved the GA's ability to find high-fitness CAs on any given run. (That is, we could use 50% fewer generations per GA run and still find high-performance CAs.) The results we report here are from experiments in which density-uniform distributions were used.

The experimental parameters we used were $M = 100$, $I = 100$, $E = 20$, $N = 149$, $T_{\max} = 2N$, $p_c = 1.0$ (i.e., crossover was always performed), $p_m = 0.016$, and $G = 100$. Experiments using variations on these parameters did not result in higher performance solutions or faster convergence to the best performance solutions.

We used $\mathcal{P}_N^{10^4}$ with $N \in \{149, 599, 999\}$ to test the quality of the evolved CAs. This performance measure is a more stringent quality test than the fitness F_N^{100} used in the GA runs: under $\mathcal{P}_N^{10^4}$, the ICs are chosen from an unbiased distribution and thus have ρ_0 close to the density threshold $\rho = 1/2$. Such ICs are the hardest cases to classify. Thus, $\mathcal{P}_N^{10^4}$ gives a lower bound on other performance measures. In machine-learning terms, the ICs used to calculate F_{149}^{100} are the training sets for the CAs, and the ICs used to calculate $\mathcal{P}_N^{10^4}$ are larger and harder test sets that probe the evolved CA's generalization ability.

5 RESULTS OF EXPERIMENTS

In this section, we describe the results from 300 independent runs of this GA, with different random number seeds.

In each of the 300 runs, the population converged to CAs implementing one of three types of computational strategies. The term "strategy" here refers to the mechanisms by which the CA attains some level of fitness on the $\rho_c = 1/2$ task. These three strategy types, "default," "block expanding," and "particle," are illustrated in figures 4–6. In each figure, each row contains two space-time diagrams displaying the typical behavior of a CA ϕ that was evolved in a GA run. Thus, CAs from six different runs are shown. In each row, $\rho_0 < 1/2$ in the left space-time diagram and $\rho_0 > 1/2$ in the right. The rule tables and measured $\mathcal{P}_N^{10^4}$ values for the six CAs are given in table 1.

5.1 DEFAULT STRATEGIES

In 11 out of the 300 runs, the highest performance CAs implemented "default" strategies, which on almost all ICs iterate to all 0s or all 1s, respectively. The typical behavior of two such CAs, $\phi_{\text{def}}^{\text{a}}$ and $\phi_{\text{def}}^{\text{b}}$, is illustrated in figures 4(a) and 4(b). Default strategies each have $\mathcal{P}_N^{I}(\phi) \approx 0.5$, since each classifies one density range (e.g., $\rho < 1/2$) correctly and the other ($\rho > 1/2$) incorrectly. Since the initial CA population is generated with uniform distribution over λ, it always contains some CAs with very high or low λ. And since λ is the fraction of 1s in the output bits of the look-up table, these extreme-λ CAs tend to have one or the other default behavior.

5.2 BLOCK-EXPANDING STRATEGIES

In most runs (280 out of 300 in our experiments), the GA evolved CAs with strategies like those shown in figures 5(a) and 5(b). In figure 5(a), $\phi_{\text{exp}}^{\text{a}}$ defaults to an all-1s fixed point (right diagram) unless there is a sufficiently large block of adjacent (or almost adjacent) 0s in the IC. In this case, it expands that block until 0s fill up the entire lattice (left diagram). In figure 5(b), $\phi_{\text{exp}}^{\text{b}}$ has the opposite strategy. It defaults to the all-0s fixed point unless there is a sufficiently

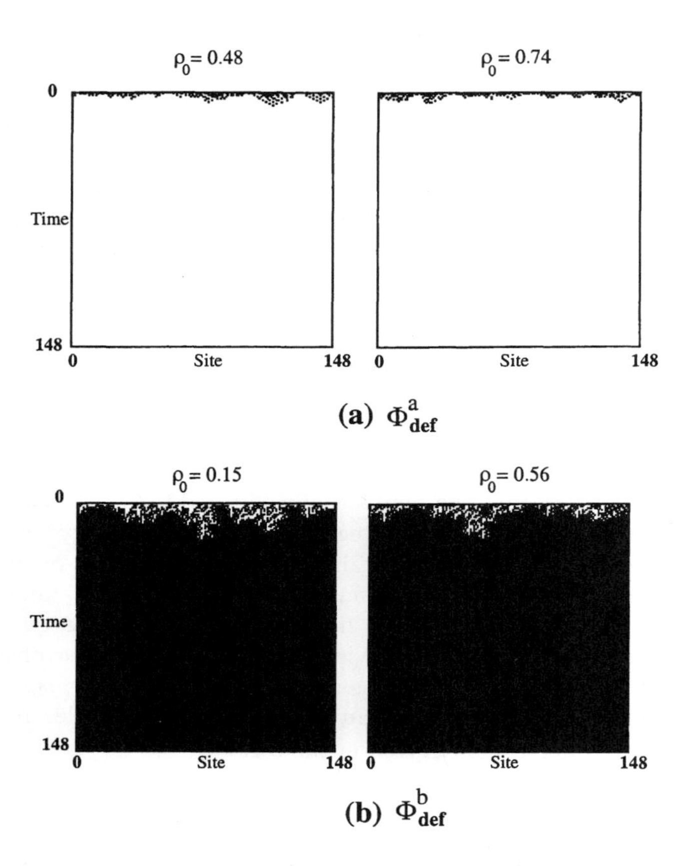

FIGURE 4 Space-time behavior of "default" strategy CAs evolved on two different GA runs. (a) $\phi^{\mathrm{a}}_{\mathrm{def}}$ with $\rho_0 = 0.48$ (left) and $\rho_0 = 0.74$ (right). On almost all ICs this CA iterates to a fixed point of all 0s, correctly classifying only low-ρ ICs. (b) $\phi^{\mathrm{b}}_{\mathrm{def}}$ with $\rho_0 = 0.15$ (left) and $\rho_0 = 0.56$ (right). On almost all ICs this CA iterates to a fixed point of all 1s, correctly classifying only high-ρ ICs.

large block of 1s in the IC. The meaning of "sufficiently large block" depends on the particular CA but is typically close to the neighborhood size $2r + 1$. For example, $\phi^{\mathrm{a}}_{\mathrm{exp}}$ will expand blocks of 8 or more 0s and $\phi^{\mathrm{b}}_{\mathrm{exp}}$ will expand blocks of 7 or more 1s.

These "block-expanding" strategies rely on the presence or absence of blocks of 1s or 0s in the IC: blocks of adjacent 0s (1s) are more likely to appear in low- (high-) density ICs. Since the occurrence of such blocks is statistically correlated with ρ_0, recognizing and then expanding them leads to fitnesses above those for the default strategy. The strength of this correlation depends on the initial

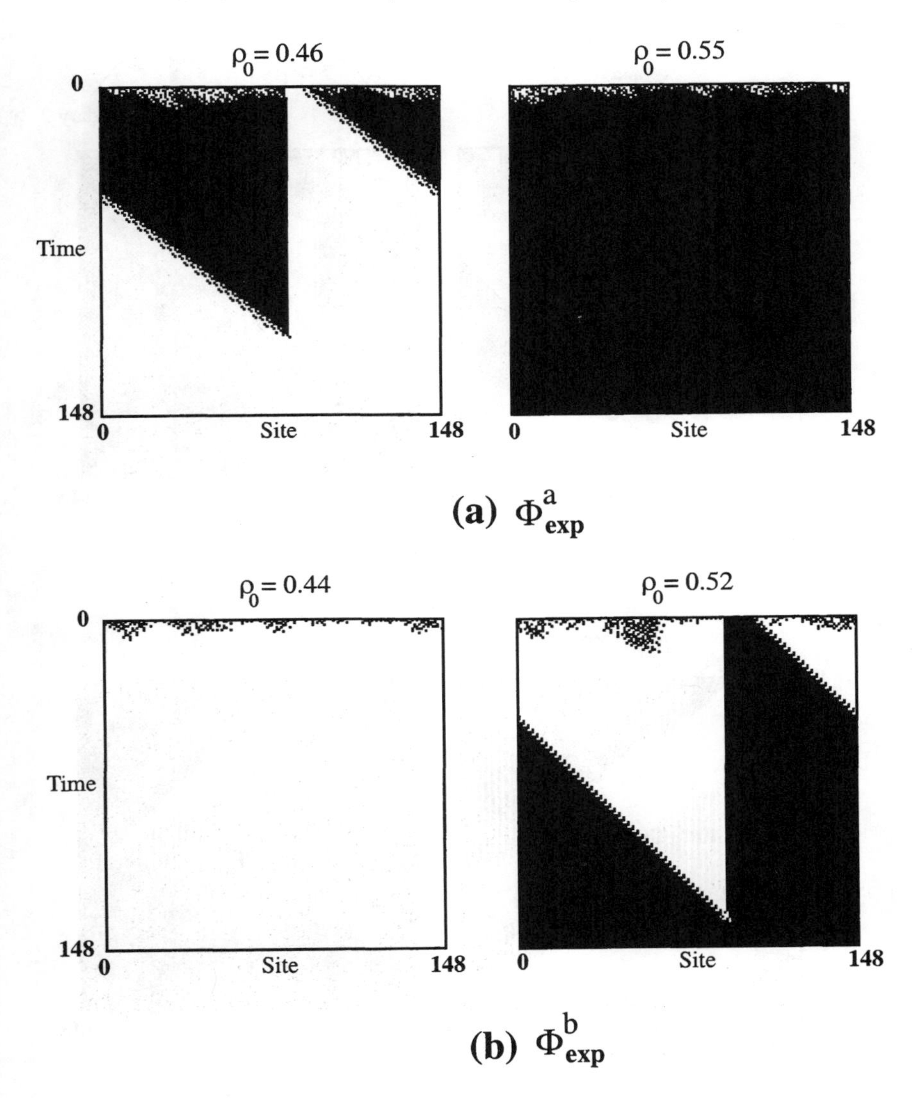

FIGURE 5 Space-time behavior of two "block expanding" CAs evolved on different GA runs. (a) $\phi^{\mathrm{a}}_{\mathrm{exp}}$ with $\rho_0 = 0.46$ (left) and $\rho_0 = 0.55$ (right). This CA defaults to a fixed point of all 1s unless the IC contains a sufficiently large block of adjacent 0s, in which case, that block is expanded. (b) $\phi^{\mathrm{b}}_{\mathrm{exp}}$ with $\rho_0 = 0.44$ (left) and $\rho_0 = 0.52$ (right). This CA defaults to a fixed point of all 0s unless the IC contains a sufficiently large block of adjacent 1s, in which case, that block is expanded. The classification of the IC is correct in each of these four cases.

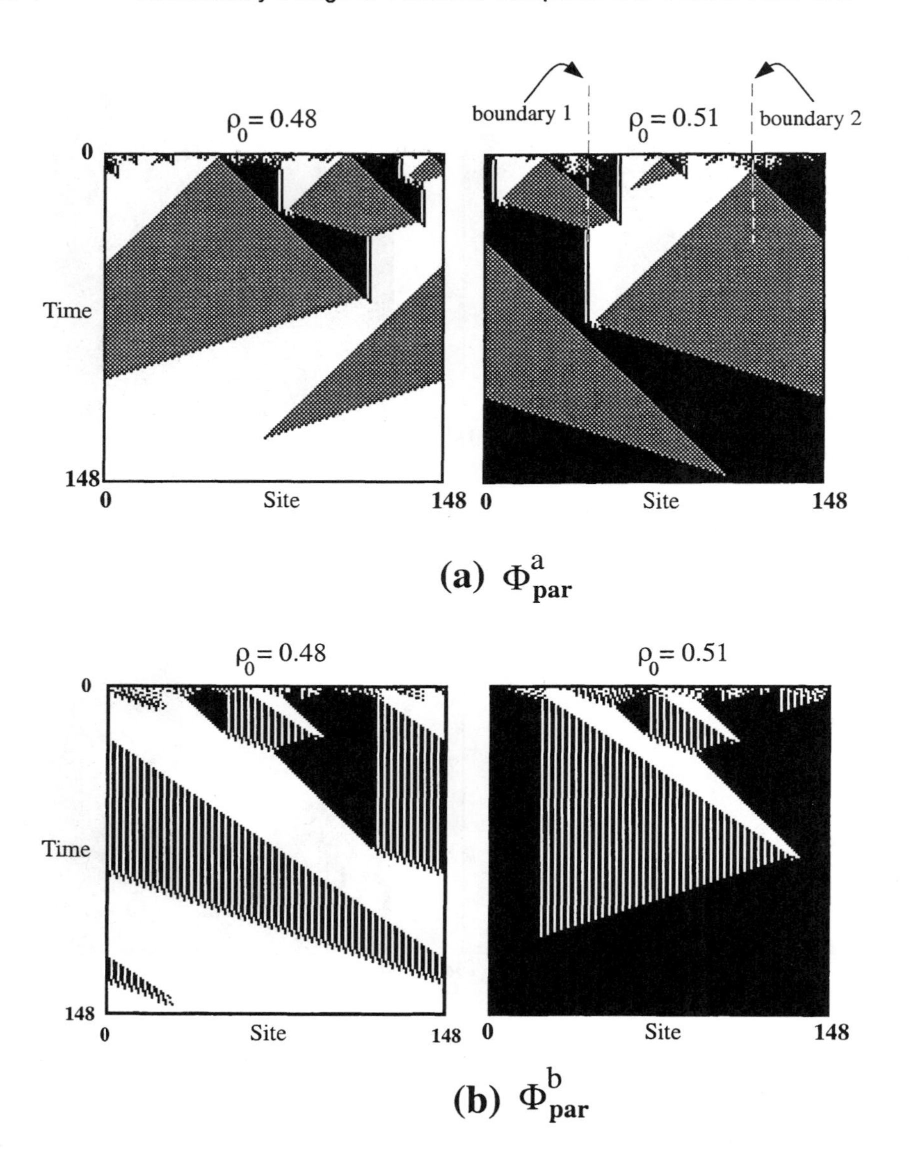

FIGURE 6 Space-time behavior of two "particle" CAs evolved on different GA runs. (a) ϕ^{a}_{par} with $\rho_0 = 0.48$ (left) and $\rho_0 = 0.51$ (right). (b) ϕ^{b}_{par} with $\rho_0 = 0.48$ (left) and $\rho_0 = 0.51$ (right). These CAs use the boundaries between homogeneous space-time regions to effect information transmission and processing. Again, the classification of the IC is correct in each of these four cases.

density ρ_0 and on the lattice size N. Typical block-expanding strategies have $F_{149}^{100} \approx 0.9$ and $\mathcal{P}_{149}^{10^4} \approx 0.6$. The block-expanding strategies designed by the GA are adapted to $N = 149$; their performances do not scale well to larger lattice sizes. This occurs since the probability of a block of, say, seven adjacent 1s appearing for a given ρ_0 increases with N, and this means that the correlation between the occurrence of this block and density decreases. This can be seen in the measured values of $\mathcal{P}_N^{10^4}$ for $\phi_{\text{exp}}^{\text{a}}$ and $\phi_{\text{exp}}^{\text{b}}$ for longer lattices given in table 1.

5.3 EMBEDDED-PARTICLE STRATEGIES

The block-expanding strategies are not examples of the kind of sophisticated coordination and information transfer that we claimed must be achieved for robust performance on the $\rho_c = 1/2$ task. Under these strategies, all the computation is done locally in identifying and then expanding a "sufficiently large" block. Moreover, the performance on $N = 149$ does not generalize to larger lattices. Clearly, the block-expanding strategies are missing important aspects required by the task. The third class of strategies evolved by the GA, the "embedded-particle" strategies, do achieve the coordination and communication we alluded to earlier. Typical space-time behaviors of two particle strategies, $\phi_{\text{par}}^{\text{a}}$ and $\phi_{\text{par}}^{\text{b}}$, are given in figures 6(a) and 6(b). It can be seen that there is a transient phase during which the spatial and temporal transfer of information about the local density takes place. Such strategies were evolved in 9 out of the 300 runs.

The behavior of $\phi_{\text{par}}^{\text{a}}$ is somewhat similar to that of ϕ_{maj} in that local high-density regions are mapped to all 1s and local low-density regions are mapped to all 0s. In addition, a vertical stationary boundary separates these regions. The set of local spatial configurations that make up this boundary is specified in formal language terms by the regular expression $(1)^+01(0)^+$, where $(w)^+$ means a positive number of repetitions of the word w [42].

The stationary boundary appears when a region of 1s on the left meets a region of 0s on the right. However, there is a crucial difference from ϕ_{maj}: when a region of 0s on the left meets a region of 1s on the right. A checkerboard region $(01)^+$ grows in size with equal speed in both directions. A closer analysis of its role in the overall space-time behavior shows that the checkerboard region serves to decide which of the two adjacent regions (0s and 1s) is the larger. It does this by simply cutting off the smaller region, and so the larger (0 or 1) region continues to expand. The net decision is that the density in the region was in fact below or above $\rho_c = 1/2$. The spatial computation here is largely geometric: there is a competition between the sizes of high- and low-density regions.

For example, consider the right-hand space-time diagram of figure 6(a). The large low-density region between the lines marked "boundary 1" and "boundary 2" is smaller than the large high-density region between "boundary 2" and "boundary 1" (moving left from boundary 2 and wrapping around). The left-hand side of the checkerboard region (centered around boundary 2) collides with boundary 1 before the right-hand side does. The result is that the collision cuts off the inner white region, letting the outer black region propagate.

In this way, $\phi^{\mathrm{a}}_{\mathrm{par}}$ uses local interactions and simple geometry to determine the relative sizes of adjacent low- and high-density regions that are larger than the neighborhood size. As is evident in figures 6(a) and 6(b), this type of size competition over time happens across increasingly larger spatial scales, gradually resolving competitions between larger and larger regions.

The black-white boundary and the checkerboard region can be thought of as signals indicating "ambiguous" density regions. Each of these boundaries has local density exactly at $\rho_c = 1/2$. Thus, they are not themselves "classified" by the CA as low or high density. The result is that these signals can persist over time. The creation and interaction of these signals can be interpreted as the locus of the computation being performed by the CA: they form its emergent "algorithm," what we have been referring to as the CA's "strategy."

In figure 6(b), $\phi^{\mathrm{b}}_{\mathrm{par}}$ follows a similar strategy, but with a vertically striped region playing the role of the checkerboard region in $\phi^{\mathrm{a}}_{\mathrm{par}}$. However, in this case, there are asymmetries in the speeds of the propagating region boundaries. This difference yields a lower $\mathcal{P}^{10^4}_N$, as can be seen in table 1.

These descriptions of the computational strategies evolved by the GA are informal. A major goal of our work is to make terms such as "computation," "computational strategy," and "emergent algorithm" more rigorous for cellular automata. In the next section, we will describe how we are using the notions of domains, particles, and particle interactions to do this. We will use these notions to answer questions such as, How, precisely, is a given CA performing the task? What structural components are used to support this information processing? How can we predict $\mathcal{P}^I_N$ and other computational properties of a given CA? Why is $\mathcal{P}^I_N$ greater for one CA than for another? What types of mistakes does a given CA make in performing the $\rho_c = 1/2$ task? These types of questions are difficult, if not impossible, to answer in terms of local space-time notions such as the bits in a CA's look-up table or even the raw space-time configurations produced by the CA. A higher-level description is needed, one that incorporates computational structures.

6 UNDERSTANDING COLLECTIVE COMPUTATION IN CELLULAR AUTOMATA

In this section, we will describe our approach to formalizing the notion of computational strategy in cellular automata and in other spatially extended systems. This approach is based on the computational mechanics framework of Crutchfield [13], as applied to cellular automata by Crutchfield and Hanson [16, 39, 40]. This framework comprises a set of methods for classifying the different patterns that appear in CA space-time behavior, using concepts from computation and dynamical systems theories. These methods were developed as a way of ana-

TABLE 1 CA chromosomes (look-up table output bits) given in hexadecimal and $\mathcal{P}_N^{10^4}$ for the six CAs illustrated in figures 4–6, on lattices of sizes $N = 149$, $N = 599$, and $N = 999$. To recover the 128-bit string giving the CA look-up table outputs, expand each hexadecimal digit (left to right, top row followed by bottom row) to binary. This yields the neighborhood outputs in lexicographic order of neighborhood, with the leftmost bit of the 128-bit string giving the output bit for neighborhood 0000000, and so on. Since $\mathcal{P}_N^{10^4}$ is measured on a randomly chosen sample of ICs, it is a random variable. This table gives its mean over 100 trials for each CA. Its standard deviation over the same 100 trials is approximately 0.005 for each CA for all three values of N. For comparison, the best known $(k, r) = (2, 3)$ CAs for the $\rho_c = 1/2$ task have $\mathcal{P}_{149}^{10^4} \approx 0.85$ (see section 9). This appears to be close to the upper limit of $\mathcal{P}_{149}^{10^4}$ for this class of spatial architectures.

CA Name	Rule Table (Hexadecimal)	$\mathcal{P}_{149}^{10^4}$	$\mathcal{P}_{599}^{10^4}$	$\mathcal{P}_{999}^{10^4}$
$\phi_{\text{def}}^{\text{a}}$	100111215030114D 01613507143B05BF	0.500	0.500	0.500
$\phi_{\text{def}}^{\text{b}}$	0BF9D97AF26F4F4B F3FF301F0B110DF7	0.499	0.499	0.501
$\phi_{\text{exp}}^{\text{a}}$	1010614604273F9B 7FD7D9DF35F53FFF	0.656	0.523	0.504
$\phi_{\text{exp}}^{\text{b}}$	02330A4B07016711 42D080C3CD877B7F	0.643	0.513	0.502
$\phi_{\text{par}}^{\text{a}}$	0504058605000F77 037755877BFFB77F	0.775	0.740	0.728
$\phi_{\text{par}}^{\text{b}}$	0012003350501123 3B77F7FFFDFFD57F	0.766	0.687	0.641

lyzing the behavior of cellular automata and other dynamical systems. They extend more traditional geometric and statistical analyses by revealing the intrinsic information-processing structures embedded in dynamical processes.

6.1 COMPUTATIONAL MECHANICS OF CELLULAR AUTOMATA

As applied to cellular automata, the purpose of computational mechanics is to discover an appropriate "pattern basis" with which to describe the structural components that emerge in a CA's space-time behavior. A CA pattern basis consists of a set Λ of formal languages $\{\Lambda^i, i = 0, 1, \ldots\}$ in terms of which a CA's space-time behavior can be decomposed concisely and in a way constrained by the temporal dynamics. Once such a pattern basis is found, those cells in space-time regions that are described by the basis can be seen as forming background "domains" against which coherent structures—defects, walls, etc.—not fitting

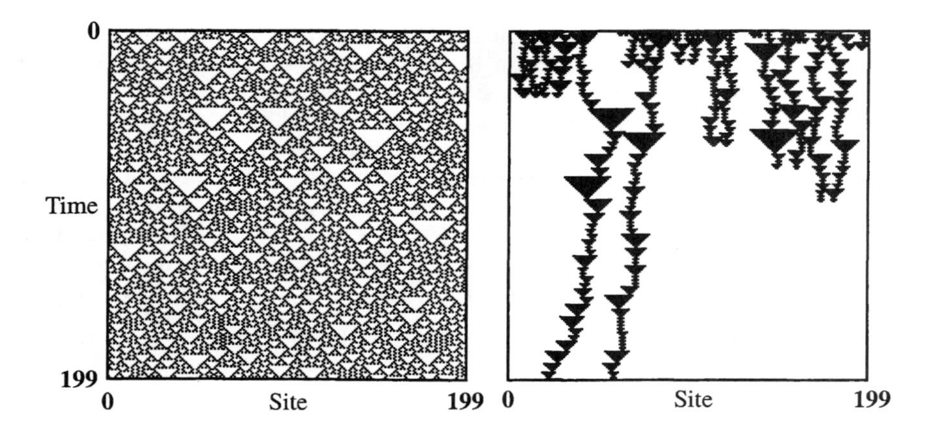

FIGURE 7 (a) Space-time diagram illustrating the typical behavior of ECA 18—a CA exhibiting apparently random behavior; i.e., the set of length-L spatial words has a positive entropy density as $L \to \infty$. (b) The same diagram with the regular domains—instances of words in Λ^0—filtered out, leaving only the embedded particles $\mathbf{P} = \{1(00)^n 1, n = 0, 1, 2, \ldots\}$. (After Crutchfield and Hanson [15]. Printed with permission of the authors.)

the basis move. In this way, structural features above and beyond the domains can be identified and their dynamics analyzed and interpreted on their own terms.

For example, consider the space-time diagram of figure 7(a), illustrating the apparently random behavior of ECA 18. This example is a useful illustration of embedded information processing since the coherent structures are not immediately apparent to the eye. The computational mechanics analysis [15, 40] of ECA 18 uses a pattern basis consisting of the single domain language $\Lambda = \{\Lambda^0 = (0\Sigma)^+\}$, where $\Sigma = \{0, 1\}$. That is, over most regions in ECA 18's configurations, every other site is a 0 and the remaining sites are wildcards, either 0 or 1. (Often this type of formal-language description of a set of configuration features can be discovered automatically via the "ϵ-machine reconstruction" algorithm [13, 39].)

Crutchfield and Hanson define a *regular domain* Λ^j as a space-time region that (i) is a regular language and (ii) is space- and time-translation invariant. Regular domains can be represented by either the set Λ^i of configurations or by the minimal finite-state machine $\mathrm{M}(\Lambda^i)$ that recognizes Λ^i. More specifically, let $\{\Lambda^i\}$ be the pattern basis for CA ϕ. Then the regular domain Λ^i describes the space-time regions of $\{\Phi^t(\mathrm{s}^0) : t = 0, 1, 2, \ldots\}$ whose configurations are words in Λ^i. Formally, then, a regular domain Λ^i is a set that is

1. temporally invariant: the CA always maps a configuration in Λ^i to another configuration in Λ^i: $\Phi(\mathbf{s}) = \mathbf{s}'$, $\mathbf{s}, \mathbf{s}' \in \Lambda^i$; and
2. spatially homogeneous: the same pattern can occur at any site; the recurrent states in the minimal finite automaton $\mathrm{M}(\Lambda^i)$ recognizing Λ^i are strongly connected.

Once a CA's regular domains are discovered, either through visual inspection or by an automated induction method, and proved to satisfy the above two conditions, then the corresponding space-time regions are, in a sense, understood. Given this level of discovered regularity, the domains can be filtered out of the space-time diagram, leaving only the "unmodeled" deviations, referred to as domain "walls," whose dynamics can then be studied in and of themselves. Sometimes, as is the case for the evolved CA we analyze here, these domain walls are spatially localized, time-invariant structures and so can be considered to be "particles."

In ECA 18, there is only one regular domain Λ^0. It turns out that it is stable and so is called a regular "attractor"—the stable invariant set to which configurations tend over long times, after being perturbed away from it by, for example, flipping a site value [15, 40]. Although there are random sites in the domain, its basic pattern is described by a simple rule: all configurations are allowed in which every other site value is a 0. If these fixed-value sites are on even-numbered lattice sites, then the odd-numbered lattice sites have a wild card value, being 0 or 1 with equal probability. The boundaries between these "phase-locked" regions are "defects" in the spatial periodicity of Λ^0 and, since they are spatially localized in ECA 18, they can be thought of as particles embedded in the raw configurations.

To locate a particle in a configuration generated by ECA 18, assuming one starts in a domain, one scans across the configuration, from left to right say, until the spatial, period-2 phase is broken. This occurs when a site value of 1 is seen where the domain pattern indicates a 0 should be. Depending on a particle's structure, it can occur, as it does with ECA 18, that scanning the same configuration in the opposite direction (right to left) may lead to the detection of the broken domain pattern at a different site. In this case, the particle is defined to be the set of local configurations between these locations.

Due to this, ECA 18's particles are manifest in spatial configurations as blocks in the set $\mathbf{P} = \{1(00)^n1, n = 0, 1, 2, \ldots\}$, a definition that is left-right scan invariant. Figure 7(b) shows a filtered version of figure 7(a) in which the cells participating in Λ^0 are colored white and the cells participating in $\mathbf{P}$ are colored black. The spatial structure of the particles is reflected in the triangular structures, which are regions of the lattice in which the particle—the breaking of Λ^0's pattern—is localized, though not restricted to a single site.

In this way, ECA 18's configurations can be decomposed into natural, "intrinsic" structures that ECA 18 itself generates—viz., its domain Λ^0 and its particle $\mathbf{P}$. These structures are summarized for a CA in what we call a *particle*

TABLE 2 ECA 18's catalog of regular domains, particles, and particle interactions. The notation $p \sim \Lambda^i \Lambda^j$ means that p is the particle formed by the boundary between domains Λ^i and Λ^j.

Regular Domain
$\Lambda^0 = \{(0(0+1))^*\}$
Particle
$\alpha \sim \Lambda^0\ \Lambda^0 = \{1(00)^n 1, n = 0, 1, 2, \dots\}$
Interaction (annihilation)
$\alpha + \alpha \rightarrow \emptyset\ \ (\Lambda^0)$

catalog. The catalog is particularly simple for ECA 18; cf. table 2. The net result is that ECA 18's behavior can be redescribed at the higher level of particles. It is noteworthy that, starting from arbitrary initial configurations, ECA 18's particles have been shown to follow a random walk in space time on an infinite lattice, annihilating in pairs whenever they intersect [25, 40]. One consequence is that there is no further structure, such as coherent particle groupings, to understand in ECA 18's dynamics. Thus, one moves from the deterministic dynamics at the level of the CA acting on raw configurations to a level of stochastic particle dynamics. The result is that ECA 18 configurations, such as those in figure 7(a), can be analyzed in a much more structural way than by simply classifying ECA 18 as "chaotic."

In the computational mechanics view of CA dynamics, embedded particles carry various kinds of information about local regions in the IC. Given this, particle interactions are the loci at which this information is combined and processed and at which decisions are made. In general, these structural aspects—domains, particles, and interactions—do not appear immediately. As will be seen below, often there is a initial disordered period, after which the configurations condense into well-defined regular domains, particles, and interactions. To capture this relaxation process, we define the *condensation time* t_c as the first iteration at which the filtered space-time diagram contains only well-defined domains in Λ and the walls between them. In other words, at t_c, every cell participates in either a regular domain, of width at least $2r + 1$, in a wall between them, or in an interaction between walls. (See Crutchfield et al. [17] and Hordijk et al. [43] for a more detailed discussion of the condensation phase and its consequences.)

6.2 COMPUTATIONAL MECHANICS OF EVOLVED CELLULAR AUTOMATA

This same methodology is particularly useful in understanding and formalizing the computational strategies that emerged in the GA-evolved CAs. Fortunately, in the following exposition, most of the structural features in the evolved CAs are apparent to the eye. Figure 6(a) suggests that an appropriate pattern basis for $\phi_{\text{par}}^{\text{a}}$ is $\Lambda = \{\Lambda^0 = 00^+, \Lambda^1 = 11^+, \Lambda^2 = (01)^+\}$, corresponding to the all-white, all-black, and checkerboard regions. Similarly, figure 6(b) suggests that for $\phi_{\text{par}}^{\text{b}}$ we use $\Lambda = \{\Lambda^0 = 00^+, \Lambda^1 = 111^+, \Lambda^0 = (011)^+\}$, corresponding to the all-white, all-black, and striped regions.

Note that a simple shortcut can be used to identify domains that are spatially and temporally periodic. If the same "pattern" appears to repeat over a sufficiently large ($\gg r$ cells by r time steps) space-time region, then it is a domain. It is also particularly easy to prove such regions are regular domains. Exactly how the pattern is expressed as a regular language or as a minimal finite-state machine typically requires closer inspection.

Once identified, the computational contributions of these space-time regions can be easily understood. The contributions consist solely of the generation of words in the corresponding regular language. Since this requires only a finite amount of spatially localized memory, its direct contribution to the global computation required by the task is minimal. (The density of memory vanishes as the domain increases in size.) The conclusion is that the domains themselves, while necessary, are not the locus of the global information processing.

Figure 8 is a version of figure 6 with $\phi_{\text{par}}^{\text{a}}$'s and $\phi_{\text{par}}^{\text{b}}$'s regular domains filtered out. The result reveals the walls between them, which for $\phi_{\text{par}}^{\text{a}}$ and $\phi_{\text{par}}^{\text{b}}$ are several kinds of embedded particles. The particles in figure 8 are labeled with Greek letters. This filtering is performed by a building a transducer that reads the raw configurations and can recognize when sites are in which domain. The transducer used for figure 8(a), for example, outputs white at each site in one of $\phi_{\text{par}}^{\text{a}}$'s domains and black at each site participating in a domain wall. (The particular transducer and comments on its construction and properties can be found in the Appendix. The general construction procedure is given in Crutchfield and Hanson [16].)

Having performed the filtering, the focus of analysis shifts away from the raw configurations to the new level of embedded-particle structure. The questions now become, Are the computational strategies explainable in terms of particles and their interactions? Or, is there as yet some unrevealed information processing occurring that is responsible for high performance?

Tables 3 and 4 list all the different particles that are observed in the space-time behavior of $\phi_{\text{par}}^{\text{a}}$ and $\phi_{\text{par}}^{\text{b}}$, along with their velocities and the interactions that can take place between them. Note that these particle catalogs do not include all possible structures, for example, possible three-particle interactions. The computational strategies of $\phi_{\text{par}}^{\text{a}}$ and $\phi_{\text{par}}^{\text{b}}$ can now be analyzed in terms of the particles and their interactions as listed in the particle catalogs.

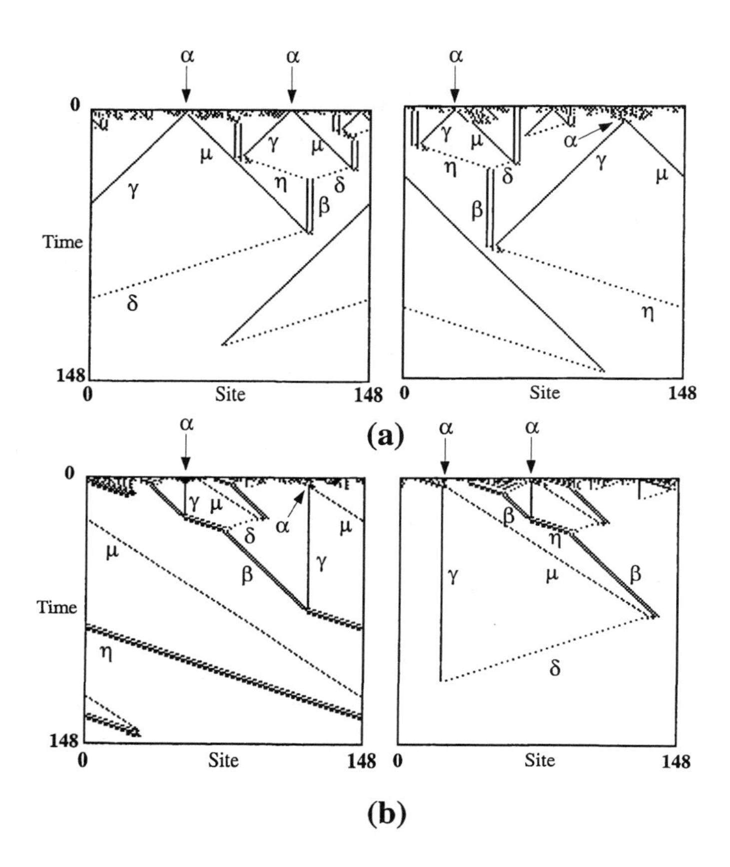

FIGURE 8 (a) Version of figure 6(a) with the regular domains filtered out, revealing the particles and their interactions. (b) Filtered version of figure 6(b).

6.3 COMPUTATIONAL STRATEGY OF ϕ^{A}_{PAR}

In a high-performance CA such as ϕ^{a}_{par}, particles carry information about the density of local regions in the IC, and their interactions combine and process this information, rendering a series of decisions about ρ_0. How do these presumed "functional" components lead to the observed fitness and computation performance?

Referring to table 3 and figure 8(a), ϕ^{a}_{par}'s β particle is seen to consist of the zero-velocity black-to-white boundary. β carries the information that it came from a region in the IC in which the density is locally ambiguous: the density of 1^k0^k, when determined at its center, is exactly ρ_c. The ambiguity cannot be

TABLE 3 $\phi^{\rm a}_{\rm par}$'s catalog of regular domains, particles (including velocities in parentheses), and particle interactions. Note that this catalog leaves out possible three-particle interactions.

Regular Domains		
$\Lambda^0 = \{0^+\}$	$\Lambda^1 = \{1^+\}$	$\Lambda^2 = \{(01)^+\}$
Particles (Velocities)		
$\alpha \sim \Lambda^0\ \Lambda^1$ (—)	$\beta \sim \Lambda^1\ \Lambda^0$ (0)	$\gamma \sim \Lambda^0\ \Lambda^2$ (-1)
$\delta \sim \Lambda^2\ \Lambda^0$ (-3)	$\eta \sim \Lambda^1\ \Lambda^2$ (3)	$\mu \sim \Lambda^2\ \Lambda^1$ (1)
Interactions		
decay	$\alpha \to \gamma + \mu$	
react	$\beta + \gamma \to \eta$, $\mu + \beta \to \delta$, $\eta + \delta \to \beta$	
annihilate	$\eta + \mu \to \emptyset\ (\Lambda^1)$, $\gamma + \delta \to \emptyset\ (\Lambda^0)$	

TABLE 4 $\phi^{\rm b}_{\rm par}$'s catalog of regular domains, particles (including their velocities in parentheses), and particle interactions.

Regular Domains		
$\Lambda^0 = \{0^+\}$	$\Lambda^1 = \{1^+\}$	$\Lambda^2 = \{(011)^+\}$
Particles (Velocities)		
$\alpha \sim \Lambda^1\ \Lambda^0$ (0)	$\beta \sim \Lambda^0\ \Lambda^1$ (1)	$\gamma \sim \Lambda^1\ \Lambda^2$ (0)
$\delta \sim \Lambda^2\ \Lambda^1$ (-3)	$\eta \sim \Lambda^0\ \Lambda^2$ (3)	$\mu \sim \Lambda^2\ \Lambda^0$ (3/2)
Interactions		
decay	$\alpha \to \gamma + \mu$	
react	$\beta + \gamma \to \eta$, $\mu + \beta \to \delta$, $\eta + \delta \to \beta$	
annihilate	$\eta + \mu \to \emptyset\ (\Lambda^0)$, $\gamma + \delta \to \emptyset\ (\Lambda^1)$	

resolved locally. It might be, however, at a later time, when more information can be integrated from other regions of the IC.

Likewise, the α "particle" consists of the white-to-black boundary, but unlike the β particle, α is unstable and immediately decays into two particles γ (white-checkerboard boundary) and μ (checkerboard-black boundary). Like β, α indicates local density ambiguity. The particles into which it decays, γ and μ, carry this information, and so they too are "ambiguous density" signals. γ carries the information that it borders a white (low density) region and μ carries the information that it borders a black (high density) region. The two particles also carry the mutual information of having come from the same ambiguous density region where the transient α was originally located. They carry this positional information about α's location by virtue of having the same speed (± 1).

To see how these elements work together over a space-time region, consider the left side of the left-hand ($\rho_0 < 1/2$) diagram in figure 8(a). Here, α decays into a γ and a μ particle. The μ then collides with a β before its companion γ (wrapping around the lattice) does. This indicates that the low-density white region, whose right border is the γ, is larger than the black region bordered by the μ. The μ-β collision creates a new particle, a δ, that carries this information (low-density domains) to the left, producing more low-density area. δ, a fast moving particle, catches up with the γ (low density) and annihilates it, producing Λ^0 over the entire lattice. The result is that the white region takes over the lattice before the maximum number of iterations has passed. In this way, the classification of the (low density) IC has been correctly determined by the spatial algorithm—the steps we have just described. In the case of $\phi^{\mathrm{a}}_{\mathrm{par}}$, this final decision is implemented by δ's velocity being three times that of γ's.

On the right side of the right-hand ($\rho_0 > 1/2$) diagram in figure 8(a), a converse situation emerges: γ collides with β before μ does. The effective decision indicates that the black region bordered by μ is larger than the white region bordered by γ. In symmetry with the μ-β interaction described above, the γ-β interaction creates the η particle that catches up with the μ and the two annihilate. In this way, the larger black region takes over, and the correct density classification is effected.

A third type of particle-based information processing is illustrated at the top left of the right-hand diagram in figure 8(a). Here, an α decays into a γ and a μ. In this case, the white region bordered by γ is smaller than the black region bordered by μ. As before, γ collides with the β on its left, producing η. However, there is another β particle to the right of μ. Instead of the μ proceeding on to eventually collide with the η, the μ first collides with the second β. Since the μ borders the larger of the two competing regions, its collision is slightly later than the γ-β collision to its left. The μ-β collision produces a δ particle propagating to the left. Now the η and the δ approach each other at equal and opposite speeds and collide. Since η is carrying the information that the white region should win and δ is carrying the information that the black region should win, their collision appropriately results in an "ambiguity" signal—here, a β that later on interacts with particles from greater distances. But since η traveled farther than δ before their collision, a β is produced that is shifted to the right from the original α. The net effect—the net geometric subroutine—is to shift the location of density ambiguity from that of the original α particle in the IC to a β moved to the right a distance proportional to the large black region's size relative to the white region's size.

Even though this β encodes ambiguity that cannot be resolved by the information currently at hand—that is, the information carried by the η and δ that produce it—this β actually carries important information in its location,

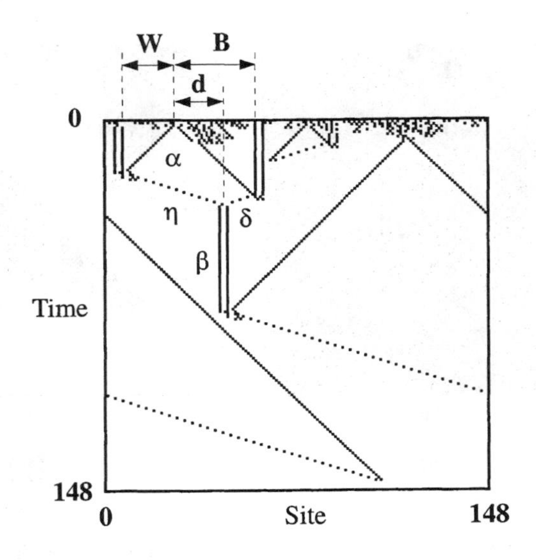

FIGURE 9 An enlargement and relabeling of the right diagram of figure 8(a) with some particle labels omitted for clarity. $\mathbf{W}$ is the length of the leftmost white region, $\mathbf{B}$ is the length of the black region to its right, and $\mathbf{d}$ is the amount by which the β produced by the η–δ interaction has been shifted from the leftmost α. Given the particle velocities listed in table 3 and using simply geometry, it is easy to calculate that $\mathbf{d} = 2(\mathbf{B} - \mathbf{W})$.

which is shifted to the right from the original α. To see this, refer to figure 9, an enlargement of the right diagram of figure 8(a) with some particle labels omitted for clarity. $\mathbf{W}$ and $\mathbf{B}$ denote the lengths of the indicated white (low density) and black (high density) regions in the IC. Given the particle velocities listed in table 3 and using simple geometry, it is easy to calculate that the β, produced by the η-δ interaction, is shifted to the right by $2(\mathbf{B} - \mathbf{W})$ cells from the α's original position. The shift to the right means that the high-density region (to the left of the leftmost β) has gained $\mathbf{B} - \mathbf{W}$ sites in size as a result of this series of interactions. In terms of relative position the local particle configuration $\beta\alpha\beta$ becomes $\beta\gamma\mu\beta$ and then $\eta\delta$, which annihilate to produce a final β. This information is used later on, when the rightmost γ collides with the new β before its partner μ does, eventually leading to black taking over the lattice, correctly classifying the ($\rho_0 > 1/2$) IC.

It should now be clear in what sense we say that particles store and transmit information and that particle collisions are the loci of decision making. We described in detail only two such scenarios. As can be seen from the figures, this type of particle-based information processing occurs in a distributed, parallel

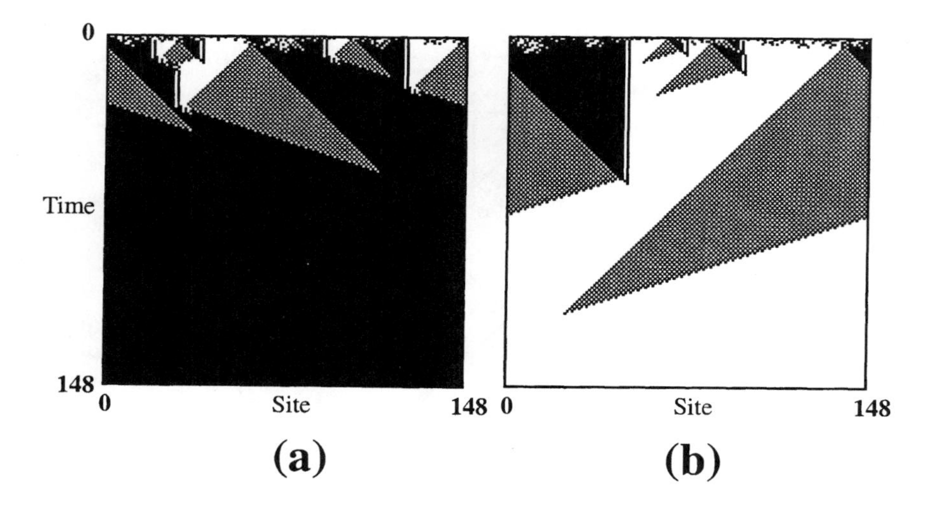

FIGURE 10 (a) Type 1 misclassification by $\phi_{\text{par}}^{\text{a}}$, with $\rho_0 = 0.48$. Even though $\rho_0 < \rho_c$, at t_c (here, $t = 8$), the lengths of the black regions sum to 65 cells and the lengths of the white regions sum to 44 cells. This leads to a misclassification of IC density. (b) Type 2 misclassification by $\phi_{\text{par}}^{\text{a}}$, starting with $\rho_0 = 0.52$. At t_c (here also, $t = 8$), the sum of the lengths of the black regions is 65 cells and the sum of the lengths of the white regions is 61 cells. However, even though these condensed lengths correctly reflect the fact that $\rho_0 > \rho_c$, the black regions in the IC's center occur within white regions in such a way that they get cut off. Ultimately, this yields a large white region that wins over the large black region, and the IC is misclassified.

fashion over a wide range of spatial and temporal scales. The functional organization of the information processing can be usefully analyzed at three levels: (i) the information stored in the particles and decisions made during their interaction, (ii) geometric subroutines that are coordinated groupings of particles and interactions that effect intermediate-scale functions, and (iii) the net spatial computation over the whole lattice and from $t = 0$ to $t = T_{\text{max}}$.

In the next section, we will argue that these levels of description of a CA's computational behavior—in terms of information transmission and processing by particles and their interactions—is analogous to, but significantly extends, Marr's "representation and algorithm" level of information processing. It turns out to be the most useful level for understanding and predicting the computational behavior of CAs, both for an individual CA operating on particular ICs and also for understanding how the GA evolved the progressive innovations in computational strategies over succeeding generations. (We will put these latter claims on a quantitative basis shortly.)

Almost always $\phi^{\mathrm{a}}_{\mathrm{par}}$ iterates to either all 0s or all 1s within $T_{\max} = 2N$ time steps. The errors it makes are almost always due to the wrong classification being reached rather than no classification being effected. $\phi^{\mathrm{a}}_{\mathrm{par}}$ makes two types of misclassifications. In the first type, illustrated in figure 10(a), $\phi^{\mathrm{a}}_{\mathrm{par}}$ reaches the condensation time t_c, having produced a configuration whose density is on the opposite side of ρ_c than was ρ_0. The particles and interactions then lead, via a correct geometric computation, to an incorrect final configuration. In the second type of error, illustrated in figure 10(b), the density ρ_{t_c} is on the same side of the threshold as ρ_0, but the configuration is such that islands of black (or white) cells are isolated from other black (or white) regions and get cut off. This error in the geometric computation eventually leads to an incorrect final configuration. As N increases, this type of error becomes increasingly frequent and results in the decreasing $\mathcal{P}_N^{10^4}$ values at larger N; see table 1.

6.4 COMPUTATIONAL STRATEGY OF $\phi^{\mathrm{B}}_{\mathrm{PAR}}$—FAILURE ANALYSIS

As noted in table 4, the space-time behavior of $\phi^{\mathrm{b}}_{\mathrm{par}}$ exhibits three regular domains: Λ^0 (white), Λ^1 (black), and Λ^2 (striped). The size-competition strategy of $\phi^{\mathrm{b}}_{\mathrm{par}}$ is similar to that of $\phi^{\mathrm{a}}_{\mathrm{par}}$. In $\phi^{\mathrm{b}}_{\mathrm{par}}$, the striped region plays the role of $\phi^{\mathrm{a}}_{\mathrm{par}}$'s checkerboard domain. However, when compared to $\phi^{\mathrm{a}}_{\mathrm{par}}$, the roles of the two domain boundaries, $\Lambda^0\Lambda^1$ and $\Lambda^1\Lambda^0$, are now reversed. In $\phi^{\mathrm{b}}_{\mathrm{par}}$, $\Lambda^0\Lambda^1$ is stable, while $\Lambda^1\Lambda^0$ is unstable and decays into two particles. Thus, the strategy used by $\phi^{\mathrm{b}}_{\mathrm{par}}$ is, roughly speaking, a 0-1 site-value exchange applied to $\phi^{\mathrm{a}}_{\mathrm{par}}$'s strategy. Particles α, β, γ, δ, η, and μ are all analogous in the two CAs, as are their interactions, if we exclude three-particle interactions; cf. tables 3 and 4. They implement competition between adjacent large white and black regions. In analogy with the preceding analysis for $\phi^{\mathrm{a}}_{\mathrm{par}}$'s strategy, these local competitions are decided by which particle, a γ or a μ, reaches a β first.

In $\phi^{\mathrm{a}}_{\mathrm{par}}$, γ and μ each approach β at the rate of one cell per time step. In $\phi^{\mathrm{b}}_{\mathrm{par}}$, although γ is now a stationary particle, it also effectively approaches β at the rate of 1 cell per time step, since β moves with velocity 1. And μ approaches β at the rate of 1/2 cell per time step, the velocity of μ minus the velocity of β. Thus, there is an asymmetry in $\phi^{\mathrm{b}}_{\mathrm{par}}$'s geometric computation that can result in errors of the type illustrated in figure 11(a). There the IC, with $\rho = 0.52$, condenses around iteration $t_c \approx 20$ into a block of 85 black cells adjacent to a block of 64 white cells. The γ particle, traveling at velocity 1 relative to β, reaches β in approximately 85 time steps. The μ particle, traveling at velocity 1/2 relative to β, reaches β in approximately 103 time steps. Thus, even though the black cells initially outnumber the white cells, the black region is cut off first and white eventually wins out, yielding an incorrect classification at time step 165. In contrast, $\phi^{\mathrm{a}}_{\mathrm{par}}$, with its symmetric particle velocities, reaches a correct classification on this same IC (figure 11(b)).

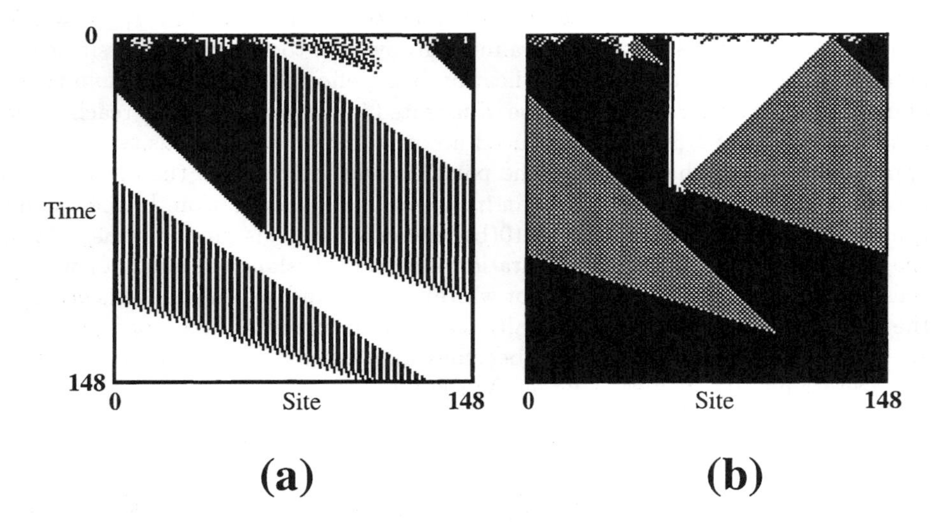

FIGURE 11 (a) Misclassification by $\phi_{\text{par}}^{\text{b}}$, with $\rho_0 = 0.52$. (By $T_{\text{max}} = 2N$, the CA reaches an all-0s fixed point.) (b) Correct classification by $\phi_{\text{par}}^{\text{a}}$ on the same IC.

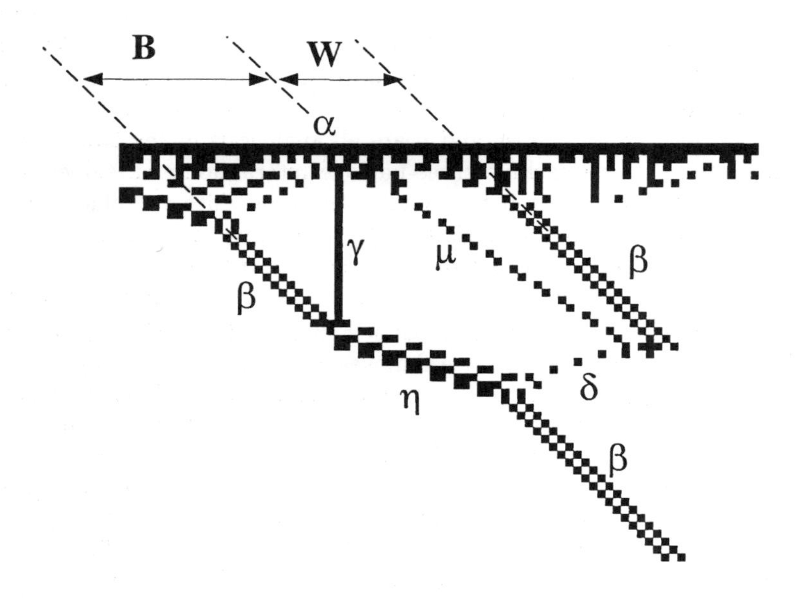

FIGURE 12 Blow-up of part of right-hand diagram of figure 8(b), illustrating asymmetries in $\phi_{\text{par}}^{\text{b}}$'s particle velocities that can result in misclassifications.

Like $\phi^{\rm a}_{\rm par}$, $\phi^{\rm b}_{\rm par}$ makes two types of classification errors—the type in which $\phi^{\rm b}_{\rm par}$ reaches t_c with a configuration whose density ρ_{t_c} is on the opposite side of ρ_c than ρ_0 and the type illustrated in figure 11(a). The first type ("type 1") is an error in how the CA condenses into domains and particles. The second type ("type 2") is due to asymmetries in particle velocities. Consider figure 12, a blow-up of part of the right-hand diagram of figure 8(b). To the left of the α (labeled) is an isolated black island and to the right is a white island. Together these two contiguous islands are bounded by two β particles on either side. Inside, as in $\phi^{\rm a}_{\rm par}$, an α decays into a γ and a μ. The resulting set of local particle interactions is such that the two islands compete for space within the two bounding β's, ending with the creation of a new β. If $\mathbf{W}$ and $\mathbf{B}$ respectively denote the lengths of the white and black islands, then after a series of interactions—$\beta\gamma\mu\beta \rightarrow \eta\delta \rightarrow \beta$—the white region (to the left of the original leftmost β) gains $2\mathbf{W} - \mathbf{B}$ sites in size. Thus, to increase this region's size, the internal white island must be at least half the size of its adjacent black island.

It is evident, therefore, that unlike $\phi^{\rm a}_{\rm par}$, there are asymmetries in $\phi^{\rm b}_{\rm par}$'s particle "logic," and these are biased in favor of classifying high densities. These asymmetries are what make $\mathcal{P}^{10^4}_N(\phi^{\rm b}_{\rm par})$ lower than $\mathcal{P}^{10^4}_N(\phi^{\rm a}_{\rm par})$. (See table 1.)

7 SIGNIFICANCE OF THE PARTICLE-LEVEL DESCRIPTION

There are several alternative ways in which cellular automata such as $\phi^{\rm a}_{\rm par}$ and $\phi^{\rm b}_{\rm par}$ can be described as performing a computation. Marr anticipated some of these in delineating the various levels of information processing in vision [52]. In principle, our CAs are completely described by the 128 bits in their look-up tables. This is too low level a description, however, to be useful for understanding how a given CA performs the $\rho_c = 1/2$ task. Using this level is like trying to understand how a pocket calculator computes the square root function by examining the physical equations of motion for the electrons and holes in the calculator's silicon circuitry.

Moreover, attempting interpretation at this level also violates, in a sense, one of the central tenets of the century-long study of dynamical systems, namely, that for nonlinear systems (e.g., most CAs), the local space-time equations of motion do not *directly* determine the system's long-term behavior. In the case of CAs, it is not the individual look-up table neighborhood-output-bit entries acting over a single time step that directly give rise to the observed computational strategy. Instead, it is the interaction of *subsets* of CA look-up table entries that over a *number of iterations* leads to the emergence of domains, particles, and interactions.

A second possibility for describing computational behavior in CAs is in terms of its detailed space-time behavior—i.e., the series of raw configurations of 0s and 1s. Again, this description is too low level for understanding how the solutions to the task are implemented. This approach is like trying to understand how a calculator's square root function is performed by taking a long series snapshots of the positions and velocities of the electrons and holes traveling through the integrated circuits. This prosaic view is analogous to Marr's "hardware implementation" level of description [52].

A third possibility is to describe the CA in terms related to the task's required input/output mapping and the task's computational complexity. For example, on a particular set of 10^4 random ICs, half with $\rho < 1/2$ and half with $\rho > 1/2$, $\phi_{\rm par}^{\rm a}$ correctly classified 81% of the $\rho < 1/2$ ICs and 74% of the $\rho > 1/2$ ICs. On average $\phi_{\rm par}^{\rm a}$ took 81 time steps to reach a fixed point; the maximum time was 227. The computational complexity of the $\rho_c = 1/2$ task on a serial architecture is $\mathcal{O}(N)$. This kind of operational analysis is roughly at Marr's "computational theory" level.

None of these levels of description gives much insight into *how* the task is being performed by a particular CA in terms of what information processing is being done and how it leads to a particular measured performance. What is needed is an intermediate-level description whose primitives are informationally related to the task at hand. This is what the computational mechanics level of particles and particle interactions gives us. However one might detect the primitives at this level, it is analogous to Marr's "representation and algorithm" level, in which particles can be seen as representing aspects of the IC, and their actions and interactions can be seen as the CA's emergent algorithm.

Representations, in the form of data structures and algorithms, have been studied extensively for von Neumann-style computers, but there have been few attempts to define such notions for decentralized spatially extended systems such as CAs. One can, of course, in principle implement any standard data structure and algorithm in a computation-universal CA, such as the game of Life CA [4], by simulating a von Neumann-style computer. However, this is not a particularly useful notion of information processing if one's goals are to understand how nonlinear systems in nature compute. It is even more problematic if one wishes to design computation in complex decentralized spatially extended architectures. We believe that it will be essential to develop new "macroscopic-level" vocabularies in order to explain how collective information processing takes place in such architectures. (One benefit of this development would be an understanding of how to program these architectures in genuinely parallel ways.)

A close reading shows that Marr's analysis of the descriptional levels required for visual processing misses several key issues. These are the facts that (i) representations emerge from the dynamics (i.e., are *intrinsic* to the dynamics), (ii) a clear formal definition is required to remove the subjectivity of detecting these intrinsic representations, and (iii) their functionality is entailed by a new level of dynamics, also intrinsic, which describes their interactions. As illustrated above in several cases, the computational mechanics framework that we are employing here makes these distinctions and provides the necessary concepts and methods to address these issues [13, 14]. The result is that we can analyze in detail the emergent computational strategies in the evolved CAs.

Our particle-level description forms an explanatory vocabulary for emergent computation in the context of one-dimensional, binary-state CAs. As was described above, particles represent various kinds of information about the IC and particle interactions are the loci of decision making that use this information. The resulting particle "logic" gives a functional description of how the computation takes place that is neither directly available from the CA look-up table nor from the raw space-time configurations produced by iterating the CA. It gives us a formal notion of "strategy," allowing us to see, for example, how the strategies of $\phi^{\mathrm{a}}_{\mathrm{par}}$ and $\phi^{\mathrm{b}}_{\mathrm{par}}$ are similar and how they differ. One immediate consequence of this level of analysis is that we can say why $\phi^{\mathrm{b}}_{\mathrm{par}}$'s strategy is weaker.

The level of particles and interactions is not only a qualitative description of spatial information processing, it also enables us to make quantitative predictions about computational performance. Crutchfield et al. [17] and Horkijk et al. [43], describe how to model a CA using its particle catalog and statistical properties at the condensation time. Each of several different CAs ϕ are compared the model's prediction for $\mathcal{P}^{I}_{N}(\phi)$ as well as for the average time taken to reach a fixed point with the values measured for the actual ϕ. Some of these comparisons will be summarized in section 8.7. The degree to which a model's predictions agree with the corresponding CA's behavior indicates the degree to which the particle-level description captures how the CA is actually performing the computation. Since, as we will show, the model's predicted performance and the observed performance are very close, we conclude that the particle-level description accurately captures the intrinsic computational capability of the evolved CAs.

8 EVOLUTIONARY HISTORY OF ONE PARTICLE STRATEGY: INNOVATION, CONTINGENCY, AND EXAPTATION

The structural analysis of CA space-time information processing that we have just outlined also allows us to understand the evolutionary stages during which the GA produces CAs. Here we will show how the functional components—domains, particles, and interactions—arise and are inherited across the evolutionary history of a GA run. We will also demonstrate a number of evolutionary dynamical phenomena, such as the historical contingency of functional emergence and the appearance of initially nonfunctional behaviors that later are key to the final appearance of high performance CAs.

To begin, figures 13(a) and 13(b) illustrate $\phi^{\mathrm{a}}_{\mathrm{par}}$'s evolutionary history. Figure 13(a) gives a partial tree of the parent-child relationships between some of $\phi^{\mathrm{a}}_{\mathrm{par}}$'s ancestors, each numbered by its generation of birth. Note that, since elite CAs can survive for more than one generation, parents and offspring, e.g., ϕ_{10} and ϕ_{13}, can have nonconsecutive generation labels. The CAs listed are those with the best fitness in the generation in which they arose. Table 5 lists the look-up tables, F^{100}_{149}, and $\mathcal{P}^{10^4}_{149}$ for the six ancestors of $\phi^{\mathrm{a}}_{\mathrm{par}}$ described below.

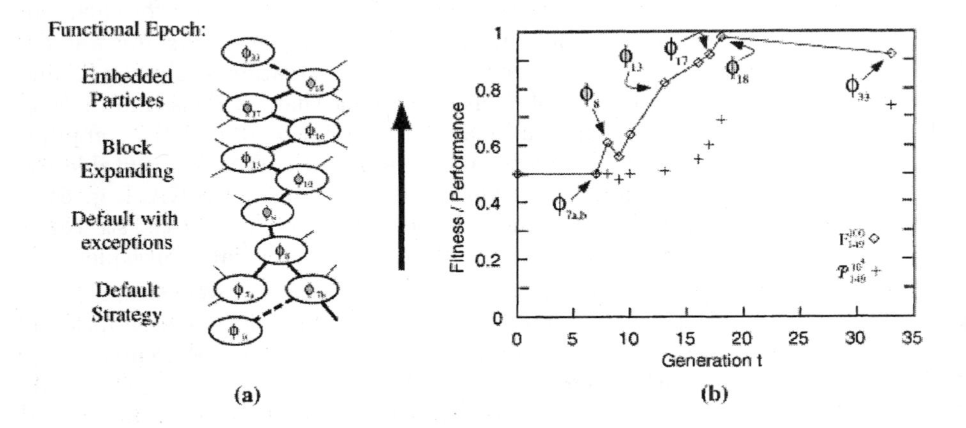

FIGURE 13 (a) Partial ancestral tree of $\phi_{\mathrm{par}}^{\mathrm{a}}$. (b) $F_{149}^{100}(\phi)$ (diamonds) and $\mathcal{P}_{149}^{10^4}(\phi)$ (crosses) of the CAs ϕ_i in (a). Six of the data points are marked with the name of the corresponding CA.

TABLE 5 CA chromosomes (look-up table output bits) given in hexadecimal, F_{149}^{100}, and $\mathcal{P}_{149}^{10^4}$ for the six ancestors of $\phi_{\mathrm{par}}^{\mathrm{a}}$ described in this section. (See figure 5.3 for directions on how to recover the CA rule table outputs from the hexadecimal code.) The F_{149}^{100} values in this table are those calculated in the CA's generation of birth by the GA; the $\mathcal{P}_{149}^{10^4}$ values given are the means over 100 trials of the performance function, calculated after the run was complete. When tested over 100 trials, the standard deviation of F_{149}^{100} is approximately 0.02, and the standard deviation of $\mathcal{P}_{149}^{10^4}$ is approximately 0.005 for each CA.

CA Name	Rule Table (Hexadecimal)	F_{149}^{100}	$\mathcal{P}_{149}^{10^4}$
ϕ_{7b}	F6EFFFFFFFFFFFFF	0.50	0.500
	6B9F7F93FFFFBFFF		
ϕ_{8}	0400448102000FFF	0.61	0.500
	6B9F7F93FFFFBFFF		
ϕ_{13}	0400458100000FFF	0.82	0.513
	6B9F77937DFFFF7F		
ϕ_{17}	0500458100000FBF	0.92	0.595
	6B9F75937FBFFF5F		
ϕ_{18}	0500458100000FBF	0.98	0.691
	6B9F75937FBDF77F		
ϕ_{33}	0504058100840FB7	0.97	0.735
	4BBF55837FBDF77F		

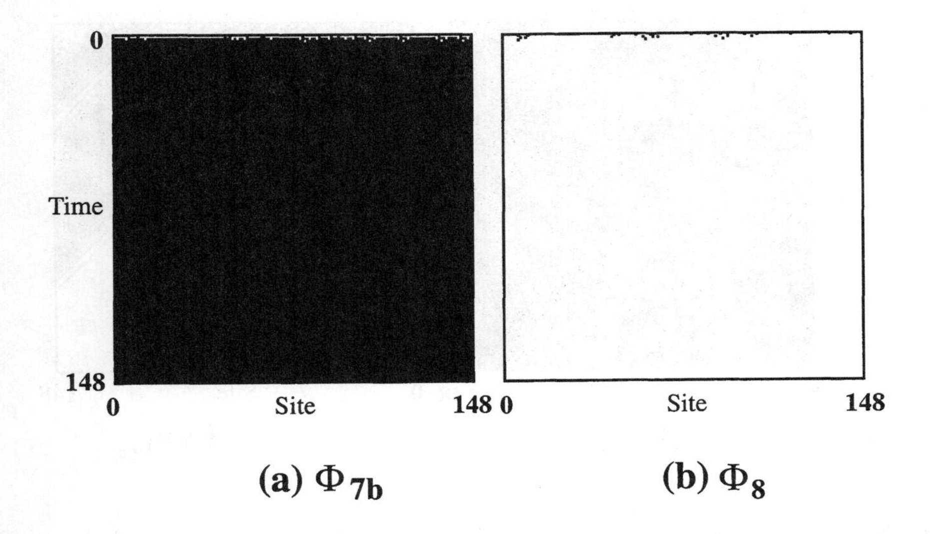

FIGURE 14 Space-time behavior of generation 7 and 8 ancestors, ϕ_{7b} and ϕ_8, of $\phi^{\mathrm{a}}_{\mathrm{par}}$. Both start from the same IC with $\rho_0 = 0.11$.

Figure 13(b) plots F^{100}_{149} (diamonds) and $\mathcal{P}^{10^4}_{149}$ (crosses) versus generation of birth for each of these ancestors. In generations 0–7, the best CA in the population has $F^{100}_{149} = \mathcal{P}^{10^4}_{149} = 0.5$, achieved by a "default" strategy like those of figure 4. Starting at generation 8, evolution proceeds in a series of abrupt increases in F^{100}_{149}. More gradual increases are seen in $\mathcal{P}^{10^4}_{149}$; of course, this statistic is not available to, and thus is not used by, the GA. The occasional small decreases result from the stochastic nature of the fitness and performance evaluations.

The goal now is to use the functional analysis to understand why these increases come about. To do so, we present a series of space-time diagrams, in figures 14–19, that compare space-time behaviors of CAs along the ancestral tree of figure 13(a). In each figure, space-time behavior with the same IC is given for two ancestrally related CAs to highlight the similarities and evolutionary innovations.

8.1 ANCESTORS ϕ_{7b} AND ϕ_8

Here the IC has very low density: $\rho_0 = 0.11$, ϕ_{7b}, a "default" CA, always iterates to all 1s and in figure 14(a) misclassifies the IC. ϕ_{7a} (not shown) is a default CA that always iterates to all 0s. ϕ_{7b}'s look-up table contains mostly 1s (see table 5), and ϕ_{7a}'s look-up table contains mostly 0s. They crossed over at locus 52 to produce ϕ_8: therefore, the first part of ϕ_8's look-up table contains mostly

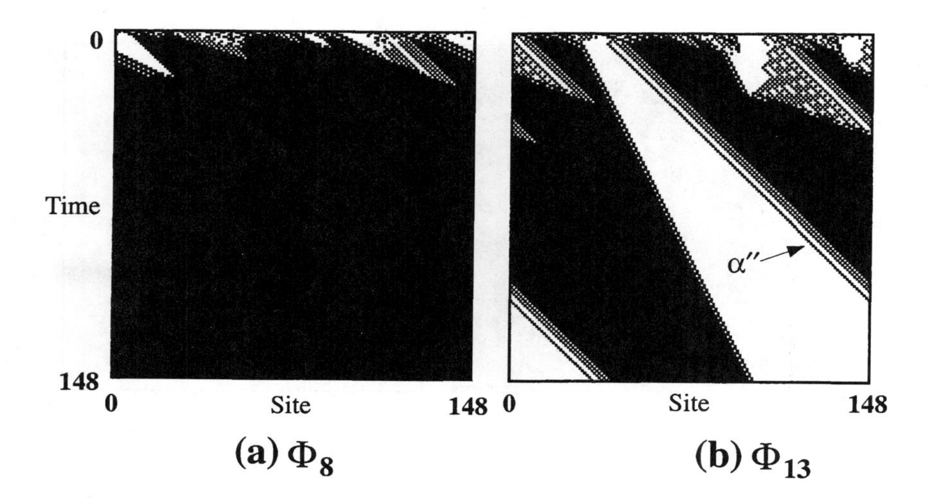

FIGURE 15 Space-time behavior of generation 8 and 13 ancestors, ϕ_8 and ϕ_{13}, of $\phi^{\mathrm{a}}_{\mathrm{par}}$. Both start from the same IC with $\rho_0 = 0.47$. (The significance of α'' is explained below, when ϕ_{13} and ϕ_{17} are compared.)

0s, and the rest is mostly 1s. In figure 14(b), ϕ_8 iterates to all 0s and correctly classifies the IC.

8.2 ANCESTORS ϕ_8 AND ϕ_{13}

Here $\rho_0 = 0.47$. In figure 15(a) ϕ_8 quickly iterates to all 1s. This is its more typical behavior than that shown in figure 14(b); very small regions of black quickly grow to take over the entire lattice. In this way, ϕ_8 is only slightly better than a default CA like ϕ_{7b}; it correctly classifies all high-density ICs and only a small number of very low density ICs. Note that while $F^{100}_{149}(\phi_8) > 0.5$, $\mathcal{P}^{10^4}_{149}(\phi_8)$ remains at 0.5. So ϕ_8 can be said to be carrying out a "default-with-exceptions" strategy. All runs that produced such strategies went on to converge on either block-expanding strategies or embedded-particle strategies.

Interestingly, the checkerboard domain $\Lambda^2 = \{(01)^+\}$ is produced by ϕ_8 on some ICs (figure 15). However, Λ^2 does not contribute to ϕ_8's fitness or performance. It is a functionally neutral feature. To determine this, we modified ϕ_8's rule table to prevent the checkerboard domain from propagating. The two relevant entries are 0101010 $\rightarrow$ 0 and 1010101 $\rightarrow$ 1. Flipping the output bit on either or both of these entries produces CAs with $F^{100}_{149} = 0.61$ and $\mathcal{P}^{10^4}_{149} = 0.5$, that is, with fitness and performance identical to those of ϕ_8. (The standard deviations of F^{100}_{149} for this and the other variant CAs discussed in this section were approximately 0.02. The standard deviations of $\mathcal{P}^{10^4}_{149}$ were approximately

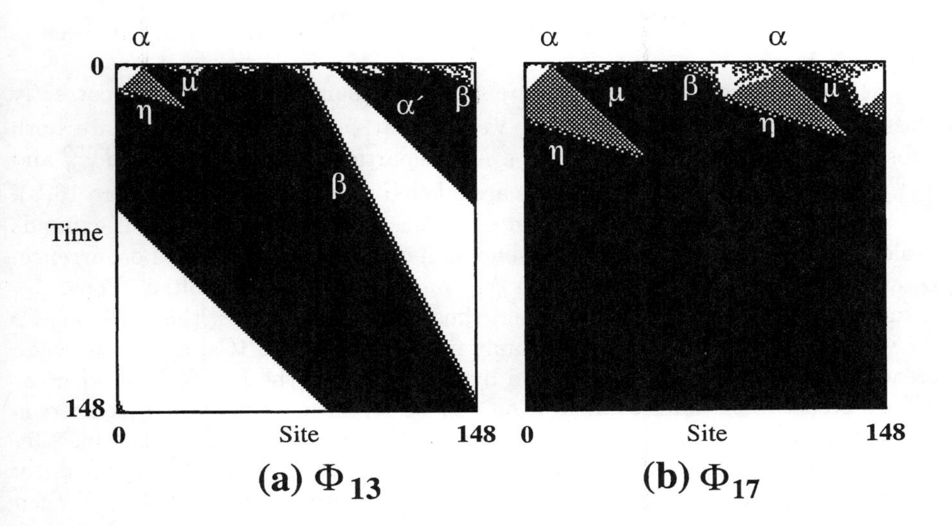

FIGURE 16 Space-time behavior of $\phi^{\mathrm{a}}_{\mathrm{par}}$ ancestors, ϕ_{13} and ϕ_{17}, arising in generation 13 and 17. Both start from the same IC with $\rho_0 = 0.58$.

0.005.) Appropriating biological terminology, we can consider the checkerboard domain, at this generation, to be an adaptively neutral trait of ϕ_8.

Ancestor ϕ_{13} represents a steep jump in fitness over ϕ_8, as seen in figure 13(b). ϕ_{13} is a block-expanding CA. It maps ICs to all 1s unless there is a sufficiently large block of adjacent 0s in the IC, in which case that block expands to eventually fill up the entire lattice, as in figure 15(b), which is a correct classification by $t = T_{\mathrm{max}}$. On some ICs, ϕ_{13} also produces a checkerboard domain and a similar but less ordered region; the latter can be seen in figure 15(b). We determined, in a fashion similar to that just explained above, that these traits also were adaptively neutral.

8.3 ANCESTORS ϕ_{13} AND ϕ_{17}

Here $\rho_0 = 0.58$. ϕ_{13} expands blocks of 0s on many ICs with $\rho > 1/2$, including the one in this figure, resulting in misclassifications. In fact, many high-density ICs with $\rho \approx 0.5$ are misclassified and, while ϕ_{13} has markedly higher F^{100}_{149} than its ancestors, its performance $\mathcal{P}^{10^4}_{149}$ is only marginally improved (see table 5).

Ancestor ϕ_{13} creates three types of boundaries between white and black domains. Two of them are shown in figure 16(a), labeled α and α'. The α, like $\phi^{\mathrm{a}}_{\mathrm{par}}$'s α, exists for only a single time step and then decays into η and μ; whereas, α' remains stable. A third type, α'', does not appear for this IC but can be seen in figure 15(b). Both α' and α'' support the block-expanding strategy; whereas,

α leads to a competition between white and black regions similar to that seen in $\phi^{\mathrm{a}}_{\mathrm{par}}$.

In contrast, consider figure 16(b), where the same $\rho = 0.58$ IC is correctly classified by ϕ_{17}. Recalling table 5, we see that ϕ_{17}'s F^{100}_{149} and $\mathcal{P}^{10^4}_{149}$ are both substantially higher than those of ϕ_{13}. At the particle level, ϕ_{17}'s higher F^{100}_{149} and $\mathcal{P}^{10^4}_{149}$ can be explained. The particles are labeled in figure 16(a) and figure 16(b).

Ancestor ϕ_{17} creates the same set of particles as ϕ_{13} (on some ICs it expands 0-blocks, not shown in figure 16(b)) but with different frequencies of occurrence: α' and α'' appear less often than in ϕ_{13}, and α appears more often. Thus, ϕ_{13} is more likely to expand 0-blocks, and thus make more errors, than ϕ_{17} on ICs for which $\rho_0 > 0.5$. Given 100 randomly generated $\rho > 0.5$ ICs, α' and α'' were created by ϕ_{13} in 86% of the ICs and by ϕ_{17} in 12% of the ICs. Whenever α' or α'' are created, the final configuration will be all 0s regardless of whether α is created. That is, block expanding dominates other behaviors. This explains why flipping output bits to suppress the checkerboard domain does not significantly affect ϕ_{13}'s F^{100}_{149} and $\mathcal{P}^{10^4}_{149}$, but does significantly affect these values for ϕ_{17}. When the checkerboard domain was suppressed in ϕ_{17}, F^{100}_{149} decreased only to 0.86, but $\mathcal{P}^{10^4}_{149}$ decreased to 0.54.

Following Gould and Vrba [37], we consider the checkerboard domain Λ^2 to be an example of an "exaptation"—a trait that has no adaptive significance when it first appears but is later co-opted by evolution to have adaptive value. According to Gould and Vrba, such traits are common in biological evolution. In the evolutionary innovation that goes from ϕ_{13} to ϕ_{17}, the exaptation of Λ^2 in ϕ_{13} makes just this transition to functionality associated with a marked increase in fitness and performance. This, in turn, leads to the change in dominant computational strategy away from block expansion.

8.4 ANCESTORS ϕ_{17} AND ϕ_{18}

Here $\rho_0 = 0.45$. The misclassification by ϕ_{17} (illustrated in figure 17(a)) is compared with the correct classification by ϕ_{17}'s higher fitness and performance child ϕ_{18} (figure 17(b)). Both CAs create similar particles, but in ϕ_{17}, the velocity of the β particle is 1/3; whereas, in ϕ_{18} its velocity is zero.

In figure 17(a), the white region (marked **W**) is larger than the black region to its right (marked **B**). Since the β particles have positive velocity, the black regions to **W**'s left and right both expand to the right. Coming in from the left, this decreases the size of **W**. On the other side, the (rightmost) β particle moves away from the **W** region. This asymmetry allows the **B** region to win the size competition, when the **B** region should not.

The asymmetry between black and white regions is corrected in ϕ_{18} by the change in β's velocity to zero. This makes the size competition between black and white regions symmetric. The result, seen in figure 17(b), is that the smaller **B** region is now cut off by the μ and β, the **W** region is allowed to grow, and the correct classification is made.

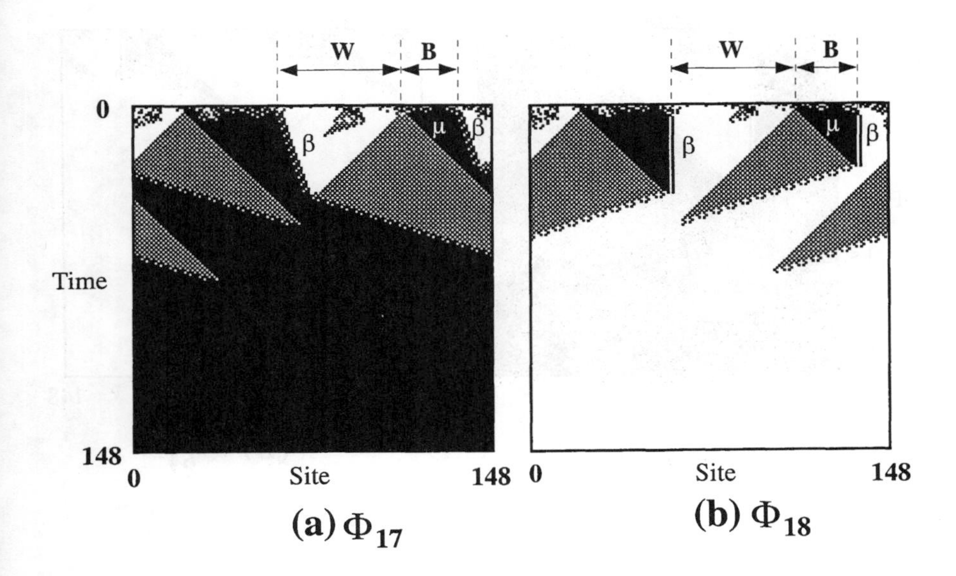

FIGURE 17 Space-time behavior of generation 17 and 18 ancestors, ϕ_{17} and ϕ_{18}, of $\phi^{\mathrm{a}}_{\mathrm{par}}$. Both are shown starting from the same IC that has $\rho_0 = 0.45$.

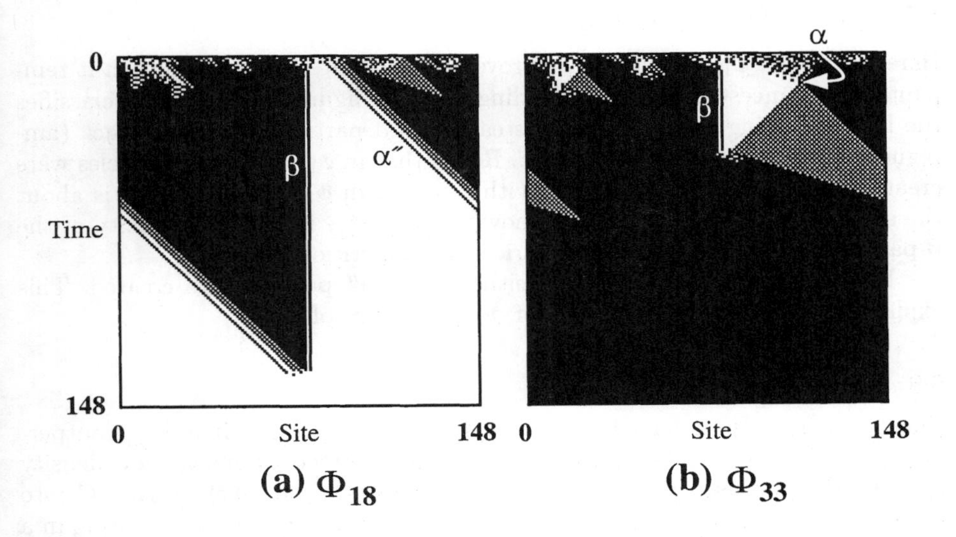

FIGURE 18 Space-time behavior of generation 18 and 33 ancestors, ϕ_{18} and ϕ_{33}, of $\phi^{\mathrm{a}}_{\mathrm{par}}$. Both start from the same IC with $\rho_0 = 0.61$.

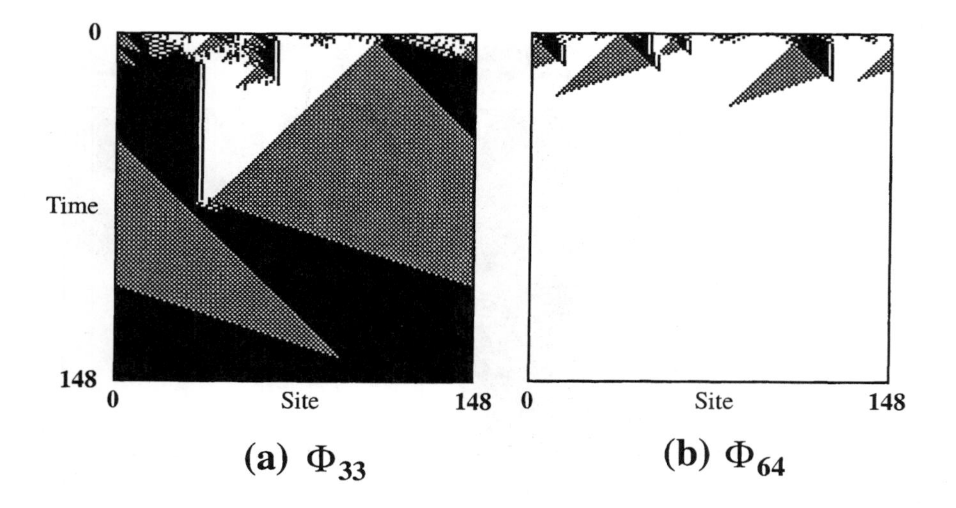

FIGURE 19 Space-time behavior of generation 33 ancestor ϕ_{33} of ϕ^{a}_{par} and ϕ_{64} (itself ϕ^{a}_{par}). Both start from the same IC, which has density $\rho_0 = 0.45$.

8.5 ANCESTORS ϕ_{18} AND ϕ_{33}

Here $\rho_0 = 0.61$. ϕ_{18}, though an improvement over ϕ_{17}, still carries with it remnants of its ancestors' block-expanding past. In figure 18(a), ϕ_{18} misclassifies the IC by creating an α'' particle instead of an α particle at a white-black (ambiguous density) boundary in the IC. Recall that in ϕ_{17}, α' or α'' particles were created in 12% of the random ICs with $\rho > 1/2$. In ϕ_{18} this frequency is about the same, 13%. Thus, ϕ_{18}'s main innovation over ϕ_{17} is the zero velocity of the β particle and the resulting symmetric size-competition strategy.

In ϕ_{33}, a descendant of ϕ_{18}, neither α' or α'' particles are created. This explains the higher F^{100}_{149} and $\mathcal{P}^{10^4}_{149}$ of ϕ_{33} over those of ϕ_{18}.

8.6 ANCESTORS ϕ_{33} AND ϕ_{64}

Here, $\rho_0 = 0.45$. Here named ϕ_{64} to denote its generation of birth, ϕ^{a}_{par}, outperforms ϕ_{33} since it classifies more low-density ICs correctly. On some low-density ICs like the one used in figure 19(a), ϕ_{33} condenses too much of the IC into black regions, a type 1 error. These then win the size competition, resulting in a misclassification. ϕ_{64} makes type 1 errors on low-density ICs less often (e.g., as seen in figure 19(b)), it correctly classifies the same IC as in figure 19(a). On the same set of 10^4 ICs, 62% of ϕ_{33}'s errors were on low-density ICs, whereas only 43% of ϕ_{64}'s errors were on low-density ICs.

TABLE 6 The CA and model performances $\mathcal{P}_{10^4}^{149}$ of ϕ_8, ϕ_{13}, ϕ_{17}, ϕ_{18}, ϕ_{33}, and ϕ_{64}. (After Crutchfield et al. [17].) Note $\phi_{\text{par}}^{\text{a}}$ has been referred here as ϕ_{64}. The CA rule tables are given in table 5.

CA Name	$\mathcal{P}_{149}^{10^4}(\phi)$	$\mathcal{P}_{149}^{10^4}(\mathcal{M}_\phi)$
ϕ_8	0.500	0.500
ϕ_{13}	0.513	0.524
ϕ_{17}	0.595	0.601
ϕ_{18}	0.691	0.747
ϕ_{33}	0.735	0.765
ϕ_{64}	0.775	0.775

8.7 PARTICLE MODELS OF EVOLVED CELLULAR AUTOMATA

The "natural history" of $\phi_{\text{par}}^{\text{a}}$'s evolution given above demonstrates how we can understand the jumps in F_{149}^{100} and $\mathcal{P}_{149}^{10^4}$ in terms of regular domains and particles—functional components in the CA's dynamical behavior. The GA's actions can be described at a low level as manipulating bits in CA rule tables via crossover and mutation, but a better understanding of the evolutionary process emerges when we describe its actions at the higher level of manipulating particle types, velocities, and interactions. An important component of this viewpoint is that the particles and their interactions lead to higher fitness. To test the hypothesis more quantitatively, we ask to what extent the CAs' observed fitnesses and performances can be predicted from the particle and interaction properties alone. To this end, in collaboration with Wim Hordijk, we have constructed "ballistic" particle models of the CAs ϕ_8 through ϕ_{64}. These models are intended to isolate the particle-level mechanisms and in so doing allow us to determine how much of the CA behavior this level captures.

A ballistic particle model $\mathcal{M}_\phi$ of a CA ϕ consists of the catalog of particle types, velocities, interactions, and their frequencies of occurrence at t_c. Model $\mathcal{M}_\phi$ is "run" by first using the particle frequencies to generate an initial configuration $\mathbf{s}_{t_c}$ of particles at the condensation time and then using the catalog of velocities and interactions to calculate the initial particles' ballistic trajectories and the products of subsequent particle interactions. The final configuration is reached when either all particles have annihilated or when $T_{\max} - t_c$ steps have occurred. This configuration and the actual time at which it was reached gives us $\mathcal{M}_\phi$'s prediction of what ϕ's classification would be for an IC corresponding to $\mathbf{s}_{t_c}$ and the time it would take ϕ to reach it. Particle models and their analysis are described in detail in Crutchfield et al. [17] and Hordijk et al. [43].

A comparison of the performances of the six CAs just analyzed and their particle models are given in table 6. As can be seen, the agreement is within a few percent for most cases. In these and the other cases, small discrepancies are due to simplifications made in the particle models. These include assumptions such as the particles being zero width and interactions occurring instantaneously. These error sources are analyzed in depth in Crutchfield et al. [17]. For ϕ_{18} the error is higher, around 8%, due to a long-lived transient domain that is not part of the particle catalog used for the model. The main effect of this is that the condensation time is overestimated on some ICs that generate this domain. This, in turn, means that the model describes only the last stages of convergence to the answer configurations, which it gets correctly and so has a higher performance than ϕ_{18}. For ϕ_{33} the error is around 4%. This appears to be due to errors in estimates of the distribution of particle types at the condensation time.

The conclusion is that the particle-level descriptions can be used to quantitatively predict the computational behavior of CAs and so also the CA fitnesses and performances in the evolutionary setting. In particular, the results support the claim that it is these higher-level structures, embedded in CA configurations, that implement the CA's computational strategy. More germane to the preceding natural history analysis, this level of description allows us to understand at a functional level of structural components the evolutionary process by which the CAs were produced.

9 RELATED WORK

In section 3, we discussed some similarities and differences between this work and other work on distributed parallel computation. In this section, we examine relationships between this work and other work on computation in cellular automata.

It should be pointed out that $\phi^{\mathrm{a}}_{\mathrm{par}}$'s behavior (and the behavior of many of the other highest-performance rules) is very similar to the behavior of the so-called Gács-Kurdyumov-Levin (GKL) CA. This CA was invented, not to perform the $\rho_c = 1/2$ task, but to study reliable computation and phase transitions in one-dimensional spatially extended systems [34]. More extensive work by Gács on reliable computation with CAs is reported in Gács [35].

The present work and earlier work by our group came out of follow-on research to Packard's investigation of "computation at the edge of chaos" in cellular automata [61]. Originally, Wolfram proposed a classification of CAs into four behavioral categories [82]. These categories followed the basic classification of dissipative dynamical systems: fixed-point attractors exhibiting equilibrium behavior, limit-cycle attractors exhibiting periodic behavior, chaotic attractors exhibiting apparently random behavior, and neutrally stable systems at bifurcations exhibiting long transients. Wolfram suggested that the latter category was particularly appropriate for implementing sophisticated (even universal) computation.

Following this with a more quantitative proposal, Langton [50] hypothesized that a CA's λ—the fraction of "nonquiescent states" (here, 1s) in its look-up table's output states—was correlated "generically" with the CA's computational capabilities. In particular, he hypothesized that CAs with certain "critical" λ values, which we denoted λ_c, would be more likely than CAs with λ values away from λ_c to be able to perform complex computations, or even universal computation. Packard's goal was to test this hypothesis by using a genetic algorithm to evolve $(k, r) = (2, 3)$ CAs to perform the $\rho_c = 1/2$ task, starting from an initial population chosen from a distribution that was uniform over $\lambda \in [0, 1]$. He found that after 100 generations, the final populations of CAs, when viewed only as distributions over λ, tended to cluster close to λ_c values. He interpreted this clustering as evidence for the hypothesized connection between λ_c and computational ability.

Mitchell et al. [58] were able to show, via theoretical arguments and empirical results, however, that the most successful CAs for the $\rho_c = 1/2$ task must have $\lambda \approx 1/2$. This value of λ is quite different from Packard's quoted λ_c values. We argued that Packard's results were due to an artifact in his particular implementation of the GA. Using more standard versions and his version of GA search, we obtained results that disagreed with Packard's findings and that were roughly in accord with our theoretical predictions that high-performance CAs were to be found at $\lambda \approx 1/2$, far from λ_c, and not in, for example, Wolfram's fourth CA category. We were also able to explain the deviations of our results from the theoretical predictions. The current work came out of the discovery of phenomena, such as embedded-particle CAs [18, 21], that were not found in Packard [61]. Moreover, according to Langton the $\lambda = 1/2$ value for our high-performance CAs corresponds to CAs in Wolfram's chaotic class. The space-time diagrams shown earlier demonstrate that they are not "chaotic"; their behavior, in fact, puts them in the first (fixed-point) category.

Later, other researchers performed their own studies of evolving cellular automata for the $\rho_c = 1/2$ task. Sipper and Ruppin [66, 67] used a version of the GA to evolve "nonuniform CAs"—CA-like architectures in which each cell uses its own look-up table to determine its state at each time step. For a lattice of size N, the individuals in the GA population are the N look-up tables making up a nonuniform CA. Sipper and Ruppin used this framework to evolve $r = 1$ nonuniform CAs to perform the $\rho_c = 1/2$ task, as well as other tasks. They reported the discovery of nonuniform CAs with $\mathcal{P}_{149}^{10^4}$ values comparable to that of $\phi_{\text{par}}^{\text{a}}$. They did not report $\mathcal{P}_N^{10^4}$ results for any other value of N, nor did they give statistics on how often high-performance nonuniform CAs were evolved. Moreover, no structural analysis of CA space-time behavior or GA population dynamics was given. Thus, it is unclear how the high fitnesses were obtained, either dynamically or evolutionarily.

Andre et al. used a genetic programming algorithm to evolve $(k, r) = (2, 3)$ CAs with $N = 149$ to perform the $\rho_c = 1/2$ task [2]. This algorithm discovered particle CAs with higher $\mathcal{P}_{149}^{10^4}$ than that of $\phi_{\text{par}}^{\text{a}}$ (e.g., 0.828 versus 0.776). We obtained the look-up table for one such CA, ϕ_{GP} [1] and found that on larger lattices, the performance of ϕ_{GP} was close to that of $\phi_{\text{par}}^{\text{a}}$ ($\mathcal{P}_{599}^{10^4}(\phi_{\text{GP}}) = 0.765$ and $\mathcal{P}_{999}^{10^4}(\phi_{\text{GP}}) = 0.723$; cf. table 5). It is not clear whether the improvement in $\mathcal{P}_{149}^{10^4}$ was due to the genetic programming representation CA look-up tables or some other factor related to increased computational resources. For example, their runs had a 500-fold larger population size M and 10-fold larger number of ICs over our GA runs. Their runs did, however, find high-performance CA in average numbers of generations that were half those in our GA. Thus, the computational resources they used in their evolutionary search were approximately 2500 times larger than in our GA runs.

Paredis [63] and Juillé and Pollack [46] experimented with coevolutionary learning techniques to improve the GA's search efficiency to find embedded particle CAs for the $\rho_c = 1/2$ task. The latter work specifically rewarded or penalized ICs of particular densities, depending on the amount of information ICs of those densities provided for distinguishing fitnesses between CAs in the population. This resulted in a higher percentage of GA runs in which high-performance embedded-particle CAs were discovered and in the discovery of higher-performance CAs than in any of the non-coevolutionary runs. The highest performance CA discovered had $\mathcal{P}_{10^5}^{149} = 0.863 \pm 0.001$, $\mathcal{P}_{10^5}^{599} = 0.822 \pm 0.001$, and $\mathcal{P}_{10^5}^{999} = 0.804 \pm 0.001$. Unfortunately, the performance of this coevolved CA, although high on small lattices (e.g., $N = 149$), decays more rapidly with lattice size than the GKL CA, which happens to have lower performance than the coevolved rule on small lattices. This appears to be the result of the more complex domains that preclude, through additional persistent particles, convergence to the answer configurations, 0^N or 1^N. Compared to the coevolved CAs, the GKL CA is one of the CAs that maintains high performance on larger lattices.

Our own work has been extended to other tasks, most thoroughly to a global synchronization task for which we have performed similar analyses to those given in this chapter [20].

Our notion of computation via particles and particle interactions derives from that introduced by the computational mechanics framework [13, 39, 40] and so differs considerably from the notions used in most other work on designing CAs for computation. For example, propagating particlelike signals were used in the solution to the Firing Squad Synchronization Problem [53, 59, 78], in Smith's work on CAs for parallel formal-language recognition [69], and in Mazoyer's work on computation in one-dimensional CAs [54]. However, in all these cases, the particles and their interactions were designed by hand to be the explicit behavior of the CA. That is, the particles are explicitly coded in each cell's local state, and their dynamics and their interactions are coded directly into the CA lookup table. Typically, their interactions were effected by a relatively large number of states per site. Steiglitz, Kamal, and Watson's carry-ripple adder [71] and the universal computer constructed in the Game of Life [4] both used binary-state signals consisting of propagating periodic patterns. But, again, the particles were explicitly designed to ride on top of a quiescent background and their interaction properties were carefully hand coded. In Squier and Steiglitz's "particle machine" [70] and in Jakubowski, Steiglitz, and Squier's "soliton machine" [44], particles are the primitive states of the CA cells. Moreover, their interaction properties are explicitly given by the CA rule table. These machines are essentially kinds of lattice gas automata [24] that operate on "particles" directly. (Other work on arithmetic in cellular automata has been done by Sheth, Nag, and Hellwarth [65] and Clementi, De Biase, and Massini [9], among others.)

In contrast to these, particles in our system are embedded as walls between regular domains. They are often apparent only after those domains have been discovered and filtered out. Their structures and interaction properties are emergent properties of the patterns formed by the CAs. Notably, although each cell has only two possible states, the structures of embedded particles are spatially and temporally extended, and so are more complex than atomic or simple periodic structures. Typically, these structures can extend over spatial scales larger than the CA radius. For example, the background domain of the elementary CA (ECA 110) shown in figure 2 has a temporal periodicity of 7 time steps and a spatial periodicity of 14 sites, markedly larger than the $r = 1$ nearest-neighbor coupling.

10 CONCLUSION

Our philosophy is to view CAs as systems that naturally form patterns (such as regular domains) and to view the GA as taking advantage—via selection and genetic variation—of these pattern-forming propensities so as to shape them to perform desired computations. Within this framework, we attempt to understand the behavior of the resulting CAs by applying tools, such as the computational mechanics framework, formulated for analyzing pattern-forming systems. The result gives us a high-level description of the computationally relevant parts of the system's behavior. In doing so, we begin to answer Wolfram's last problem from "Twenty problems in the theory of cellular automata" [83]: "What higher-level descriptions of information processing in cellular automata can be given?" We believe that this framework will be a basis for the "radically new approach" that Wolfram claimed will be required for understanding and designing sophisticated computation in CAs and other decentralized spatially extended systems.

Our analysis showed that there are three levels of information processing occurring during iterations of the evolved high-performance CAs. The first was the type of information storage and transmission effected by the particles and the type of "logical" operations implemented by the particle interactions. The second, higher level comprised the geometric subroutines that implemented intermediate-scale computations. We analyzed in detail two of these that were important to the size competition between regions of low and high density. We also showed how variations in the particles led to several types of error at this level. The third and final level is that of the global computation over the entire lattice up to the answer time. This is the level at which fitness is conferred on the CAs.

We analyzed in some detail the natural history that led to the emergence of such computationally sophisticated CAs. The evolutionary epochs typically proceed in a set sequence, with earlier epochs setting the (necessary) context for the later, higher performance ones. Often the jumps to higher epochs were facilitated by exaptations—changes in adaptively neutral traits appearing in much earlier generations.

There are a number of fruitful directions for future work. The first is to extend the lessons learned here to more general evolutionary search algorithms and pattern-forming dynamical systems. The problem of choosing a genetic representation of dynamical systems that helps, or at least does not hinder, the search will play an important role in addressing this. The evolution of CAs that operate on two-dimensional images rather than one-dimensional strings will also help address this issue and also open up application areas, such as iterative nonlinear image processing [12].

We also need to develop substantially better analytical descriptions of the search's population dynamics and of how the intrinsic structures in CAs interact with those dynamics. Although the evolution of CAs is a very simplified problem from the biological perspective, the evolutionary time scale of the population dynamics and the development time scale of the CAs result in a two-time-scale stochastic dynamical system that is difficult to analytically predict. Such predictions, say, of how to set the mutation rate or population size for effective search, are centrally important both for basic understanding of evolutionary mechanisms and for practical applications. Progress on quantitatively predicting the population dynamics occurring during epochal evolution has been made [74, 75]. The adaptation of the "statistical dynamics" approach introduced there to the evolution of CAs will be an important, but difficult, step toward understanding complicated genotype-to-phenotype maps. The latter is highly relevant for using such search methods on complex problems.

Another quantitative direction is the estimation of computational performance of distributed systems based on higher-level descriptions. The results, reported here and described in detail in Crutchfield et al. [17] and Hordijk et al. [43], on predicting CA computational performance are encouraging. Constructing a more accurate model along with a quantitative analytical model of higher-level computation in CAs will help us understand how much the embedded CA structures contribute individually to overall fitness. And this, in turn, will allow us to monitor the evolutionary mechanisms that lead to the emergence of collective computation in coordinated groups of functional units.

ACKNOWLEDGMENTS

The authors thank Wim Hordijk for calculating the CA particle-model performances and Silas Alben for calculating the coevolved CA performances. They also thank Jim Hanson, Wim Hordijk, Cris Moore, Erik van Nimwegen, and Cosma Shalizi for helpful discussions. This work was supported at UC Berkeley by ONR grant N00014-95-1-0524 and at the Santa Fe Institute by ONR grant N00014-95-1-0975, DARPA contract F30602-00-2-0583, National Science Foundation grant IRI-9705830, and AFOSR via NSF grant PHY-9970158.

APPENDIX: DOMAIN FILTER

In this appendix we describe the properties and construction of $\phi^{\mathrm{a}}_{\mathrm{par}}$'s domain-recognizing and filtering transducer.

The transducer, shown in figure 20, reads in binary CA configurations and outputs strings of the same length, the lattice size N, in the domain-wall alphabet $\{\lambda, 0, 1, 2, w\}$. In this alphabet, λ indicates that the transducer has not yet "synchronized" (see below) to the domain or wall structures in the configuration, $\{0, 1, 2\}$ label each of the three domains, respectively, and w indicates a wall between domains. In the filtered space-time diagrams, w is mapped to black, and all other output symbols are mapped to white.

Briefly, $\phi^{\mathrm{a}}_{\mathrm{par}}$'s domain-wall transducer is constructed as follows. $\phi^{\mathrm{a}}_{\mathrm{par}}$ has three domains, each of which can be described by simple finite-state machines. These machines form the recurrent states of the transducer. When the transducer first begins to read in the configuration, it may take several steps to disambiguate the site values and identify the appropriate domain in which they are participating. Working through the transitions and transient states that lead to the recurrent (domain) states determines the transitions from the start state. When the transducer is reading site values consistent with one of these domains, but then encounters site values that are not consistent with it (e.g., values indicating walls), then some number of additional site values must be read in to determine the domain type into which the transducer has moved. Such transitions determine the transducer's domain-to-domain transitions.

Note that due to the steps required to initially read in a sufficient number of site values to recognize the domains and walls, a process that we call synchronization, the transducer may have to read some portion of the configuration that it has already read, as it wraps around due to the lattice's periodic boundary conditions. This takes at most one additional pass over the configuration.

The general construction procedure for domain-wall transducers is given in Crutchfield and Hanson [16].

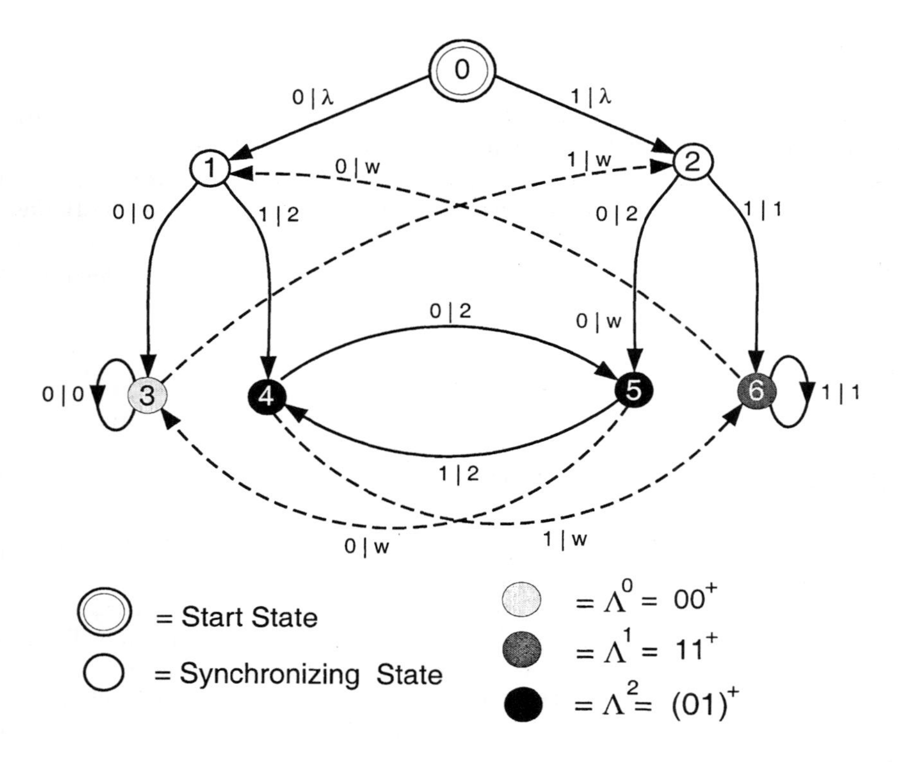

FIGURE 20 ϕ^{a}_{par}'s domain-recognizing and filtering transducer. The edge labels $s|t$ indicate that the transition is to be taken on reading configuration site value $s \in \{0,1\}$ and then outputting structural label $t \in \{\lambda, 0, 1, 2, w\}$.

REFERENCES

[1] Andre, D. Personal communication, 1998.

[2] Andre, D., F. H. Bennett, III, and J. R. Koza. "Evolution of Intricate Long-Distance Communication Signals in Cellular Automata using Genetic Programming." In *Artificial Life V: Proceedings of the Fifth International Workshop on the Synthesis and Simulation of Living Systems*, 513–520. Cambridge, MA: MIT Press, 1996.

[3] Bäck, T. *Evolutionary Algorithms in Theory and Practice: Evolution Strategies, Evolutionary Programming, Genetic Algorithms.* Oxford: Oxford University Press, 1996.

[4] Berlekamp, E., J. H. Conway, and R. Guy. *Winning Ways for Your Mathematical Plays*, vol. 2. New York, NY: Academic Press, 1982.

[5] Burks, A. W., ed. *Essays on Cellular Automata.* Urbana, IL: Univerity of Illinois Press, 1970.

[6] Capcarrere, M. S., M. Sipper, and M. Tomassini. "Two-State, $r = 1$ Cellular Automaton that Classifies Density." *Phys. Rev. Lett.* **77** (1996): 4969–4971.

[7] Chau, H. F., K. K. Yan, K. Y. Wan, and L. W. Siu. "Classifying Rational Densities using Two One-Dimensional Cellular Automata." *Phys. Rev. E* **57** (1998): 1367–1369.

[8] Churchland, P. S., and T. J. Sejnowski, eds. *The Computational Brain.* Cambridge, MA: MIT Press, 1992.

[9] Clementi, A., G. A. De Biase, and A. Massini. "Fast Parallel Arithmetic on Cellular Automata." *Complex Systems* **8** (1994): 435–441.

[10] Colorni, A., M. Dorigo, and V. Maniezzo. "Distributed Optimization by Ant Colonies." In *Toward a Practice of Autonomous Systems: Proceedings of the First European Conference on Artificial Life*, edited by F. J. Varela and P. Bourgine, 134–142. Cambridge, MA: MIT Press/Bradford Books, 1992.

[11] Crichlow, J. M. *An Introduction to Distributed and Parallel Computing.* London: Prentice Hall, 1997.

[12] Crutchfield, J. P. "Spatio-Temporal Complexity in Nonlinear Image Processing." *IEEE Trans. Circ. Sys.* **37** (1988): 770–780.

[13] Crutchfield, J. P. "The Calculi of Emergence: Computation, Dynamics, and Induction." *Physica D* **75** (1994): 11–54.

[14] Crutchfield, J. P. "Is Anything Ever New? Considering Emergence." In *Complexity: Metaphors, Models, and Reality*, edited by G. Cowan, D. Pines, and D. Melzner, 497–514. Santa Fe Institute Studies in the Sciences of Complexity, proc. vol. XIX . Reading, MA: Addison-Wesley, 1994.

[15] Crutchfield, J. P., and J. E. Hanson. "Attractor Vicinity Decay for a Cellular Automaton." *CHAOS* **3** (1993): 215–224.

[16] Crutchfield, J. P., and J. E. Hanson. "Turbulent Pattern Bases for Cellular Automata." *Physica D* **69** (1993): 279–301.

[17] Crutchfield, J. P., W. Hordijk, and M. Mitchell. "Predicting the Behavior of Evolved Cellular Automata." Manuscript in preparation.

[18] Crutchfield, J. P., and M. Mitchell. "The Evolution of Emergent Computation." *Proc. Natl. Acad. Sci. USA* **92** (1995): 10742–10746.

[19] Das, R. *The Evolution of Emergent Computation in Cellular Automata.* Ph.D. thesis. The Colorado State University, Fort Collins, CO, 1998.

[20] Das, R., J. P. Crutchfield, M. Mitchell, and J. E. Hanson. "Evolving Globally Synchronized Cellular Automata." In *Proceedings of the Sixth International Conference on Genetic Algorithms*, edited by L. J. Eshelman, 336–343. San Francisco, CA: Morgan Kaufmann, 1995.

[21] Das, R., M. Mitchell, and J. P. Crutchfield. "A Genetic Algorithm Discovers Particle-Based Computation in Cellular Automata." In *Parallel Problem Solving from Nature—PPSN III*, edited by Y. Davidor, H.-P. Schwefel, and

R. Männer, vol. 866, 344–353. Lecture Notes in Computer Science. Berlin: Springer-Verlag, 1994.

[22] Deneubourg, J. L., S. Goss, N. Franks, A. Sendova-Franks, C. Detrain, and L. Chrétien. "The Dynamics of Collective Sorting: Robot-like Ants and Ant-like Robots." In *From Animals To Animats: Proceedings of the First International Conference on Simulation of Adaptive Behavior*, edited by J.-A. Meyer and S. W. Wilson, 356–363. Cambridge, MA: MIT Press, 1991.

[23] Dolev, D., and H. R. Strong. "Authenticated Algorithms for Byzantine Agreement." *SIAM J. Comp.* **12** (1983): 656–666.

[24] Doolen, G., ed. *Lattice Gas Methods for Partial Differential Equations.* Santa Fe Institute Studies in the Sciences of Complexity, Proc. Vol. IV. Reading, MA: Addison-Wesley, 1990.

[25] Eloranta, K. "The Dynamics of Defect Ensembles in One-Dimesional Cellular Automata." *J. Stat. Phys.* **76** (1994): 1377–1398.

[26] Farmer, J. D., N. H. Packard, and A. S. Perelson. "The Immune System, Adaptation, and Machine Learning." *Physica D* **22** (1986): 187–204.

[27] Farmer, J. D., T. Toffoli, and S. Wolfram, eds. *Cellular Automata: Proceedings of an Interdisciplinary Workshop.* Amsterdam: North Holland, 1984.

[28] Farrag, A. A., and R. J. Dawson. "On Designing Efficient Consensus Protocols." In *Distributed Processing*, edited by M. H. Barton, E. L. Dagless, and G. L. Reijns, 413–427. Amsterdam: Elsevier Science Publishers, 1988.

[29] Fiesler, E., and R. Beale, eds. *Handbook of Neural Computation.* New York: Oxford University Press, 1997.

[30] Fogel, D. B. *Evolutionary Computation: Toward a New Philosophy of Machine Intelligence.* New York: IEEE Press, 1995.

[31] Fogelman-Soulie, F., Y. Robert, and M. Tchuente, eds. *Automata Networks in Computer Science: Theory and Applications.* Manchester, UK: Manchester University Press, 1987.

[32] Forrest, S., S. Hofmeyr, and A. Somayaji. "Computer Immunology." *Comm. ACM* **40** (1997): 88–96.

[33] Fukś, H. "Solution of the Density Classification Problem with Two Cellular Automata Rules." *Phys. Rev. E* **55** (1997): R2081–R2084.

[34] Gács, P. "Nonergodic One-Dimensional Media and Reliable Computation." *Contemp. Math.* **41** (1985): 125.

[35] Gács, P. "Reliable Computation with Cellular Automata." *J. Comp. & Sys. Sci.* **32** (1986): 15–78.

[36] Goldberg, D. E. *Genetic Algorithms in Search, Optimization, and Machine Learning.* Reading, MA: Addison-Wesley, 1989.

[37] Gould, S. J., and E. S. Vrba. "Exaptation: A Missing Term in the Science of Form." *Paleobiology* **8** (1982): 4–15.

[38] Gutowitz, H. A., ed. *Cellular Automata.* Cambridge, MA: MIT Press, 1990.

[39] Hanson, J. E. *Computational Mechanics of Cellular Automata.* Ph.D. thesis, University of California at Berkeley, Berkeley, CA, 1993.

[40] Hanson, J. E., and J. P. Crutchfield, "The Attractor-Basin Portrait of a Cellular Automaton." *J. Stat. Phys.* **66** (1992): 1415–1462.

[41] Holland, J. H. *Adaptation in Natural and Artificial Systems,* 2d ed. Cambridge, MA: MIT Press, 1992. (First edition, 1975.)

[42] Hopcroft, J. E., and J. D. Ullman. *Introduction to Automata Theory, Languages, and Computation.* Reading: Addison-Wesley, 1979.

[43] Hordijk, W., J. P. Crutchfield, and M. Mitchell. "Mechanisms of Emergent Computation in Cellular Automata." In *Parallel Problem Solving from Nature-V*, edited by A. E. Eiben, T. Bäck, M. Schoenauer, and H.-P. Schwefel, 613–622. Springer-Verlag, 1998.

[44] Jakubowski, M. H., K. Steiglitz, and R. K. Squier. "When Can Solitons Compute?" *Complex Systems* **10** (1996): 1–21.

[45] Jesshope, C., V. Jossifov, and W. Wilhelmi, eds. *International Workshop on Parallel Processing by Cellular Automata and Arrays (Parcella '94).* Berlin: Akademie Verlag, 1994.

[46] Juillé, H., and J. B. Pollack. "Coevolutionary Learning: A Case Study." In *ICML '98 Proceedings of the Fifteenth International Conference on Machine Learning*, 251–259. San Francisco, CA: Morgan Kaufmann, 1998.

[47] Karapiperis, T., and B. Blankleider. "Cellular Automaton Model of Reaction-Transport Processes." *Physica D* **78** (1994): 30–64.

[48] Kung, H. T. "Why Systolic Architectures?" *Computer* **15** (1982): 37–46.

[49] Land, M., and R. K. Belew. "No Perfect Two-State Cellular Automata for Density Classification Exists." *Phys. Rev. Lett.* **74** (1995): 5148–5150.

[50] Langton, C. G. "Computation at the Edge of Chaos: Phase Transitions and Emergent Computation." *Physica D* **42** (1990): 12–37.

[51] Lindgren, K., and M. G. Nordahl. "Universal Computation in a Simple One-Dimensional Cellular Automaton." *Complex Systems* **4** (1990): 299–318.

[52] Marr, D., ed. *Vision: A Computational Investigation into the Human Representation and Processing of Visual Information.* San Francisco: W. H. Freeman, 1982.

[53] Mazoyer, J. "An Overview of the Firing Squad Synchronization Problem." In *Automata Networks*, edited by C. Choffurt, 82–94. New York: Springer-Verlag, 1988.

[54] Mazoyer, J. "Computations on One-Dimensional Cellular Automata." *Annl. Math.& Art. Intel.* **16** (1996): 1–4.

[55] Mitchell, M. *An Introduction to Genetic Algorithms.* Cambridge, MA: MIT Press, 1996.

[56] Mitchell, M. "Computation in Cellular Automata: A Selected Review." In *Nonstandard Computation*, edited by T. Gramss, 213–241. Weinheim: VCH Verlagsgesellschaft, 1998.

[57] Mitchell, M., J. P. Crutchfield, and P. T. Hraber. "Evolving Cellular Automata to Perform Computations: Mechanisms and Impediments." *Physica D* **75** (1994): 361–391.

[58] Mitchell, M., P. T. Hraber, and J. P. Crutchfield. "Revisiting the Edge of Chaos: Evolving Cellular Automata to Perform Computations." *Complex Systems* **7** (1993): 89–130.

[59] Moore, E. F. "The Firing Squad Synchronization Problem." In *Sequential Machines: Selected Papers*, edited by E. F. Moore, Reading, MA: Addison-Wesley, 1964.

[60] Nagel, K. "Particle Hopping Models and Traffic Flow Theory." *Phys. Rev. E* **53** (1996): 4655–4672.

[61] Packard, N. H. "Adaptation Toward the Edge of Chaos." In *Dynamic Patterns in Complex Systems*, edited by J. A. S. Kelso, A. J. Mandell, and M. F. Shlesinger, 293–301. Singapore: World Scientific, 1988.

[62] Packard, N. H. "Intrinsic Adaptation in a Simple Model for Evolution." In *Artificial Life*, edited by C. G. Langton, 141–155. Santa Fe Institute Studies in the Sciences of Complexity, Proc. Vol. VI. Reading, MA: Addison-Wesley, 1989.

[63] Paredis, J. "Coevolving Cellular Automata: Be Aware of the Red Queen!" In *Proceedings of the Seventh International Conference on Genetic Algorithms*, edited by T. Bäck, 393–400. San Francisco: CA: Morgan Kaufmann, 1997.

[64] Rumelhart, D. E., G. E. Hinton, and J. L. McClelland. "A General Framework for Parallel Distributed Processing." In *Parallel Distributed Processing*, edited by D. E. Rumelhart, J. L. McClelland, and the PDP Research Group, vol. 1, 45–76. Cambridge, MA: MIT Press, 1986.

[65] Sheth, B., P. Nag, and R. W. Hellwarth. "Binary Addition on Cellular Automata." *Complex Systems* **5** (1991): 479–486.

[66] Sipper, M. *Evolution of Parallel Cellular Machines: The Cellular Programming Approach.* Berlin: Springer, 1997.

[67] Sipper, M., and E. Ruppin. "Co-evolving Architectures for Cellular Machines." *Physica D* **99** (1997): 428–441.

[68] Smith, A. R. "Simple Computation-Universal Cellular Spaces." *J. ACM* **18** (1971): 339–353.

[69] Smith, A. R. "Real-Time Language Recognition by One-Dimensional Cellular Automata." *J. Comp. Sys. Sci.* **6** (1972): 233–253.

[70] Squier, R. K., and K. Steiglitz. "Programmable Parallel Arithmetic in Cellular Automata using a Particle Model." *Complex Systems* **8** (1994): 311–323.

[71] Steiglitz, K., I. Kamal, and A. Watson. "Embedding Computation in One-Dimensional Automata by Phase-Coding Solutions." *IEEE Trans. Comp.* **37** (1988): 138–145.

[72] Toffoli, T., and N. Margolus. *Cellular Automata Machines: A New Environment for Modeling.* Cambridge, MA: MIT Press, 1987.

[73] Turing, A. M. "Computing Machinery and Intelligence." *Mind* **59** (1950).

[74] van Nimwegen, E., and J. P. Crutchfield. "Optimizing Epochal Evolutionary Search: Population-Size Independent Theory." Special issue on Evolutionary and Genetic Algorithms in Computational Mechanics and Engineering,

edited by D. Goldberg. *Comp. Meth. App. Mech. & Eng.* **186** (2000): 171–194.

[75] van Nimwegen, E., J. P. Crutchfield, and M. Mitchell. "Statistical Dynamics of the Royal Road Genetic Algorithm." *Theoret. Comp. Sci.* **229** (1999): 41–102.

[76] von Neumann, J. *The Computer and the Brain.* New Haven, CT: Yale University Press, 1958.

[77] von Neumann, J. *Theory of Self-Reproducing Automata,* edited and completed by A. W. Burks. Urbana, IL: University of Illinois Press, 1966.

[78] Waksman, A. "An Optimum Solution to the Firing Squad Synchronization Problem." *Infor. & Control* **9** (1966): 66–78.

[79] Waldspurger, C. A., T. Hogg, B. A. Huberman, J. O. Kephart, and W. S. Stornetta. "Spawn: A Distributed Computational Economy." *IEEE Trans. Software Eng.* **18** (1992): 103–117.

[80] Weiner, N. *Cybernetics, or Control and Communication in the Animal and the Machine.* New York: John Wiley and Sons, 1948.

[81] Wellman, M. P. "A Market-Oriented Programming Environment and Its Application to Distributed Multicommodity Flow Problems." *J. Art. Intel. Res.* **1** (1993): 1–23.

[82] Wolfram, S. "Statistical Mechanics of Cellular Automata." *Rev. Mod. Phys.* **55** (1983): 601–644.

[83] Wolfram, S. "Twenty Problems in the Theory of Cellular Automata." *Physica Scripta* **T9** (1985): 170–183.

[84] Wolfram, S., ed. *Theory and Applications of Cellular Automata.* Singapore: World Scientific, 1986.

Index

J

P